ON THE ORIGIN OF PHYLA

ON THE ORIGIN OF

PHYLA

JAMES W. VALENTINE

The University of Chicago Press   *Chicago and London*

The University of Chicago Press, Chicago 60637
The University of Chicago Press, Ltd., London
© 2004 by The University of Chicago
All rights reserved. Published 2004
Paperback edition 2006
Printed in the United States of America

13 12 11 10 09 08 07 06   2 3 4 5

ISBN: 0-226-84548-6 (cloth)
ISBN: 0-226-84549-4 (paperback)

Library of Congress Cataloging-in-Publication Data

Valentine, James W.
    On the origin of phyla / James W. Valentine.
      p.   cm.
    Includes bibliographical references (p.) and index.
    ISBN 0-226-84548-6 (cloth)
    1. Phylogeny.   2. Evolution (Biology)   I. Title.
    QH367.5 . V26 2004
    591.3′8—dc22

                           2003016601

♾ The paper used in this publication meets the minimum requirements of the American
National Standard for Information Sciences—Permanence of Paper for Printed Library
Materials, ANSI Z39.48-1992.

To Diane

Contents

Preface

The title of this book, modeled on that of the greatest biological work ever written, is in homage to the greatest biologist who has ever lived. Darwin puzzled over but could not cover the ground that is reviewed here, simply because the relevant fossils, genes, and their molecules, and even the bodyplans of many of the phyla, were quite unknown in his day. Nevertheless, the evidence from these many additional sources of data simply confirm that Darwin was correct in his conclusions that all living things have descended from a common ancestor and can be placed within a tree of life, and that the principle process guiding their descent has been natural selection. And he was correct in so much more.

The data on which this book is based have accumulated over the near century and a half since Darwin published *On the Origin of Species,* some gradually, but much in a great rush in the last several decades. I have been working on this book for well over a decade, and much of that time has been spent in trying to keep up with the flood of incredibly interesting findings reported from outcrops and laboratories. I am stopping now, not because there is any lull in the pace of new discoveries (which if anything is still picking up), but because there never will be a natural stopping place anyway, and because the outlines of early metazoan history have gradually emerged from mysteries to testable hypotheses.

The purpose of this book is to bring together the findings from a number of quite separate fields that have contributed to our growing understanding of early metazoan evolution, and to explain their evidentiary bases. There are three main parts. Part 1 details the sorts of evidence that have been garnered from fossils, molecules, and morphology. Part 2 is a review of the basic morphology and development of metazoan phyla, with emphasis on the major design elements of the bodyplans of living clades and of the many distinctive taxa, known only as fossils, that were the major elements in the early metazoan faunas. Part 3 examines the evolutionary patterns displayed by metazoan clades through the Phanerozoic, and presents models for the evolutionary patterns of the origin and early diversification of Metazoa, and of the establishment of the living phyla.

It is usual to thank one's editor last, but mine must be acknowledged right up front. Christie Henry, always supportive and full of good advice, made this book possible in its present form. She also is responsible for subjecting me to repeated waves of somewhat painful but extremely useful critiques from knowledgeable reviewers in the diverse fields involved. These reviewers include Tomasz Baumiller (Department of Geological Sciences and Museum of Paleontology, University of Michigan), Ferdinando Boero (Dipartimento di Biologia, Universita di Lecce), Sandy Carlson (Department of Geology, University of California, Davis), Michael Foote (Department of Geophysical Sciences, University of Chicago), Michael LaBarbera (Department of Organismal Biology and Anatomy, University of Chicago), Peter Wagner (Department of Geology, Field Museum of Natural History), Greg Wray (Department of Biology, Duke University), and one anonymous soul. I am a paleontologist, and my attempts to understand and summarize information from other relevant disciplines, and from areas within paleontology in which I have little expertise, were greatly enhanced by the quality of those reviews. My warmest thanks to all.

A lot of help has also come from colleagues who have read chapters in their areas, sometimes more than once, and whose advice I have almost always followed; these are Allen Collins, Doug Erwin, Healy Hamilton, Dave Lindberg, Kevin Padian, Pete Sadler, and Pete Wagner. And several colleagues have tendered various other sorts of advice and encouragement, such as patiently fielding naïve questions, providing unpublished data, or guiding me to useful information; these include Sean Carroll, Simon Conway Morris, Mischa Fedonkin, John Gerhart, Jim Lake, Mike Levine, Chris Lowe, Ben Waggoner, and Adam Wilkins. Finally, there are my collaborators in various attempts to make sense of some aspects of the puzzles of early metazoan life; these are Allen Collins, Doug Erwin, Healy Hamilton, Dave Jablonski, Jere Lipps, Cathleen May, and Chris Meyers. Many other individuals, far too numerous to list, have also provided help during the last decade or so, all of which has been deeply appreciated, and I sincerely thank them all. David K. Smith (Scientific Visualization Center, University of California, Berkeley) has been responsible for the lion's share of the graphics. Finally, the librarians of the BioSciences Library at Berkeley have been most helpful over the many years.

Most of my own research has been funded by the National Science Foundation and by faculty research grants from the University of California, and belated thanks are due the Guggenheim Foundation for supporting the studies that began my interest in metazoan phyla per se.

PART ONE

Evidence of the Origins of Metazoan Phyla

A phylum should consist of closely allied animals distinguishable from any other phylum by well-defined positive characteristics, some of which do not exist in other phyla or not in that particular combination. LIBBIE HYMAN

The bodyplans of phyla have been much admired as representing exquisite products of evolution, in which form and function are combined into architectures of great aesthetic appeal. While the phyla seem relatively simple in their basic designs, most contain branches that form important variations on their structural themes, and some body types display remarkable embellishments in their morphological details. Presumably these variants reflect something of the ranges of ecological roles and environmental conditions in which the various phyla have evolved and functioned. The origins of phyla are very remote; it is consistent with present evidence that all of them appeared over half a billion years ago. The geologic record of events during those ancient times has been obscured during the passage of so many years. Nevertheless, we have some interesting fossils, and they are very helpful indeed, especially when they can be interpreted in the light of biological information from their living relatives. There are few fields within biology that do not contribute directly to the understanding of some aspect of the origin of phyla.

Much work on the origins of the phyla has focused on their phylogenetic interrelations, with evidence classically drawn chiefly from comparative studies of development and of adult anatomy. Yet many of the developmental and anatomical patterns do not correlate well with one another across the phyla, so that the relationships that are inferred among the phyla depend upon the particular characters that are selected. Proponents of different phylogenies are often in different fields or belong to different schools of thought, and they stress different features as being the more reliable indicators of relationships. As a result, a great number of phylogenetic schemes have been proposed. Thus for most phyla the nature of the ancestral form is actively disputed. Indeed, for many systematists the problem of the origin of a phylum is the problem of its ancestral root. I am not using *origin* in such a restricted sense, but more in the spirit in which Darwin used it. He was not concerned with the ancestry of any given species, but in the processes that were responsible for how species could evolve. In somewhat the same vein, it is not the main purpose here to establish a phylogenetic tree for metazoan phyla, but to try to understand how the wonderfully disparate bodyplans of the major branches have arisen. Nevertheless, it is clear that, without some knowledge of phylogenetic branching patterns, there can be no consensus on the pathways of change that have produced the bodyplans of the phyla. And without some understanding of the interrelationships among phyla, the problems surrounding the evolutionary processes involved in their origins are difficult to confront.

Fortunately, three recent advances help in understanding which of the phylogenetic schemes is more likely to be correct, and thus permit some insights into the evolutionary processes that produced phyla. One advance is the establishment of molecular techniques of phylogenetic reconstruction. The molecular methods are independent of development and morphology per se and thus provide a new data set to aid in discovering metazoan relationships. The molecular phylogenetic trees have produced a few surprises, separating some phyla that were commonly believed to be closely related, and bringing into close proximity others that were thought to be on quite distant branches. These new placements require that the ancestral bodyplans of the last common ancestors of many of the phyla be reinterpreted. It is still early in the application of molecular techniques, however, and many of the relations indicated in molecular trees cannot yet be considered definitive. Nevertheless, there are a couple of reasons to believe that the major molecular branching patterns will stand the test of time. Some of the assemblages of phyla long regarded as valid on developmental or morphological evidence are well supported by the molecular evidence. And some of the major new assemblages created by molecular branches seem, on review, to be plausible. Also, the sequence of bodyplans implied by the molecular trees is consistent with evidence from the early fossil record. Still, the chance that every branching now suggested by molecular evidence is in fact correct is essentially nil.

The second advance has been the continuing discovery, description, and interpretation of early metazoan fossils. The abrupt early appearance in the fossil record of the remains of numbers of animal phyla has been a famous phenomenon since it was first emphasized by Darwin as a difficulty to his theory. Continued work during the following 140 years has only verified this pattern; most of the major metazoan phyla appear within a geologically narrow window of time, during the Cambrian "explosion" about 520–530 million years ago (Ma). Furthermore, a variety of unusual fossils appear within this window, indicating that numbers of distinctive major branches of living phyla, and perhaps even some additional phyla, also arose during this explosion, but have become extinct. While the fossil record preceding the explosion is still too poor (or too poorly known) to permit explicit reconstructions of the forms ancestral to the Cambrian phyla, it does provide evidence of some of their behavioral repertoires and grades of organization. This evidence produces important constraints on the sorts of organisms that were present, and thereby significantly restricts the possible phylogenetic schemes.

The third advance is in the rapid growth of knowledge of genetic aspects of development, particularly of the processes of bodyplan patterning and morphogenesis on the molecular level. These studies are leading to a deeper understanding of how the bodyplans develop from the information contained in the zygote. This is quite a young field and an exceedingly complicated subject, but even to a paleontological observer standing well behind the sidelines, the significance of the main findings are apparent. Some sorts of genes—regulatory genes—act to mediate the transcription of other genes, operating within a regulatory network to produce a programmed series of gene expressions. Usually this regulatory system is set in motion by fertilization, and the regulatory cascade mediates the differentiation of cell and tissue types as development unfolds. Different cell types arise because different fractions of the genome are expressed in them, and a spatially coordinated pattern of cell differentiation and tissue formation then leads to an integrated pattern of organogenesis to produce a particular body architecture. The genes that produce the molecules that form physical parts of organisms and carry out their physiological activities are targets for the regulatory gene systems; they are the structural genes.

One of the more surprising findings of developmental genetic research is that similar suites of regulatory genes control patterns of gene expression in organisms with very different architectures—as different as nematodes, insects, and mammals, for example. Clearly, those regulatory genes were not much changed even during the evolution of such disparate phyla. Nevertheless, the patterns of expression mediated by the regulatory genes have clearly had to undergo major remodeling to produce new bodyplans during the evolution of phyla. There must have been considerable evolution in the relations between given transcriptional regulators and their target genes. It also seems plausible that the number of gene-expression events increased as more complex organisms arose.

Findings from these three rapidly advancing fields, then—molecular phylogenetics, early metazoan paleontology, and molecular evolutionary aspects of developmental biology—can be integrated with classical information on developmental and anatomical features, hopefully to throw light on the problems of the origin of phyla. The first part of this book reviews the nature of phyla and the problems in their classification, major features of metazoan architecture, major features of development, phylogenetics, and the nature and quality of evidence from the fossil record. The second part is a morphological tour of the molecular tree of phyla, major branch by major branch, to examine the bodyplans of the phyla and to explore the range of likely possibilities for their kinships. The third part is an attempt to examine the subsequent fossil record of phyla for clues as to their early histories, and to review scenarios of the adaptive bases of bodyplan origins. Although resolution of many problems involved with the evolution of phyla is clearly still beyond our reach, it is at least possible to grapple with competing hypotheses with the evidence that is now available.

CHAPTER I

The Nature of Phyla

Phyla Are Morphologically Based Branches of the Tree of Life

Concepts of Animal Phyla Have Developed over Hundreds of Years

Classifications of organisms in hierarchical systems were in use by the seventeenth and eighteenth centuries. Usually organisms were grouped according to their morphological similarities as perceived by those early workers, and those groups were then grouped according to *their* similarities, and so on, to form a hierarchy. Thus species, which form the lowest level in the hierarchy, have membership in each of a series of increasingly inclusive hierarchical levels or categories. By an international agreement first formalized in 1901, the tenth edition of Linnaeus's *Systema Naturae* (1758) was taken as the starting point for priority in zoological nomenclature for taxa in the lower categories. However, there are no international rules for the nomenclature of orders, classes, or phyla. Linnaeus himself used five hierarchical levels for animals: *regnum, classis, ordo, genus,* and *species.* His principal subdivisions of the animal kingdom, that is, the classes, were Mammalia, Aves, Amphibia, Pisces, Insecta (essentially arthropods), and Vermes (essentially everything else). Thus four of the principal animal taxa were vertebrates, and only two classes embraced the entire spectrum of invertebrate animals.

Cuvier (1812) is sometimes credited with erecting a level that corresponds roughly with modern phyla; he called the principal subdivisions of animals *embranchements,* of which he recognized only four: Vertébrés, Mollusques, Articulés (annelids and arthropods), and Zoophytes (echinoderms, cnidarians, and just about everything else). This is a slightly more balanced classification, putting three major invertebrate taxa on equal footing with the vertebrates. Nevertheless, Cuvier's embranchements are clearly on the same level as the classes of Linnaeus. As Cuvier called the subdivisions of the embranchements "classes," it appears that he had defined a higher taxonomic level, but that was not really the case; he simply demoted Linnaeus's term. Cuvier based his embranchements on his perceptions of functional unity within each division (see Appel 1987). Von Baer (1828) found that the same four divisions could be recognized on developmental criteria. The embranchements were used by many eminent naturalists into the second half of the nineteenth century. For example, a classification by Agassiz (1857) used four "branches," and one by Owen (1860) used four "provinces," all essentially the embranchements of Cuvier.

Haeckel (1866) introduced the term *phylum,* based on the Greek word *phylon,* a tribe or stock. The phyla were principal subdivisions of the kingdom Animalia, and thus on a level with Linnaeus's classes and Cuvier's embranchements. Haeckel used the term in two ways. In one usage, phyla denoted major branches in his famous trees of life. Six such branches of animals were recognized, although one, Spongiae, was placed in Protista rather than Animalia; the other five were Coelenterata, Echinodermata, Articulata, Mollusca, and Vertebrata. A second usage was as a level in a hierarchical classification. Haeckel (never one to underdo things) used twelve principal categories from species to phylum, each further subdivided by a subcategory, for a total of twenty-four hierarchical levels.

The number of recognized living phyla in hierarchical classifications has grown significantly since Haeckel's time. The classifications used in recent textbooks differ in detail, splitting some phyla and lumping others, but the number of phyla recognized usually lies in the low to middle thirties. Three representative synopses of the animal kingdom employ thirty (Parker 1982), thirty-two (Margulis and Schwartz 1982), and thirty-four phyla (Barnes 1984). Since those synopses were assembled, two additional phyla have been described (Loricifera Kristensen 1983 and Cycliophora Funch and Kristensen 1995), some phyla have been sunk, and a few groups have been raised to phylum rank by some workers.

A century after Linnaeus supplied what became the foundation of zoological nomenclature, Darwin (1859) proposed that the diversity of life had arisen through evolutionary processes, a hypothesis that was thoroughly tested and corroborated in the following century. Organisms are related to one another and can be arranged in a genealogy of life, as humans can be arranged in family trees. Darwin himself produced a treelike diagram of the relations among some hypothetical species (fig. 1.1), and Haeckel (1866) produced a number of trees of life, many complete with bark and gnarled branches (fig. 1.2). Once it was accepted that we owe the diversity of life to evolution, the Linnean hierarchy became a way of expressing relatedness as well as morphological similarity per se. Today it is accepted that each taxon should be monophyletic—that it contains only species that have a common ancestor that is the founding member of that taxon (fig. 1.3). Despite the acceptance of a tree of life, systematists continued to employ the Linnean hierarchy. Until the second half of the twentieth century, phyla were nearly always regarded as composing a hierarchical rank that represented a principal subdivision of the animal kingdom, whether or not they were represented in a tree.

In phylogenetic trees the position of a species (or of another taxon) depends upon its ancestors and descendants rather than its morphology per se. The branches of the tree represent actual entities, and (if the tree is correct) they have historical reality as lineages. The landmarks that are most easily identified in a tree structure are the branch points or nodes. As the nodes in a phylogenetic tree usually represent the onset of independent evolutionary paths for organisms along each branch, systematists often use nodes as the basis for their taxonomic nomenclature (node-based

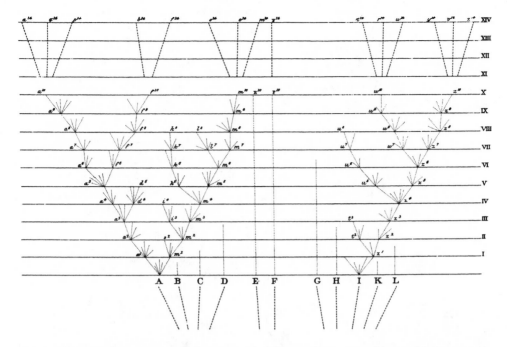

FIG. 1.1 Darwin's hypothetical tree of species, used to illustrate the patterns of descent that he expected. Note that some extinctions are indicated. From Darwin 1859.

taxa). Identifying the nodes and determining their sequences, or identifying discrete branches in some other way, produces a classification with a treelike structure rather than a hierarchical one. The methodology in common use for determining such a treelike structure was founded by Hennig (1950, 1966), and is termed cladistics.

A branching creates two (or more) sister species; each sister founds a monophyletic taxon—a branch that includes the sister and all descendant branchings, which form a clade. Branches of a phylogenetic tree that include all descendants of a founding species are termed holophyletic (fig. 1.3A); they are also monophyletic, of course. In Linnean taxonomy, monophyletic branches may be divided into several taxa of equivalent rank if they form distinctive morphological clusters. Thus the founding species of a branch, and its cluster, may form a taxon that does not include all of the species descending from it—the taxon is not holophyletic. Such a taxon is termed paraphyletic and can be termed a paraclade (fig. 1.3B; as defined by Raup [1985], the term *paraclade* included holophyletic clades; Wagner [1999] used it for strictly paraphyletic clades, and that is how it is used here). Such a taxon is also monophyletic since all of its members are descended from its founding species. In Hennigian classifications paraphyletic taxa are not permitted, and thus all monophyletic clades must be holophyletic, which becomes a redundant term. In Linnean classifications a monophyletic clade may be either paraphyletic or holophyletic, so the distinction is not redundant. With either classification system mistakes are

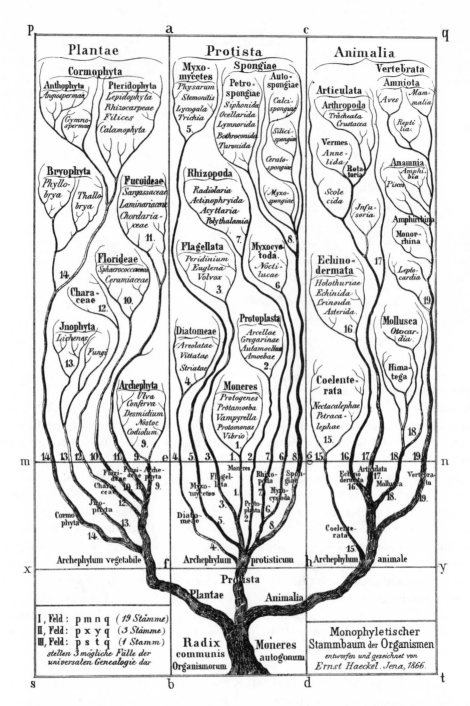

FIG. 1.2 Haeckel's tree of life, bark and all, presenting his ideas as to the phylogeny of the major groups of organisms, including many groups now recognized as metazoan phyla. Here several hierarchical levels are represented in the same tree, their importance indicated roughly by limb thicknesses. From Haeckel 1866.

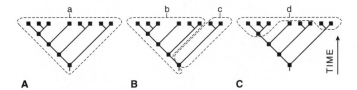

FIG. 1.3 Trees of species for hypothetical taxa that include various combinations of ancestors and descendants. (*A*) Taxon a (within the dashed line) includes an ancestral species and all of its descendants, and is therefore monophyletic and holophyletic. (*B*) Taxon b includes an ancestral species and only some of its descendants, and is therefore monophyletic and paraphyletic. Taxon c includes an ancestral species and all of its descendants, just like taxon a, and is therefore both monophyletic and holophyletic. (*C*) Taxon d does not include the last common ancestor of its species, and is therefore polyphyletic. Squares are species.

sometimes made, and species or other taxa may be included within a higher taxon or clade with which they do not in fact share a founding common ancestor. Groups resulting from such erroneous placements are termed polyphyletic (fig. 1.3C) and should be emended whenever such an error is detected.

The Concept of Homology Is Basic to Determining Animal Relationships

A major source of phylogenetic error lies in confusing homologues with analogues. In evolutionary terms, homologues are similar features in separate organisms or taxa that have evolved from a similar ancestral feature, and owe their similarity to this common inheritance—dog and lizard forelimbs, for example. Analogues are similar features in separate organisms or taxa that have not evolved from a common ancestral feature, that have evolved independently, but that share a common or similar function—bird and bat wings, for example. The distinction between homology and analogy was first developed in preevolutionary times, when homologues were commonly represented as reflections of the characters in some "ideal" type (see Appel 1987). It was Owen (1843, glossary) who originated this usage of the terms *homologue* and *analogue;* for him a homologue was "the same organ in different animals under every variety of form and function." Good treatments of homology are in Wiley 1981; Patterson 1988; and papers in Hall 1994 and Bock and Cardew 1999. Morphological features used as criteria of relatedness must have a common evolutionary origin—must be homologous—whether one is constructing a Linnean or a Hennigian classification. Classing birds and bees together because they both have wings is clearly not appropriate. If there were a way to identify morphological homologies unambiguously, it would be much easier to construct an accurate tree of life. Recognition of homologues among phyla has proven difficult. The branching of any two phyla from their last common ancestor has always occurred well before we can identify the phyla morphologically, by which time homologous features have usually diverged significantly. Furthermore, features that have evolved independently for similar functions commonly resemble each other and have often been mistaken for homologues.

The most uncomplicated case of homology, sometimes termed static homology, occurs when a feature is inherited in sister lineages and has remained recognizably similar during subsequent divergence. A complication arises if the feature undergoes significant change during the divergence to create a transformational homology; the feature is then recognized as a distinct character in the transformed lineage (see Streidter and Northcutt 1991). If the transformed characters can be connected by intermediates, they may still fall under the definition of homology, but when the features cease to resemble each other, the concept loses much of its utility. When similar features are repeated serially within the same organism, as with the segments of an earthworm, they are sometimes regarded as serial homologues. In evolutionary terms this usage would imply that they have descended by multiplication of a single ancestral feature, but the history of serial homologues is usually not understood. In many cases so-called serial homologues tend to diverge in time, becoming morphologically and functionally specialized according to their positions along a body to produce a pattern of regionation. Some features exist in multiple similar copies within individuals but not in a serial arrangement, as for example many cell types; such copies are termed homonyms.

Linnean and Hennigian Taxa Have Different Properties
All of the processes that introduce, conserve, and change morphological features are significant components of evolution. Hierarchies and trees are the two structures in which taxa are normally classified, the former in a Linnean system, the latter in a Hennigian one. The hierarchical approach to phyla, as in Linnean classifications, emphasizes patterns of morphological similarity, based on characters that arise and spread among a lineage and its collateral relatives. As a result, a paraclade can be created when a distinctive new morphological group originates from within a taxon and is ranked equally with its parent group. The cladistic approach, as in Hennigian classifications, emphasizes descent, and is based on characters that indicate the sequence of their introduction from ancestor to descendant. As a result, only true clades that include all descendants of the clade ancestor are recognized as taxa. Using a tree or a hierarchy is not simply a choice among different methods of assorting taxa, however, for when treated in these different systems, taxa, including phyla, are defined differently. Each of these approaches provides a distinctive perspective on the phyla and on their roles in the history of life, and to ignore either is to risk missing some of the more basic features of evolution. Understanding the differences in the properties of trees and ranked hierarchies is therefore essential in evaluating the natural relations within and among phyla.

In fact, hierarchy and tree structures occur in most of the biological systems that are involved in the origin and evolution of phyla. Evidently all natural hierarchies are formed by treelike processes. For example, take a metazoan body. Cells proliferate from an egg by subdivision and then duplication, and if the ancestral-descendant relations among cells are traced, they form a treelike branching system. The resulting body is not organized as a tree, however, but as a hierarchy of cells, tissues, organs, and so forth. Another example is the hierarchy of ecological entities forming the planetary biota, from individuals to populations to communities to regional faunas to provinces and realms. The biota is produced by a tree—the tree of life. Furthermore, trees and hierarchies are both found within metazoan genomes (chap. 3), the former producing the latter. And finally, the Linnean hierarchy is clearly formed by a treelike branching process that can be described in Hennigian terms. Before coming to grips with the evolution of these entities, it is most useful to discuss the distinctive properties of hierarchies and trees, indispensable to understanding the way in which the biological world is organized.

Genealogical Histories Can Be Traced in Trees, Which Are Positional Structures

The various family trees, trees of life, and other branching diagrams used in the biological sciences include structures that have often been termed hierarchical, but that do not have the properties of hierarchies (see below). They have other properties, however, that hierarchies do not possess. Trees usually proceed from a single founding unit or "root," tracing the courses of lineages through time; they can order entities in ancestor-descendant relationships. The entities in trees may be clearly individuated, as people in a family tree, or they may grade into each other, as do evolving populations. Gradational lineages can be delineated by one or more criteria, such as, in evolving lineages, by using branch nodes, or periods of rapid morphological change, or the first appearance of some particular feature, or gaps in the fossil record, to define taxonomic entities within the continua. The entities in a tree are all alike—in a family tree, they are all people—while in a hierarchy they are nested into new entities from rank to rank (see below).

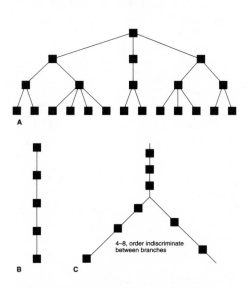

FIG. 1.4 Some types of positional structure used in biology and paleontology. (A) A tree such as a family tree, in which all units occupying the positions are of the same kind (persons), but may be ranked by some criterion, such as generations. (B) An unbranched serial structure. (C) A branched serial structure in which there are no criteria by which to establish ranks between branches. From Valentine and May 1996.

Three sorts of positional structure are shown in fig. 1.4. One type contains tiers (fig. 1.4A). This type is exemplified by family trees that are positioned by generations, as discussed above. A second sort of positional structure is exemplified by a series of entities that can be placed in a series, one ahead of another (fig. 1.4B). A single lineage of ancestral to descendant fossil populations would form an example. This structure is analogous to a tree that lacks branches. A third positional structure has branches, but the relations between entities on separate branches are indeterminate (fig. 1.4C). This is the way that taxa are generally positioned within the tree of life.

Morphological Entities within Metazoan Bodies, Such as Cells, Can Be Positioned in Trees

Developmental cell trees trace cellular descent during the ontogeny of a given individual, beginning with the zygote. Evolutionary cell trees trace the evolution of cell types in a clade through geologic time. Most cell types have been identified morphologically (cell morphotypes), at first by techniques of light microscopy and more recently by those of electron microscopy. The simplest living sponge, for example, has perhaps five cell morphotypes, and the most complex may have twelve or more. In the sponges as a whole, however, over thirty cell morphotypes have been described (Simpson 1984), and there were probably additional cell morphotypes in extinct sponges that are not represented today. For any individual sponge there is a tree that indicates the ontogenetic sequence of its cell lineages. For all sponges there is a tree that indicates the evolutionary derivations and relations of all sponge cell types.

Trees of tissue or organ types can be imagined, either ontogenetic trees for individual organisms, or phylogenetic trees for all tissues or organs, or trees of individual tissues or organs in all metazoans. As used here, a tissue need contain only one cell type and an organ only one tissue type. Thus the structurally simplest free-living metazoans, the placozoans (chap. 6), might be considered to be composed of only a single tissue, or perhaps two, but here they would also be considered to be composed of a single organ and a single organ system, and of course each represents an organism. A tissue usually includes several cell types, however, and an organ commonly includes more than one tissue, and so forth. Trees of individual organs have been used to suggest the phylogeny of the taxa in which they appear, but it can be difficult to separate similarities owing to function from those owing to ancestry.

Trees Composed of Individual Organisms Can Be Incredibly Complicated

A significant complication at the individual level is that, among sexually reproducing forms at least, each individual has two parents, and they may come from quite different genealogical backgrounds. When, say, males are added to a female family tree, each brings his own family tree, and the structure becomes too complicated to describe easily. Genealogical trees of sexual species become simple branching structures only when descent is traced by either male or female lines alone. The rich consequences of the complications imposed on evolution by sexuality are extensively examined in the literature of population genetics and microevolution.

Trees Composed of Species Are Much Simpler

Species can be defined as systems of populations within which gene exchange may occur and between which gene exchange is quite limited or prevented by intrinsic reproductive isolating mechanisms (see Dobzhansky 1937; Avise 1994). A tree of species resembles a simple branching tree, with minor complications. One complication is that branchings may be multiple rather than binary; another is that some separate species lineages may reunite (especially among plants), so that some anastomosing occurs within the tree structure. There are numerous modes of speciation, which may have different morphological consequences. For example, a species may be founded by only a few individuals (two genetic individuals is the usual minimum in sexual species). When the founding population is only a minor fraction of the parental population, its isolation need not affect the biology or the future of the parental population in any way. The morphology of the small founding population may well change significantly, however, perhaps in response to its radically new situation. In other cases, say when a species' population becomes split more nearly in two (vicariates), each of the sister populations might diverge morphologically from the parental condition, depending upon a host of factors that vary from case to case, as is made clear in the immense literature on speciation. At any rate, species do not proceed from other species in toto but from some sample of the parental species. As species may form by very different modes, they represent a wide array of biological entities (for an excellent review incorporating molecular data see Avise 1994).

Trees Can Be Formed of Linnean Taxa above the Species Level

The literature contains many trees of Linnean taxa above the species level. In a tree of, say, genera, each branch will contain a (nondepicted) species tree for the genus, as within each branch of the imaginary species trees there are implicit trees of individuals. At the points where genera branch, new genera are derived from only one parental species within the parental genus. The same principles apply to familial trees, ordinal trees, and so on to the highest taxonomic level. Hennigian-based taxonomies do not permit the formation of trees of Linnean taxa above the species level, as such taxa represent ranks, which are not permitted.

Molecular Information Can Position Morphologically Based Taxa in a Tree

With seemingly minor exceptions, DNA sequences in animals must trace to ancestral sequences that were present in the last common ancestor of all living metazoans. Each of the genes in that common ancestor (that did not become extinct immediately) gave rise to a gene tree within Metazoa, branching whenever they were duplicated, or when modifications of the original DNA sequence arose within separately evolving lineages as metazoans diversified. As the homologous genes of organisms that branched earlier along the phylogenetic tree have been separated longer than those of organisms that branched later, it would be expected that their DNA sequences would diverge more. The pattern of sequence similarities of a gene (or its product) among living representatives of a number of taxa should, in principle, reflect the sequence of branching among those taxa (Zuckerkandl and Pauling 1965). Hence, determining this branching pattern in living taxa should provide the basis for a tree that shows their interrelationships. Unfortunately, evolutionary changes within DNA sequences do not always produce clear divergent patterns that reflect their branching histories—for example, homologous DNA sequences may evolve at such different rates that their true relationships are obscured (chap. 4). Nevertheless, trees based on interpretations of sequence comparisons of individual genes or their product molecules have become of great importance in understanding evolutionary patterns of morphologically based taxa. Molecular phylogenetic studies provide evidence that is independent of morphology per se, and thus should help to avoid some of the problems with morphological criteria of relatedness, such as establishing which characters are homologues and which are not.

Natural Biological Hierarchies Are Nested Structures of Functional Entities That Emerge When Complex Systems Are Organized

Nearly everywhere within the complex world of biology, whatever perspective we choose, we find entities aggregated into hierarchies. Why this may be so was elegantly described by Simon (1962, 1973), who suggested that hierarchies are the only useful architecture for the organization of complexity. Given complex systems, as biological systems certainly are, it is no surprise that they are organized as hierarchical structures. The entities that compose biological hierarchies should show enough spatiotemporal integrity to be considered, in logic, as individuals, and not as types or classes (see discussions in Ghiselin 1974; Salthe 1985; Eldredge 1985). Several natural hierarchies are associated with the origin and diversification of phyla. These include a somatic hierarchy (Eldredge 1985), used to describe the structure of organisms; an ecological hierarchy, used to describe biotic structure in the biosphere (see Valentine 1973b); and, evidently, systems within metazoan genomes associated with the regulation of development and of its evolution. Additionally there is the Linnean taxonomic hierarchy, which is real enough but only partly natural (discussed later).

There Are Four Major Types of Hierarchical Structure
The biological hierarchy that best displays the utility of hierarchical arrangements is the quasi-artificial Linnean taxonomic hierarchy, which is a perfect example of a purely aggregative hierarchy (see Mayr 1982). The basic entities are discrete species, composed of individuals related by common descent, and identified by various biological attributes involving gene flow, and by morphological similarities. The

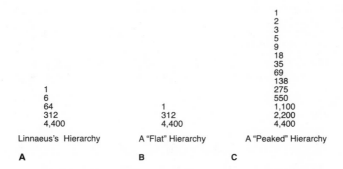

FIG. 1.5 Taxonomic diversity in three hierarchies. (*A*) Linnaeus's hierarchy, with about 4,400 species arranged in 312 genera, 64 orders, and 6 classes within the kingdom Animalia—a total of 4,783 taxa. (*B*) A "flat" hierarchy of 4,400 species arranged in three levels; each second-level taxon has the same number of species as Linnaeus's hierarchy, and although there are 4,713 taxa, not many fewer than Linnaeus used, there is no guide to the patterns of similarity among the higher taxa. (*C*) A "peaked" hierarchy of 4,400 species arranged in fourteen levels so that the number of subtaxa in each higher taxon is about 2; there are 8,805 taxa.

ranks (categories) are formed on the basis of increasingly generalized morphological resemblances, given monophyly of the entities (taxa). Linnaeus himself recognized about 4,400 species of animals organized into five levels (fig. 1.5A). The number of levels useful for simple classification in an aggregative hierarchy depends very much upon the number of entities—species in the present example—that lie at the base. If there are 40 species, no hierarchy is really needed if the only purpose is to organize the pattern of morphological variation, which can be handled by "brute force" memory. With 4,000 species, however, it greatly helps to associate them in a hierarchy, as many early naturalists did. A two-level hierarchy merely names the collective of all 4,000 species, providing no real organization. Three hierarchical levels is a bit better, but still, if one used 312 second-level units (Linnaeus's number) in a three-level hierarchy, one would have to remember not only their names but also their patterns of resemblance, which would not be displayed further in that hierarchy (fig. 1.5B). Linnaeus's five levels seem just about right; zealous students could master all of the taxa from the second level and up. On the other hand, if one used many more levels, say fourteen, there would be an average of about two taxa nested in each taxon of the next higher level, and while the patterns of resemblance would be explicitly portrayed, there would be over 4,400 taxonomic names in the hierarchy above the species level—the addition of levels is not free (fig. 1.5C). A similar problem occurs when naming all of the branches in a tree. The more nodes or levels, the more information is required to keep track of the entities, and in a Linnean-style hierarchy that means adding many more taxa above the species level.

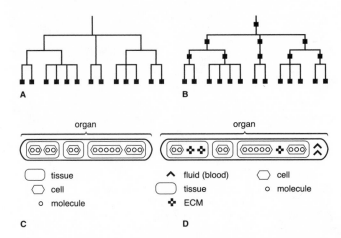

FIG. 1.6 Some types of hierarchies used in biology and paleontology. (*A*) An aggregative hierarchy; the only units present are the
basal ones, and increasingly higher ranks are formed of increasingly
inclusive collectives of these. (*B*) A cumulative aggregative hierarchy.
New units may join the hierarchy at ranks above the basal level, but
once joined, they are aggregated like other units in the ranks. (*C*) A
constitutive hierarchy. The only units present are the basal ones, but
they are physically joined to produce units in the next higher rank,
and these units are physically joined in turn to produce units in the
next higher rank, and so on. (*D*) A cumulative constitutive hierarchy.
New units may join the hierarchy at ranks above the basal level, but
once joined, they are constituted like the other units in the ranks.
ECM, extracellular matrix. From Valentine and May 1996.

In the taxonomic hierarchy, nothing new joins the hierarchy at any given rank.
Organisms are aggregated into species, these same species are aggregated into genera, and so forth (fig. 1.6A). However, there is a second sort of aggregative hierarchy
in which new entities *do* join the hierarchy at levels above the fundamental one,
and are subsequently aggregated in higher levels just as are the entities that they
join (fig. 1.6B)—a cumulative aggregative hierarchy (Valentine and May 1996). An
example of this sort of hierarchy can be drawn from the military. Individuals are aggregated into squads. Squads are aggregated into platoons, which are commanded
by lieutenants; the lieutenants are not members of any of the squads, but *are* members of the platoons. Ascending the hierarchy, platoons (lieutenants now included)
are aggregated in companies, commanded by captains. Captains join the hierarchy
only at the company level but are aggregated in higher levels, and so forth. Part
of the regulatory genome that mediates the development of metazoans seems to be
organized as a cumulative aggregative hierarchy (chap. 3).

A third sort of hierarchy is the constitutive hierarchy (Mayr 1982); the units
in each level form physical parts of the units on the next higher level (fig. 1.6C).
The usual biological example of a constitutive hierarchy is the somatic hierarchy

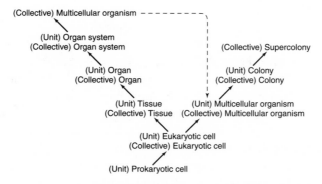

(Collective) Multicellular organism

(Unit) Organ system
(Collective) Organ system

(Collective) Supercolony

(Unit) Organ
(Collective) Organ

(Unit) Colony
(Collective) Colony

(Unit) Tissue
(Collective) Tissue

(Unit) Multicellular organism
(Collective) Multicellular organism

(Unit) Eukaryotic cell
(Collective) Eukaryotic cell

(Unit) Prokaryotic cell

FIG. 1.7 Two constitutive hierarchies evolved during metazoan
evolution. Bodies along the left branch are constructed by elaboration of collectives *within* individuals; those along the right branch
are constructed by elaboration of collectives *of* individuals.

that underlies the construction of an individual organism. Using elementary particles as the fundamental units, the levels that can be recognized in a somatic hierarchy include the following: atoms/molecules/organelles/cells/tissues/organs/organ
systems/individual organisms. Through evolution, a unicellular ancestor has originated and has been elaborated into multicellular organisms, some of which contain
many scores of cell types, but which (usually) start their ontogenies as a single cell,
a zygote. The more complex organisms are not huge single cells, nor associations of
identical cells, but are hierarchies of components that may be treated as units within
hierarchical ranks. Indeed, some accounts of the various grades found in animals
treat increasingly complex forms as being at the cell, tissue, and organ levels of
construction (e.g., Hyman 1940). In the more complex organisms there are systems
of organ systems, such as the human cardiopulmonary systems.

Actually, in the evolutionary history of metazoans there has been more than
one sort of somatic hierarchy (fig. 1.7; see McShea 2001). Eukaryotic cells have
evolved through the amalgamation of two or three types of cells (Margulis 1993),
so eukaryotic unicellular organisms may technically be termed multicellular (see
Taylor 1974), collectives formed from those more primitive cell types as units.
However, what we now call multicellular organisms are composed of differentiated
and integrated assemblages of eukaryotic cells. It would be perfectly logical to integrate such multicellular organisms into a collective to form the next level in a
somatic hierarchy; this would produce a somatically joined colony (i.e., not a so-
cial colony). Numbers of phyla contain such colonial levels of organization. Some
such colonies are so well integrated as to be virtual superorganisms, as among
the siphonophore cnidarians (Mackie 1963). Furthermore, in some siphonophores,
integrated collections of colonies exist during early stages of their life cycles
(Beklemishev 1969). Thus there is a hierarchy of prokaryotic cell/eukaryotic
cell/multicellular individual/multi-individual colonies/colonies of such colonies (see

McShea 2001). However, this is clearly not the route that evolution followed in achieving the most complex organisms. Metazoan somatic coloniality exists only among groups with relatively simple individual morphologies (cnidarians, bryozoans, entoprocts, hemichordates, urochordates). The higher levels of complexity found in invertebrates result from an elaboration of the hierarchy within a single multicellular organism—cell/tissue/organ and so forth—all forming components within the individuals rather than among them (fig. 1.7). One can imagine a somatic hierarchy evolving without going through steps from prokaryotes to eukaryotes or from individuals to colonies, but when such a step occurs, it probably greatly shortcuts the time required to achieve greater complexity.

A fourth hierarchy type is the cumulative constitutive hierarchy (Valentine and May 1996). This structure is analogous to the cumulative aggregative hierarchy in that new units may join the hierarchy at a level above the fundamental one, but then become constituted into the succeeding levels (fig. 1.6D). Despite having just considered the various levels in an organism's construction as a simple constitutive hierarchy, such a structure is not actually sufficient to contain neatly all of the physical parts of an organism. For example, the extracellular matrix of metazoan bodies, which forms the "glue" that holds most of them together, is not present in cells, and though partly manufactured by them, it is partly manufactured extracellularly. Extracellular matrix is a constituent of tissues, however, and therefore it is present in all higher levels of the hierarchy also—organs and so forth. Some fluids may be considered tissues, forming parts of hydrostatic organs, and hence are present in organ systems also. Thus an actual metazoan body is constructed as a cumulative constitutive hierarchy.

The taxonomic and somatic hierarchies can be logically joined at the level of the individual organism, where the constitutive somatic hierarchy gives way to the purely aggregative taxonomic hierarchy. Although the general properties are the same in both these hierarchies, whether units can or cannot exist independently is obviously of considerable biological importance.

Hierarchies Help Sort Out Relations among Biological Features

Novel Phenomena Emerge at Successive Hierarchical Levels. One reason that hierarchies have been so widely used by biologists to classify entities is that they are able to simplify and systematize complexity, ordering complicated entities so that they are more tractable to study. The ordering is possible because hierarchies forgive the differences among the entities within units; the entities within a given unit are assumed to be equivalent for purposes of classification at that level. Hierarchies also forgive differences among units at a given level. As described with great clarity by Medawar (1974), when moving from level to level within a constitutive hierarchy from the basic constituents to realized units on higher levels, the scope of possible phenomena is progressively reduced, yet the richness of the realized phenomena is progressively increased. For example, elementary particles may form the ultimate base of a constitutive hierarchy. The potential number of things that can be formed from elementary particles is unimaginably large, presumably comprising everything in the universe and a lot that is not, and might involve unimaginably many potential hierarchies. As one proceeds up an actual hierarchy, however, the scope of the potential becomes limited. Cells are composed of elementary particles, but the scope of the possible entities on the cellular level and of the things that can be constructed from them is clearly greatly reduced from all the potential things that can be composed of elementary particles. On the other hand the complexity and richness of form and function encountered on the cellular level far outshines the drab sameness of collections of elementary particles. And so it goes up the hierarchy. In multicellular animals such as ourselves, the actual cells are only a narrow subset of potential cells, yet animals based on this subset are richly diverse in plan and detail.

The structure and function of cells are clearly made possible by the properties inherent in elementary particles and in the units on hierarchical levels below the cell (molecules, atoms, etc.). The cells, it is assumed, do not violate any of the laws that govern the particles. On the other hand, no one is smart enough, when contemplating only a collection of particles and the laws governing them, to predict the existence of the myriad processes that occur on the cellular level, out of all the possible processes that could occur. Nor, given the properties of cells, could one predict the morphologies that we find among animals, though we can "explain" them once we find them. Higher-level laws do not exist in lower-level theory because the higher levels do not exist there. The opposite is not true, however, for lower levels *do* exist in and in fact make up the higher levels. The properties and laws that first appear at higher levels are said to be emergent; of course, they must conform to the laws of all the lower levels of their hierarchy. Some emergent properties can be predicted by ordinary humans, while others seem to be total surprises to everyone.

The Effects of Levels upon One Another Are Quite Asymmetrical. Salthe (1985) discusses interlevel interactions in terms of (1) a focal level on which interest is centered, (2) the level immediately below, and (3) the level immediately above, together forming a "basic triadic system." Properties on the focal level are derived from a subset of the potential properties of the units on the level below, and include emergent properties not found on the lower level, but their ultimate importance is determined on the level above. The lower levels are the generators of focal-level properties, while the upper levels constrain them. In a very real sense, then, even the ultimate in reductionist explanation of a given level is incomplete, for it does not take into account the narrowing range of possibilities imposed at each higher level. In a somatic hierarchy, asymmetry implies that, for example, the functional requirements of a tissue constrain the ranges of cell types that compose it. While this sort of constraint is clear enough, ecological and evolutionary processes can alter the requirements of tissues and permit novel cell types to arise—that is, the constraints themselves are subject to evolutionary change. Biological hierarchies evolve on all levels.

Natural Hierarchies Are Formed by Trees

It seems likely that all natural biological hierarchies consist of functional entities and are constructed by trees (see Woodger 1952; Gregg 1954). The somatic hierarchy, for example, is created from a tree of cells. As cells proliferate in their treelike pattern, they differentiate and become positioned in tissues. In the general case, however, cells within a given tissue are not recruited from a single cell "clade," but may be assembled from many different branches, and commonly include more than one cell type. Thus from the standpoint of a tissue, the cells are conjoined through a network; the cells are recruited from distinct ancestors, and the tissues are genealogically "polyphyletic." Organs may be composed of more than one tissue and are polyphyletic in the same way, and so on up the hierarchy. To transform a tree into the very different structure of a hierarchy requires a network of processes. The network responsible for the transformation of a cell tree into a somatic hierarchy is created by the portion of the regulatory genome that mediates the developmental patterning of metazoan bodies (chap. 3). Other examples of a relationship between complexity and the emergence of hierarchies within systems created by trees are to be found within the basic elements of the biosphere, created by evolutionary processes.

An Ecological Hierarchy of Biotic Entities Is Formed by the Tree of Life

The ranks of another biological hierarchy, the ecological hierarchy, are shown in column C of table 1.1. This is an aggregative hierarchy, beginning with the individual organism as the basic unit. Ecology is usually defined as dealing with the interrelations between organisms and the physical and biological aspects of their environments. Organisms are associated in populations, populations in communities, communities in bioregions or bioprovinces, and these in realms (e.g., terrestrial and marine); the most inclusive level (for Earth at least) consists of the entire planetary biota. The entities in this hierarchy can be associated with terms that include the environment that is involved in the ecological system as well as the organisms themselves: each individual has a range of tolerances and basic requirements; populations

TABLE 1.1 Three aggregative hierarchies. Column A is a taxonomic hierarchy, and column C an ecological hierarchy. Column B gives the ranks of a genetic hierarchy, which would be different when involving taxonomic rather than ecological units. For example, to correspond to genes in a genus, the collection of gene pools would be those of all the species within the genus; to correspond to genes in a community, the collection of gene pools would be those of all the species within the community. After Valentine and May 1996.

A	B	C
Order	Collection of collected gene pool collections	Realm
Family	Collection of gene pool collections	Bioprovince
Genus	Collection of gene pools	Community
Species	Gene pool	Population
Organism	Genome(s)	Organism
	Gene	

have niches; communities have trophic pyramids and other interpopulation inter-actions; bioregions, bioprovinces, and realms have ecosystems at larger scales; and the global ecosystem or, more usually, the biosphere, comprises the planetary biota.

The tree that has produced this ecological hierarchy is the phylogenetic tree of life, and the network that transforms this tree into a hierarchy is formed by ecolog-ical interactions between organisms and between organisms and the environment. Plotting the tree in space as well as time, the tips of the branches form the present biosphere, within which the biogeographic-ecological subdivisions form a hierar-chy. If one takes a slice of ecological time right across the tree at some previous time, one finds the biosphere of that age. The physical environment has formed a template upon which evolutionary processes have fashioned biotic assemblages, adapted to the mosaic of ambient ecological conditions. Of course environmental conditions are in constant change, producing a kaleidoscopic pattern of hierarchi-cal arrays of biotic assemblages through time. This pattern does not imply that the entities are highly integrated; indeed, the unique character of each population or species suggests that their patterns of association are largely based on overlapping tolerances and requirements.

The tattered remnants of sequences of past biospheres, preserved in the fos-sil record (chap. 5), form the basic data of paleoecology. And it is the connection of branches between temporal successions of fossils that forms the direct evidence for morphological reconstruction and interpretation of metazoan history. Unfor-tunately, the relations of community and provincial settings to the origin of phyla have not been much explored (though see papers in Zhuravlev and Riding 2001).

Hierarchies of Genes Can Be Mapped onto the Somatic and Ecological Hierarchies

Perhaps because genes have an especially important place in the biological world (on the one hand in biosynthesis and development, on the other in heredity and evolution), they have frequently been included in somatic hierarchies at a level below the cell, as gene/cell/tissue, and so on. Cells are not composed of genes, however, even though their contents do chiefly derive from gene activity. Thus genes as molecules do not form a level of the somatic hierarchy, but are merely among the many molecular constituents of cells. Genes have also been included with taxa in aggregative hierarchies, as gene/individual organism/species/higher taxa; but they do not belong there either—individual organisms are not composed of genes.

Genes lie within a constitutive hierarchy, so far as their physical presence is concerned on the molecular level. However, genes are commonly physically dis-persed, sometimes overlapping, and are subjected to complicated regulation, so that their recognition as forming part of a single constitutive level requires many glitches in the structure. If the rubric "gene" is used to indicate a unit of hereditary information, then a hierarchy may be formed as gene/genome/gene pool/collection of gene pools/collection of gene pool collections/and so forth. The genes form the basic units, aggregated at the genomic level; the subsequent pattern of aggregation depends upon the hierarchy of interest (table 1.1). If it is a taxonomic hierarchy, the genomes are aggregated into gene pools associated with species, these into genera, these into families, and so forth. If it is an ecological hierarchy, the genomes are aggregated into gene pools associated with populations, these into communities (aggregating the gene pools of all the populations in the community), these into bioprovinces, and so forth. These sorts of genetic hierarchies are probably the least useful (for our purposes) of the hierarchies discussed.

The Linnean Hierarchy Is Quasi-Natural

Artificial hierarchies are quite common, for their organizing principles are useful in any complicated system. Armies have been mentioned above, but churches, large businesses, universities, governments and their bureaus, and many other institutions also harbor hierarchies of functional entities. The composition of artificially organized hierarchies is usually quite regular; organizational charts of businesses may consist of neatly arranged boxes in tiers within rather symmetrical triangles. Natural hierarchies are far more complex and far, far messier. The sizes of entities within a level vary enormously, and it is not always clear what are the boundaries of an entity or even to which level a given entity should be assigned. Constitutive hierarchies seem a bit more regular and less puzzling in these regards than do aggregative ones, for the physical entities are helpful guides, but ambiguities in assignments to entities and to levels are still present.

The Linnean hierarchy is a sort of hybrid between a tree and a true biological hierarchy. Linnean taxa are produced by a tree, as natural hierarchies are. Membership in an entity is defined by the possession of common morphological features, and such entities can be arranged in levels (which produce paraclades). If this hierarchy is natural, the taxa should be functional entities. Commonalities in the functions of allied taxa certainly exist; many of the structures used to define morphological entities have distinctive functions. Taxa recognized on morphological criteria often have other features in common, such as geographic range sizes (Jablonski 1987) or taxonomic turnover rates (Van Valen 1985; Erwin et al. 1987), that suggest functional similarities or overlaps in aspects of their biology. Presumably the functional commonality of Linnean taxa arises from their inheritance and is expressed in the way they become arrayed within the ecogeographic template. Thus they are molded into rough hierarchical structures. Linnean taxa are required to be monophyletic, however, so that membership in an entity is not open to taxa in outgroups, and in that respect the taxa are not exclusively functional. A Linnean structure is thus a hybrid, formed as a tree between levels but ordering the units within levels in a hierarchical manner. Perhaps it can be concluded that the Linnean hierarchy is "quasi-natural." Nevertheless, like all hierarchies, artificial or natural, it provides the organizing powers and the insights available for comparative studies that are characteristic of such a structure.

Trees and Hierarchies Have Very Distinct Properties

Trees are positional structures, in which each entity is present at only one place in the tree (unlike hierarchies, in which all of the founding entities are present at every level). The ordering criterion in many biological trees is descent, as in family trees (fig. 1.8). Such trees require that entities on higher levels precede those on lower levels, as fathers precede sons (just the opposite of hierarchies, in which units on lower levels are assembled to form the higher levels). When family trees are ranked by generations, they have the appearance of a hierarchy. However, fathers and sons do not represent hierarchical levels—fathers are neither aggregated nor constituted from their sons. The higher tiers are not collectives of entities in lower tiers, but are merely similar entities positioned at a higher level—the units are all individual people. The point of placing ancestors in tiers is simply to enumerate the generations. Phylogenetic trees are also positional structures, with taxa forming the entities, but as generations do not have the same meaning for species or other taxa as they do for families, even the generational tiers found in family trees, which lend them a hierarchy-like aspect, are absent.

There are two ways in which systematic hierarchies may be formed from trees, one being a Linnean solution, and the other a Hennigian one. In order to pursue comparative studies, as of evolutionary rates, ecological deployments, or biodiversities among taxa, different branches or branch segments of the phylogenetic tree must be contrasted. Taxa that are to be enumerated with respect to some property must be considered a group; they must become members of a nest, and thus they no longer have the positional properties of entities in a tree, but have the properties of an entity within a hierarchical rank. Obviously, morphological criteria are used to form such an entity by Linnean methods, with the constraint set by requiring monophyly. If phylogeny were ignored, purely morphological hierarchical entities could be joined by nonmonophyletic taxa owing to similarities provided by suites of convergent characters, would tend to approximate a functional

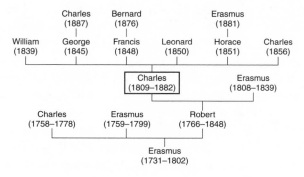

FIG. 1.8 A patrilineal family tree that includes Charles Darwin.

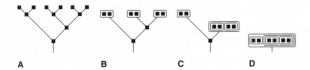

FIG. 1.9 Nesting of positional structures. (*A*) A tree of species. (*B*) Collapse of most distal row of six species into their three ancestors, each of which is a species. (C) Collapse of the most distal remaining sisters into their ancestor, which is a single species. (*D*) Collapse of the remaining sisters into their common ancestor, which is a species. This is a binary "Russian doll" sort of nesting, but it does not produce aggregated ranks.

hierarchy, and would be a natural hierarchy if plotted in ecological time. With their constraints, monophyletic Linnean hierarchies trace the evolutionary deployment of suites of functions across the environmental mosaic of adaptive zones, subzones, and niches.

Hennigian hierarchies are different. When clades are collapsed into a rank, they have a pattern like a series of boxes of decreasing sizes that fit one inside another to form a sort of serial nest. This pattern is easiest to envision in the unbranched serial structure; if nesting is to occur, it must come about through the collapse of the later entities into the earlier, like a collapsing series of increasingly large Chinese boxes or Russian dolls (fig. 1.9). In the branched serial structure, each branch must collapse into its founding entity, which then fits within its founding entity on the earlier branch, if new branches arise from a split that produces two equivalent sisters within each larger one. Thus the branching relations among the units are retained inside the collapsed structure. While thus embedded in trees, Hennigian taxa do not form functional assemblages, nor can they be compared; for those purposes they must be removed from their positions and associated or compared with other taxa.

If in comparative studies the groups of interest are holophyletic, Hennigian taxa, removed from their tree, are precisely as useful as Linnean ones. However, if it is desired to study one or more groups that have descendants that are excluded from the study—paraclades—then there are no Hennigian taxa available for those groups. When dealing only with lower taxa in crown groups that do not have deep roots, paraclades are not usually required. The fact that when these groups are being compared they are stripped from the tree and placed in entities with the properties of hierarchical ranks can (and usually does) go unnoticed; it seems to be a mere technical detail. When dealing with clades that are rooted more deeply in geologic time, though, and that have undergone significant morphological changes, the use of paraphyletic taxa permits a more accurate depiction of their evolutionary history than does use of holophyletic taxa, which cannot evaluate important aspects of morphological change.

Cladistics Is a Systematics Based on Trees

The methodology of Hennigian systematics has been elaborated in a large literature (e.g., see accounts in Ridley 1986; Panchen 1992; Smith 1994; Edwards 1996; Kitching et al. 1998). A major aim of cladistic methodology is to establish the most parsimonious hypothesis of relations among taxa, on the available evidence. In cladistics, as relationships are strongest along lines of descent, all of the taxa descending from a clade founder are considered to be more closely related to the founder than to any of the collateral relatives of the founder. Commonly, collateral relatives resemble each other more closely than descendant relatives. Therefore, morphological similarity per se cannot be used to establish degrees of relationship. Rather, the distribution of morphological characters is used to infer the most parsimonious or likely branching pattern among taxa that could have produced their observed distribution.

Some Cladistic Terms Are Hypotheses as to the Evolutionary Status of Characters

The most likely evolutionary branching order of a group of species is determined by the distribution of homologous characters (morphological, molecular, or of some other attribute), which can be indicated in a figure called a cladogram (fig. 1.10). Structurally, cladograms are trees, but they are not phylogenetic trees. Consider a group of three species. A character that is held in common by all of the species is

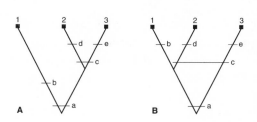

likely to have been inherited by them from a common ancestor that possessed that character—it is a shared ancestral character. Ancestral characters are termed plesiomorphies, and shared ancestral characters are symplesiomorphies. Such ancestral characters are of no use in determining the branching order among these three species, as all of the species possess the characters, which provide only some weak evidence of a relationship. If only two of the species have a character in common, however, it may be that they have a common ancestor that is not shared by the species that lacks this character. If so, the character is a derived character with respect to the last common ancestor of the three species. Derived characters are termed apomorphies, and shared derived characters, as possessed by the two species

FIG. 1.10 Two cladograms based on the same distribution of five characters (a–e) among three species (1–3). (A) The cladogram that implies the simplest history, which assumes neither loss nor duplication of characters: a is plesiomorphic for 1, 2, and 3 (thus a shared primitive character); c is apomorphic for 2 and 3 (thus a synapomorphy or shared derived character); and b, d, and e are apomorphic for 1, 2, and 3, respectively. (B) A more complicated cladogram. To account for the distribution of c, one has to infer either that it is plesiomorphic for 1, 2, and 3 but lost in the ancestry of 1 after 2 branched off, or that it evolved independently in 2 and 3. In either case the interpretation involves more evolutionary steps than are required for cladogram A.

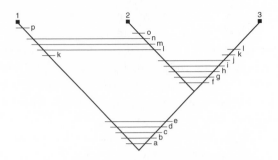

FIG. 1.11 Most parsimonious cladogram of sixteen characters (a–p) distributed among three species (1–3) that display numerous homoplasies. See text for discussion.

here, are synapomorphies. If one of the two species with this synapomorphy displays an additional character not found in the other, then it is likely that this character is an apomorphy that has originated after branching between these two species. The cladograms in fig. 1.10 indicate the distribution of five characters among three species. The lines in a cladogram are not lineage traces within a phylogenetic tree, but rather they represent a hypothesis about the relationships among the species according to the relative derivation of their characters. A cladogram can be used to inform the construction of a phylogenetic tree, and it can be used as a basis for a classification.

Interestingly, the preparation of cladograms has revealed the great extent to which the distribution of characters commonly used in classification can be at odds with evolutionary history. In fig. 1.11, species 2 and 3 have evidently five shared derived characters. However, 1 and 2 share three characters not found in 3, and 1 and 3 share one not found in 2. Interpretation of this pattern is uncertain without further data. It is possible that the characters shared only by 1 and 2 and by 1 and 3 have arisen independently in the species lineages where they are now found—such nonhomologous but similar characters are termed homoplasies (Lankester 1870; see discussions in Sanderson and Hufford 1996; the term is now used for either morphological or molecular cases). If two features are similarly but independently transformed, they produce a parallel homoplasy, and when two features converge on a similar character from different ancestral characters, they produce a convergent homoplasy. Most morphological homoplasies are due to the independent evolution of similar solutions to common functional problems. Homoplasies may also be caused by character reversals. For example, in fig. 1.11 it is possible that one of the characters common only to species 1 and 2 may be plesiomorphic for all three species but has reverted to an ancestral condition in species 3. The cladograms of the characters depicted in figs. 1.10A and 1.11 are the most parsimonious of the possible cladograms—that is, they show the arrangements that require the fewest

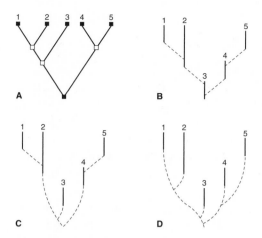

FIG. 1.12 (*A*) A cladogram of five species (1–5) assumed to be unambiguous for the characters studied. (*B–D*) Three trees that can be constructed from the cladogram in A, each with a unique branching pattern of lineages ancestral to the species. From Schaeffer et al. 1972.

character changes. It does not necessarily follow that the historical branching order produced the most parsimonious character arrangement, however. It is possible that all five apparent synapomorphies shared by species 2 and 3 in fig. 1.11 are actually plesiomorphies that have been lost in species 1, for example.

Computer programs are available that determine the preferred arrangement of taxa according to the characters that are input, depending on the criteria used, such as parsimony (see list in Swofford et al. 1996). Some programs produce arrays of cladograms, arranged for example in order of their relative degrees of parsimony. However, caution is clearly indicated when interpreting character distributions as phylogenies. Even correct cladograms do not automatically lead to an unambiguously correct tree. Fig. 1.12 depicts a cladogram (A) and three different trees (B–D) that it represents equally well. If there is, for example, good fossil evidence to indicate the order of lineage branching with respect to the origin of the characters used in the cladogram, the correct tree might be identified.

One way to reduce uncertainty as to the status of characters is to use two or more allied taxa as outgroups. These allied forms are not positioned inside the cladogram, but are used "outside" to help decide the polarity of characters, for if characters appear in the outgroup it would strongly suggest that they were plesiomorphic with respect to the taxa in the cladogram (Watrous and Wheeler 1981). However, even outgroup comparisons may mislead. A feature may have evolved independently in the outgroups, or a feature that is truly plesiomorphic may have been lost from the outgroups as well as from some of the taxa in the cladogram. Choosing multiple outgroups that are as closely allied as possible with the taxa under study will minimize such problems.

In Cladistic Classifications, Branch Points May Define Sister Taxa That Are Holophyletic

Clades can be based on any of three criteria: node-based, being founded by sister taxa at their common branching; stem-based, founded by the earliest member of one branch; or apomorphy-based, founded when a novel feature makes its first appearance (see de Queiroz and Gauthier 1992). As the definition of phyla used here involves morphology, the apomorphy-based definition is preferred, and can be used for groups in both Linnean and Hennigian classifications if they happen to be holophyletic.

Cladistic classifications exploit the characteristics of a positional structure. When branching occurs to produce two lineages in place of one, the daughter clades are termed sister taxa regardless of their eventual diversity or distinctiveness. One sister may fail to produce any daughter lineages and quickly become extinct, while the other sister may produce thousands of descendant lineages that include distinctive morphological novelties. Thus although sister taxa come as close to representing a rank as can occur in cladistics, they form quite a different sort of entity than a rank in a hierarchy.

Within higher taxa, a distinction can be made between clades that have living representatives and those that are entirely extinct (Jefferies 1979). All of the living representatives of a higher taxon have a last common ancestor, and this ancestor and all its descendants form a crown group, a clade that includes all branches leading to the living "crown" of the phylogenetic tree (fig. 1.13A). However, the higher taxon may have numbers of fossil representatives that branched off earlier than the last common ancestor of the living taxa. These earlier branches form a stem group. The stem and crown groups together form a total group, the last common ancestor of which founded the clade, and which is holophyletic. Because of the nesting pattern of cladograms, nodes that lead to inclusive crown groups may have stem groups of their constituent sub-taxa intercalated between their earlier nodes (fig. 1.13B). Stem-group taxa expand the morphological features and, usually, the morphological disparity of a phylum beyond that of the crown group, while at the same time they lack crown-group synapomorphies. Indeed, if the total group is based on the sister species that first founded the clade, total groups that include stem groups cannot be defined morphologically (see Runnegar 1996; Valentine 1996).

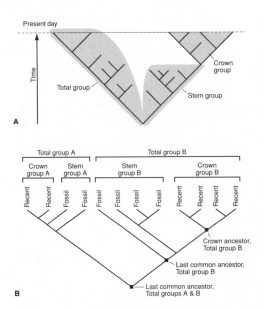

FIG. 1.13 Stem, crown, and total group from a cladistic standpoint. (A) A hypothetical tree. (B) A hypothetical cladogram that identifies a number of stem and crown groups within a major taxon that is divided into two total groups. From Smith 1994.

In dealing with the classification of crown and stem groups, many of the differences between Linnean and Hennigian approaches to classification are exposed, and some remain unresolved. In practice, the limits of most total groups at very high taxonomic levels are based on the distinctive features possessed by the living members of the clades. In the case of living phyla, their distinctive architectures mostly occupy unique regions of morphological space (morphospace). In other words there is in general a morphological gap between the crown groups of phyla, and this Linnean-like criterion is used to define the crown ancestor, thus creating a ranking criterion. The two crown ancestors of sister phyla, which contain features that are derived with respect to each other but are shared by their respective living taxa, are also separated by a significant morphological gap. However, there is no morphological gap of any significance between the sister species that found stem-based clades in which separate phyla evolve, and early branches of both clades will share many morphological features for a period of time. Similarly, there is no morphological gap between the founder of a node-based clade and its ancestral species, both sharing the same bodyplan. The same may be said for the last ancestor of, and the founder of, an apomorphy-based clade, although the apomorphy at least provides a morphological distinction between them.

Consider the changing morphospace within the stem group of a phylum when it is traced back through successively earlier nodes from the ancestor of the crown group (the crown ancestor) to its last common ancestor with a sister phylum. The crown synapomorphies are lost immediately. Some stem-group branches may be quite diverse, and their derived features may be so morphologically striking that in Linnean classification these branches would rank as extinct classes or orders of the phylum, ranks they cannot be assigned in cladistic classifications. Proceeding toward the last common ancestor with a sister phylum, the constellation of features that constitute the unique morphology of the phylum continues to erode. Eventually the morphologies at the nodes on the main line toward the common ancestor become dominated by features that are plesiomorphic. Finally, unique features of the phylum can no longer be recognized, and by definition the phylum is no longer present.

As phyla appear in the fossil record quite abruptly, the earliest branches of their stem groups, which would present the most difficulty in classification, are in general not known. However, the branches of stem groups share many characters with the crown group and thus can be allied with them on criteria of general resemblance. Smith (1994) discusses these and allied problems clearly and in far greater detail. Neither the earliest prestem descendants of common ancestors between phyla, nor the common ancestors themselves, are known for any phylum.

Phyla Have Split Personalities

Like photons and other entities that can be simultaneously waves and particles, phyla can be simultaneously branches in a phylogenetic tree and ranks in a hierarchy. Although trees do not have inherent ranks, ranks can be imposed by the use of nonpositional criteria. Phyla, then, are monophyletic branches whose members have morphological themes, which may be used to define a rank. Linnean-style nomenclature is ingrained and familiar and is quite useful in communicating clade characteristics, so it is common for workers examining some clade feature such as its diversity to base their studies on nodes within cladograms that correspond to the appearance of synapomorphies that define Linnean taxa. As Linnean taxa are ranked, comparative studies of clade richness are commonly conducted in what are technically hierarchies, although clades are being compared. The transformation is a consequence of the differences in the properties of positional and hierarchical structures, and exploits the best features of each type. In general, positional structures are best for defining the unique features of the branching system of the tree of life, while hierarchical structures are best for comparative studies involving taxa within or between levels that are defined on some morphological criteria.

The lag in appearance of taxon-defining synapomorphies after the origin of sister groups presents a significant difficulty in transformations between taxonomic hierarchies and trees. In a hierarchy based on morphologically similar monophyletic taxa, the segments of a tree between the origin of sister groups and the origin of defining morphological synapomorphies belong to an ancestral taxon (or to paraphyletic taxa), while in a cladistically based tree, they are basal and near-basal members of their clades. This situation is as true for species, which commonly originate cladistically within conspecific populations, as it is for phyla. However, this difference need not be a deterrent to the use of either sort of structure when it is useful to attack or illustrate the problem at hand. As phyla are Linnean taxa, they will be defined morphologically here, but their positions in clades clarify their histories and form a major component in any explanation of their origins.

Molecular Branchings Can Define Clades, while Morphological Features Define Linnean Taxa

There is a significant disconnection between trees of morphologically based taxa and the molecular trees that are used to determine their phylogenetic relationships. This disconnection becomes increasingly great at higher taxonomic levels. The lineage splits that first isolate molecules occur initially within populations of the same species. A molecule may become polymorphic before speciation, and thus diverge even before the lineage splitting, or, if the molecule evolves slowly, molecular-sequence divergence may not occur until some time after the lineages have branched. The sequences of different genes may diverge at different times and be sorted into different lineages within parental species. With these caveats, however, the time of final branching of the ancestral lineages should usually mark the approximate geologic time of divergence of ancestral molecules, though not necessarily of the lineage morphologies. After splitting, the daughter lineages may or may not be morphologically identical at first, but they will certainly share a conspecific similarity.

In the general case the speciation will have nothing to do morphologically with the remote futures of the lineages. As phyla are defined morphologically in Linnean classifications, they do not appear until their characteristic features evolve, perhaps millions or tens of millions of years after the founding of the molecularly defined clades. The molecular tree branches at the split, thus creating sister clades; a phylum originates after significant morphological evolution along a branch.

Bodyplans Consist of Evolutionarily Disparate Features

Bodyplans Are Polythetic

As Linnean taxa share lots of morphological features—more general features at higher and higher taxonomic ranks—members of phyla were grouped together because they share some very general features. The existence of such groups of organisms with characteristic features led to attempts to encapsulate their designs as sort of ideal architectures, of which the various subtaxa within the phyla displayed modified versions. These idealized architectures have been variously called Baupläne, ground plans, ground patterns, and body plans, and in some cases they have been used as a basis for rather philosophical hypotheses of how evolution proceeds. Here I will use the term bodyplan for an assemblage of morphological features shared among many members of a phylum-level group. Not all of the features that characterize a given architecture need be present in every member of the phylum, however, and many of the features may be found in other phyla; it is the assemblage of numbers of these features that is unique. Classifications based on unique combinations of features drawn from a pool of characterizing (but not necessarily unique) features are termed polythetic (see Beckner 1959; Sokal and Sneath 1963). All Linnean taxa are polythetic in principle.

Important Bodyplan Features May Be Plesiomorphies or Synapomorphies, and May Be Homoplasies

In any given phylum, some elements of the bodyplan evolved in ancestral forms and are not unique to the phylum (plesiomorphies). Other elements evolved at the founding of the phylum, and are unique to the phylum (synapomorphies). For example, the arthropod bodyplan involves mesodermal body-wall muscles and jointed appendages. The mesodermal muscles originated in remote ancestors with a different bodyplan and are shared with many phyla, while the jointed appendages are derived features of arthropods. Defining a particular phylum involves the transformation of a treelike evolutionary pattern into a hierarchy. To establish the origin of the phylum, one must pick a position on the tree at which a particular set of morphological features are first present. In the case of arthropods, the defining features would involve the evolution of jointed appendages, presumably from lobopods of some sort (chap. 7). This innovation would not arise overnight, but rather there would be a series of transitional forms that show a sequence of change from an ancestral bodyplan. The point at which one could say that the arthropods had branched from their sister group would have to be subjectively chosen, although once chosen it could be defined objectively.

Following the establishment of a phylum by the appearance of one or more apomorphies, additional characters appear as the phylum diversifies. Judging from the fossil record, it is a common pattern for morphological disparity to increase rapidly during the early history of a phylum (chaps. 5, 12). Some of the features that evolve after the defining morphology has appeared may be present in many or all of the major clades within the phylum, and form important elements in their bodyplans. For example, the sort of regionation called tagmosis in arthropods (chap. 2) may represent such a feature. Furthermore, a branch of a phylum may, during the course of its evolutionary history, lose some of the pool of characteristic bodyplan features. This situation commonly arises when organisms are miniaturized or when they become parasitic, in either case adapting to highly distinctive ways of life that permit them to dispense with many ancestral morphological characters. The pentastomes (tongue worms) form an impressive example; these are wormlike parasites that lack jointed appendages and have nonchitinous cuticles. The earliest known pentastomes are from near the Cambrian-Ordovician boundary (Walossek, Repetski, and Müller 1994). Pentastomes have frequently been considered to form a phylum of their own, so distinctive is their morphology, but there is morphological and molecular evidence to indicate that they are descended from crustaceans (Wingstrand 1972; Abele et al. 1989) and therefore branched within the phylum Arthropoda. Thus in this case one has a choice between the topology of the tree and the nature of the bodyplan per se to determine the taxonomic rank accorded the pentastomes; if they are considered a phylum, their ancestral arthropod groups become paraphyletic. Some workers using Linnean taxonomy tend to downplay the significance to be given any such "secondary modifications" and are content to permit organisms with even such extreme morphologies to remain within their parent phylum, although they hardly meet the received morphological criteria, even polythetic ones, for inclusion. In cladistic classifications, such morphologically simplified branches are recognized by character losses and do not raise any special complications, as morphology per se is not a concern in classification.

With higher taxa the criteria for severing the continuum of descent are not codified, and, indeed, dissatisfaction with subjective judgments as to their limits has prompted more than one attempt to alter or abandon the Linnean system (see de Queiroz and Gauthier 1992). The best possible situation for the transformation of a phylogenetic tree into a Linnean hierarchy would be if evolution had provided discrete clusters of closely allied taxa, between which there were rare, rapidly evolving lineages, and if the clusters were themselves clustered, with the most rapidly evolving lineages separating the clusters of the highest order. The continuum of descent could then be broken into the series of clusters at the intervals of rapid evolution; by convention, the rapidly evolving intermediate lineages could be assigned to the ancestral taxa. Phylogenetic trees are often sketched as if this situa-tion were the historical norm, with congeneric species clustered within families, and the families well separated by long branches, commonly with abundant question marks that indicate uncertain connections between families.

There would seem to be some truth to this pattern, for evolution has certainly not produced a regular correlated sequence of changes among or within major phy-logenetic branches. Rates of evolution have varied significantly among and within branches throughout life's history, and many of the branches, large as well as small, are cryptogenetic (cannot be traced into ancestors). Some of these gaps are surely caused by the incompleteness of the fossil record (chap. 5), but that cannot be the sole explanation for the cryptogenetic nature of some families, many invertebrate orders, all invertebrate classes, and all metazoan phyla. Furthermore, phyla for which we have good early records usually achieve great morphological disparity early in their histories (chap. 13), for example, Arthropoda (Briggs et al. 1992), Bryozoa (Taylor and Curry 1985; Anstey and Pachut 1995), and Echinodermata (Paul 1977; Campbell and Marshall 1987).

Many of the disparate lineages arising from such early morphological radia-tions are recognized as Linnean classes or orders and are cryptogenetic themselves. It is not uncommon for these classes and orders also to achieve great morphological disparity early in *their* histories. Particularly well quantified examples are in the echinoderm taxa Blastozoa (Foote 1992a) and Crinoidea (Foote 1994a, 1994b), and there is qualitative evidence of similar patterns within other invertebrate and vertebrate groups, including the dinosaurs (Sereno and Novas 1992; Benton 1993). This pattern, though common, is not universal; Trilobita, for example, displays greatest morphological diversity well after its initial radiation (Foote 1991). Lower taxa such as families and genera, which can sometimes be connected to an ancestral Linnean taxon of the same or lower rank, often display a gradual rise to their peak morphological diversities rather than an abrupt one.

This situation is somewhat counterintuitive, for if the origin of higher taxa (as used here, orders and up) were similar to that of lower taxa (as used here, families and down), one would expect that there would be fewer unbridged gaps between ancestral and descendant taxa at higher ranks. The period of divergence required to produce a higher taxon would naturally be far longer than that required for the origination of a new species or to accumulate differences that merited recognition of a new genus, and intermediates should be found during that period. The abruptness of the first appearance of taxa at increasingly higher levels strongly suggests that a different tempo of morphological evolution was commonly operative in higher taxa than in lower taxa. Whatever the cause of this phenomenon, the record of abrupt appearance of higher taxa provides a basis for the erection of morphological criteria for their recognition and for the severing of the phylogenetic continuum for purposes of classification. The pattern of differences in key developmental genes within the genomes of phyla also suggests that phyla are "real" evolutionary entities (chap. 3).

Even if the origin of novelties has proceeded in bursts at several levels, there is still the problem of ranking. Evolution has certainly not given us a neat series of ordered clusters; cladogenesis has been reasonably independent and perhaps largely random along different branches (though bursts of originations seem to be correlated among some branches), producing idiosyncratic clusters within the various taxa. Clearly, their rankings have commonly been resolved qualitatively and somewhat subjectively.

The number of ranks at which disparity may be recognized within phyla varies greatly. Within the Phoronida, known only from a dozen or so living species, the species can be divided into genera, but these are so similar morphologically that only a single family is used (Emig 1979), with an implied monotypic order and monotypic class. Thus the pattern of morphological disparity in phoronids is recognized in such a manner that phoronid taxa are somewhat commensurate with taxa in the same nominal ranks in other phyla.

A contrasting situation is presented by Arthropoda, today the most diverse phylum overall. In order to contain this record of arthropod diversity, hierarchical classifications commonly employ more levels than are usual for other phyla. However, it should be possible to relate the principal levels (phylum, class, order, family, etc.) to some normative range of morphological disparity displayed at similar ranks in other groups. The use of multiple ranks intercalated between the principal ranks organizes the staggering diversity into manageable entities and can obviously be of great utility (assuming that the entities are monophyletic, though not necessarily holophyletic). The major arthropod clades have as defining features characteristic patterns of body subdivisions, and appendages of different types in characteristic patterns on segments within the body subdivisions (chap. 7). Such patterns are sufficiently consistent to contain nearly all living arthropods and most extinct ones (trilobites also have a characteristic pattern). There are exceptions, however, especially in the Cambrian fauna, which has a variety of organisms that are clearly arthropods but that have patterns of segments and/or appendages not found among living taxa. These are stem forms that fall morphologically and presumably evolutionarily outside of living subphyla or classes (some that can be placed in a living class cannot be placed in a living order), but do not seem to have represented highly diverse clades. It appears that, early in arthropod history, many lineages evolved that displayed significant variation in characters that define the major living groups, and that at the time were on a more or less equal footing with the surviving clades. By the logic that permits classification by monophyletic taxa that are arranged hierarchically and ranked by morphological disparity—Linnean taxonomy—these extinct forms should be subphyla or classes. By the logic of cladistic classification, not recognizing higher and lower taxa but rather earlier and later ones, these forms should be positioned according to their ancestral branching patterns (which unfortunately are very poorly known; but see Briggs and Fortey 1989). Budd and Jensen (2000) argue against including the features of stem taxa when assessing bodyplans of phyla. However, stem taxa should provide more evidence of the ancestral features of a phylum than do the crown taxa. Restriction of bodyplan features to crown taxa confuses factors associated with the origins of taxa with those associated with their extinctions (Collins and Valentine 2001).

Systematic Hierarchies and Trees: A Summary

Organisms with the characteristic bodyplans that we identify as living phyla appear abruptly in the fossil record, many within a narrow window of geologic time—perhaps 5 to 10 million years, beginning about 530 Ma (chap. 5). Nearly all of these are stem taxa. Some fossils are known that may represent extinct phyla, and these appear chiefly during this same window. It is consistent with the fossil record that all of the characteristic animal bodyplans had evolved by the close of this period, but none of them can be traced through fossil intermediates to an ancestral group. The organisms belonging to many of the phyla can be further subdivided into distinctive morphological groups that possess characteristic body subplans. In no case is a morphological continuum found across a broad range of bodyplan morphologies, nor do phyla resemble each other more closely during their early fossil histories. It is possible to aggregate the organisms within each phylum into a hierarchy of morphologically based subdivisions. At less inclusive subdivisions—lower taxa— intermediate morphologies are commonly found. The Linnean hierarchy employs the subdivisions as nominal taxa. Phyla are thus taxa representing a high-level subdivision or rank within the animal kingdom, and they are conceptually polythetic.

The advantages of the hierarchical aspect of phyla are that it produces a system that renders the complexity of relationships among organisms tractable, and that it permits comparative studies among taxa within and between ranks. For taxa, there is more subjectivity involved in constructing a hierarchical classification than a cladistic one, because the irregularities and messiness of the natural hierarchy must be organized into elements and ranks, requiring decisions as to their limits and equivalence. On the other hand, there is more information in the hierarchy than the tree, for it includes estimates of morphological similarities at various levels. When the morphologies within several diverse phyla are each aggregated into a series of, say, six ranks, within each of which the morphological disparities represented by related taxa are kept within a similar range, then each rank may represent a reasonably similar morphological product, and the use of ranks in exploring evolutionary dynamics seems quite legitimate; in doing so, one is not implying that the ranks are precise equivalents. Because dynamics within hierarchies display regular trends (chap. 12), the reassignment of a taxon to a higher or lower rank will not ordinarily provoke a major deformation of dynamic models.

Evolutionary theory, bolstered by a wide array of evidence, predicts that all metazoans have descended from a common ancestor, and molecular evidence indicates that the last common ancestor was probably a metazoan itself, and the

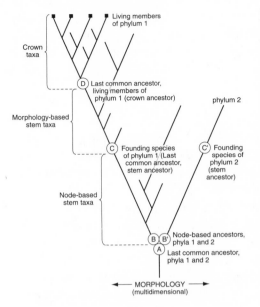

FIG. 1.14 Relations between the branchlike (phylogenetic) and the ranklike (morphological) aspects of phyla. Parental species A gives rise to two daughters, species B and B′, each of which founds a clade that produces a new bodyplan, phylum 1 and phylum 2, respectively. Species A is the last common ancestor of the two descendent phyla. The node-based ancestors B and B′, with the same bodyplans as species A, are not members of the descendant phyla. Phylum 1 is established when a number of characters give rise to a new bodyplan at species C. If species C is unknown (the universal case), its bodyplan can be hypothesized by reference to homologies among descendant taxa and to the nature of the ancestral bodyplan, if known. Species C and its descendants, except for species D and its descendants, are stem taxa; species D and its descendants are crown taxa. Modified from Valentine and Hamilton 1997.

kingdom is therefore likely to be holophyletic. The phyla are assumed to represent branches on a single diversifying tree of metazoan life. Some of the phyla, such as Porifera, must be paraphyletic; it is not certain whether any bilaterian phyla are paraphyletic, though it seems likely (e.g., Acoelomorpha, chap. 6, and Onychophora, chap. 7). Tracing these branches back along their continua of descent, they must converge and unite in some particular order, until finally the single root lineage is reached. As no fossils representing the union of any phylum-level branches are known, the bodyplans present at the branch points must be inferred from indirect evidence.

Fig. 1.14 portrays something of the dual nature of phyla. Species A gives rise to daughters (B, B′) that each happen to found a clade from which a new bodyplan evolves (phylum 1 and phylum 2). Species A is the last common ancestor of those phyla. Species A, B, and B′ are about as closely related as species can be, and so all must have the same bodyplan, even though species B and B′ are each the ancestor of a clade that eventually represents a separate phylum. Although it is possible that the morphological differences between the daughters and their parent species are steps toward their eventual distinctive bodyplans, this would be unusual. Morphological changes arising during any given speciation event are likely to be associated with adaptations to some local conditions and would rarely open up opportunities for the evolution of a new bodyplan. Eventually, however, the descendants of one of the clade ancestors, species B, accumulate derived characters that form the basis of a novel architecture, which is first realized after an indefinite period of time (it could be many tens of millions of years) in species C. It is a successful bodyplan and is modified into a number of subplans, and species C becomes the founder of phylum 1, in keeping with its morphological definition. Only species

C and its descendants can be considered to belong to phylum 1, in keeping with its phylogenetic definition. If the interval between species B and C was very long, numerous branchings may occur there, which would not belong to phylum 1. During the early history of phylum 1, between C and D, numbers of other branches are likely to arise, which become extinct. These are by definition stem branches, and species C is the stem ancestor. Branches that survive to the present day are likely to have a last common ancestor that lived well after the phylum was founded. In fig. 1.14, the last common ancestor of living species, the crown of the phylogenetic tree of this phylum, is species D; this is the crown ancestor. The members of phylum 1 that branched between species C and D are stem taxa, while species D and all subsequent members of the phylum are crown taxa. For this book, there are three classes of ancestors: node-based clade ancestors (which have sister species in another phylum); stem ancestors (which first achieve the morphological criteria for inclusion in the phylum); and crown ancestors, defined as depicted in fig. 1.14.

The species that found the clades within which the two phyla evolve, B and B′ in fig. 1.14, are important in that they represent an ancestral bodyplan from which phyla have descended, thus establishing the morphological trajectories that were traversed by evolution. Hypotheses about their morphologies may be constructed by reasoning from homologies among their descendants and from the nature of the ancestral bodyplan. In a morphologically based taxonomy, intermediates along or branching from the trajectory between the node-based ancestor and the stem ancestor of a phylum should, logically, be assigned to an ancestral phylum. For example, if species A and B were onychophorans, the lineages between B and C would represent a (rather derived) branch of that phylum. It would be wonderful if there were lots of these intermediates to puzzle over, but, unfortunately, there seem to be few to trouble us.

Each of the incarnations of a phylum, as rank or branch, provides not only a different perspective but permits a different body of evidence to be used to assess their natures and to explore the evolutionary mechanisms that were in play, both during the rise of these great morphological novelties, and during their subsequent histories. These two ways of observing phyla are complementary.

CHAPTER 2

Design Elements in the Bodyplans of Phyla

The wonderfully rich comparative studies of classic morphologists have provided an invaluable account of the design elements that underlie invertebrate structure and function. The growth in the number of phyla recognized between the mid-nineteenth century and today, from four to over thirty, occurred chiefly because those studies revealed distinctive differences in invertebrate bodyplans. For the most part, relatively simple biomechanical principles underlie the architectures of invertebrates. Even when different lineages exploit similar principles, however, differences in their ancestral morphologies and in the conditions that they have encountered during their evolution have produced distinctive morphological adaptations—although the principles incorporated in the bodyplan designs are sometimes similar, the designs themselves are distinctive. Furthermore, organisms are adapted to a spectrum of particular conditions and requirements that produce adaptive compromises. Many lineages of organisms have encountered conditions that have evoked entirely new (for them) sets of biomechanical or physiological adaptations, which have been achieved by rather tortuous modifications of ancestral morphology, as the organisms were not "designed" from scratch for the new conditions (barnacles certainly form a case in point).

At any rate there is a relatively restricted number of design elements associated with the major developmental stages and bodyplans of phyla. General accounts of these elements organized by phyla can be found in invertebrate biology texts and in more detailed or specialized accounts in Hyman 1940, 1951a, 1951b, 1955, 1959; Beklemishev 1969; and Nielsen 1995, 2001. Stachowitsch (1992) has produced a very useful glossary of morphological terms for invertebrates, group by group. Especially valuable reviews are by Bullock and Horridge (1965) on invertebrate nervous systems and by Bereiter-Hahn et al. (1984) on invertebrate integuments. Many other important contributions, to which reference should be made for more detailed descriptions and documentation, are cited below and in part 2.

Cells Are the Basic Building Blocks of Metazoan Bodies

Following the founding of cell theory in 1838, and especially after Schwann's famous treatise (1839), the notion that cells are building blocks of all living things has become absolutely fundamental to our understanding of how life is organized. The

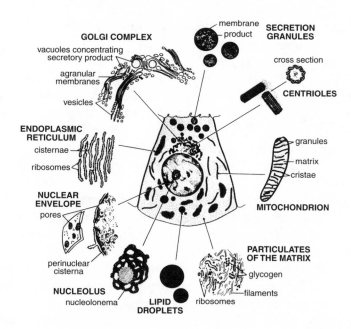

FIG. 2.1 A metazoan cell with its complement of major organelles. Cell as visualized by light microscopy, details of organelles as visualized by electron microscopy. From Fawcett 1994.

properties of metazoan tissues and organs are derived from the properties of the cells of which they are composed, and therefore the evolution of differentiated cell types is a most important part of the origin of metazoan bodyplans. A generalized metazoan cell is depicted in fig. 2.1. The earliest multicellular animals must have inherited their basic organelles, cytoarchitectures, and cell functions from their unicellular ancestors. Detailed accounts of cell types of invertebrate phyla may be found in Harrison et al. 1991–1999. A general account of (chiefly metazoan) cells is in Alberts et al. 1989.

Cytoskeletons Provide the Framework for Cytoarchitectures

Metazoan cells have a cytoskeleton that comprises three main sorts of filaments: microtubules (25 nm in diameter, composed of the globular protein tubulin); microfilaments ("thin" microfilaments are 8 nm in diameter, composed of the globular protein actin); and intermediate filaments (8–10 nm in diameter, composed of a variety of fibrous proteins differing from cell to cell; keratin fibers are a common kind). Together these three filament types form the protein scaffolding involved in the skeletal and motor functions required for cell division and movement. Parts of the cytoskeleton are highly dynamic; the average life of a microtubule, for example, is about 10 minutes. During the life of a cell, many structural elements of the cytoskeleton are assembled and disassembled as new elements are generated, particularly during active phases of the cell cycle.

Intermediate filaments are more stable. These filaments form a ropy cage around the nucleus and extend outward to the cell periphery, where they form an open network beneath the plasma membrane. A network lining the interior of the nuclear envelope or lamina is also formed by a family of intermediate filaments. As intermediate filaments vary among cell types, the evolution of their genes is presumably implicated in the evolution of cellular differentiation within the metazoans. Intermediate filaments have not been identified in protistans and are evidently a metazoan synapomorphy.

By contrast, microfilaments have quite similar compositions and structures throughout the eukaryotes, and metazoans have clearly inherited them from protistan ancestors. The genes coding for actin molecules have been relatively well conserved during metazoan evolution, and the evolutionary history of the actin molecule is chiefly one of changes in the pattern of its expression. Microfilaments are developed in different amounts, in different geometries, and in association with different protein varieties in different cell types and different taxa. An important function of microfilaments is displayed by contractile fibers, wherein actin filaments are interlayered with filaments of the protein myosin to produce "thick" microfilaments. These filament stacks are bundled into sarcomeres—segments of muscle fiber. The action of muscle fibers is particularly well described in cross-striated muscles, found only in vertebrates and some arthropods. During contraction the different filament types slide past each other, so that, although the length of the filaments has not changed, the sarcomere is shortened. When numbers of sarcomeres form segments of a long fiber, their contractions can result in significant shortening. The energy for the filament sliding is derived from ATP hydrolysis, and is used to promote "walking" of myosin molecules along microfilaments. The myosin walks are unipolar, so that the contracted sarcomeres must be "stretched" back into their elongated states—their action must be antagonized, usually through contraction of other muscles, with the force sometimes transmitted hydrostatically.

Microtubules are generated at foci that have been termed, collectively, microtubule organizing centers. In unicellular eukaryotes, the number of microtubule organizing centers per cell varies from one to many, depending upon the taxon (see Margulis 1981). Spindle formation and chromosome movement during cell division are the best-known behaviors in which microtubules are implicated.

Microtubules form cilia and flagella. Cilia are commonly disposed in rows or tracts and usually beat much as oars stroke, either to carry fluids across a cell surface or to move a cell through a fluid. Flagella undulate longitudinally to send waves along their axes, either to propel a cell or to move fluid, usually along the length of the flagellum. Both these structures operate at very low Reynolds numbers (see below), where viscous forces are quite important. Fig. 2.2 depicts the role of microtubules in a flagellum. Nine microtubule doublets run the length of the

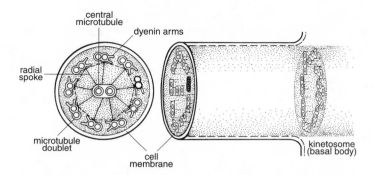

FIG. 2.2 Structure of a eukaryotic flagellum, cross section at left. Microtubules run the length of the flagellar axis; dyenin arms would "walk" pairs
of microtubules past each other, but spokes and struts prevent differential
microtubule movement, so the flagellum bends. After Margulis 1984.

flagellum, and are connected to a central microtubule assembly by radial spokes.
To bend the flagellum, molecules of the protein dyenin that are interspersed between
doublets change shape (employing energy from ATP), and this exerts a shear stress
between adjacent doublets. Doublets are prevented by the spokes and associated
struts from relieving this stress via differential sliding, and so the flagellum flexes
instead.

Cytoskeletal filaments and tubules are responsible in large part for establishing
the shape and internal geometry of cells. Microtubules are involved in altering cell
geometry during important activities or cell phases; for guiding the movements
of vesicles, organelles, and some other features within cells; and for effecting cell
locomotion, as in amoeboid crawling or ciliary swimming.

Metazoan Cells Have Descended from Protistans, Probably Choanoflagellate-Like

Both molecular and morphological evidence indicate that the earliest metazoans descended from unicellular protistan ancestors, though perhaps from colonial forms.
Protistan cells ingest materials ranging from "large" particles to small molecules.
Small molecules may enter the cell by infiltrating the plasma membrane, commonly aided by transport proteins. Large molecules or particles are ingested by
endocytosis—a pit forms to contain the material, and the membrane invaginates
and pinches off, forming an internal vesicle or vacuole containing the ingested matter. Particles (perhaps a small protistan ingested by an amoeba) are digested by
enzymes introduced into the vacuole through fusion with a lysosome. Fluids may
be ingested in vesicles or vacuoles as well. Materials can be removed from cells
by reversing the ingestion processes. Waste-filled vesicles may fuse with the plasma
membrane and then open to the exterior to release their contents into the extracellular environment (exocytosis).

Sponges (chap. 6) contain an important cell type called a choanocyte, which has a terminal flagellum surrounded by a collarlike structure and which is strikingly similar to the cells of the protistan group Choanoflagellata (see Fjerdingstad 1961; Laval 1971; Leadbeater 1985). Feeding in both groups is by ingestion of particles trapped at the collars of these cells. These morphological and functional similarities are taken as strong evidence that sponge ancestry is to be found among that group or a close ally. Other collar cells are known among metazoans, but none are so similar to choanoflagellates, particularly as regards the details of the collar and flagellar apparatus (Hibberd 1975). Furthermore, molecular phylogenetic data support a close relationship of choanoflagellates to metazoans (Sogin 1991, 1994; Wainright et al. 1993; Collins 1998). Molecular data also suggest that choanoflagellates and metazoans are allied to the Mesomycetozoa (Herr et al. 1999), a group of parasites and pathogens that have some fungi-like characteristics. In some species of choanoflagellates, the cell is encased by a lorica, which may be fibrillar or may be formed of siliceous costae. About 140 living choanoflagellate species have been described, including both pelagic and benthic forms (see Norris 1982; Leadbeater 1985). In some colonies cells are evidently undifferentiated and form benthic mats. Reproduction is by fission (Leadbeater 1985). At least one pelagic species forms spherical colonies of cells, attached by loricas, flagella

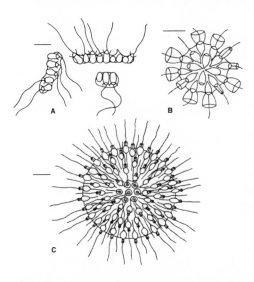

pointing inward (Thompsen 1976). Other forms display some cell differentiation. The chiefly freshwater genus *Proterospongia* includes species with biphasic life cycles (Leadbeater 1983): a unicellular stage is sedentary, while a colonial stage has peripheral collar cells, flagella outward, surrounding a number of amoeboid cells in a gelatinous or mucilaginous matrix (fig. 2.3). Margulis (1981) has suggested a plausible basis for such a combination of cell types (see also Buss 1987). A choanoflagellate cell has only a single microtubule organizing center and cannot produce a spindle and divide while maintaining a flagellum. Before division, the cell resorbs the microtubules of the flagellum, and the tubulin is then employed in spindle formation. Colonial choanoflagellates therefore contain both dividing cells and flagellar cells. Growth and maintenance

FIG. 2.3 Colonial stages of choanoflagellates of the genus *Proterospongia*. Cells are embedded within mucilaginous (A) or gelatinous (B, C) substances; presumptive dividing cells lack flagella. (A) *P. dybsoeensis*, marine; scale bar 10 μm. (B) *P. haeckeli*, freshwater; scale bar 10 μm. (C) *P. volvox*, freshwater; scale bar 20 μm. After Leadbeater 1983.

of a choanoflagellate colony might be promoted by the presence of a nonflagellar cell type that could divide to produce a flagellar daughter, without requiring any reduction in the number of flagellar cells that provide propulsive and feeding currents. In effect, such a cell type would be a stem cell. Like choanoflagellates, metazoan cells do not divide while bearing flagella or similar microtubule-based structures, such as axons. The origin and the character of metazoan cell differentiation may be rooted in this inheritance from choanoflagellate-like ancestors.

It seems plausible that many metazoan cell types arose by specialization of particular features of more generalized cells. For example, muscle cells may have arisen from generalized cells that exploited their microfilament complements for contraction and in which cellular elements that were employed for other purposes became reduced. At present, however, there is not much concrete evidence as to the detailed paths of descent of many cell types.

Cells Are Integrated into Tissues by Protein Bindings or Matrices

In metazoans, cells are organized into tissues, functional elements that may include one or many cell types. General accounts of tissues are by Welsch and Storch (1976), for animals in general, and by Fawcett (1994), emphasizing humans. More specific accounts of the tissues found in each invertebrate phylum can be found in the multivolume series edited by Harrison and others (1991–1999).

Extracellular Matrix Supports Metazoan Tissues
Metazoan tissues display novel features that are not present in unicellular organisms. In addition to cells, metazoan tissues contain an extracellular matrix that provides structural support for the cells, sometimes elaborated into a virtual armature on which thin tissue layers rest. Extracellular matrix has many physiological and developmental functions as well. Extracellular matrix consists of a gelatinous matrix of glycosaminoglycans, usually proteoglycans, infused with fibers, chiefly of collagen, and often containing an array of other macromolecules. Many of these molecular species appear to have arisen through a recombining of a number of protein domains (see Engel et al. 1994). Although the matrix materials are secreted within cells, the final synthesis of many of these molecules is accomplished extracellularly. Most metazoan extracellular-matrix proteins in general, and collagens in particular, have not been found in nonmetazoan kingdoms, and are evidently metazoan synapomorphies (Towe 1981; Morris 1993; Engel et al. 1994).

In Most Metazoan Tissues, Cells Are Connected or Anchored by Protein Molecules
In tissues, cells are connected to each other and to extracellular matrix via specialized molecules or molecular structures. These connections form attachments and/or permit cell-cell communication, and most are termed cell junctions (table 2.1; see Green 1984; Alberts et al. 1989). Nonjunctional adhesion proteins or receptors also

TABLE 2.1   Cell junction types common in metazoans.

Occluding junctions
   Septate junctions (unknown in urochordates and cephalochordates)
   Tricellular junctions (widespread among invertebrates)
   Tight junctions (in urochordates, chordates, and some specialized arthropod tissues)

Adherens junctions
   Adhesion belts ("belt desmosomes," actin filaments, connect to cells or extracellular matrix)
   Desmosomes ("spot desmosomes," intermediate filaments, connect to cells)
   Hemidesmosomes (intermediate filaments, connect to extracellular matrix)

Communicating junctions
   Gap junctions (unknown in placozoans or sponges)
   Synapses (unknown in placozoans or sponges)

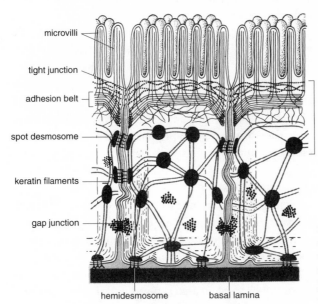

FIG. 2.4 Cells in an epithelial tissue (mammalian small intestine); the complex of junctions and the attachments to the basal lamina produce an integrated tissue sheet. From Alberts et al. 1989.

connect cells to other cells or to matrix. In some tissues there are several junction types (fig. 2.4). The various junctions, together with some nonjunctional adhesion systems, form complexes to produce a support system that helps with the mechanical integration of cells into tissues.

Occluding junctions prevent the intercellular passage of materials across cell sheets, though some sheets are more "leaky" than others in this regard. Septate junctions are a class of occluding junction that is widespread among invertebrates. Usually cells are separated by a 15–18 nm space, sometimes wider, that is crossed near the apex by a band of septa that encircles the cell; the septa include intramembrane components from the apposing cells. In some phyla (in most protostomes) the septa may be pleated and may sport peglike structures. Urochordates and chordates do not have septate junctions, and their roles are taken by tight junctions, in which apposed plasma membranes are sealed by belts of intermeshed strands and grooves that encircle the cells near their apical ends. Tight junction–like structures are also found in some specialized arthropod tissues. Tricellular junctions are found in a variety of invertebrates where three cells join; they consist of series of intercellular diaphragms and attachment structures affixed to the septate junctions of the three cells; they seal what would otherwise be open "corners" between cells.

Other junctions function as adhesion structures. Adherens junctions are either belts of contractile actin filaments ("belt desmosomes") that are attached to the plasma membranes of apposed cells, or are localized concentrations of actin filaments that anchor a cell to extracellular matrix. Other anchoring junctions involve intermediate filaments that may run cell to cell, termed desmosomes, or cell to extracellular matrix, termed hemidesmosomes; within the cell these filaments are attached to the cytoskeleton.

Additionally, there are communicating junctions. Gap junctions are clusters of tubes made from protein molecules that penetrate the plasma membranes of adjoining cells and permit the passage of small molecules (effective diameter to about 1.5 nm), and therefore of chemical and electrical signals, between cells. Another signaling system is found where neurotransmitters operate across synapses. Finally, there are some unusual junctions, such as the pore-bearing plug that connects cells with syncytia in hexactinellid sponges (unique to that taxon; see chap. 6).

Metazoans Have Several Major Types of Tissues

Most Tissues Are Epithelial or Connective
Tissues are usually grouped into two chief categories, epithelial and connective. Epithelial tissues (fig. 2.4) are sheets of cells that are polarized into apical and basal portions and interconnected in some fashion, usually by occluding junctions. Epithelia rest on a basal lamina of extracellular matrix to which they are attached, a feature that distinguishes them from connective tissue. Epithelia with basal laminae are present in living metazoans above the sponge grade. In connective tissues (fig. 2.5), cells are usually somewhat separated and embedded in extracellular matrix (although in acoel flatworms there appears to be no matrix in connective tissues; Rieger et al. 1991). When cells are encased in matrix, they may be free or attached by hemidesmosomes, and many are joined to other cells by various types of junctions. Both epithelial and connective tissues may incorporate a number of cell types (fig. 2.6).

Muscle Tissue May Be either Epithelial or Connective
Muscle tissue consists of elongate contractile cells (myocytes), commonly all of the same type. The cells are connected to each other or to neurons (via synapses), are attached to supporting structures (extracellular matrix, cuticle, or skeleton) by

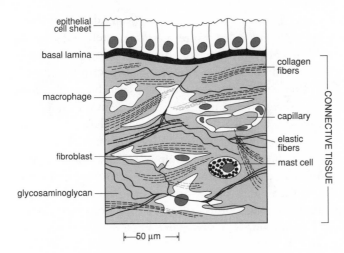

epithelial
cell sheet

basal lamina

macrophage

fibroblast

glycosaminoglycan

collagen
fibers

capillary

elastic
fibers

mast cell

CONNECTIVE TISSUE

|← 50 μm →|

FIG. 2.5 Cells in a verte-brate connective tissue be-neath an epithelial sheet; in this tissue most of the volume is taken up by glycosamino-glycan, which is invested by fibers. From Alberts et al. 1989.

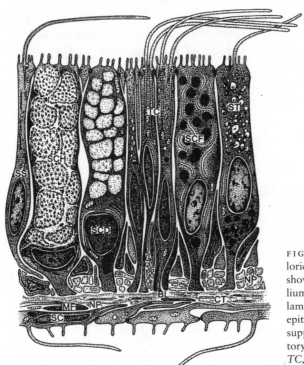

FIG. 2.6 Sketch of the wall of the py-loric cecum of an asteroid echinoderm, showing several cell types in the epithe-lium, and adjacent features. *BL,* basal laminae; *CT,* connective tissue; *MF,* myo-epithelial cell; *NP,* nerve plexus; *SC,* supporting cell; *SCD, SCF, SCH,* secre-tory cells; *SS,* sensory cell; *ST,* storage cell; *TC,* supporting cells in a transportation channel. From Chia and Koss 1994.

hemidesmosomes, and may be interconnected by adhesion belts. Smooth muscle tissue commonly consists of uninucleate spindle-shaped cells. Striated muscle tissue (such as the skeletal muscle of vertebrates) consists of multinucleate cylinder-shaped syncytia arranged in fascicles. The striations are caused by alternating anisotropic (A) and isotropic (I) bands within the muscle fibers that are in transverse register

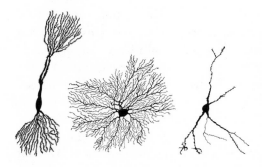

FIG. 2.7 Some neurons from various sites in monkey brain, a small sampling of the wide diversity of nerve-cell forms found in vertebrates. After Truex and Carpenter 1969.

across fiber bundles. Additional tissue types found in invertebrates include obliquely striated and double-obliquely striated muscles, in which A and I bands are staggered obliquely from filament to filament; cells in these tissues are uninucleate, at least in annelids (Lanzavecchia et al. 1988). Some invertebrates have several muscle types (e.g., the minute rotifers have smooth, striated, and obliquely striated muscle tissues), while others appear to have only one or two of these cell morphotypes (e.g., crustacean muscle fibers are all striated, although they display much functional diversity). It is not likely that all the morphotypes are homologous among phyla. Most metazoan muscle tissues are bounded by extracellular matrix and are classed as epithelia, though muscles in some forms, such as in some ctenophores and flatworms, are organized more like connective tissues.

Nervous Tissues Are Not Organized either as Epithelia or as Connective Tissues

Nervous tissues arise from ectodermal epithelia but invest most body tissues and are so specialized that they are usually treated as a separate tissue category. They form the main integrating and coordinating system of the body, and are present in all free-living phyla except sponges and placozoans. Sensory receptors, neuromotor systems, and central nervous systems are composed of nerve cells (neurons; fig. 2.7). Neurons exhibit an incredible array of morphologies, with lower invertebrates tending to display only simpler forms while vertebrates have the greatest variety and complexity—it is speculated that in humans the number of nerve cell types may be as great as all our other cell types combined. Despite their specializations, nerve cells possess a normal range of cellular machinery and junctions and have specialized communicating junctions as well.

Multinucleate (Syncytial) Tissues Are Found in Many Disparate Phyla

Many phyla have some tissues that are syncytial (containing multiple nuclei but not partitioned by cell membranes), particularly well known in hexactinellid sponges and in striated muscles throughout the Metazoa. They also occur as important tissues in acanthocephalans, rotifers, and flatworms.

Organs and Organ Systems Are Formed of Tissues

Organs are usually defined as a structure, such as a heart, that may be composed of more than one tissue, integrated so as to perform a specific function that serves the whole organism. Organ systems are composed of more than one organ, integrated so as to contribute to a specific function; they may involve multiple organs of the same type, such as multiple nephridia in an excretory system, or organs of different types, such as a cardiopulmonary system, which performs a larger but distinctive function. These definitions apply fairly well to vertebrates or complex invertebrates but are not so useful in lower invertebrates. Whole organisms that are composed of only a few cell types might be regarded as composed of single tissues. Indeed, Hyman (1940) considered Porifera to be at the "cellular grade of construction," little more than a "loose aggregation of cells hardly formed into tissues" and lacking organs and organ systems. However, I will follow sponge workers in recognizing that sponges possess tissues (such as pinacoderm and choanoderm; chap. 6), and, further, will regard structures such as the flagellated chambers of sponges as organs, and collections of flagellated chambers within the same organism as organ systems.

Organisms Are Best Understood as Developmental Systems

The design of multicellular metazoan bodies unfolds during growth and development from a single cell. The appearance of permanent novelties must be underlaid by heritable changes, and while some phenotypic changes may be due to changes in structural genes, those that affect the patterns of features in bodyplans must be due to changes in the gene regulatory system. Similarities and differences in early developmental features have been widely used as phylogenetic criteria, but, as with adult features, some of their patterns of distribution among taxa are inconsistent. Alternative phylogenetic schemes are sometimes suggested by different developmental features.

Classical observations of developmental phenomena are beginning to be informed by a growing understanding of the mechanisms that underlie them, thanks to the rise of developmental studies on the molecular level (chap. 3). It appears that many of the questions raised by the fossil record of bodyplan origination and diversification will eventually be answered by evolutionary developmental studies. For invertebrates there is a detailed treatment of the early nonmolecular literature by Kume and Dan (1968), a clear, concise summary of classic features of embryology by Brusca (1975), and a multiauthored survey that includes up-to-date reviews of the general embryology of some major phyla edited by Gilbert and Raunio (1997). Mechanistic approaches have been discussed by Davidson (1986, with supplementary reviews of 1990, 1991, and 1993; 2001) and others cited. Fine reviews of molecular aspects of developmental biology are by Gilbert (1997), which is more encyclopedic, and Wolpert et al. (1998), which is more general. Evolutionary aspects of development are treated by Raff (1996), Gerhart and Kirschner (1997), Carroll et al. (2001), and Wilkins (2002).

Cleavage and Cell Differentiation Are Linked in Most Metazoan Ontogenies

Development in metazoans usually begins with an egg or oocyte that is fertilized, although in some cases development may begin from cells that have not been sequestered as gametes, or from unfertilized gametes as in parthenogenesis. In any case, the adult metazoan body is composed of a number of somatic cell morpho-types (between three and several hundred in living metazoans), all of which have descended from a zygote or other founding cell, and which (with a few exceptions; review in Davidson 1986) have identical genomes despite the fact that they have be-come morphologically differentiated. However, different cell types express different fractions of the genome. The metazoan oocyte contains an assortment of organelles and many kinds of molecules that are provided during oogenesis, and which are unequally distributed—the eggs are not isotropic. Most oocytes are polarized, with reference to the fate of their cytoplasm, into vegetal and animal poles. The vegetal pole is so called because the cytoplasm there contributes to the development of di-gestive organs. The cytoplasm at the animal pole contributes to much of the nervous system (among other things), an exclusively animal attribute, hence the name.

Early cell divisions are mediated by some of the maternal gene products in the oocyte, such as elements of a complex known as MPF (maturation promoting factor) composed of a large subunit, cyclin, and a small subunit, a cyclin-dependent kinase (see accounts in Gilbert 1997; Gerhart and Kirschner 1997; Wilkins 2002; and references therein). As is true for many molecular systems, the workings of these molecules in cleavage and in cell proliferation have been carefully studied in only a few organisms (notably in yeast; review in Nigg 1995); however, the general mitotic controls must be universal among metazoans, although details vary among clades. In at least many invertebrates, cyclin concentration builds up through time within the zygote and its blastomeres, owing to the continuing translation of maternal cyclin mRNAs (fig. 2.8). At a threshold concentration the cyclin ac-tivates the kinase, which proceeds to phosphorylate and otherwise affect a variety of proteins to produce mitosis. During these processes the cyclin is de-graded, deactivating the kinase. Cyclin concentration then gradually rebuilds as translation continues, and the next mi-totic cycle gets under way. Embryonic nuclear activity eventually starts up and produces the molecules necessary for further mitoses.

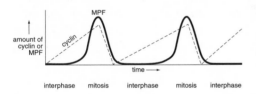

FIG. 2.8 Cyclicity in mitosis. During cleavage, the protein cyclin is produced from maternal mRNA and forms a complex with a kinase (Cdc, cell-division cycle). As the concentration of the complex increases, the kinase is activated and the complex becomes MPF (maturation promot-ing factor), which sets in motion the myriad pro-cesses of mitosis. One of these processes results in cyclin degradation, deactivating the kinase; the next mitotic cycle must await the return of cyclin to high concentrations, the timing of which is con-trolled in some organisms by the translation rate of the maternal mRNA. Mitosis associated with cell proliferation following cleavage relies on tran-scription of nuclear cyclin genes and may involve controls associated with cell growth. After Gerhart and Kirschner 1997.

As development begins, the egg cleaves into a series of smaller and smaller cells. Usually the egg cleaves first into two cells and these into four by cleavages that are longitudinal with ref-erence to cell polarity, then these into eight by meridional cleavages, and so on

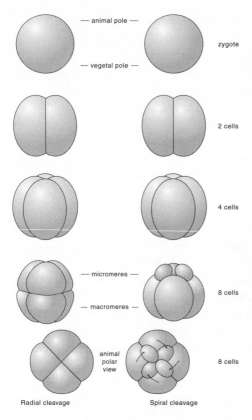

FIG. 2.9 Two common cleavage patterns in metazoans through the eight-cell stage. Blastomere spindles in radially cleaving embryos are orthogonal with respect to their polar axis, but in spirally cleaving forms are tilted in the eight-cell stage and thereafter. From Brusca and Brusca 1990.

(fig. 2.9); sometimes the cleavage is in synchronous steps, sometimes not. Cells produced by cleavages are termed blastomeres. Cell division may produce daughter "equivalence cells" similar to the parent (proliferative mode), one daughter similar to the parent and one that is differentiated (stem-cell mode), or two daughters, each of which is distinct from the parent (diversifying mode) (Stent 1985). In some cases maternal factors are not evenly distributed throughout oocytes, or become localized within the zygote during fertilization. Thus when cleavages occur, differentiation of cells is under way; blastomeres may have different constituents, and their fates may differ right from the first cleavage. On the other hand, despite such differences, early blastomeres retain the ability to produce a normal adult in some cases. Cells that retain the ability to differentiate into any cell, including gametes, are totipotent; those that can differentiate into any of a number of somatic cell types are pluripotent. Once a cell can continue to differentiate into a given phenotype, it is said to be committed. If a committed cell will develop to a given phenotype without further signaling, it is specified (and if it will develop to a given phenotype even in the face of contradictory signals, it is determined). Cells may be specified autonomously (only by their internal constituents) or conditionally (by signaling from other cells). Cleavages in which cell lineages are committed to a given fate or range of fates autonomously are termed determinate. Cleavages in which cell lineages may remain uncommitted for some time, and thus must be conditionally specified, are termed indeterminate. Both determinate and indeterminate cleavages can occur in the same developmental systems.

Two famous patterns of cleavage, radial and spiral, are shown by invertebrates (fig. 2.9), most commonly in species that have planktonic feeding larvae (indirect developers). Cleavage planes pass between chromosome sets, and at mitosis are normal to the axis of the mitotic spindle. In radial cleavage, the axes of the spindles are either parallel or orthogonal to the polar axis of the dividing cells, and the cells form vertical rows from tier to tier. Cnidarians, deuterostomes, and the lophophorate phyla, among others, show radial cleavages. In some radially cleaving forms, such as echinoids, the spindle axes are offset from the cell center during some meridional cleaving cycles to produce tiers that have blastomeres of different sizes; small blastomeres are termed micromeres, large ones, macromeres.

In spiral cleavage, spindle axes are tilted with respect to the polar axis from the eight-cell stage onward, so that cells in each tier are centered above the sutures of the previous cells. Blastomeres in subsequent tiers alternate between lying over the suture to the right or to the left. The cells of the first four blastomeres cleave so as to produce micromeres toward the animal pole, the four vegetal blastomeres being macromeres. Classic spiral cleavage is found among many of the Eutrochozoa, which include mollusks, annelids, and their allies. Fig. 2.10A–D gives the conventional cell coding, developed by Wilson (1892), describing the order and position of the four-, eight-, sixteen-, and thirty-two-cell stages of a perfectly spiralian embryo.

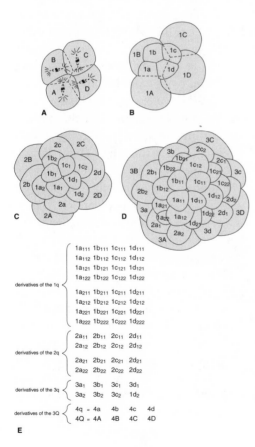

FIG. 2.10 The coding system designed by Wilson (1892) for the first sixty-four cells in a perfectly regular spiral cleavage. (*A*) Division along the first two cleavage planes produces four cells, the macromeres, coded A, B, C, and D and collectively termed Q (for quartet). (*B*) The next cleavage produces four micromeres; for simplicity we will follow just the products of the A macromere. This cell A has divided, so the daughter macromere is now 1A; the micromere arising from A is 1a; the macromeres are now collectively 1Q, the micromeres 1q (quartets arising from the first macromere division). (*C*) The next cleavage produces micromeres from 1Q, 1A becomes 2A, and the micromere arising from 1A becomes 2a; the macromeres are now collectively 2Q, the new micromeres 2q. During this round of cleavage the 1a micromeres also divide to produce cells coded $1a_1$ and $1a_2$; the lower number indicates the cell that is nearer to the animal pole. (*D*) The numbering continues at the thirty-two-cell stage in the same system, so that the code for any blastomere indicates both its parentage and position. (*E*) The codes for all cells in a sixty-four-cell blastula. Note that 4d is the mesentoblast for most spiralians (see text). From Wilson (1892) modified by Brusca and Brusca 1990.

Note that the cell designations change to indicate the number of cleavage events. The next cleavage cycle would result in a sixty-four-cell embryo, with blastomeres labeled as in fig. 2.10E. In most spirally cleaving forms the first four cells (A–D) define developmental quadrants of the adult. Normally, the A and C quadrants give rise to left and right regions, respectively, while the B and D quadrants give rise to the ventral and dorsal regions, respectively (see Henry et al. 1995). Other developmental features, such as the origin of mesoderm from the 4d blastomere (see below), are also shared among spirally cleaving eutrochozoans, strongly indicating that they form a clade, often called Spiralia.

Species that grow into an adult without a planktonic feeding stage deposit large amounts of yolk in their eggs in order to nurture the embryo as it develops into a juvenile. In these cases blastomeres may be proliferated in caps or layers around the yolk periphery, so that spiral or radial patterns displayed by yolk-poor species in the same taxon are not shown. Many arthropods have unusual cleavage patterns, but some have patterns that can be interpreted as radial while others can be interpreted as spiral. The homology of spiral arthropod cleavages with those of eutrochozoans is uncertain, however, and may represent a separate evolution of spiral cleavage, for arthropods do not share the constellation of developmental characters found in Spiralia (chap. 7). A number of phyla that develop directly into adults without a larval stage show neither spiral nor radial cleavages, but have distinctive, unique cleavage patterns. These are discussed phylum by phylum in part 2.

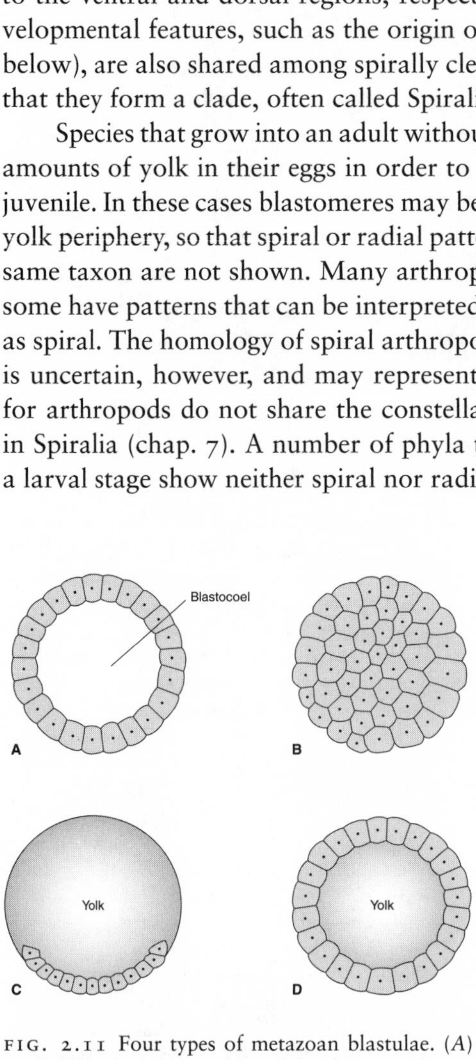

When cleavage terminates, the early embryo has between about thirty and some hundreds of cells, and is termed a blastula (fig. 2.11). If hollow, the blastula is a coeloblastula; if solid, a stereoblastula. The interior cavity of a blastula, if present, is the blastocoel. In some mosaic embryos all blastomeres are committed at this stage (and, commonly, well before), although in some regulative embryos many cells are still pluripotent. In fig. 2.12 the eventual fates of cell lines arising in various regions of a polychaete annelid blastula are mapped. The presumptive areas are similar for all polychaetes, although the cell lineages included in the areas often vary among taxa.

FIG. 2.11 Four types of metazoan blastulae. (A) Coeloblastula; sectioned to show blastocoel. (B) Stereoblastula; the embryo is a solid ball of cells at this stage. (C) Discoblastula; cells form a cap at the animal pole over a yolk mass. (D) Periblastula; sectioned to show cells forming a single layer around central yolk mass. From Brusca and Brusca 1990.

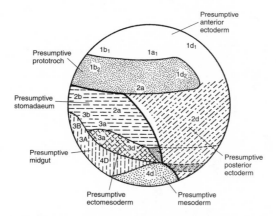

FIG. 2.12 Map of the developmental fates of some blastomeres of an annelid of the genus *Scoloplos*. From Anderson 1973.

Gastrulation Gives Rise to Ectodermal and Endodermal Germ Layers

Sponges have a unique development beyond the blastula (chap. 6). In other metazoans in which development is known, the next stage is gastrulation, which creates an embryo with two sheets of cells in outer and inner layers (fig. 2.13) that will produce particular tissue types. The outer cell sheet, termed ectoderm, is slated to produce tissues of the integument, including exoskeletons, and the nervous system. The inner cell sheet, termed endoderm, forms the embryonic gut, or archenteron, which communicates to the exterior through the blastopore, and later forms tissues of the adult stomach, intestine, and digestive organs. Gastrulation involves the appropriate early positioning of cells that give rise to these different tissues; cells whose lineages form the gut are internalized, for example. In annelids, as a particular example, the presumptive endoderm cells 4A, 4B, 4C, 4D, 4a, 4b, and 4c (as well as the presumptive mesoderm cell 4d) are internalized either via invagination or epiboly (fig. 2.13). The gastrula may or may not retain a remnant of a blastocoel.

Middle Body Layers Range from Simple Sheets of Extracellular Matrix to Mesodermal Germ Layers

In sponges the various surface cell types of both the exterior and interior of the body wall and its chambers rest upon a middle layer of extracellular matrix, termed the mesohyl (fig. 2.14A–C); tissues that can be identified as ectoderm and endoderm are lacking, however (chap. 6). In adults of so-called diploblastic organisms—cnidarians and ctenophorans—sheets of ectodermal and endodermal cells form epithelia, lining the outer and inner body surfaces, respectively. Basal laminae composed of extracellular matrix support the tissue sheets, and these laminae can be considered to

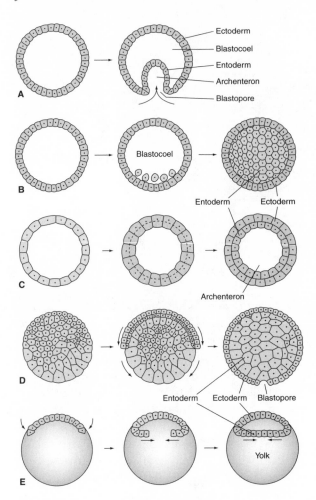

FIG. 2.13 Five types of metazoan gastrulation, each involving a different pattern of presumptive endoderm specification. (*A*) Coelogastrula formed by invagination of a coeloblastula, internalizing the presumptive endodermal blastomeres and producing an early gut (archenteron) with opening to the exterior (blastopore). (*B*) Ingression of cells, descended from cells in coeloblastula wall, into the blastocoel, filling it to produce a solid stereogastrula. (*C*) Delamination, the division of cells in the coeloblastula wall to produce an inner layer of endoderm that then defines an archenteron. (*D*) Epiboly, the growth of a layer of dividing ectodermal cells from the animal pole over the vegetal pole, the cells thus internalized forming endoderm. (*E*) Involution, growth of discoblastula margins in a very yolky egg to form a discogastrula. From Brusca and Brusca 1990.

occupy the blastocoel compartment. This extracellular matrix may be thickened to form a middle body layer, termed mesoglea, which is the dominant body layer in many Cnidaria, especially in medusae (fig. 2.14D–E). An external cuticle of extracellular matrix, sometimes elaborated into a complex layered structure that may include an exoskeleton, is secreted atop the epidermal cell sheet. The properties of extracellular matrix sometimes vary from place to place within individuals and vary among taxa, not only in Cnidaria but throughout Metazoa, presumably in response to their varied mechanical and physiological requirements. In Cnidaria, muscular contractions are furnished by cells within epithelial tissues (myoepithelial cells), some of which are derived from endoderm and form muscular tissues. In Ctenophora there are endodermally derived muscle cells, sometimes bundled, lying beneath the ectodermal epithelium and thus occupying a position within the former blastocoel compartment (see Martindale and Henry 1997a). These muscles are

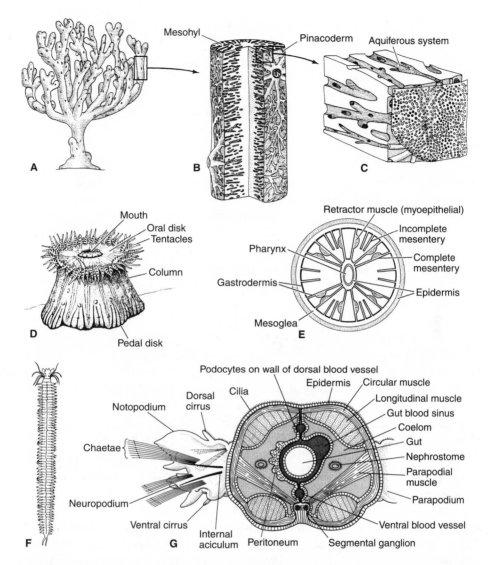

FIG. 2.14 Forms of middle body layers. (*A–C*) The mesohyl layer in a sponge (*Microciona*), overlaid by a thin pinacoderm and containing flagellated cylinders lined by choanoderm. (*D, E*) The mesogleal layer in a cnidarian (sea anemone), overlaid by epidermis (ectoderm) and underlain by gastrodermis (endoderm). *D* is *Calliactis*. *E* is a transverse section through a generalized anemone pharynx. (*F, G*) The endomesodermal muscular system displayed in an annelid, overlaid by epidermis with chaetae and invaded by nerves (ectoderm) and encasing an intestine (endoderm). *F* is the polychaete *Nereis*. *G* is a transverse section through a trunk segment. A–E after Brusca and Brusca 1990; F after Barnes et al. 1993; G after Ruppert and Barnes 1994.

thus at least analogous in position with the mesodermal muscles of triploblastic forms.

In the so-called triploblastic organisms the extracellular matrix of the middle layer is usually not dominant but is largely replaced by cellular tissue, which is thus elaborated within the compartment of the blastocoel to form a compartment of its own. This medial tissue is called mesoderm, which is regarded as a third germ layer because it gives rise to a suite of organs. In most triploblastic forms the dominant mesodermal derivatives are muscles (fig. 2.14G), and perhaps mesoderm first evolved from muscle cells in early bilaterians. Extracellular matrix may remain abundant within mesodermal tissue or may be restricted to basal laminae (and sometimes mesenteries), and is evidently nearly missing in some acoel flatworms. Mesodermal tissues that arise from ectodermal or endodermal cells are distinguished as ectomesoderm or endomesoderm, respectively. When ectomesoderm is not developed into epithelial sheets, it is termed mesenchyme, acting as connective tissue; mesenchyme is commonly infused with extracellular matrix. Epithelial mesoderm is termed mesothelium and is usually endomesoderm. However, muscles and other organized tissues are sometimes also derived from ectomesoderm, especially in nemerteans, mollusks, and annelids. In flatworms, ectomesodermal muscle layers lie along the outer part of the blastocoel compartment, with the more internal mesodermal tissue being termed the parenchyma. Larval muscles are commonly derived from ectodermal sources in spiralians (see Boyer et al. 1996a). Furthermore, most spiralians derive mesoderm from cell 4d, the mesentoblast, which may divide to produce endodermal and mesodermal lineages. It is not clear that the mesoderm formed in this manner should be termed endomesoderm, although it is usually so characterized. Mesenchymal and mesothelial tissues arise from each other in some cases. Thus, as Ruppert (1991b) has emphasized, mesoderm is derived from a variety of sources, both ectodermal and endodermal, in a variety of modes, and arises at different times during ontogeny in different organisms.

Endomesoderm *sensu lato* is the source of most muscles; of blood cells, kidney, and gonadal structures; and of the mesothelial linings of secondary body cavities in higher invertebrates. Mesodermal tissues are invaded by nerve processes from the ectoderm and are pervaded by channels or lumina that carry blood, which arise from the topographic site of the blastocoel to form the blood vascular system. In some groups (such as arthropods and chordates) the blood vascular system that penetrates mesodermal tissues is lined by a mesothelium. Endoskeletons are produced from mesoderm, most notably in echinoderms and chordates, and in some other phyla as well (e.g., some skeletal elements in brachiopods).

Endomesoderm may originate as cells internalized from endodermal precursors or as evaginations of epithelial endoderm or of mesenchyme. In taxa with spiral cleavage, the mesentoblast is usually internalized near the blastopore during gastrulation (fig. 2.15A). This pattern is found in flatworms, nemerteans, mollusks, annelids, and a number of phyla that seem related to these, and thus is likely to be

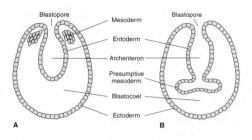

FIG. 2.15 The formation of endomesoderm. (A) In spiralians, including flatworms, annelids, and mollusks, mesoderm proliferates from a mesentoblast that lies near the blastopore following gastrulation. (B) In triploblastic radialians, including echinoderms and hemichordates, mesoderm proliferates from cells in the wall of the archenteron. After Brusca and Brusca 1990.

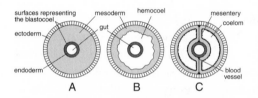

FIG. 2.16 Topology of the pseudocoelomic and coelomic spaces relative to mesodermal tissue (gray). (A) An acoelomate with blastocoel compartment entirely filled by mesoderm. (B) A pseudocoelomate with fluid-filled blastocoel space between the endoderm and somatic mesoderm. (C) A coelomate, coelom lined by a mesothelium. Former blastocoel compartment compressed to an outer lining between somatic mesoderm and ectoderm, an inner lining between splanchnic mesoderm (lining the gut) and endoderm, and a lining between mesentery tissues. Note that the blood vessels run on the site of the blastocoel compartment and thus, in forms such as annelids with mesodermal septa, bypass the coelom. After Willmer 1990.

a plesiomorphic trait for these forms. Groups with idiosyncratic cleavage employ other endomesodermal precursors. In triploblastic taxa with radial cleavage (such as hemichordates, echinoderms, and lophophorates) endomesoderm usually originates from endodermal cells that line the archenteron of the gastrula (fig. 2.15B). These cells divide to produce a middle body layer either by pouching out to line a sort of "bubble" that becomes a coelomic space (see below), or by proliferation of solid packets of cells into the blastocoel compartment.

Pseudocoels and Hemocoels Are Body Cavities That Lie on the Site of the Blastocoel

In some triploblastic bodyplans there are no body cavities (acoelomate, as in platyhelminths), and any body fluid is dispersed within a meshwork of interstitial spaces in the parenchyma; this meshwork is simply considered part of the mesenchyme (fig. 2.16A). In some cases there are small fluid-filled regions within the meshwork, termed lacunae (as in kinorhynchs and gastrotrichs). In most bilaterians, however, there are fluid-filled spaces that are lined by an extracellular matrix that is the basal laminae of tissues representing one or another of the germ layers (fig. 2.16B). These spaces lie in the topological position of the blastocoel and form the primary body cavity. Simple primary body cavities are termed pseudocoels, and they commonly function as hydrostatic skeletons, though many physiological activities certainly occur in pseudocoelomic fluid. Some pseudocoels are bounded by mesodermal tissues along the body wall, but not usually along the complete length of the gut. In most larger invertebrates there is a blood vascular system, which also develops on the topological position of the blastocoel, ramifying into channels, sinuses, or vessels that penetrate the tissues and that contain circulating fluids (fig. 2.16C). Bilaterians with only pseudocoels as body cavities, lacking a circulatory system, tend to be small-bodied. When the blood vascular system functions structurally, as a fluid space, it is a hemocoel. Here, two sorts of hemocoels will be recognized, hydrostatic hemocoels that serve a significant hydrostatic function, and organ hemocoels, such as pericardial sinuses, that surround particular organs. These terms parallel the usage for coelomic spaces discussed below.

The small-bodied phyla with acoelomate, lacunate, and pseudocoelomate bodyplans are rather comparable in size and complexity, and all of these states are sometimes found within a single phylum. Major functional differences among these design elements are that the blastocoel compartment is filled mostly with fluid or mostly with tissue, and hydrostatic processes are performed chiefly by one or the other. Inglis (1985) has suggested the term paracoelomate to describe all these small-bodied, noncoelomate bodyplans. The term is used throughout this book for animals with these bodyplans, without any systematic implications.

Coeloms Are Body Cavities That Lie within Mesoderm

In most triploblastic organisms there are fluid-filled body spaces that are cut off from topological connections with any pseudocoel or hemocoel spaces by a lining of mesoderm, and so form compartments within the mesoderm compartment. Such intramesodermal spaces are lined by cells (rather than by extracellular matrix as in most pseudocoels) and are termed secondary body cavities, or coeloms (fig. 2.16C). In many phyla a coelom is developed periviscerally, separating the muscles of the body wall (somatic muscles) from those lining the digestive tract (splanchnic or visceral muscles). The coelomic lining may consist of muscle tissue forming a myoepithelium, or of a peritoneum—a mesothelial tissue with its own basal lamella that coats the muscle tissue. In the coeloms of annelids and their allies it is common for the somatic lining to have a peritoneum while the splanchnic lining is myoepithelial. In some annelids the nature of the coelomic linings varies even among different regions of the body (see Fransen 1988 and especially Bartolomaeus 1994). In forms in which the coelom is divided into isolated compartments by transverse septa, as in annelids, blood vessels can bypass the septa since they lie between mesoderm and either endoderm or ectoderm, on the site of the blastocoel (fig. 2.16C; Ruppert and Carle 1983).

Among the radially cleaving deuterostomes in which mesoderm arises from cells along the wall of the archenteron, the coelom commonly develops from outpockets of the gut wall that produce spaces, which are then captured within the elaborating mesoderm; these are enterocoels (fig. 2.17A). In some taxa with radial cleavage, such as vertebrates, mesoderm that arises from the archenteron does so as solid bands or packets, and coelomic spaces are then produced by splitting within these tissues. Although these spaces are literally split coeloms, the origin of the mesoderm from the archenteron wall is taken as evidence that enterocoely was the primitive

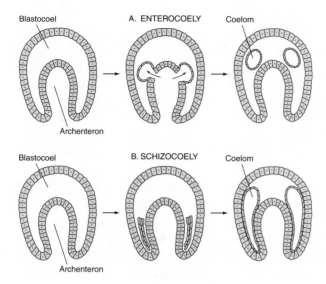

FIG. 2.17 Diagrammatic representation of early development of the coelom. (*A*) Enterocoely, mesoderm arising from archenteron wall and coelomic space being captured from the enteron. (*B*) Schizocoely, mesoderm arising from a mesentoblast and coelomic spaces arising by splitting within mesodermal tissue. After Brusca and Brusca 1990.

condition, and this pattern of coelom formation is considered a modified form of enterocoely. In classic spiralians the mesoderm is elaborated from blastomere 4d (the mesentoblast) as bands that invade the blastocoel compartment, and as cells are not recruited from the archenteron, enterocoely is not possible. Coelom formation must occur by splitting within these bands, termed schizocoely (fig. 2.17B–F).

Some Coeloms Function as Hydrostatic Skeletons. Coeloms have a variety of functions. When they are perivisceral or form tentacular lumina, coeloms act as hydrostatic skeletons that provide turgor and transmit muscular forces. In some cases coelomic fluid contains circulating cells; in adult sipunculans and probably in nemerteans there is no primary body cavity and no fluid system except for the coelomic system. Unless the context clearly indicates otherwise, the unmodified term coelom is used here to refer to perivisceral or tentacular coeloms—to coeloms with hydrostatic functions—and only organisms with such body cavities are considered to be full-fledged coelomates (eucoelomates) structurally. Phyla with perivisceral hemocoels in which intramesodermal spaces can be found at some developmental stage are still considered eucoelomates by some workers; this usage evidently derives from an early belief that arthropods were derived from annelids (chap. 7).

Some workers have specified that true coelomic spaces should be lined by a peritoneum, but other workers have required only the presence of a simple mesothelial lining, especially as ongoing work has revealed that many spaces regarded as coeloms lack peritoneums. The evolution of a mesothelial lining, in which the cells form a sheet with apices directed to the coelom, would seem to be a simple adaptation that could arise whenever intramesodermal spaces appeared. However, some organisms have intramesodermal spaces that are entirely lined by mesenchymal muscles. Some of these forms are minute members of phyla in which, in larger taxa, the coelom is lined by a peritoneum, and as the minute forms are not necessarily primitive, their intramesodermal linings have presumably been lost. Such spaces are not pseudocoels, and might as well be considered coelomic. In other organisms, lining of perivisceral spaces with myoepithelial cells alone is thought to be the primitive condition (Bartolomaeus 1994). Interstices and lacunae within mesoderm that are parts of fluid networks within those tissues, such as are found in flatworm parenchyma, are best regarded as internal spaces that are neither pseudocoel nor coelom. The utility of spaces within the mesoderm is clear from even this brief account, and it is quite likely that they have had multiple origins (though see Budd and Jensen 2000). The homology of coeloms in different phyla should not be assumed; rather, evidence of homology should be required in each case where it is asserted.

Some Coeloms Are Adjuncts to Organs. Intramesodermal spaces that do not function primarily as parts of a hydrostatic system include ducts that traverse mesoderm, and spaces that lodge organs, which undergo volume changes. The chief organ coeloms are associated with excretory organs, with gonads, and with the heart. Intramesodermal spaces are found in the excretory organs of many coelomates. These are metanephridia, multicellular mesodermal structures with a proximal ciliated funnel (the nephrostome) or similar structure that leads to an excretory duct. Gonocoels are spaces containing reproductive cells, connecting to the exterior via ducts that arise from mesothelial tissues, and therefore are coelomoducts. Such ducts commonly meet distally with an invaginated ectodermal tube or collar, thus sealing off space within the ducts from primary body spaces. Gonocoels may be enlarged during and following gonadal discharge. Gonoducts and excretory ducts are commonly fused. Pericardial coeloms surround the heart in some phyla, such as Mollusca, and provide a space for heart pulsations.

Larval Stages Commonly Possess Bodyplans of Their Own
All metazoan phyla contain at least some species that develop within egg capsules and emerge as juveniles from which growth and some further development produce adults. These species are termed direct developers. But nearly half of metazoan phyla produce one or more free-living larval stages that disperse, chiefly by flotation in the water column, as plankton, before developing into juveniles. Some of these larvae do not feed (lecithotrophic development), their eggs usually being provided with yolk. Many other planktonic larvae do feed (planktotrophic development) and thus can sustain themselves in the plankton for days to weeks or longer. Most of the adults in these phyla are benthic, and thus their larvae occupy an entirely different ecological realm, eventually settling to the bottom to metamorphose into juveniles. Evolution has therefore produced larval bodyplans, adapted to life in the plankton, that represent separate phases in the life cycles of these forms. Biphasic life cycles of this sort are found in the most basal living metazoans—sponges—and in some groups in each of the major metazoan alliances (chap. 4). The larval bodyplans have been given names; fig. 2.18 shows the bodyplans of a number of invertebrate larvae

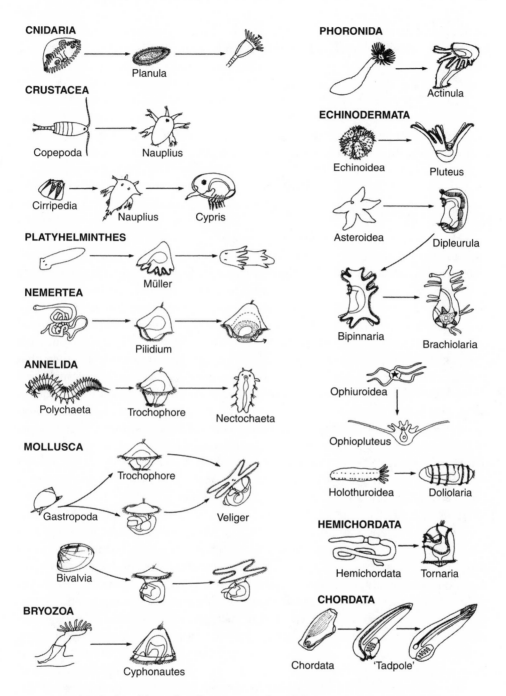

CNIDARIA

Planula

CRUSTACEA

Copepoda   Nauplius

Cirripedia   Nauplius   Cypris

PLATYHELMINTHES

Müller

NEMERTEA

Pilidium

ANNELIDA

Polychaeta   Trochophore

Nectochaeta

MOLLUSCA

Trochophore

Gastropoda   Veliger

Bivalvia

BRYOZOA

Cyphonautes

PHORONIDA

Actinula

ECHINODERMATA

Echinoidea   Pluteus

Asteroidea   Dipleurula

Bipinnaria   Brachiolaria

Ophiuroidea

Ophiopluteus

Holothuroidea   Doliolaria

HEMICHORDATA

Hemichordata   Tornaria

CHORDATA

Chordata   'Tadpole'

FIG. 2.18 Marine larval forms found among a number of phyla. Adults on the left, their larvae on the right. Most larvae are quite small, in the range of a millimeter or so. From Barnes et al. 1993.

with their adults. In many cases the larval bodyplans vary among classes or other higher taxa, showing larval body subplans, so to speak. The larvae are discussed in part 2, and their evolutionary significance is assessed in part 3.

Many Bodyplan Features Reflect Locomotory Techniques

In Soft-Bodied Forms, Locomotory Devices Range from Cilia to Limbs

Many of the major features of metazoan bodyplans are organized as benthic locomotory mechanisms or assist in locomotion in some way, especially among the bilaterians. The earliest Bilateria are likely to owe their anteroposterior axial organization to selection for directional locomotion, and their dorsoventral organization to locomotion on a substrate. The earliest bilaterian locomotory mechanism probably involved a two-layered mucociliary system. In this locomotory method, a mucous ribbon is secreted on the substrate beneath the organism; the outer mucous layer is the denser and adheres to the substrate, while cilia beating within a watery inner layer impinge on the outer layer and propel the animal (Collins et al. 2000). Such methods are used by a wide variety of small metazoans, but at larger body sizes, such relatively weak ciliary transport systems are inadequate, and muscular contractions become involved (see Clark 1964).

Skeletons are particularly important in locomotion and obviously contribute greatly to body shape. Fluid (hydrostatic) skeletons are the most widespread of skeletal types, being found in most animals even when a solid skeleton is present as well (Chapman 1958; Clark 1964). Even in humans, the fluid (coelomic) skeleton is an important element in many mechanical functions. Sealed within a compartment by tissues, a liquid cannot be compressed but will flow to any shape permitted by deformation of the compartment walls. As the total volume of the compartment must be constant, reduction in the volume in one region of a fluid compartment must cause a compensatory enlargement of some other region. Thus muscular contractions that alter the shape of a compartment will stretch muscles that lie along enlarging dimensions, and an antagonizing muscle system is created to return contracted muscles to a stretched condition. Fluid skeletal compartments are often surrounded by two (or more) layers of muscles with fibers oriented in different directions—often circular and longitudinal muscles in vermiform taxa, such as annelids. Pseudocoelomic fluid, blood, coelomic fluid, and seawater (trapped in the gut of anemones, for example) are used in such skeletons. Tissues are used in lieu of other skeletons in some cases (e.g., mesoderm in flatworms), for they are also incompressible and can be deformed, but they do not flow like liquids and probably are less efficient.

Benthic locomotion commonly involves waves of body contractions. Peristaltic locomotion, with waves of contraction around the circumference of vermiform bodies, is used in numbers of burrowing animals (see Annelida, chap. 8), and pedal locomotion, involving similar waves but restricted to a ventral sole in contact with

TABLE 2.2 Locomotory gripping mechanisms employed as accessories to peristaltic progression (including pedal locomotion). Modified from Elder 1980.

Mechanism	Major Taxon Containing Exemplar Organisms
Mucus	Platyhelminths, nemerteans, annelids, mollusks
Adhesive rugae	Anthozoans
Setae and chaetae	Annelids
Suckers	Asteroids, holothurians
Cuticular flanges	Priapulids
Calcareous spicules	Holothurians
Foot processes	Polychaetes, holothurians
Mouth or jaws	Nemerteans

the substrate, is used in a number of creeping forms (see Platyhelminthes, chap. 8). Ancillary structures such as integumental annuli and chaetae are sometimes used in conjunction with peristaltic mechanisms (table 2.2). Such structures grip or adhere to the substrate to prevent backsliding, and may also provide purchase for burrowing. Another technique used in moving over the substrate or, more commonly, in burrowing, employs a proboscis of some sort, usually an introvert, to probe forward (or to probe the direction of locomotion) and then to anchor; the body may then be pulled forward and the probing resumed. Anchoring devices such as cuticular flanges and spines are particularly common in such forms.

Ambulatory locomotion in soft-bodied forms is provided by limbs or limblike devices that are manipulated by muscles. In these forms the limb skeleton is fluid, as in onychophoran limbs with hemocoels, and in annelid parapodia with coeloms. Serpentine locomotion, in which the body is thrown into lateral snakelike waves as in some annelids and cephalochordates, is sometimes used in elongate forms. Still other locomotory modes occur, such as the swimming techniques adopted by a number of groups, but they are derived modes and the associated morphological modifications are unlikely to have been involved in the origins of the bodyplans of phyla, except for a few cases.

"Hard" Skeletons May Complement or Replace the Biomechanical Functions of Fluid Skeletons

Solid skeletons can add rigidity and strength to metazoan bodies. Fortunately, many solid skeletons are quite durable and thus have a fossilization potential high enough to provide paleontologists with their bread and butter. About a third of the phyla have at least fairly durable skeletal parts. Exoskeletons are secreted by ectoderm and are chiefly external to or form the outer part of the integument; endoskeletons are secreted by mesoderm and therefore lie inside tissues. Some solid skeletons are entirely of organic materials, most commonly of chitin, a tough polysaccharide that is usually strengthened by cross-linkages with proteins. Other skeletons, the ones most easily preserved as fossils, are mineralized. Mineralized invertebrate skeletons are chiefly of calcium carbonate, in the form of calcite or aragonite, and most vertebrate skeletons are of calcium phosphate, as the mineral dahllite, but there are many exceptions. Lowenstam and Weiner (1989) review the distribution of minerals in skeletal and other metazoan tissues, and discuss the processes of mineralization in several major phyla.

Solid skeletons are used extensively as muscle attachments. Durable skeletons commonly provide the scaffolding on which muscles may be arranged for manipulation of feeding or locomotory organs and the working of other body parts. The rigidity of more or less encompassing shells, carapaces, or tests, such as in clams and lobsters, permits the external body shape to be controlled without muscular effort, with body turgor maintained by a fluid skeleton. Hydrostatic systems may thus be braced by rigid walls and made more efficient, and their biomechanical potential is thus expanded by such skeletons. Many skeletal elements form levers, as in arthropod or vertebrate limbs. The mechanical advantage of the levers permits relatively small muscular contractions to be translated into much larger movements. Even skeletal elements that are scattered through tissues, such as granules or spicules, provide significant stiffening and thus can have a structural role in the bodyplan (see Koehl 1982).

The adaptive pathways that led to the origin of skeletonization have been much debated. The two most common hypotheses are that skeletons evolved either for protection or for support. Given the presence of a solid skeleton, evolutionary processes would seem likely to exploit both those functions so long as they did not interfere with one another. In a few groups, however, a protective function has clearly become of major importance. Marine gastropod exoskeletons, for example, provide heavy armor and are ingeniously sculptured to secure protection from predators (see Vermeij 1978). Bivalves exhibit both protective and supportive skeletal functions; many could not function (in burrowing, for example) without a solid exoskeleton, which provides protection as well. All phyla that have solid skeletons appear to have evolved them independently, and in some cases independently in different classes or possibly in even lower taxa within phyla. Given this evident multiplicity of originations, it is likely that different adaptive advantages were of importance in different cases. The coadaptations of skeletal plans with the bodyplans of their phyla are reviewed in chapters 6–11.

Symmetry and Seriation Are the Principal Descriptors of Body Style

The arrangement of organs, organ systems, and other design elements within the organism imparts symmetry to and characterizes the architectural style of the bodyplan. Organisms are rarely if ever ideally symmetrical but can nevertheless be referred to ideal symmetries, which they more or less approximate. Symmetries are described in relation to planes that (ideally) separate the body into mirror-image parts, and to orthogonal axes related to the symmetry planes that describe the bodies' dimensions. A few metazoans (e.g., some sponges) lack such planes and are essentially asymmetrical.

Symmetry Is Imparted by Repetition of Parts across Planes or along Radii

In perfect radial symmetry, symmetry planes occur along all diameters normal to an axis of symmetry; each divides the body into identical halves, equal in architecture and dimensions. Among adult metazoan bodyplans, some cnidarians come closest to being radially symmetrical (see fig. 6.18A). Their usual symmetry axis is oral-aboral. Because their mouths and associated structures are elongated rather than round in cross section, sea anemones have only two symmetry planes that divide the body into mirror images, a condition termed biradial; most ctenophores are also biradial. Jellyfish have four gastric pouches arranged symmetrically around their oral-aboral axis. There are thus four symmetry planes that divide the body into mirror images; this is tetraradiate symmetry. Many echinoderms have five radii, which define a pentaradiate symmetry (fig. 11.8A–B), while others have even more symmetry planes. However, echinoderms, like most organisms, have some features that destroy the precision of their symmetries, such as the singular madreporite and its adjuncts, which are offset from any potential symmetry plane (chap. 11). There are few examples of triradiate body symmetry (although mouthparts of some forms with pharyngeal pumps, such as leeches and nematodes, form a partial exception), but some enigmatic fossils from the Neoproterozoic evidently had triradiate bodyplans (see fig. 6.16). Varieties of radial symmetry are found chiefly in floating forms, for which the environment is more or less similar on all sides, and in sessile benthic forms that feed from the water column and may find prey in more or less any direction. Body parts that are similar (usually across some symmetry plane), such as the gastric pouches of medusae or the arms of starfish, are termed antimeres.

Most metazoans have bilateral symmetry, being divided into nearly mirror-image halves by a dorsoventral (midsagittal) plane down the anteroposterior (sagittal) axis—their antimeres are laterally disposed. Planes orthogonal to the dorsoventral symmetry plane and that divide bilaterian bodies into dorsal and ventral parts are frontal planes. Planes that are orthogonal to both dorsoventral and frontal planes are transverse planes. These last two sorts of planes separate body sections that are quite different. Frontal and transverse planes, while not symmetry planes, permit visualization and description of anatomical arrangements. Many bilaterians are markedly bilaterally asymmetrical in some features; the one large claw of fiddler crabs is a flamboyant example, and it is common for singular internal organs to be well off center, such as the human heart, liver, and spleen. The cnidarians and ctenophores, presumably primarily radial, have commonly been classed together as the taxon Radiata. Most of the bilaterians are cephalized at least weakly (though not all, e.g., echinoderms and bivalve mollusks).

Anteroposterior Regionation Involves Seriation, Segmentation, and Tagmosis

In bilaterians the anteroposterior axis is usually elongated and anatomically differentiated, often organized into distinctive regions, such as a head, trunk, and tail, each serving a distinctive function. In many forms there is an anteroposterior repetition of one or more organs, such as of gills in chitons (chap. 8) or of gonads in some enteropneusts (chap. 11). In some cases the serial repetition of some organs is so regularized as to create a series of similar body parts—markedly seriated organisms include some flatworms, kinorhynchs, and most arthropods and annelids. A terminology has sprung up to express these arrangements, and the descriptive terms are commonly given phylogenetic significance.

Similar body segments that are arranged serially are termed metameres, and organisms that exhibit numbers of such segments are metameric. A major taxon termed Metameria was once in common use; it usually included arthropods and annelids, and sometimes onychophorans and tardigrades, and in some schemes still other phyla that exhibit seriation in some organs. However, the serial arrangement of organs or organ systems is such a felicitous solution to the physiological and mechanical problems presented by an elongate body that there is no reason to believe that it could not have arisen independently in any number of lineages. The homology of serial architectures among phyla should not be assumed but, as with the coeloms, should be supported by evidence in each case wherein it is hypothesized. Because the metameric terminology has been bound up with phylogenetic hypotheses, it will be avoided here. Seriation will be used to indicate repetition of organs or organ systems along the anteroposterior axis of bilaterians (e.g., gonads in some flatworms; fig. 2.19A). Segmentation will be used to designate a variety of seriation in which several organs or organ systems are repeated in a coordinated fashion along that anteroposterior axis (e.g., gonads, nephridia, parapodia, and various muscle groups in annelids; fig. 2.19B). The terms are meant to be descriptive only. In some organisms, such as arthropods, segments are specialized within distinctive regions, such as the cephalon, thorax, and pygidium of trilobites. Such a regionalized condition is termed tagmosis, the regions being tagmata (fig. 2.19C–D).

Another common form of regionation in bilaterians, seen especially in those that are sessile suspension feeders, is a two- or threefold anteroposterior compartmentalization of the body to produce functionally distinct, hydrostatic skeletomuscular systems based on coeloms (e.g., in hemichordates and phoronids). When three such regions are present, they are termed (from anterior to posterior) the prosoma, mesosoma, and metasoma (fig. 2.20), and their coelomic compartments are the protocoel, mesocoel, and metacoel. In pterobranch hemichordates, the prosoma is a locomotory organ, the mesosoma is elaborated into a tentacular feeding system, and the metasoma is the trunk (chap. 11). In keeping with the terminology based on the root *mere* (part or segment), this architecture is commonly termed trimeric or, since some phyla seem to have only two coelomic compartments, oligomeric. As "meres" are supposedly somewhat similar units, and as the whole point of

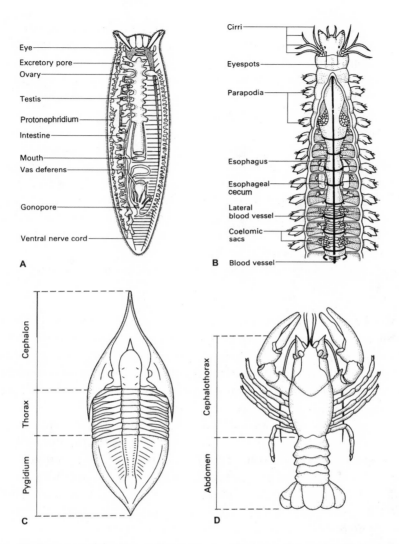

FIG. 2.19 Regionation in bilaterians. (*A*) Seriation in a flatworm, *Procerodes*. (*B*) Segmentation in the anterior of an annelid, *Nereis*. (*C, D*) Tagmosis in segmented animals, a trilobite and a crustacean, respectively. From Boardman et al. 1987.

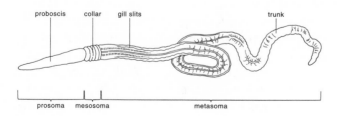

FIG. 2.20 A triregionated coelomate, the enteropneust hemichordate *Balanoglossus;* the three body regions reflect an internal coelomic regionation.

regionation in these forms is to create dissimilar units, this terminology has been roundly criticized (see Hyman 1959, 137). Accepting this criticism, the terms biregionated and triregionated will be used here for "oligomerous" architectures as appropriate, and are meant to be descriptive only.

Evolutionary Changes in Body Size Occur throughout Metazoan History

Size is associated with many trends in animal functions and plays an important role in animal design, for both mechanical and physiological reasons (e.g., see Thompson 1942; Schmidt-Nielsen 1984). Two of the more significant features in the history of metazoans are the increases in size from their very small unicellular or colonial protistan ancestors to the metazoans of the Neoproterozoic, and from very small Neoproterozoic bilaterians to the bodyplans of many living phyla. Many of these changes in body size have been associated with and may have engendered changes in architectural complexity.

Area/Volume Ratios Are Sensitive to Most Size Changes
One of the major effects of size difference or change is the concomitant change in the ratio between the volume and surface area; in a sphere the surface area increases as a square, the volume as a cube of the radius. Some or even all of this disparity in volume-to-surface-area change can be compensated for by shape changes; in an organism that grows as an expanding disk, for example, one of the dimensions is not changing at all and both volume and surface area increase proportionately. Most organisms increase in size in all dimensions, however; as their body size increases, their ratio of surface area to volume falls. Therefore, service to interior cells that depends in some way upon surface area, such as upon oxygen diffusion surfaces or the size of ciliary feeding tracts, must be proportionately enhanced as body size increases. On the other hand, if body size decreases, such services may be downsized, and may be simplified or even discarded in extreme cases.

In organisms with two tissue layers, whether sponges or diploblastic forms, the tissues are arranged so that each layer is bathed by seawater, and although the organisms are three-dimensional, the tissue layers are essentially two-dimensional and escape most problems of internalization and associated volume effects. Some pelagic cnidarians are as large as small rooms, and giant benthic sponges are legendary. In triploblastic forms, however, mesodermal cells are cut off from the exterior either by ectoderm or endoderm. In flatworms, body size is generally small, or at least the body is quite thin dorsoventrally, permitting some oxygen (and perhaps some nutrients) to reach the interior cells from the body surface, and nutrients (and perhaps some oxygen) to reach interior cells from the gut; in some forms the gut is elaborated into diverticula that evidently serve as distributaries. Fluid-filled interstices and lacunae appear within the flatworm mesoderm, perhaps having hydrostatic functions associated with "pedal" locomotion or swimming but perhaps also serving a primitive circulatory function as muscular activity sloshes the fluid around. In triploblastic organisms with thicker or larger bodies, a more elaborate circulatory system is required.

Life Is Different at High versus Low Reynolds Numbers

For aquatic organisms, size is of importance to morphology because it is a parameter of Reynolds number (Re). This dimensionless variable is a ratio of inertial forces divided by viscous forces, and indicates whether the fluid currents with which an organism deals are turbulent or laminar, and whether the fluid medium is experienced as having more or less viscosity. Reynolds number can be expressed as

$$Re = \frac{lU}{v},$$

where l is a linear dimension such as the length of an organism, U is the velocity of the fluid (or of the organism through the fluid), and v is the kinematic viscosity (the ratio of dynamic viscosity to density). Vogel (1981), who gives an eminently readable account of the biomechanical effects of Reynolds number, has estimated Re for several marine organisms. For a large whale swimming at 10 m/s (meters per second), $Re = 300,000,000$; for a tuna swimming at the same speed, $Re = 30,000,000$; for a copepod in a pulse of 20 cm/s, $Re = 300$; for an invertebrate larva moving at 1 mm/s, $Re = 0.3$; and finally, for a sea urchin sperm moving at 0.2 mm/s, $Re = 0.03$ (these figures, obviously, are only approximate). As large marine organisms tend to move faster than small ones, both l and U decline at smaller body sizes. At very small Reynolds numbers, the effects of viscosity dominate movement within the fluid. Thus when active swimming ceases, a large whale has a lot of inertia and can glide a long way, but a larva will be brought to an abrupt halt, as if trapped in molasses, by the viscosity.

This contrast creates size-related differences between organisms. Larvae and other small metazoans can utilize minute food items that are essentially embedded in the (to them) highly viscous fluid and can be drawn to them en bloc via ciliary activity. However, invertebrates measured in centimeters filter food items from ambient currents, and large animals that are planktivores engulf their food items by swimming through populations of their planktonic prey or by sucking them in when opening a voluminous oral cavity—both are methods of inertial capture. The change from larval life at very low Reynolds numbers to the higher numbers associated with adult body sizes must help account for the change in life mode that commonly accompanies growth and maturation even in free-living marine invertebrates. There is remarkably little data on this point. However, Koehl (1995) has presented experimental results and reviewed the literature on feeding in copepods, which feed via setae, for which the operational l can be taken as setal diameter. She found that at very low Reynolds numbers (such as $\leq 10^{-2}$), setal rows function as paddles, and by alternate "fling and squeeze" motions (flinging outward and squeezing back together) they draw blocks of water and their contained food items to the mouth. At a Reynolds number near 1, setal rows function as leaky sieves, creating currents from which food items could be filtered. The border between dominantly viscous and dominantly inertial feeding depends in part on the shape of the appendage and on the size of setal gaps, and is affected by changing the speed of setal motion or by positioning the effective part of the setae away from the body wall. It would be most interesting to learn what, if any, key developmental changes in organisms growing from larvae to large-bodied adults coincide with critical values of Re, and if evolution from small-bodied to larger adults, such as is evidently found among early bivalves for example, involved changes in life mode that are correlated with such values.

Morphological Complexity Is Not a Simple Topic

Although all biologists recognize that some organisms are more complex than others, differences in their complexities have been difficult to define and measure. A useful approach to a concept of complexity that can be applied to morphology was proposed by Hinegardner and Engleberg (1983): the complexity of an object can be taken as the minimum amount of information required to describe it. Factors that increase the size of the minimum description of an organism include an increase in the number of parts, their increasing disorder, and their iteration (see Wicken 1979; McShea 1991). An object with many different kinds of parts is more difficult to describe than an object with just one kind, and is more complex. An object whose parts are disarrayed and jumbled is more difficult to describe than one whose parts are neatly ordered, and is more complex. And if parts are iterated, even though all copies are identical, it requires specification of the number for each part and thus a longer description. Most metazoans have a great many kinds of parts—for example, human beings have many organs composed of numerous tissues containing hundreds of different types of cells. Furthermore, these parts are not arranged in any easily described fashion. Organs are asymmetrical, tissues are penetrated by cell types, such as neurons, belonging to other tissues, and by conduits for circulating fluids, and by ducts; humans are quite disordered. And parts—cells, for example—are present in a great range of iterations. This rich and messy anatomy qualifies humans as quite complex. Placozoans, on the other hand, with only four rather neatly arranged somatic cell types and no differentiated organs, are extremely simple morphologically as metazoans go.

Whether relatively simple or relatively complex, metazoans generally function quite well, but it is clearly not because they are well ordered or are complex. A pile of trash can be more complex than any organism if it is large enough; the kinds and numbers of its parts, and their disorder, can be increased indefinitely. But no particular organizational plan is required of the trash, which does not function. The functionality of organisms is possible because they are highly organized.

A number of workers have suggested using the number of cell morphotypes as a measure of the complexity of an organism (Sneath 1964; Bonner 1965, 1988; Raff and Kaufman 1983; Valentine 1991; Valentine et al. 1994). This simple metric should provide a rough index of comparative complexity among bodyplans; it counts parts on the level of the basic building blocks of animal bodies. However, it does not take account of the distribution of cell morphotypes into parts at higher levels of the somatic hierarchy—tissues and organs, and so forth—or of the topology of their arrangements. Generally speaking, differences in cell morphotype counts greatly underestimate the complexity differentials between bodyplans. Thus these counts may provide a rough method of ranking bodyplans along a scale (although differences of a few cell types in these crude estimates should not be given much weight) but do not place them at intervals appropriate to their relative complexities. On the other hand, it is possible in principal that bodyplans with significantly different levels of complexity could be constructed from the same number of cell morphotypes. One bodyplan might use combinations of the cell morphotypes in many more organs or other structures than another one, and thus have more and more different parts at higher levels of the somatic hierarchy. Inspection of cell-morphotype numbers in living phyla suggests that this possibility is not a problem. One case where the assessment of relative complexity by cell-type numbers may be a problem is comparing some sponges and cnidarians: some sponges have nearly as many cell morphotypes as some cnidarians and are more disordered. However, the cnidarians do have more parts at higher constructional levels—nerve nets, gonads, and so forth—and thus may qualify as more complex on that basis.

One difficulty in turning cell-morphotype counts to practical use is in defining and identifying the cell types. The cell morphotypes recognized through techniques of light or electron microscopy are based on visible morphological distinctions among which there is relatively little intergradation. Variations in cell size or shape that seem to intergrade are usually lumped; usually it is discrete variations in cell structure or visible cell contents that are split (see discussion in Alberts et al. 1989, 995). Molecular studies have revealed that the differentiation of cells with similar morphotypes is often mediated by diverse genetic signals and that the cells express different fractions of their genomes. These cells, though similar morphologically, differ developmentally. If cell types were differentiated on the basis of their expressed genes, many more would be recognized. However, it can be argued that these differences, as they are not morphological, are not consequential to measuring the morphological complexity of the adult, though they are indications of developmental complexity.

In this book I shall follow the conventions of cell-morphotype determination that are in wide use by histologists today, represented for example by papers in Harrison et al. 1991–1999. The numbers of cell morphotypes clearly vary among taxa within phyla, which is not unexpected since most phyla contain a range of forms representing variations of their bodyplans. Nonsensory nerve cells are so difficult to deal with that they are simply lumped here. It may well be that practices

TABLE 2.3 Cell-morphotype numbers estimated for stem ancestors of metazoan phyla, based on descriptions of the fine-structure anatomy of living
groups (as in Harrison et al. 1991–1999). Neurons have been lumped; distinctive syncytial tissues such as those in Rotifera and Acanthocephala are counted
as one cell type each. Numbers in parentheses are cell morphotypes estimated
for a derived crown group. Cell-morphotype identifications are somewhat
subjective, as are my assessments of the descriptions, though every attempt
has been made to avoid bias among organisms.

Phylum	Cell-Morphotype Number
Placozoa	4
Porifera (Cellularia)	5
Nematomorpha	8
Cnidaria	10
Acanthocephala	12
Entoprocta	13
Nematoda	14
Cycliophora	15
Rotifera	15
Gnathostomulida	16
Kinorhyncha	17
Ctenophora	18
Loricifera	18
Tardigrada	18
Priapulida	20
Echiura	20
Platyhelminthes (Turbellaria)	20
Pogonophora	20
Chaetognatha	21
Gastrotricha	23
Phoronida	23
Hemichordata (Pterobranchia)	25
Bryozoa	30
Onychophora	30
Brachiopoda (Articulata)	34
Nemertea	35
Sipuncula	35
Arthropoda	37 (90)
Mollusca (Polyplacophora)	37 (60)
Urochordata (Ascidiacea)	38
Chordata (Cephalochordata)	39
Annelida (Polychaeta)	40
Echinodermata (Crinoidea)	40
Chordata (Agnatha)	60 (215)

in cell-type splitting and lumping are somewhat uneven among the phyla or among
the researchers, a common problem in taxonomy at all levels. Nevertheless, an
attempt has been made to achieve a standardized count of cell morphotypes for those
phyla or classes for which the literature is adequate. Table 2.3 lists cell-morphotype
numbers for metazoan phyla; these are a naïve lumper's estimate for the number
of cell morphotypes required of an individual with the characteristic bodyplans

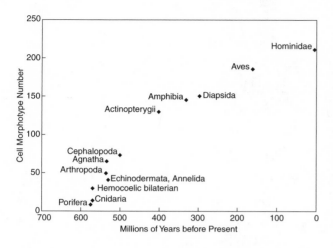

FIG. 2.21 Cell-morphotype numbers estimated for a number of phyla, plotted against the time of their first appearance in the fossil record. Only taxa presumed to have the highest cell-morphotype counts at any time are plotted. Complexity appears to have risen more rapidly early in metazoan history, though as taxa that appear later are increasingly more complex than can be represented by this simple index, the slowdown in complexity increase suggested by the figure is misleading. Modified from Valentine et al. 1994.

of the phyla at a primitive level—the hypothetical stem ancestor of each phylum. For three phyla, arthropods, mollusks, and chordates, the estimated number of cell morphotypes for a derived crown group is also shown.

Note that marine invertebrate bodyplans inferred for stem ancestors, based on the more basal living clades of the phyla, are not really very complex compared with derived terrestrial vertebrates and arthropods. The more complex of the basic invertebrate bodyplans have only a fifth or so of the number of cell morphotypes found in higher vertebrates, and as mentioned above, the actual difference in complexity is certainly far greater. The fossil record suggests a rising upper bound of metazoan complexity through the Phanerozoic Eon—the last 543 million years—that probably extended back to the origin of Metazoa (fig. 2.21). Thus the envelope of metazoan complexity has enlarged. Of course unicellular forms are still with us, and indeed dominate the biosphere in many ways, and primitive invertebrates abound. Still, the cell-morphotype number of the most complex metazoans has grown at rates that average on the order of 0.3 to 0.4 cell types per million years (Valentine et al. 1994). However, some clades have clearly undergone reductions in cell-morphotype numbers. Cell-type numbers appear to have been reduced in Acanthocephala (evolving from within Rotifera, chap. 10), in Pogonophora (evolving from within Annelida, chap. 8), and in such obligate parasitic phyla as Rhombozoa, Orthonecta, and Myxozoa (so reduced that their ancestries are uncertain, chap. 6). Several other phyla (such as the minute Tardigrada) have suspiciously low cell-morphotype numbers.

CHAPTER 3

Development and Bodyplans

The Evolution of Developmental Systems Underpins the Evolution of Bodyplans

If the interpretations of the morphological patterns found in the fossil record are roughly correct, they have implications for the sorts of evolutionary processes that must have been involved. Evolution is evidently capable of producing geologically short bursts of morphological novelties, such as reflected in the Cambrian explosion, as well as of preserving the basic bodyplans of phyla and other higher taxa for hundreds of millions of years. The genome underlying the morphology of metazoan bodyplans must be capable of both great innovation and great conservatism.

Genetic studies of evolutionary change were at one time devoted chiefly to the processes involved in substituting one allele for another within a population, as identified when the substitution produced some recognizable morphological change. Alleles could be studied in nature, and mutational changes in alleles could also be induced experimentally by the use of mutagens. Most of the changes that could be considered viable and potentially useful to a population were small ones. As new techniques were developed, it became possible to study variation in allele products, whether they produced recognizable morphological differences or not, and it became apparent that in general there were large reservoirs of allelic variation within the gene pools of populations. Selection did not operate to reduce all variation to a single "most fit" gene, but balanced the frequency of alleles in various ways to provide a rich variety of alleles and allelic combinations that could be used to adapt to environmental changes. Some workers concluded that even the evolution of phyla proceeded only by this sort of allele substitution. For those workers, macroevolution was microevolution writ large: the small changes brought about by allele substitutions, accumulated over long periods of time, could eventually produce disparate bodyplans in different lineages. Whether the high rates of morphological change inferred from portions of the fossil record could plausibly be achieved by these microevolutionary processes was not clear. Alternative suggestions, such as that a morphological jump arising from a "macromutation" could produce a viable new bodyplan in one abrupt event, did not seem to have any evidential basis.

With the rise of molecular biological techniques, it has become possible to explore the structure and function of genes in great detail. This work has redefined the concept of "the gene," and has been particularly important in increasing our understanding of the genetics of ontogeny—how the genome is involved in the production of a given bodyplan by cell multiplication, differentiation, and positioning as the processes of development unfold. Progress in unraveling the molecular basis of developmental genetics is rapid and rather overwhelming, and a revolution in the understanding of genomic processes is in full swing. Comparative studies have revealed surprising similarities and differences among the developmental systems in different phyla, thereby permitting inferences as to the sort of genome evolution that must have taken place. Unfortunately, such comparative data are as yet available for only a few phyla and for only a sample of molecular systems. Nevertheless, the evolutionary history of metazoan development seems to be open to discovery, and in the very near future.

From what is already known, it is evident that the evolution of regulatory gene systems, rather than of structural alleles, has been chiefly responsible for the sorts of major morphological innovations revealed by the fossil record. As knowledge of these systems increases, it will permit a synthesis that rationalizes morphological aspects of the history of life with the evolution of the developmental processes that produce them. This is not to belittle the role of fitness within populations in defining the pathways of change and in monitoring the sort of evolutionary changes that may occur. The evolution of development must proceed through the processes and under the constraints of population biology. As it is clear that differences in developmental instructions in the genome are responsible for differences in bodyplan morphologies, however, it is appropriate to concentrate here on the sorts of changes in the organization and regulation of the developmental genome that are most likely to be implicated in bodyplan evolution. We begin by reviewing the salient features of the organization and regulation of metazoan genomes. Good accounts of genome organization with reference to developmental mechanisms are by Gilbert (1997) and Davidson (2001), and accounts by Raff (1996), Gerhart and Kirschner (1997), Carroll et al. (2001), and Wilkins (2002) discuss at some length the evolutionary implications of developmental processes.

The English Language and Genomes Both Have Combinatorial, Hierarchical Structures

In Narrative English the Immensity of Combinations Inherent in the Alphabet Is Constrained within a Hierarchy

Genes may be said to form a sort of metaphorical language, both of evolution and of development. Indeed, the analogies between aspects of the structure of, say, written English, and of the metazoan genome are striking. Both written English and genomes involve methods of preserving and conveying information, and both are combinatorial systems which are organized into hierarchies capable of essentially infinite variation. In English, for example, there are only twenty-six letters, yet with combinations of these and a small array of punctuation marks one can create, nearly effortlessly, entirely original sequences of words that convey information. For example, this simple paragraph has never been written by anyone before in the history of English writing (for an interesting account of combinatorics in language see Pinker 1995). The possible combinations are immense even though letter combinations are constrained to form actual words, and the word combinations are constrained to "make sense."

To briefly trace the written English hierarchy, a set of letters is aggregated at several levels to produce a meaningful narrative. First is the level of words, the meanings of which are more or less well defined, and which fall into a number of classes such as nouns and verbs. Most letter combinations do not form words, so the essentially infinite potential of combinations is reduced to the vocabulary of the language. Nevertheless, we have gone from a basic level of twenty-six letters to the vocabulary of an entire language, one that is particularly word-rich. If we consider an individual writer, a working vocabulary seems to be in the range of 10,000 to 30,000 words. In the next round of aggregation, the words are combined into phrases or sentences, the meanings of which are mediated by grammatical rules. The grammatical machinery chiefly involves word order and word modification (such as the changing of word endings or other motifs, much more common in many other languages than in English). Random associations of words must be ruled out as not making sense, but the possible number of associations that do make sense is so immense that it can be considered infinite for practical purposes. Think of all the different sentences that occur in all of English literature, and of your own ability to write a vast number of perfectly proper sentences, none of which have ever been composed before. In three hierarchical levels we have gone from tens, to tens of thousands, to some unspecified but exceedingly large number of units permitted by English.

There are still other levels that are involved in making sense of written English. The next level involves the combining of phrases or sentences so as to produce a rational narrative. Truly random associations of sentences may have a certain postmodernist appeal, but they do not carry much information beyond proclaiming an attitude. Thus the number of possible sentence combinations is greatly constrained, although it must nevertheless exceed the number of possible sentences. Then there are paragraphs and chapters, or, in poetry, stanzas and cantos, the arrangements of which are basic to an understanding of an author's intent. Randomizing paragraphs might suggest experimental writing but would not provide a coherent narrative. The levels from letters to finished work form an aggregative hierarchy which, when constraints are taken into account, produces a message that makes sense.

The constraints form the organizing principles that permit the construction of a meaningful narrative, which is thus anything but a random process, involving as it does selection of "meaningful" units at each level. The inability of random processes to produce meaningful English without constraints is classically demonstrated by having a monkey at a typewriter, selecting letters at random. It can be conjectured that the monkey would eventually produce something intelligible and important, perhaps a poem by Marvell, perhaps a play by Shakespeare, perhaps a book by Darwin. The truth of this supposition depends on how much leeway one gives to the word "eventually," because, on average, writing any of those works by monkey would take far, far longer than our solar system will survive, and even far longer than the universe has been around. Take just the first sentence of Darwin's *Origin of Species*: "When on board H.M.S. 'Beagle,' as naturalist, I was much struck with certain facts in the distribution of the inhabitants of South America, and in the geological relations of the present to the past inhabitants of that continent." Given a monkey with a thirty-key typewriter, randomly striking a key per second, about 10^{180} years are needed to explore all possible combinations of keystrokes for that sentence. Even with billions of monkeys typing, composing the sentence would still require much more time than we know about. Composing a beautiful poem at random is well beyond imagination; there is simply not world enough and time.

Hierarchical Constraints Also Operate within Metazoan Genomes

Genomes are composed of aggregative molecular hierarchies, and they display many of the properties of English (and many other languages) that are associated with that sort of structure. So far as information content is concerned, the basic genetic unit can be taken as the nucleotides of DNA, of which there are four kinds in each genome. These units are then aggregated into sixty-four triplets that form codons, which in a given organism code for twenty different amino acids and include "punctuational" codes, such as stop signals. The codons are, in turn, aggregated into genes. Nearly all codons in the translated portions of a gene code for polypeptides, which are proteins or combine with other polypeptides to form proteins. This part of the genomic hierarchy is constitutive (chap. 1). The number of possible genes—codon combinations—must be very great. Elsasser (1987) has classed numbers into three size categories: ordinary sizes, from 1 to 10^{10} members; very large sizes, from 10^{10} to 10^{100} members; and immense sizes, above 10^{100} members. The number of possible codon combinations is clearly immense, but many of them must be biologically inert or otherwise uninteresting. Just how many possible codon combinations can produce biologically active or useful molecules is unknown, but may well be very large in Elsasser's sense. However, most of this imagined potential genetic vocabulary is not realized. Counts and estimates of gene numbers in metazoans are available for only a few organisms (see below). Judging by these few, the average invertebrate probably has between 13,000 and 28,000 genes, about equal to the number of words in a person's vocabulary.

Of course, to construct a metazoan body requires additional levels of organization of genes within the genome. In terms of the somatic hierarchy, at the cellular level, some fraction of the genome is expressed in a given gene order and dosage to create biosynthetic pathways that produce a given cell type. Other fractions of the genome, which overlap greatly with each other in expressed genes, produce still other cell types. To make developmental "sense" the cell types must be spatially and temporally integrated to form tissues, those to form organs, and those to form systems or organs to function as a viable metazoan organism. Some of the genes act as a sort of grammatical machinery to create meaningful sequences of gene activities from strings of genes, producing cell types, and some act as a sort of rhetorical machinery to create a developmental narrative that makes organismal sense, topologically and functionally. In these respects the genome is behaving as a classic hierarchy, with the entities at each higher level being more complicated, but with the sorts of entities possible at any level restricted by the nature of the entities available from the next lower level.

The Metazoan Gene Is a Complex of Regulatory, Transcribed, and Translated Parts

Transcribed Gene Regions Are Processed to Produce mRNA

Within eukaryotes, transcribed DNA is scattered within the genome like islands in a sea of nontranscribed or silent DNA, to use the metaphor of Loomis and Gilpin (1986). The genes themselves include a number of nontranscribed elements (fig. 3.1). There are nontranscribed regulatory regions, usually upstream of the transcribed regions, which bind molecules that control transcription (see below). Even within a transcribed gene sequence there are both coding elements (exons) and noncoding elements (introns). Introns are known that range from about 80 to over 10,000 nucleotides in length. In some cases the introns are longer than the exons, which are thus archipelagoes within the transcribed genes. Many exons code for protein domains, with introns lying at the boundaries between them. Evidently the

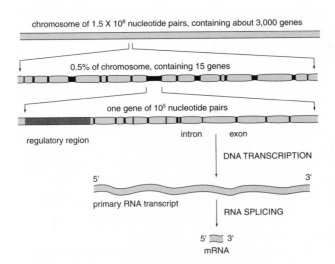

chromosome of 1.5×10^8 nucleotide pairs, containing about 3,000 genes

0.5% of chromosome, containing 15 genes

one gene of 10^5 nucleotide pairs

regulatory region   intron   exon

DNA TRANSCRIPTION

5'   3'

primary RNA transcript

RNA SPLICING

5'   3'

mRNA

FIG. 3.1 Gross structure of genes on a eukaryotic chromosome (a vertebrate). The DNA between genes does not directly function in gene activities. Regulatory regions are not necessarily restricted to upstream sites as depicted here. Modified from Alberts et al. 1989.

sequences within many introns are not usually conserved except at the ends, where splicing signals occur; the sequences within exons tend to be somewhat to highly conserved, presumably depending partly upon the functional importance of the nucleotides. The intron transcripts are excised from the RNA between transcription and translation, and the exon transcripts are spliced to produce messenger RNA (mRNA) that specifies the sequence of amino acids in the gene product.

The stage of exon splicing provides an opportunity for the creation of new sequences. Splicing is accomplished by molecular machines, the spliceosomes, assembled for the purpose from a variety of small molecules. Alternative splicings that include one or more of the putative introns produce alternative codings; parts of the introns of one gene can become the exons in another gene that is manufactured from the same transcribed DNA. Alternative splicings may form a family of genes whose functions are rather similar, but in some cases are distinctive; the alternative genes are usually expressed at different times and in different cell types. The result of alternative gene splicings is to increase the number of genes available to perform different functions, without increasing the number of DNA sequences associated with those genes. Alternative gene expressions must rely on regulatory signals of some sort, but events that lead to the production of alternative sequences from the same gene are not understood in detail.

Cis-*Regulatory Elements Mediate Transcription*

DNA sequences in the regulatory regions of genes form binding sites that are complementary to sites on binding domains of the proteins. These regions lie on the same chromosome and adjacent to the transcribed portion of the genes (fig. 3.1)— in *cis* position (Latin meaning "on this side"). Genes are transcribed through the activity of RNA polymerase II. By convention, directions along genes are based on the RNA transcripts, which are read from 5' to 3'. Features lying above the 5' end are upstream, those below the 3' end downstream. Usually just upstream from the transcribed region of a gene is a sequence known as the promoter, where a molecular complex of transcriptional machinery is assembled. The transcription complex commonly binds to a short sequence, the TATA box, which lies about thirty base pairs upstream of the transcribed region (the complex may also bind to other promoter sequences still farther upstream). For promoters that lack a TATA box, the transcription complex may bind to a protein bound in turn to another promoter sequence upstream of the transcribed region. The transcription complex chiefly consists of the RNA polymerase II molecule, with a TATA-binding protein associated with a number of transcription factors. This complex binds to the promoter, positions the RNA polymerase II molecule appropriately, and releases it from the promoter; the DNA helix unwinds as the polymerase proceeds downstream. The processes surrounding the activation of the transcription complex may require thirty or more different proteins, and vary among taxa. In some cases the complex is bound to the promoter and is activated by one or more transcription factors, while in other cases the complex may be recruited to the promoter by a factor that is itself already bound or tethered to the promoter (Ptashne and Gann 1997). When activated, the polymerase produces a complementary RNA copy of one of the DNA sequences, beginning at a start codon and ending at a termination codon. With the aid of many more enzymes the transcribed sequence is then processed to remove the introns, capped at the 5' end, and given a tail at the 3' end to become mRNA that can be translated into the gene product on a ribosome.

The expression or repression of a gene can be regulated by molecules that bind to a sequence within the promoter. However, expression is also mediated by factors that bind to other *cis*-regulatory sites that usually lie upstream of the promoter; these sites are sometimes known as enhancers. Enhancers may also lie downstream of the transcribed gene, or even inside the transcribed gene, within introns (and therefore their RNA products are excised after transcription). Most enhancers are modular elements that contain binding sites for numbers of factors—protein molecules—that can influence the expression of the gene. As the name implies, enhancers were originally named for their ability to affect gene expression positively. The effects of the transcription factors are combinatorial, however, and depending on which and how many factors are bound to an enhancer module, can act to repress as well as to express a gene. There can be as many as 100 or more binding sites associated with the regulatory complex of a gene, though usually there are fewer. Most enhancers lie relatively close to the transcribed portion of the gene, perhaps a few hundred to a few thousand base pairs away, though instances are known of enhancers that exert influence from over fifty thousand base pairs away. Transcription factors commonly recruit additional molecules to act as coregulatory proteins (Mannervik et al. 1999; and see Wilkins 2002). The mechanical details of enhancer activities have not been completely elucidated. As most enhancer mutations are dominant, many with useful effects may be expected to spread through a population if they manage to escape early extirpation by drift.

The *cis*-regulatory regions of only a few genes are known at all well (see reviews by Arnone and Davidson 1997; Davidson 2001). A gene that has been particularly well analyzed is *Endo16*, which encodes a protein involved in development of the embryonic and larval midgut of sea urchins (fig. 3.2). In the late blastula stage, *Endo16* is expressed in cells at the vegetal pole (in the vegetal plate) and repressed in nearby cells fated to become ectoderm or skeletogenic mesenchyme. As the archenteron develops, *Endo16* is expressed in all midgut cells, and additionally in cells around the blastopore that invaginate during gastrulation to form the hindgut. In larval stages *Endo16* becomes restricted to differentiating midgut cells. The *cis*-regulatory element contains scores of enhancer sites that accommodate nine different binding factors (fig. 3.2). Yuh and Davidson (1996) and Yuh et al. (1998) have shown the sites to be organized into a series of modules (A to G) upstream from the promoter. Module A controls transcription; it is the only module that communicates directly with the promoter. Module A is evidently responsible

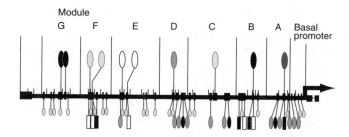

FIG. 3.2 Organization of the *cis*-regulatory sequence of a sea urchin gene, *Endo16*. Binding sites upstream from the basal promoter are organized into functional modules (*A* to *G* proceeding upstream). *Heavy horizontal line*, the DNA sequence; *dark boxes on line*, binding sites; *upper balloons*, proteins at binding sites; *lower balloons and rectangles*, proteins with multiple binding sites. The heavy arrow indicates where transcription starts; black boxes downstream of initiation site represent exons. From Yuh et al. 1994.

for the earliest expression of *Endo16*. Module B is active later, in larval midgut; it interacts with module A so that, together, a synergistic amplification of expression is obtained. Modules CD, E, and F act to repress *Endo16* activity, CD in skeletogenic mesenchyme and E and F in ectoderm, but require mediation by module A to accomplish this. Module G expression is weak, but it evidently persists throughout *Endo16* expression and is amplified by interactions with module A. A complicated *cis*-regulatory apparatus such as that of *Endo16* is capable of expressing the transcribed portion of the gene at a number of different times and places and dosages during ontogeny, and of repressing it at other times and places. Some genes are expressed scores of times, and presumably have even more complex regulatory elements.

Regulatory Signals Are Produced by *Trans*-Regulatory Systems

Transcription Factors Bind to Enhancers

The molecules that bind to the promoter complex, and that may produce regulatory effects, are proteins that act as *trans* (Latin meaning "across") regulators, produced by genes that need not be adjacent to the *cis*-regulatory system of a gene, or even on the same chromosome. These *trans*-regulators are transcription factors that arrive at their binding sites either through diffusion or transport. Transcription factors contain DNA binding domains, stretches of amino acids that can bind to particular sites, such as enhancer sites, on target genes. The binding domains are coded by nucleotide sequence motifs, commonly highly conserved, in the *trans*-regulatory genes that produce them. In the target genes, transcription is initiated when appropriate factors are bound to promoters or enhancers. Presumably, binding of a factor is affected by the number of DNA binding sites, the relative affinity of the factor to the binding sites, and the abundance of factor molecules. There are commonly multiple sites within the regulatory portions of the target gene that will bind a given transcription factor. In some cases there appears to be redundancy in transcriptional signals, while in other cases the binding of multiple molecules of a given *trans*-regulatory factor may be necessary to produce an appropriate result, initiating activity or increasing the amount of gene product in the target gene.

In any given metazoan cell, many of the genes are repressed, for they are dedicated to producing other cell types; evolution of the pattern of repression is at least as important as the pattern of expression. In general the packaging of DNA in chromatin within chromosomes provides a basic repression of the genome. For repressions provided by transcription factors, Gray et al. (1995) recognized two major types, quenching and silencing. In quenching, a transcription factor acts to repress the activity of neighboring enhancer sites over short ranges. Thus even in the presence of the quencher, the target gene may be expressed in some cells, but not in cell types requiring activity in quenched sites. In silencing, a transcription factor acts to repress all of the enhancer sites of the target gene, which therefore cannot be expressed in any cell in the presence of the silencer. Another source of gene quenching and/or silencing has recently come to light; at least two classes of small RNAs (RNAi) interfere with gene transcription or translation and thus can regulate a pattern of gene expression. The evolutionary consequences of interfering RNAs have not been explored (see summary by Couzin 2002).

Some of the genes activated or repressed by transcription factors produce regulatory molecules that affect transcription of still other genes. Many *trans*-regulatory genes act at many different times at different ontogenetic stages and in different cell types to affect numerous target genes. Enhancer sites that bind transcription factors tend to be quite short, commonly from about six to twelve nucleotides long. In some instances only a couple of nucleotide changes suffice to switch the binding affinity of an enhancer site from one transcription factor to another. With so few nucleotides involved, it should not be uncommon for enhancer sites to be switched among transcription factors, and it seems possible that enhancer sites may occasionally arise de novo through mutation. Thus new patterns of gene expression should be readily accessible, so far as their mechanics are concerned, through the evolution of gene regulatory elements.

One would expect that changes that affect the binding of specific transcription factors would be most likely to occur in the *cis*-regulatory enhancer sequence rather than in the binding site on the transcription factor, since transcription factors are commonly used many times in a variety of contexts during development. A change in the binding sites on the transcription factors themselves would thus tend to have widespread effects, would be likely to be deleterious in many functional settings, and would tend to be selected against. Evolutionary changes have nevertheless clearly occurred in *trans*- as well as in *cis*-regulatory elements during the origin and subsequent evolution of metazoan bodyplans.

Trans-*Regulators Are Controlled by Signals That Ultimately Arise from Other Regulatory Genes*

The metazoan genome, then, is composed of genes that can be switched on and off through signals received within their regulatory regions from other genes. During differentiation of the egg, many species of maternal mRNAs are accumulated, and after fertilization (or in some cases at some parthenogenetic event) development of an individual metazoan begins as these maternal factors are translated. Thus at this early stage, regulation of gene expression is translational so far as the genome of the embryo is concerned. The blastomeres produced by cleavages act as compartments within which the cell types of different cell lines may be specified by different maternal factors that are positioned in different parts of the egg. Eventually, nuclear transcription is initiated, so that differentiation proceeds under the aegis of the individual's own genome. Differentiating cell types continue to be produced through different patterns of gene expression, much of which is regulated by differences in transcription signals.

As new cells arise by mitosis, they commonly receive signals to alter their gene-expression pattern from that of their parental cell. Such signals arrive via ligands, molecules that bind to cell-surface proteins acting as receptors; the ligands may be diffusible substances or may communicate directly cell-to-cell. The signals are then transduced from the receptor to the nucleus, where they initiate transcription or repression of target genes by the binding of transcription factors.

The routes of signal transduction vary widely. It is common for the ligand receptor to be a transmembrane protein, with an extracellular element that binds the ligand and an intracellular domain that responds to the reception of the ligand at the cell surface by activating target proteins, commonly through phosphorylation. Activation of the target protein may begin a chain of events along a signaling pathway involving a series of proteins, until binding of a transcription factor occurs on the regulatory complex of a nuclear gene. In some chains seven or more proteins are linked to deliver a signal that causes a promoter to initiate or repress transcription. These signaling pathways are sometimes complicated by interactions with other molecules that are required to enable signaling at many links, forming a maze of chemical bondings and splicings along the pathway. Multiple linkages in a signaling chain thus provide opportunities for many forms of regulation; by inhibiting the formation of one or more links, a signal may be prevented from reaching its target. In some, evidently rare, cases a signaling molecule may bypass a transduction pathway altogether, penetrate the cell membrane, penetrate the nucleus, and activate transcription by itself (He and Furmanski 1995). If a transcribed gene has a regulatory function, the gene product may initiate or suppress the expression of many other genes, either directly, or in a series of regulatory activities. The gene product may act as a ligand, or may influence the production of ligands, which thus signal other cells. Such systems may produce cascades of regulatory activity within a cell, and may spread patterns of transcriptional activity through other cells, producing numbers of differentiated cell populations through overlapping realms of gene expression.

As cells multiply in a growing organism, specialized areas appear within populations of multiplying cells to produce differentiated subregions of equivalence cells, and still more specialized subdivisions may appear within those. Thus there is continuing developmental refinement to produce increasingly specific features. After earliest cell differentiations that are based on maternally positioned factors, and following largely intercellular signals during later cleavage, diffusible signals, such as hormones, may continue the progressive cell differentiations. A likely aspect of this progression is that the transcription factors associated with the earliest signals condition some gene(s) within a cell cluster for expression by a subsequent signal (e.g., Carmena et al. 1998; Halfon et al. 2000).

The relative timing of developmental events is closely controlled, although development can certainly proceed at different rates and at different times in different fields. Early cell differentiations are chiefly devoted to pattern formation—the blocking out of the general body architecture—which, of course, is the essence of bodyplans. Later differentiations are devoted to organogenesis—body parts are developed in appropriate forms at locations specified in the pattern. Organisms with more than one bodyplan, such as those with larvae that lead lives quite different from the adults, must have more than one set of developmental instructions.

Genomic Complexity Is a Function of Gene Numbers and Interactions

The discussion of complexity in chapter 2 suggests that the complexity of a genome can be defined by its minimum description. This description should involve the number of different genes present, and aspects of their deployment. Genome size, as measured by the amount of DNA present, does not correlate with morphological complexity by any measure (Jahn and Miklos 1988). Metazoan genomes contain much DNA that neither codes for genes nor is directly involved in the regulation of gene expression, and that varies independently of morphological complexity. However, one might well expect that the number of functional genes would correlate well with complexity. Gene number is difficult to discover, and has been estimated from extensive sequencing projects in only a few metazoans (table 3.1)—a nematode (*Caenorhabditis*), an arthropod (the fly *Drosophila*), a urochordate (the sea squirt *Ciona*), and vertebrates (the fish *Fugu* and the mammal *Homo*). Additionally, two species of unicellular fungi (yeasts) have been sequenced. Attempts have been made to estimate gene numbers in some of these and in other organisms by several more indirect methods. These methods have not all proven to be very reliable, although some estimates have fallen between 10% and 20% of the gene number subsequently estimated from sequencing. An estimate for an echinoderm (*Strongylocentrotus*) is also available. While keeping possible errors in mind, one must still conclude that there is a rough correlation between gene number and general grade

TABLE 3.1   Some gene number estimates. "Sequenced" estimates are the most reliable, but all are somewhat incomplete. CpG, "islands" of DNA anomalously high in GC content, commonly found upstream of open reading frames. STC, sequence tag connectors. Data from Bird 1995; Simmen et al. 1998; *C. elegans* Sequencing Consortium 1998; Chervitz et al. 1998; Adams et al. 2000; Cameron, Mahairas, et al. 2000; Ventner et al. 2001; Wood et al. 2002; Aparicio et al. 2002; Dehal et al. 2002.

Organism	Approximate Gene Number	Method
Eubacteria		
Escherichia (Eubacteria)	4,000	Open reading frames
Fungi		
Saccharomyces cerevisiae (Ascomycota)	6,340	Sequenced
S. pombe (Ascomycota)	4,824	Sequenced
Metazoa		
Caenorhabditis (Nematoda)	17,800	Open reading frames
	18,424	Sequenced
Drosophila (Arthropoda)	12,000	Reassociation
	16,000	Open reading frames
	13,601	Sequenced
Ciona (Urochordata)	15,500	Homologue sampling
	16,000	Sequenced
Strongylocentrotus (Echinodermata)	27,350	STC sampling
Fugu (Chordata)	30,000	Sequenced
Mus (Chordata)	80,000	CpG islands
Homo (Chordata)	60–70,000	cDNA tagging
	80,000	CpG islands
	26–38,000	Sequenced

of morphological complexity (table 3.1). The yeast genomes certainly contain significantly fewer genes than the invertebrate genomes, by factors of two or three, and invertebrates sequenced to date have fewer genes than vertebrates, but not as many fewer as had been expected (although the gene estimates need further study). It has been speculated that large increases in gene number result from duplications of entire genomes, or at least of major portions of them (e.g., Holland and Garcia-Fernàndez 1996; Sharman and Holland 1996; Patton et al. 1998; Pébusque et al. 1998; Lundin 1999). However, the relatively small number of genes reported in the human genome implies that there have been massive gene losses after any such duplication.

Within the invertebrate phyla, complexity and gene number do not appear to be correlated, judging by the sparse data that are available. For example, flies are many times more complex than nematodes, yet they have significantly fewer genes (table 3.1). Yet the complex development of flies must have involved many more gene-expression events than the simpler development of nematodes. Either each gene sequence is used in many more contexts—at more places and times during development—or the transcribed sequences have many alternate translations that have not been counted as separate genes, or, in all likelihood, both. In either or both cases, each different expression event would presumably involve a distinct usage of the promoter complex, a unique combinatorial signal from enhancers to promote translation. Thus, while gene number is not a good measure of developmental complexity, it is hypothesized that the number of gene-expression events should be (Valentine 2000). In effect, each usage of a gene may count as a separate gene when considering complexity. While the number of genes represents genomic diversity, the number of gene-expression events represents genomic richness, and probably reflects developmental complexity more closely than any other single measure.

The inference from these considerations is that there is more than one way for a lineage to become morphologically complex. An evolving lineage may add more genes and thus enzymes and structural proteins with which to construct a more complex body; or it may use a given set of genes over and over again in different developmental contexts, adding, instead of genes, enhancer sites or modules. It is interesting that the most complex organism yet sequenced, humans, has fewer genes than expected considering how morphologically complex they have generally been judged to be. Humans seem to have chiefly followed the path of adding regulatory elements rather than of greatly enlarging the number of genes in the genome. It is possible that many of the regulatory events are associated with alternative gene splicings, but these require still more informational signaling to (upstream) enhancers.

Although developmental and morphological complexity must be related, they do not necessarily correlate precisely. For example, some fairly simple adults have complicated ontogenies, such as parasitic forms that have a number of distinct bodyplans, and these forms probably require more gene-expression events than direct-developing forms of similar adult complexities. To fairly compare adult bodyplan complexities by numbers of gene-expression events would require the comparison of species with similar ontogenetic patterns. If all stages of an ontogeny could be considered in assessing complexity, the extent of correlation between developmental and morphological complexity could be evaluated, but neither of these variables can be measured at present.

Metazoan Genomes Display Surprising Patterns of Similarities and Differences among Taxa

The variety of early developmental patterns among the phyla, reviewed in chapters 6–11, lead to distinctive larvae or juveniles, indicating that separate developmental routes to the different bodyplans of phyla are largely instituted from early stages of embryogenesis. As these distinctive developmental patterns must be underlaid by distinctive genetic patterns, it might be assumed that evolution has created novel suites of distinctive genes to produce such great morphological diversity. While it is true that the genomes of the phyla harbor many genetic novelties, they are not universally composed of different genes. On the contrary, one of the surprises stemming from comparative studies of molecular developmental processes is that many types of genes that are associated with bodyplan development are

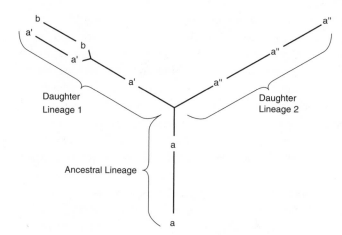

FIG. 3.3 Orthologs and paralogs in diverging daughter lineages. Letters are genes: *a*, an ancestral gene; *a'*, ortholog of *a* in daughter lineage 1; *a''*, ortholog of *a* in daughter lineage 2; *b*, gene originating by duplication of *a'*, and paralogous to *a'*.

homologous throughout Metazoa. The homologues are not necessarily identical, but they have such high sequence identity that they must be derived from a common and very similar ancestral gene. Some of these homologous genes that code for transcription factors can be artificially transferred from a member of one phylum to a member of a second phylum, in which they function normally, even though the phyla have been evolving separately for over half a billion years. There are also many gene "families" composed of numerous genes that have descended from an ancient ancestral gene through multiple duplications, and which have diverged to the point where they are considered separate gene-family members.

Gene homologies—that is, relational properties of genes themselves and not of morphological or developmental features that they help to produce—thus have special complications. There are two main types of gene homologies (fig. 3.3; Fitch 1970). Genes that have arisen in an ancestral lineage and are inherited in daughter lineages, quite like classical morphological homologues, are called orthologs. Genes that have arisen by duplication of genes that are in the same genome are paralogs. In a sense, paralogs are repetitions of the same structure, much like serial morphological homologues (Patterson 1987). Once paralogs have appeared, the direct "descendants" of each are their orthologs. The fate of the genes descending from sister paralogs may be quite varied. With respect to their parental gene, their sequences may remain about the same, or one may remain about the same while the other diverges, or both may diverge. Their expression patterns may also vary; evolution within their *cis*-regulatory elements could produce divergence in some or all of their expression events with respect to each other and with respect to their parental gene. The functional evolution and translated sequence evolution of genes are not necessarily congruent. And of course some genes may have a common ancestry but have diverged within metazoans so that their homology is no longer apparent—they have become transformational homologues.

Some Functional Classes of Genes Are Broadly Similar across Metazoan Phyla

The functions of numbers of genes and gene families have been worked out, so although our data are still quite incomplete, it is possible to get some idea of the general functions to which major portions of the genome are allotted. For example, one group of genes is responsible for the general functioning of eukaryotic cells, carrying out the basic metabolic processes that sustain cell life, permit cell replication, and so forth. These "housekeeping" genes must be subjected to strong stabilizing selection, and are found in nearly or quite all metazoan cells. Eukaryotic housekeeping genes are also widespread among protistans, and metazoans must have inherited most, perhaps nearly all, housekeeping genes from their unicellular ancestors. For example, there appear to be nearly 3,000 such genes in common between the unicellular yeast cell and cells in metazoans (Rubin et al. 2000). That this category of gene is taxonomically widespread across metazoans is not surprising.

Another group of metazoan genes is associated with the general functioning of metazoan multicellularity, that is, with the integration of many cells into a single organism. These genes include those that produce proteins that bind cells to each other or to extracellular structures or that serve as junctions to create tissue sheets or act as channels of communication between cells. Such genes did not arise all at once to produce the first metazoan, of course—some arose in protistans, where they must have served different purposes, and not all of them have been found in sponges—but they are characteristic of metazoans.

The biggest surprise has come with genes that are concerned with cell differentiation. To produce differentiation not only requires some of the intercellular communication systems mentioned above, but also involves regulating the transcription and translation of genes responsible for the particular phenotype of a differentiating cell, and the repression of genes that are used only in other cell phenotypes. Furthermore, there must be instructions that produce or mediate the correct topologies of tissues and organs. As the distinctive developmental histories of the phyla produce bodyplans with disparate anatomical architectures, one might reasonably expect that the genes that regulate developmental paths in different taxa would be different. Yet many regulatory genes are broadly similar—clearly homologous—across the Metazoa. This similarity is particularly strong with regard to the nucleotide motifs that specify the binding portions of their products—the binding domains that affect transcription. In some cases the nonbinding portions of proteins concerned with regulation have not been much conserved; presumably so long as the shape of these proteins permits access for binding to targets and cofactors, these nonbinding sites are not important for protein function. In other cases, however, changes in the sequence of nonbinding regions of transcriptional regulators are associated with the evolution of new expression patterns that have been important in the evolution of morphological patterning (Galant and Carroll 2002; Ronshaugen et al. 2002). Not only do many regulatory genes occur in very morphologically disparate phyla, but they tend to form a pattern of expression—a map of expression domains—that is largely conserved in metazoans despite great differences among phyla in morphological elements in each map domain. This observation has led to the suggestion that Metazoa can be characterized and indeed defined by the characteristic map of expression domains of patterning genes, termed the zootype (Slack et al. 1993), which is different from those of other kingdoms. It appears that the zootype concept holds up well for bilaterians, may work for radiates, and is unclear in its application to sponges.

It seems that at least half of metazoan genes are going to prove to have homologues in a number of phyla and to have had common ancestors at least in stem bilaterians; in fact, most will have had common ancestors in prebilaterians (chap. 14). That is an amazing similarity after the many rounds of radiations and diversifications during the long eras of metazoan history. On the other hand, there are significant numbers of genes that are not known to be widespread at the level of phyla. The relevance of those genes to the evolution of phyla is not yet clear. It seems likely that some of them will be found to be associated with the synapomorphies that characterize each phylum, or with apomorphies within major branches such as classes or orders. However, Rubin et al. (2000) report that about 30% of the proteins in organisms that have been sequenced seem to be unique, without known homologues. Some of these may be products of genes that are evolving at high rates and have diverged rapidly within lower taxa (Rubin et al. 2000). If this is the case, such genes are presumably not involved in specification of bodyplans.

Bodyplans Are Patterned by Sequential Expressions of High-Level Regulatory Genes

Following fertilization and, commonly, the beginning of cell differentiation under the aegis of maternal mRNA transcripts, the progressive specification of cell types and positions that pattern the bodyplan is mediated by a succession of transcriptional regulators. These genes belong to gene families that are widespread among metazoans, but many were named according to their contributions to development in *Drosophila,* where they were first studied. In *Drosophila,* as in other metazoans, a number of transcriptional regulators are expressed in a tightly regulated progression to mediate bodyplan patterning. There are about 700 such genes in *Drosophila* (Adams et al. 2000); they do not directly specify any particular characters, but rather affect the expression of other genes. Among the more important types of patterning genes are those with homeobox and zinc-finger motifs, which produce characteristic DNA binding domains in their protein products. Homeodomains, produced by homeobox motifs, consist of three α helices, while zinc-finger domains include a zinc complex and a binding extension or "finger." There are whole families of genes that produce these (and other) binding domains. Many of these genes (and their homologues, which are widespread in other metazoans) are employed many times and in different developmental contexts within the same organism, in aspects of organogenesis as well as pattern formation. The presence of homologous regulatory genes in different phyla does not mean that they are associated with homologous functions, although they are in some cases (see below).

Anteroposterior Axis Specification and Patterning Genes Are Found throughout Eumetazoa

Embryogenesis in *Drosophila* is relatively well known but is highly derived and thus not representative of forms that were primitive among phyla. The early embryonic tissues are syncytial, and regional specification of the bodyplan precedes their cellularization. Nevertheless, many aspects of the regulative structure in *Drosophila* are quite informative of the developmental system in metazoans as a whole. Within the oocyte, mRNA from the maternal gene *bicoid* is localized at the future anterior of the embryo. When translated in the zygote, the bicoid protein diffuses to produce a concentration gradient that contributes to anteroposterior differentiation, eventually causing the transcription of different target genes in nuclei that lie across key concentration thresholds (fig. 3.4A). The mRNA of a second maternal gene, *hunchback*, is evenly distributed throughout the oocyte. Hunchback protein represses genes that are responsible for patterning the posterior part of the body. The mRNA of a third maternal gene, *nanos*, is localized posteriorly, and nanos protein represses *hunchback* there, so the posterior patterning genes are expressed only in posterior nuclei (i.e., where *hunchback* is repressed). Early development at the anterior and posterior extremes of the embryo is also mediated by other maternal genes.

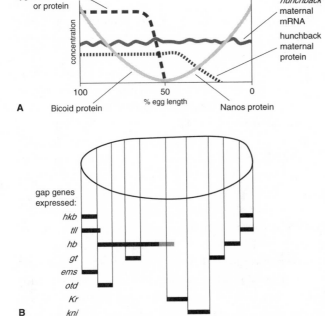

FIG. 3.4 (*A*) Concentration gradients of bicoid and nanos proteins regulate *hunchback* expression in early *Drosophila* embryos. (*B*) Gap genes are activated by their positions along the concentration gradients. *ems, empty spiracles; gt, giant; hb, hunchback; hkb, huckleberry; kni, knirps; Kr, Kruppel; otd, orthodenticle; tll, tailless.* From Gerhart and Kirschner 1997.

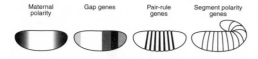

| Maternal polarity | Gap genes | Pair-rule genes | Segment polarity genes |

FIG. 3.5 Progressive regionation of the *Drosophila* embryo during the chiefly successive expression of maternal, gap, pair-rule, and segment polarity genes. Modified from Tautz and Schmid 1998 and Gilbert 2000.

These regulatory tactics, employing gradients and repressors and quenching of repressors, are common in developmental processes, which become quite byzantine.

The *Drosophila* bodyplan is of course segmented in the arthropod style; the specification of the segments is achieved through progressive regionalization of the anteroposterior axis (figs. 3.4B, 3.5). Groups of regulatory genes are expressed in temporal cascades, those groups expressed earlier usually being required for proper development of structures mediated by those that are transcriptionally downstream. The early expression patterns of the cascade that produces anteroposterior differentiation in *Drosophila* are shown in fig. 3.5; the chief gene groups involved are maternal, gap, pair-rule, and segment polarity, usually expressed in that order. These genes specify bodyplan axes, and then selector genes set up specialized body regions where organs and other structures will be developed. In effect, the body is divided into a series of developmental modules.

Gap-gene products are transcription factors with zinc-finger binding domains, largely regulated by maternal gene products. The eight gap genes are expressed in partly overlapping portions of the syncytial *Drosophila* embryo along the anteroposterior axis. The borders between regions of different gap-gene expression are partly regulated by interactions among their proteins. The gap genes in turn help to regulate the expression of the pair-rule genes, which have homeobox motifs and further serve to subdivide the rather broad regions of gap-gene expression.

Just as cells begin to form, producing an embryonic tissue, pair-rule genes are expressed in stripes that presage a segmented body pattern. The first three pair-rule genes expressed under gap-gene regulation are "primary" pair-rule genes (*hairy, evenskipped,* and *runt*), and their products regulate the expression of five other "secondary" pair-rule genes. All eight of these genes are expressed in each of seven anteroposterior stripes. The position of any given stripe is regulated chiefly by different combinations of gap or primary pair-rule gene products binding to pair-rule gene enhancers; therefore there must be seven regulatory configurations that cause transcription of a given pair-rule gene, one expression event in each different location. Here we see repeated independent expression of a single gene in different developmental-morphological modules, a tactic that has been found in all eumetazoans whose molecular development has been studied extensively.

The enhancer system governing the expression of *evenskipped* in the second anteroposterior stripe in *Drosophila* (fig. 3.6) is particularly well characterized (Stanojevic et al. 1991). The protein products of the maternal *bicoid,* and of the gap genes *hunchback, Kruppel,* and *giant,* are expressed along the anteroposterior

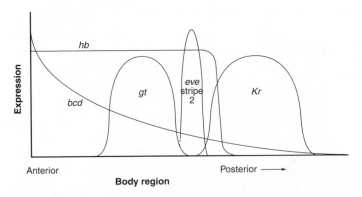

FIG. 3.6 Activation and repression patterns in the stripe 2 enhancer of the gene *evenskipped* (*eve*). This gene is expressed by products of the genes *bicoid* (*bcd*) and *hunchback* (*hb*). Expression is repressed in cells where products of the genes *giant* (*gt*) and *Kruppel* (*Kr*) are present, however; their binding to the enhancer module overrides the signals from *bcd* and *hb*. The sharp boundaries of the stripe are produced by the repressors. From Stanojevic et al. 1991.

axis, and bind to the *evenskipped* stripe 2 enhancer module. Bicoid and hunchback proteins mediate the activation of *evenskipped*, while Kruppel and giant proteins repress expression. Repression signals override the activation signals, and establish the borders of the stripe.

Because of the complicated patterns of anteroposterior overlap of gap and primary pair-rule gene products, the regions of pair-rule striping are somewhat offset from gene to gene, creating zones of overlap. The next set of segmentation genes to be expressed, the segment polarity genes, have combinations of enhancers that permit their activation in a fourteenfold repeating pattern, exploiting the overlaps of the pair-rule expression. Some segment polarity gene products have homeodomains and are transcription factors, while others are ligands or receptors. The morphogen-based system of gradients, which worked so beautifully in the syncytial embryo, has to be replaced by a ligand-based system in the cellular embryo. The fourteen compartments finally defined by segment polarity genes are termed parasegments, because their boundaries do not fall at the boundaries of the morphological segments, but near their centers.

Gap and pair-rule gene products regulate the expression of Hox genes, homeobox genes that were first recognized as being the class of gene responsible for homeotic mutations in *Drosophila* (McGinnis et al. 1984; Scott and Weiner 1984). Homeotic mutations, as defined by Bateson (1894), transform a given structure into some other structure; among Bateson's examples were mutations in insects that produced legs in place of antennae, and mutations in crustaceans that produced antennae in place of eyes. To explain a pattern of rather similar homeotic mutations known in *Drosophila*, Lewis (1978, 1985) proposed that a series of homologous genes mediates the development of structures in the successive body segments along the anteroposterior axis. In these *Drosophila* Hox genes the homeobox motif was first discovered. Homeobox-type genes are present in fungi and vascular plants, so a homeobox-type motif originated in a premetazoan ancestor. In organisms in which the physical locations of Hox genes have been determined (nematodes, arthropods, and chordates), they are clustered. In *Drosophila* the Hox gene cluster is split into two, though other flies have a unified cluster (Powers et al. 2000; Davenport et al. 2000), and a unified cluster is present in most phyla; several are shown in fig. 3.7. In these organisms the anterior boundaries of the regional expression of the Hox genes occur in sequence along the anteroposterior body axis in the same order in which the genes occur within the cluster(s), a condition termed colinearity. Clustering presumably evolved by tandem duplication of an ancestral (ProtoHox) gene, for it seems unlikely that clusters would be assembled from previously dispersed genes. Breaking up a cluster would seem to be more likely, but just why most clusters have

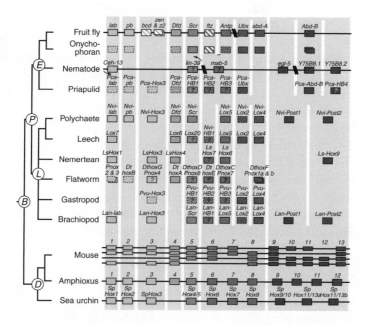

FIG. 3.7 Hox genes identified in a number of phyla. In nearly all cases they are expressed from anterior (to the left) to posterior (to the right) along the body. Note that, except for Onychophora and Arthropoda, none of these phyla have exactly the same Hox gene assemblage, and none of the vertebrate clusters are identical either. In some phyla, positions within Hox clusters have evidently been filled independently by paralogs rather than by inheritance of orthologs. The three great bilaterian alliances are characterized best by posterior genes. *B*, Bilateria; *D*, Deuterostomia; *E*, Ecdysozoa; *L*, Lophotrochozoa; *P*, Protostomia. From de Rosa et al. 1999.

stayed together for hundreds of millions of years is not understood. Hox genes can function as selector genes, which regulate the identity of structures that develop within their domains of expression in ectodermal and mesodermal tissues.

In *Drosophila*, regions of Hox gene expression begin in the posterior head lobe and are mostly broad and overlapping. The anterior boundaries of the regional expression of the Hox genes occur in colinear sequence along the anteroposterior body axis. The borders of Hox gene expression most commonly correlate with segmental boundaries, unlike the parasegmental expression of segment polarity genes. If the regulatory portion of a Hox gene is mutated, the segment(s) in which it is normally expressed or repressed may be altered, and the mutation may cause the expression of a downstream gene cascade that produces structures in one segment that are normally developed in another segment. For example, a mutation that alters a signal from a combination that mediates leg expression to one that mediates antenna expression can result in producing antennae in the segment where legs belong, a classic homeotic mutation. The antenna-determining genes have been identified as the homeobox genes *extradenticle* and *homothorax* (Casares and Mann 1998); they are regulated by the Hox gene *Antennapedia*. *Antennapedia* is expressed in segments where legs are to appear, repressing *homothorax*, which is required for nuclear expression of *extradenticle*. Transcriptionally downstream of *extradenticle* is a potential cascade of gene expression that mediates leg development. Without *homothorax* expression, this gene cascade is not activated, but rather an alternate "default" cascade is expressed that results in antenna development.

In most bilaterian phyla the anteroposterior axis is specified in the oocyte, but not always by the same molecular events. However, differentiation of structures along the anteroposterior axis does involve some of the same genes in all bilaterian phyla investigated for anteroposterior patterning (fig. 3.7). If Hox genes prove to be universally involved, it may reflect the fact that most bilaterians are elongated and rather differentiated anteroposteriorly, and a colinearly expressed gene cluster can pattern that axis so effectively that it has a selective advantage, though a cluster may be broken up or otherwise modified under some circumstances. But even in cnidarians there are Hox-type genes (Schierwater et al. 1991; Kuhn et al. 1996; Finnerty and Martindale 1997; Schierwater and Kuhn 1998; Finnerty 1998), at least some of which appear to be involved in axial patterning (e.g., Bode 2001). A typical Hox gene sequence is present in echinoderms (Martinez et al. 1999), even though they have a (secondarily) radial aspect.

Some Hox gene homeodomains have been highly conserved among different metazoan phyla, from the diploblastic phyla to highly derived, complex bilaterians. For example, the homeodomain of the product of the Hox gene *Antennapedia* in *Drosophila* differs by only a single amino acid from its homologue in mice (product of *Hox-A7*), although to be sure some homeodomains are much more divergent. Mice have not one but four Hox clusters containing thirty-nine genes in all (fig. 3.7; see Krumlauf 1994), and though each cluster is unique in the number and/or kinds of Hox genes present, their order is preserved. Presumably these four clusters were duplicated or multiplied from an ancestor with a single cluster. The zebra fish *Danio* has seven clusters, each also unique (Amores et al. 1998). The largest single cluster known is in cephalochordates; *Branchiostoma floridae* has a single cluster of fourteen Hox genes (Ferrier et al. 2000). Evidently cephalochordates possess a complete, primitive chordate cluster, for the unique clusters found among vertebrate classes are formed by differential loss of homologues of *Branchiostoma* Hox genes. The Hox gene assemblages of nearly all phyla are unique, indicating significant evolution in patterns of specification (fig. 3.7). The bodyplans are of course quite disparate, and the morphology of the features that the genes mediate may be entirely different from group to group and need not be related to particular bodyplan features such as segmentation. Priapulans, unsegmented ecdysozoans that lack limbs, have ten Hox genes, although all these genes are not necessarily orthologs of *Drosophila* genes (de Rosa et al. 1999); their developmental functions have not been worked out. In nematodes, which are also unsegmented ecdysozoans, there are five (or possibly six) Hox genes. The anatomical units whose development is mediated by given nematode Hox genes are founder cell lineages that differentiate into only a few cell morphotypes and, indeed, involve relatively few cells (Kenyon and Wang 1991; Wang et al. 1993). The nematode Hox genes are not in the same order as in flies and mice, and one of them is oriented opposite to the others.

A second Hox-type cluster of usually three genes, the ParaHox cluster, appears to be widespread among eumetazoans and may be universally represented therein. The ParaHox cluster was first identified in amphioxus (Brooke et al. 1998) and is now reported in many phyla (Ferrier and Holland 2001). The ParaHox cluster probably first arose from a Hox-type cluster at least as early as the stem ancestor of the cnidarians, when there were four genes in a ProtoHox cluster that was duplicated to form Hox and ParaHox clusters (Finnerty and Martindale 1999; Kourakis and Martindale 2000). ParaHox genes, like Hox genes, are expressed colinearly where they have been studied, but functions of ParaHox genes are not so well known as of Hox genes; they may be important in patterning endoderm but are expressed in other tissues as well.

Dorsoventral Axis Specification and Patterning Genes Are Similar across Bilateria

Just as with the anteroposterior axis, establishment of the dorsoventral axis in *Drosophila* begins in the oocyte with the expression of maternal genes—at least twenty or so—whose products interact with products of nuclear genes. Following fertilization, the zygotic nucleus migrates toward the anterior end of the zygote and expresses the gene *gurken,* the product of which represses the transcription of numbers of maternal genes in the follicle cells bordering that end of the zygote. Among the maternal genes that are repressed is *nudel,* and where *nudel* is repressed, the dorsal side of the embryo develops. The product of another maternally active gene, *Dorsal,* is a transcription factor and is found in the cytoplasm of every cell in

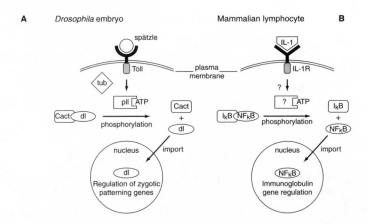

FIG. 3.8 Homologous regulatory pathways regulating nonhomologous genes. (*A*) A Toll pathway in *Drosophila*, leading from a ligand (spätzle) to the regulation of ventralizing genes in embryonic nuclei. See text for description. tub, tube, a protein associated with the early pathway; pll, pelle; Cact, Cactus protein; dl, Dorsal protein. (*B*) A pathway in *Homo*, leading from a ligand (IL-1) and its receptor (IL-1R) to the regulation of immunoglobulin genes of the human immune system. NFkB, a protein with sequence similar to dorsal; lkB, a protein with a function analogous to cactus. After Shelton and Wasserman 1993.

the developing embryo. However, Dorsal protein can be prevented from entering the nuclei when it is bound by another protein, Cactus, as it is in earliest embryogenesis. Where *nudel* is not repressed, its product is found in a layer of what will become the ventral part of the embryo. The gene *nudel* initiates a multistep cascade of gene activity that results in a signal to receptors on the surfaces of the embryonic cells— the Toll protein receptors (fig. 3.8). These receptors in turn activate a kinase that releases Cactus from Dorsal, permitting Dorsal to penetrate the nucleus, where it initiates transcription of its targets. Dorsal is most concentrated along what will become the ventral midline, and is found in a concentration gradient from there to the point where the Toll pathway is not active (i.e., where *nudel* is repressed).

Dorsal protein initiates a cascade of transcriptions and repressions that produce the phenotypes of ventral cells. Among the early effects is the expression of the gene *twist* and repression of two other key genes that regulate dorsal cell phenotype, *decapentaplegic* and *zerknüllt* (fig. 3.9). Twist protein evidently regulates a signaling pathway involving a growth factor receptor (DFR1), and further gene activity ensues to produce ventral mesodermal cell phenotypes. On the opposite side of the embryo, where Dorsal protein is absent and the key dorsal genes such as *decapentaplegic* and *zerknüllt* are not repressed, cascades of transcription that they initiate lead to dorsal cell phenotypes. A large number of gene products are involved in the many signaling and transcription systems that produce the final cell phenotypes. Between the compartments influenced by these dorsalizing and

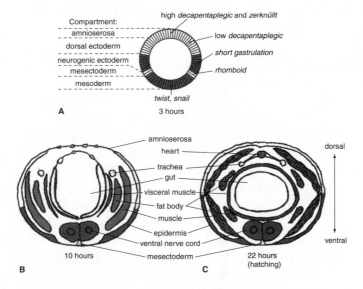

FIG. 3.9 Dorsoventral patterning during *Drosophila* development. (*A*) At 3 hours, selector genes define three main compartment types in the embryo (the *short gastrulation* compartment is repeated on each side). (*B, C*) At 10 and 22 hours, interactions among the dorsoventral and other selector domains produce successively more compartmentalized embryos. From Gerhart and Kirschner 1997.

ventralizing factors, a third compartment develops that produces what become, after gastrulation, ventral ectoderm and the ventral nerves. The gene *short gastrulation* is expressed in this third, neurectoderm compartment. Interactions among genes in these three compartments result in six dorsoventral compartments in the late embryo (fig. 3.9).

Gerhart and Kirschner (1997) give a comprehensive account of the establishment of the dorsoventral axis in *Drosophila,* which is greatly simplified below. Many of the molecular details are different in other arthropod groups and in other phyla, and it is not worth pursuing further particulars in *Drosophila.* However, key genes that regulate the dorsoventral pattern in *Drosophila* have homologues in many other phyla. It is a striking fact that, in mice, the expressions of homologues of some genes that regulate dorsoventral polarity in *Drosophila* are inverted, so that dorsal *Drosophila* factors are expressed in ventral compartments and vice versa (Arendt and Nübler-Jung 1994; DeRobertis and Sasai 1996). Thus a *decapentaplegic* homologue (*Bmp-4*) is involved in specification of the ventral compartment in mice, rather than the dorsal compartment as in *Drosophila.* Further, a *short gastrulation* homologue (*chordin*) is expressed in a dorsal compartment where it helps to block *Bmp-4* activity, rather than being active in a more ventral compartment as it is in *Drosophila.* Some other developmental genes associated with dorsoventral patterning, which are homologous in flies and mice, are expressed in inverted order as well (see chap. 11).

Organogenesis Involves Positioning by Patterning Genes and Development via Gene Cascades Controlled by Selector Genes

If anteroposterior and dorsoventral developmental specifications were complete, all positions within a body could be accounted for. The locations of internal organs and of features such as antennae, limbs, and wings are generally positioned on the body by the operation of anteroposterior and dorsoventral axial mediators. In *Drosophila* the adult body surface, and indeed most of the adult body (the imago), develops from pluripotent cells that are set aside from larval development in sac-shaped disks, four to a segment, beginning in midembryogenesis. These imaginal disks are located at the boundaries of parasegments, which places them about midway between the segmental boundaries of the adult. The disks are greatly enlarged as the larva grows, and at metamorphosis, cells in those disks that are destined to become involved in appendage development begin differentiating. Which features develop on a segment depends upon the pattern of Hox gene expression (see above). Compartments of differential gene expression are set up by the different parasegments that the disks overlap. For example, in the limb disks, the anterior part of the appendage disk overlies cells in the posterior of a parasegment that expresses the segment polarity gene *wingless*. The posterior part of the disk overlies cells in the anterior of the next parasegment back that expresses the segment polarity genes *hedgehog* and *engrailed*. Furthermore, the product of the dorsoventral selector gene *decapentaplegic* forms a gradient, most concentrated dorsally, across the site of the disk. Three disk compartments result: an anterodorsal compartment in which *hedgehog* product causes *decapentaplegic* expression; an anteroventral compartment in which *hedgehog* product causes *wingless* expression; and a posterior compartment in which *hedgehog* and *engrailed* are both expressed.

At the center of the limb disk is an area where a combination of *wingless* and *hedgehog* products mediate the expression of two selector genes, *distal-less* and *aristaless*. The homeodomain proteins of these genes interact with the proteins within the compartments set up by the segment polarity genes and with other genes that are expressed by these selectors, and by their targets, to produce distal growth and limb differentiation. Differentiation proceeds concentrically from the outer disk rim, which will be the most proximal, to the disk center, which will be the most distal, and many genes are expressed in concentric rings. Hundreds of genes are directly involved in the cascade that produces a leg. Wing disks are regulated by some of the same segment polarity genes, which establish dorsal and ventral compartments within which some different selector genes are expressed; wing growth then occurs along the compartment boundary.

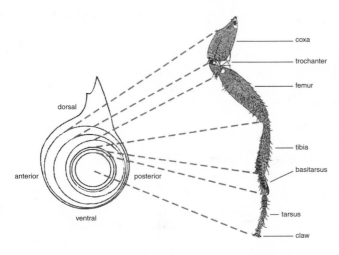

FIG. 3.10 An aspect of proxi-modistal development in *Drosophila*. *Left*, an imaginal disk in late embryogenesis, with concentric rings of cell differentiation forming the prospective limb parts. *Right*, a limb derived from the disk after metamorphosis. Cell differentiations also occur within the disk rings to produce sensory and other structures that are distributed along anteroposterior and dorsoventral limb axes. After Schubiger 1968 and Gerhart and Kirschner 1997.

The puzzling offset of the main regulatory anteroposterior gene compartments (parasegments) and the morphological-anatomical compartments (segments) of *Drosophila* is thus clarified. Appendages and other features that develop from disks naturally proceed from near the middle of segments, where their muscular systems and/or other anatomical supporting features are centered. Placing limbs at segmental junctions is hardly feasible; it would be an architectural nightmare. As limb development involves interactions of segment polarity and other regulatory genes, their regions of expression have borders that form parasegmental boundaries well away from the segmental junctions.

Many of the genes and gene assemblages involved in appendage development in *Drosophila* are also involved in vertebrate limb development. The vertebrate limb and wing axes are established by homologues of *hedgehog* (in mice *Sonic hedgehog*) and *decapentaplegic* (in mice *Bmp-2*) that act like their relatives in *Drosophila*. In the chordate wing, homologues of a number of *Drosophila* genes play similar developmental roles (fig. 3.11; reviewed in Shubin et al. 1997). Thus some homologous genes and gene-product interactions are employed in producing similar structures in arthropods and chordates, though these phyla are only distantly related.

Some homologues of *Drosophila* genes that are widespread throughout the Metazoa are expressed in structures that are neither homologous nor analogous to those in which they are expressed in flies. For example, *distal-less* is a selector gene for appendage development in *Drosophila* (see above), and many of its homologues have been studied across several phyla (Panganiban et al. 1997; I will call all these homologues *distal-less*). In onychophorans, *distal-less* is expressed in the distal portion of developing antennae and lobopods, features that are probably homologous with their arthropod analogues. *Distal-less* is also expressed in appendage development in chordates, analogues to arthropod appendages but unlikely to be homologues. Similarly, in polychaetes, *distal-less* is expressed in distal

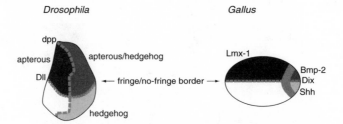

FIG. 3.11 Regulatory gene compartments in the wing imaginal disk of the fly *Drosophila* and the wing bud of the chicken *Gallus*. Some homologous genes regulate compartments with similar geometric relations in both organisms. Cells along the *fringe*/no *fringe* border lead the outgrowth of the wings in both cases. As these two organisms are only distantly related, the question is raised as to why some of their homologous genes serve similar developmental functions for morphological features that were independently evolved. After Shubin et al. 1997.

portions of developing antennae and parapodia. While the possible homology of annelid and arthropod antennae might be argued, it is quite unlikely that arthropod legs and annelid parapodia are homologous, though they are analogous as being limbs. *Distal-less* expression is also found in mollusk gills, which are not related in any obvious way to any fly organ. In urochordates, *distal-less* is expressed distally in both the developing ampullae and siphons of ascidians. These features have no obvious counterparts in arthropods. Finally, in echinoderms, *distal-less* is expressed distally during the early development of tube feet and at the tips of growing spines (sea urchins were studied). The tube feet are part of the water vascular system, a unique coelomic structure that is not represented by analogous morphological features in arthropods, and the spines are endoskeletal features and quite unlike any arthropod structure. Indeed, these echinoderm features are surely synapomorphies of that phylum and are unrelated to structures in which *distal-less* is expressed in any other phylum. Other distinctive expressions of *distal-less* in several classes of echinoderms have been described by Lowe and Wray (1997). The usages to which *distal-less* and other genes are put in different echinoderm classes are commonly class-specific and attest to widespread recruitment of these genes for a variety of developmental roles. Discoveries of other cases of individual gene homologues performing different developmental functions in different organisms are becoming commonplace.

It is interesting that, in vertebrates, a cell adhesion molecule (N-CAM) has been found to contain binding sites for numbers of homeodomains, and can be regulated by Hox and Pax gene products in vitro; there is some evidence that such regulation occurs in vivo as well. Edelman and Jones (1995), who reported this finding, point out that, if this regulatory mechanism proves to be general, it provides a connection between the molecular mechanics of morphogenesis and the gene regulatory apparatus that patterns bodyplans. Elements of this system, employing homeobox genes, may have evolved with the earliest metazoans. Thus evolution of both the differentiation of metazoan cell types and of their spatial positioning may well have involved some of the same regulatory genes.

Signaling Pathways, Like Individual Genes, Are Recruited for a Variety of Tasks

There are increasing numbers of instances known in which entire signaling pathways have been co-opted from one task to serve another. An early example that is particularly convincing involves the signaling pathway for ventralization in *Drosophila* (see above) and a pathway in the mammalian immune system that produces differentiated lymphocytes (see Shelton and Wasserman 1993; González-Crespo and Levine 1994). A similar sequence of homologous genes occurs in both systems; mammalian lymphocyte development involves a ligand and a Toll-like signaling molecule, which frees a Dorsal homologue from a Cactus homologue, permitting its translocation to the nucleus to regulate genes there (fig. 3.8). González-Crespo and Levine show that the *cis*-regulatory enhancers targeted by the Dorsal homologues are homologous as well. It is at that point that the homologous pathways end, however; the genes with which the enhancers are associated are quite different in the two systems, and their functions could hardly be less homologous. Evidently the enhancer from the target of one of these pathways (probably from a primitive immune system) has been duplicated and installed in the other target, thus recruiting that target gene (and its downstream cascade) to expression via that pathway. The entire pathway was thus co-opted for a novel function, almost certainly as a single unit (a gene cassette). There is an additional example of the use of the same pathway in still another context (Kanegae et al. 1998; Bushdid et al. 1998). In patterning of chick limb, the vertebrate homologues of the *Drosophila* genes *Dorsal, Cactus, twist,* and *decapentaplegic* form a system, essentially identical to that in *Drosophila* ventralizing, for signaling between ectoderm and mesoderm during limb growth.

Developmental Genomes May Evolve on Many, Semidecomposable Levels

Evolution of Cis-Regulatory Elements Entails Effects That Differ from the Evolution of Transcribed Genes

Evolution has been defined as consisting of changes in gene frequencies, the genes of interest being alleles of exons that produce polypeptides with a particular sequence of amino acids. Granting that the frequency changes in alleles of exons are a basic aspect of evolution, there are other exceedingly important aspects of genetic evolutionary change—those that involve changes in the organization of gene expression. It is abundantly clear from studies of development that the evolution of patterns of gene expression is of basic importance in the history of life, and has been particularly fundamental in the origin of bodyplans. For an early expression of this view see Wallace 1963; for an early hypothesis of the regulatory mechanism in eukaryotes see Britten and Davidson 1969; and for an early application to the Cambrian explosion see Valentine and Campbell 1976.

The evolution of polypeptide structure arises primarily from mutations to the transcribed DNA sequences of genes (or sometimes to their splicing mechanisms), whereas the evolution of patterns of gene expression arises largely from mutations to the regulatory sequences of genes. While mutations within a regulatory sequence do not alter functional aspects of the polypeptide, they may alter the timing, dosage, and location of its occurrence, and may quench or silence the gene altogether. A new pattern of expression caused by a mutation would then be subject to selection, the mutated regulatory sequence in effect representing a regulatory allele. As noted by González-Crespo and Levine (1994), the evolution of enhancers would seem to be as fundamental as the evolution of exons. Indeed for the origin of bodyplans, involving the patterning of novel architectures, evolution of *cis*-regulatory elements appears to have been preeminent.

There are numerous kinds of sites within the metazoan genome at which mutations can produce regulatory evolution. An obvious type of site is within modules of the enhancer sequences, where a small change may permit the binding of a different transcription factor, which might change the state of a gene, say from expression to repression or vice versa, in cells where that factor is present. A new binding site might result from mutation of a sequence that was already present, or from the introduction of an enhancer sequence or module by a shuffling of DNA among regulatory sequences. As is evident from the regulatory portion of the gene *Endo16*, reviewed above, regulatory signals are integrated. Gene activity depends upon combinatorial regulatory codes, and in *Drosophila* the combinations are at least sometimes serial, that is, a given sequence of binding events is used to achieve gene expression. Within such a system the upstream genes will tend to be expressed over broader domains, and their expression may determine the "competence" of the more downstream genes to be expressed.

If the entire developmental genetic system consisted of a set of nested domains of gene expression, the expression sequence would be treelike, with the earliest expression events near the root, and branching into progressively more localized expression domains downstream, or, in the tree metaphor, upward along the branches. In tracing expression events as if each were a separate gene, something resembling such a tree structure should be found. In tracing the pattern of all the activities of a given gene, however, the treelike pattern would not be evident, because the gene would be expressed along different branches at different times and/or places. Thus from the standpoint of all the signals associated with a gene, rather than of each independent expression event, the pattern is a network rather than a tree, for there are multiple expression domains that provide regions of competence for genes that have multiple expressions. In some cases each domain is different in that each requires a different combinatorial *cis*-regulatory signal to the gene in order for expression to occur in each of the different domains, as is true of *evenskipped* (see Frasch and

Levine 1987; Goto et al. 1989; Stanojevic et al. 1991). In other cases, as when genes are expressed farther downstream in separate modules, it seems likely that the same combinatorial *cis*-regulatory signal is simply repeated, although there still must be unique combinatorial signals from genes somewhere upstream to establish each region of competence.

Changes in the binding sites, not only of target genes, but of their transcriptional regulators, also occur, for the binding motifs display divergences. Presumably, many of such changes involve gene duplication, so that the ancestral function continues to be served by one daughter gene as the derived function evolves in the other (see below). Evolution of transcriptional regulation may also occur at sites entirely outside the DNA binding domains. One such case involves changes in a Hox gene (*Ultrabithorax;* Galant and Carroll 2002; Ronshaugen et al. 2002). One of the changes involves the carboxy-terminal domain of the molecule, a change that is evidently derived in insects, but is not found in crustaceans (Ronshaugen et al. 2002). The insect *Ultrabithorax* molecule mediates the repression of limb development in abdominal segments, thus contributing to a key element in the insect bodyplan. The insect molecule evidently achieves this regulatory effect during signal transduction.

Regulatory Variation May Be Maintained by Several Unique Mechanisms

Developmental variation is presumably accumulated within a genome by various types of balancing selection acting on alleles of regulatory gene transcripts and also on alternative *cis*-regulatory enhancer sites or enhancer modules that affect the transcription patterns of their genes. Also, duplicated regulatory genes and genetic elements may provide a pool of DNA sequences that are subject to mutations that may provide functional variation in the activity of the genes, which may then diverge to provide new functions. Gene duplications are common, evolutionarily speaking, with an estimated average rate of about 1% per gene per million years (Lynch and Conery 2000). If duplicate genes do not diverge, however, there may be no selective pressure to retain both of them, and mutations can cripple one without ill effects. Marshall et al. (1994) have estimated that such gene degradation would probably occur over a scale of 0.6 to 6 million years, and a truly redundant gene would be unlikely to remain functional for 10 million years. Nevertheless, families of genes that have duplicated and diverged, including developmentally important genes, are common. A likely reason for the success of many duplicated genes is that one of the duplicates has suffered a mutation within an enhancer module, which deleted the particular gene-expression event that the enhancer helps control. The other duplicate, with that same enhancer intact, then becomes the functional gene for that particular expression event and is retained by positive selection (Force et al. 1999). Following the establishment of positive selection for retaining both duplicates, mutations that delete other redundant events in one or the other of the genes might not be opposed by selection. As a result, both genes are retained, but each becomes specialized, being expressed for a particular fraction of the events covered by the parent gene. The class of events that produces functionally divergent duplicated genes by these methods is termed the duplication-degeneration-complementation (DDC) process (Force et al. 1999). As those authors note, degeneration of an inactivated enhancer module in one gene actually facilitates preservation of the other duplicate. There are still other ways to retain duplicates, for example, through an enhancer mutation to one of a pair of duplicated genes that creates a novel function, which is then positively selected.

Simulations of binding-site evolution within regulatory sequences suggest that fixation of new binding sites, and even combinations of two new binding sites, may evolve quickly, becoming fixed at microevolutionary time scales under neutral selection (Stone and Wray 2001). Sequence evolution within functional enhancers was studied by Ludwig et al. (2000), who examined variation in the *evenskipped* stripe 2 enhancer module in thirteen different species of *Drosophila,* all of which showed differences in the module sequence, although there were no differences in the timing or pattern of expression of the stripe in the four species for which such details were available. The stripe 2 enhancer modules of two of the species, *D. melanogaster* and *D. pseudoobscura,* were manipulated experimentally to try to understand why the sequence differences did not have developmental consequences. The last common ancestor of these species probably lived over 40 Ma, so genetic divergence is not surprising, but the conservation of the stripe is essentially perfect. In fact, when individual genetic differences in the enhancer modules were studied, they did produce developmental variations; for example, some of the changes involved impairment of *Kruppel* repression, a change that should be expressed in blurring and extending of the posterior stripe margin. It was concluded that the differences that arose from any given changes were being masked by other changes that compensated for them. Assuming that mutations are not so strongly selected against that such masking does not have time to appear, nucleotide turnover at binding sites may produce evolution within the developmental system while selection maintains a stable outcome of gene functions. Thus low-level but rather pervasive turnover in enhancer-system sequences is indicated, suggesting that genetic variations that can underpin *cis*-regulatory developmental evolution are generally available to selection.

Other sorts of potential developmental variation may be harbored in a genome at a posttranslational level. A particularly interesting finding with *Drosophila,* arising from experiments with a heat-shock protein, Hsp90, was reported by Rutherford and Lindquist (1998). Heat-shock proteins often function as "chaperones" to protect an organism from the deleterious effects of environmental stress, such as unusually high ambient temperatures. Under such conditions, proteins may become incorrectly folded or denatured. Heat-shock proteins can interact with such impaired proteins to keep them properly folded and functional until conditions return to "normal." It turns out that, in the case of proteins that are altered because their genes have suffered mutation, Hsp90 will interact with some of them to maintain their original functions also. In this way the mutant genes are hidden from selection, and a significant amount of genetic variation may build up without interfering with fitness. When Hsp90 itself is impaired or exhausted, so that its chaperone function is lost, the mutant protein structures become employed in development. In *Drosophila,* mutations unmasked by suppression of Hsp90 produced morphological changes in a great many adult structures, ranging from minor changes in wing venation to transformation of body parts. Some of these changes were maintained in subsequent generations.

Units of Selection in Developmental Evolution Include Semi-independent Modules

Debates about the most important units on which selection operates have been long-standing and intense. Among the entities on which selection may operate are the familiar levels of cells, tissues, organs, and so forth within the somatic hierarchy; individual organisms (usually considered to be the chief focus of natural selection); and higher-level "group" entities such as kinship groups, trait groups, other sorts of populations, species, and higher taxa—clades, including paraclades (see Buss 1987; Williams 1992). Selection or sorting on one or another of these entities may produce change within genes, genotypes, gene pools, and various collections of gene pools. All these levels of selection are probably associated with the evolution of novel bodyplans in one way or another.

In development, genomes are not "beanbags" of independent genes but are integrated developmental systems that display many levels. Selection is faced with genes whose products may regulate multiple developmental pathways that range from being functionally related to being functionally disparate. Constraints on the modification of these gene products must be strong. However, evolution can act individually on each pathway of expression by selection on individual regulatory elements. For developmental evolution of bodyplans, it must be the sorting out of gene-expression events (chiefly through enhancer evolution) that is most significant. The exons of the regulatory genes, particularly the binding motifs, may change hardly or not at all. Thus units of selection among these regulatory genes can involve cascades of gene-expression events that are used in the development of a feature in which change produces higher fitness.

The morphological entities that confront selection during bodyplan evolution include integrated body parts. These entities integrate biomechanical and physiological features to form selective units. From one perspective, it is these morphological units that organize development, under the aegis of selection. Distinctive body parts are commonly underpinned by developmental modules involving multigene cascades (see Raff 1996; Carroll et al. 2001). The role of modules is particularly easy to appreciate in forms that are seriated or segmented—where body parts are serially homologous. The same key genes may be responsible for initiating the

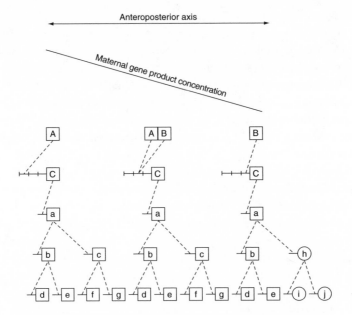

FIG. 3.12 A highly simplified cartoon of the sort of regulatory pattern that underlies some serial homologies. Transcribed genes are represented by boxes or circles, with their regulatory sequences represented by tails to the left of the boxes; dashed lines indicate an activation signal. A concentration gradient of a maternal gene product establishes broad, overlapping domains of expression of genes *A* and *B*, which have promoter complexes that are activated by different concentrations of the maternal gene product. The products of these genes are transcription factors that set up three fields, one characterized by the presence of *A* alone, one of *A* and *B*, and one of *B* alone. Trees of gene expression are shown at one site in each of these fields. There are three enhancer sites on gene *C*, each of which is activated by a different transcription factor or combination of factors produced by genes *A* and *B*. Each compartment is thus under the aegis of a separate expression event of gene *C*; the events may or may not be contemporaneous. Gene *C* products bind to enhancers in the gene labeled *a*, activating a treelike cascade of gene expression in each compartment. Each of these cascades is free to evolve independently; the cascade in the *B* compartment has in fact evolved, most likely through enhancer mutations; instead of gene *c*, gene *h* is expressed by gene *C*, which captures a new subcascade because gene *h* activates genes *i* and *j* (in circles). In cascades containing hundreds of genes, opportunities for modification of the branching patterns of expression would be many and varied.

development of each homologous module, but at least many of them are expressed in different modules by different enhancer signals (fig. 3.12). Separate but similar developmental cascades occur in each module.

The separate cascades downstream from the key genes are free to evolve independently, each developmental module corresponding to a morphologically serial homologue. As mutations produce alternative morphological outcomes of development in different modules, divergence and specialization of modular morphologies can proceed if selectively advantageous. The evolution of limbs on particular arthropod segments, changes in functional specialization within the segments, and their differentiation among segments are cases in point. Selection for a given feature must often entail a ripple effect, as for example lengthening might involve selection for strengthening, and such coadaptive changes would occur throughout the module. The evolution of limb features would occur at nodes in the tree of gene expression that are downstream of the key genes at the head of the modular cascade; in effect, subsidiary cascades may be regulated from many nodes in the gene-expression tree (fig. 3.12). As domains of expression are sometimes created by the overlapping expression of upstream genes, boundaries of expression of a given gene may not correspond to boundaries of morphological entities. The evolutionary system is certainly vastly more complex still, with complications from the activities of cofactors and of posttranscriptional modifications to proteins, and from many other sources from which a given developmental outcome may be regulated.

For organisms whose bodyplans do not contain serial homologues, modular patterns may nevertheless govern differentiation among body parts, but the systems are not as well understood and are more difficult to envision. Among mollusks, for example, torsion, loss, fusion, gain, or multiplication of body parts are shown among the different major taxa (chap. 8). Some molluscan taxa show evidence of serial expressions in the developmental genes involved in skeletogenesis (Jacobs et al. 2000), but the developmental patterning of the major morphological modules is not understood. In modules that are not serial homologues, different genes may regulate the gene tree of each morphological module. Thus in fig. 3.12, if the cascades were not mediating development of serial homologues, the role of gene C would not be required, as each module would be quite different and thus require a distinctive gene cascade, and the independent cascades could well be controlled by separate genes.

Bodyplan Evolution Commonly Uses Established Genetic Units of Selection for Novelties

Genes May Be Recruited or Captured. If it is selectively advantageous to multiply a body part, evolution does not have to build a new part from the ground up, but can simply duplicate the expression of the responsible developmental cascade(s) through the evolution of signals from enhancer modules. And when a novel sort of body part evolves, it also does so within the context of the established developmental system. Most of the cells and tissues that can be employed in the new part may already be specified, and their deployment in new topographies may be governed by growth factors and positioning systems that are already used elsewhere. The loss or reduction of "old" parts would presumably result from appropriate quenching of former gene-expression events. Thus new bodyplans may be progressively blocked out by recruitment of working gene products through new patterns in the timing, location, and dosages of gene expression. When morphological evolution is traced in fossils, and developmental evidence from representative crown groups is available, it seems likely that something of the history of such recruitment may be deciphered.

An example of a sort of normal recruitment history is the use of homologous genes in similar morphological features in different bodyplans when there seems to be no possibility of derivation from a common ancestor—the features are analogous but are not homologous. It has been a puzzle as to why the same genes seem to be recruited independently to do similar jobs. However, a plausible scenario can sometimes be suggested to account for such a pattern. One of the Pax genes found in flies, *Pax-6,* is a regulatory gene whose homologues mediate eye development not only in arthropods but in chordates, mollusks, and other phyla (Quiring et al. 1994; Tomarev et al. 1997). The eyes in, say, flies, mice, and octopuses are constructed quite differently, and none evolved from either of the others. However, Pax genes are found even in prebilaterians, and it is plausible that *Pax-6* was involved in the development of sensory cells, presumably including photosensory ones, in organisms that predated the protostome-deuterostome divergence. As increasingly complex eyes eventually evolved from such a cell type along different lines of descent, the *Pax-6* gene must have helped to regulate the growing cascades of downstream gene expressions that produced different complex eyes in the different branches. *Pax-6* now acts as one of the key developmental genes early in the cascade of gene expression that produces nonhomologous eyes in different phyla.

Another case may be represented by homologues of several key developmental genes (such as *distal-less*—see above—and *hedgehog* and *engrailed,* among others) that are used in rather similar ways in limb development in both flies and mice. The last common ancestor of those forms appears to have lacked limbs. Why then were these very genes recruited independently for limb development when limbs did evolve? Perhaps they are genes that were employed in the protostome/deuterostome ancestor to mediate body-wall extensions (Panganiban et al. 1997), and limbs have developed from such extensions independently in separate clades.

Gene-expression events may also be captured by enhancer evolution. For example, enhancer shuffling or internal mutations may provide a new target for an established signaling pathway that has no direct historical connection with a developmental subsystem that is evolving (such as the Toll pathway discussed above). One would guess that such captures are generally unlikely to improve fitness, but the use of homologous genes in unique situations suggests that there are at least occasional exceptions. It is possible that some usages of *distal-less* in association with a variety of disparate, nonlimb structures result from captures. If so, it implies that this gene is particularly fit for use in the development of structural outgrowths and is easily captured by regulatory sequences. Elucidation of the evolutionary histories of regulatory genes may solve many puzzles in bodyplan evolution.

Cases of Heterochrony and Heterotopy Are Changes in the Time or Place of Gene Expression. Comparison of the timing of the development of similar structures in the ontogenies of related organisms has shown that changes in the developmental rates or timing of appearance of bodyplan features are common; such changes are termed heterochronies. A specialized terminology has been used to describe alterations in developmental timing (see Gould 1977; Alberch et al. 1979; McKinney and McNamara 1991). In some formulations, all additions to, deletions from, and insertions within an ontogenetic system are considered to be heterochronies, in which case nearly all developmental change would qualify. However, most workers restrict heterochrony to relative changes in rate of growth, onset time of growth, and offset time (termination) of growth of some feature between an ancestor and descendant, commonly but not always with the onset of sexual maturity as a reference point. Each of those three sorts of changes may be either increases or decreases, to form a system of six basic types (fig. 3.13). When descendant adults have co-opted the characteristics of earlier ontogenetic stages, they are paedomorphic. This sort of heterochrony has been hypothesized to explain the origin of vertebrates, which were speculated to have arisen from ascidians that began reproducing as larvae and therefore shifted to the larval ("tadpole") bodyplan, discarding their adult stages, a case of neoteny (chap. 11; Garstang 1894, 1928). When descendant adults have added novel features to the end of their ontogeny, they are peramorphic. This sort of heterochrony would be involved if ascidian larvae recapitulated their ancestral adults, so that ascidian adult features would have been added to the ancestral morphology, a case of hypermorphosis (chap. 11; Jefferies 1967, 1986).

Evolutionary shifts in spatial rather than temporal relations among developing features are termed heterotopy. Both terms, heterochrony and heterotopy, were coined by Haeckel (1866). Heterotopy was used by Haeckel to refer explicitly to evolutionary changes in cell lineage fates, which seem to be common. As used here in a more general sense, heterotopy is a parallel term to heterochrony, and refers to spatial changes in growth patterns between ancestors and descendants. A classic hypothesis of heterotopy is the postulated dorsoventral inversion between

HETEROCHRONY					
PAEDOMORPHOSIS			PERAMORPHOSIS		
Progenesis	Neoteny	Post-displacement	Hyper-morphosis	Acceleration	Pre-displacement
(earlier offset)	(reduced rate of morphological development)	(delayed onset of growth)	(delayed offset)	(increased rate of morphological development)	(earlier onset of growth)

FIG. 3.13 A common terminology for heterochronies. Paedomorphosis includes three sorts of "underdevelopment" in descendant adults, while peramorphosis includes three sorts of "overdevelopment." See McKinney and McNamara 1991.

invertebrates and vertebrates first suggested by Geoffroy Saint-Hilaire (1822; chap. 11). Spatial and temporal changes in development are often correlated. It is clear from molecular developmental studies that the categories of heterochrony and heterotopy each involve numbers of distinctive genomic events. Though the categories have their uses, it seems likely that they will eventually be replaced by a taxonomy that is based on types of genomic evolutionary change.

Regulatory Gene Systems Organize Complexity

The Developmental Genome Should Be Hierarchical

Metazoan genomes are surely complex by any measure. The genomes contain thousands of parts—genes and their *cis*-regulatory systems—interacting in a bewildering pattern. Genes have evolved within a gene tree, rooted in a theoretical first gene or genes and evolving and diversifying by duplication and subsequent divergence, by combining and shuffling modules, and probably by other means. While the number of genes in the protistan lineage from which Metazoa arose is unknown, it must have been many thousands. Thus, when metazoans arrived on the scene, they inherited a large genome. While the many earlier branches of the gene tree are missing when just considering metazoans, the many genes that were represented at the dawn of Metazoa are parts of a tree that traces back to the first gene(s), and the further history of those branches within Metazoa remains treelike. The gene tree of the protistans must have been quite complex long before metazoans came on the scene, and presumably a genetic hierarchy was well established within the unicellular ancestor of Metazoa. During the establishment and elaboration of multicellularity, another hierarchy emerged as differentiated cell types multiplied and became organized into morphological features—a hierarchy devoted to metazoan development. The transformation of the developmental tree of gene-expression events into a genomic hierarchy is accomplished by the genetic regulatory apparatus. The gene-expression events accomplish this transformation through a network of interactions.

Networks That Organize the Products of a Tree into a Hierarchy Are Hypothesized to Be Scale-Free

If the hierarchical aspects of biological entities are properties of their complexities that emerge in order for them to function, then it seems possible that the networks that transform trees into hierarchies may also have a structure that is related to the origin of complexity. Rather than forming some indescribable, messy, random interaction pattern, the transforming networks might form, well, some describable messy interaction pattern. In other words, if a modular hierarchy emerges during selection for function in a complex biological system, then some sort of modular structure might be discerned in the associated network. To some extent such structures are known to exist in ecosystems, which contain key species that form nodes of functional interactions. To be sure, the functions are commonly redundant, as has been shown by the robustness of food webs in the face of local extinctions, followed by collapse when the last species capable of performing the function becomes too rare or is extirpated (see Jackson et al. 2001). And within metazoan genomes there are key developmental genes that act in multiple ways, organizing bodyplans into functional patterns, and some of the gene functions are redundant. Although complete developmental gene networks are not yet elucidated, it would appear that within their patterns of expression there are key nodes of functional interactions also.

Among the sorts of networks that would seem likely to evolve in order to organize complex biological systems are so-called scale-free networks that have been studied in complex systems (see Barabási and Albert 1999; Albert et al. 2000). In these networks many entities have relatively few functional links, but some entities have many, forming key interaction nodes or hubs (fig. 3.14A, C). The distribution of connectivities of the nodes in a scale-free network follows a power law $N(k) \approx k^{-y}$, where $N(k)$ is the number of links and the exponent $-y$ indicates the proportions of nodes with many versus few links. This is in contrast to networks in which the probability of connecting any two nodes is random (fig. 3.14B, D). There is empirical support for the existence of scale-free networks in some complex biological systems (Jeong et al. 2000; Ravasz et al. 2002 and references therein). For example, in studies using metabolic networks in forty-three disparate organisms, a model that fit the data rather well was of hierarchically nested modules of scale-free networks organized into higher-order scale-free networks (Ravasz et al. 2002). Modules in the lower levels of the hierarchy of scale-free networks are evidently

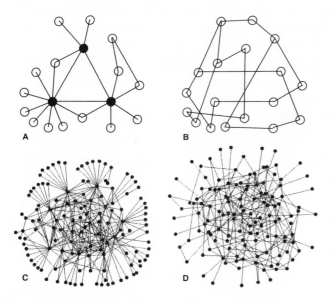

FIG. 3.14 Network structures. (*A, B*) Highly simplified neworks. (*A*) A scale-free network in which a few entities form numerous connections, while most entities form only a few. (*B*) An exponential network in which the probability of connecting any two entities is random. (*C, D*) More realistically complex networks but still vastly simpler than a genome. (*C*) A scale-free network. (*D*) An exponential network. Redrawn from Albert et al. 2000 and Jeong et al. 2000.

more coherent than modules at successively higher levels, which is a general feature of nested hierarchies (chap. 1).

If scale-free designs exist within developmental genomes, they should be shown in maps of gene expression events, and if they fit the general pattern, the maps would indicate that scale-free patterns of gene expression are stronger in the more downstream developmental modules. That is, a relatively weak scale-free pattern should be found among the genes that are expressed early in development, while modules that develop later should contain a few genes whose expressions have many connections, and thus that serve as hubs, and many genes that have relatively few connections. One of the interesting properties of scale-free networks is that they are relatively robust and tolerant of errors, such as might be introduced by mutations in a developmental network (Featherstone and Broadie 2002). The tolerance arises partly because mutations are nearly random with respect to gene loci, and most genes have few links, so that most mutations have relatively small effects, which in any event may be covered up by functional redundancy with other genes.

Regulatory Genes Are Arbiters of Developmental Narratives

To pursue the language metaphor, some of the regulatory genes function in the establishment of the orderly sequence of morphological "chapters"—the patterning of the body. Other gene-expression events are devoted to producing meaningful morphological "paragraphs" and "sentences," the organs and structures that permit survival and reproduction. In grammar such arrangements are partly governed by phrase-structure trees, and in development, by trees of gene expression, but the usages of words and of genes must form complex networks when referred to an entire vocabulary or genome. The reason that this analogy works as well as it does is surely no accident. Both the regulatory genome and the usages of grammar and narrative in language are systems to organize information so that it makes sense. There must be ways to organize information that would not appear similar to the organization of either English or development, but building up a hierarchical structure through elaboration of a tree of elements must be an easy way to reach informational complexity. The evolution of structure in many languages, and in development, has evidently produced rather similar organizational features.

CHAPTER 4

Morphological and Molecular Phylogenies

As the major metazoan bodyplans appear abruptly and without known intermediates in the fossil record, the morphological gaps between bodyplans have to be filled by hypothetical intermediates in order to connect them into a branching structure. Given this freedom to hypothesize, however, it is possible to conjure up intermediate forms that might connect almost any of the phyla with almost any other phylum. There have been many suggestions for such intermediates. Fig. 4.1 depicts various branching patterns that have been proposed in recent years by eight of the foremost comparative morphologists for twenty-six taxa that may be considered phyla, rearranged in a congruent fashion by Eernisse et al. (1992). A glance at the topology of the patterns indicates how diverse are the notions of phylogeny at this level. And in most cases, the different patterns of relationships imply different intermediates, which were not thought to be implausible creatures by the respective authorities. These eight schemes, while representative, are far from exhaustive of the variety that have been proposed. There has been more agreement on the contents of the higher-level Linnean taxa than on their interrelationships. Libbie Hyman, a most knowledgeable and acerbic commentator on invertebrate systematics, indicated her own views on some of the larger questions in metazoan phylogeny in a famous quote (Hyman 1959, 754): "Anything said on these questions lies in the realm of fantasy." Of course she was speaking in reference to data available in her time. We are most fortunate today to have a wealth of additional information that has produced testable hypotheses supported by procedures that promise to decide many of these issues in the foreseeable future. However, that future has not quite arrived.

Assumed Evolutionary Histories Affect Morphologically Based Phylogenetic Hypotheses

Many of the classical phylogenetic hypotheses are based on some overarching concept imposed as a principle from which relationships may be deduced; the assumption that evolution has always proceeded from the simple to the complex, for example, would greatly constrain the possible phylogenetic patterns. Many hypotheses combine such (often tacit) assumptions with attempts to trace in some logical way the evolution of particular, presumptively homologous features (including developmental features) or organs. As different features tend to have different patterns of

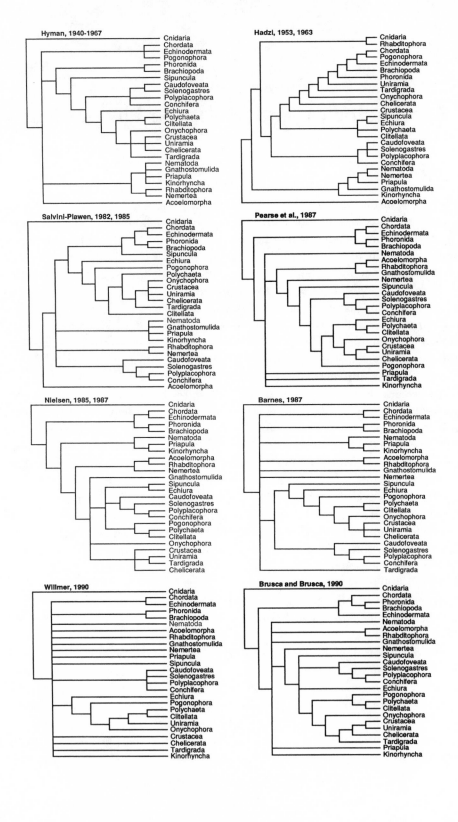

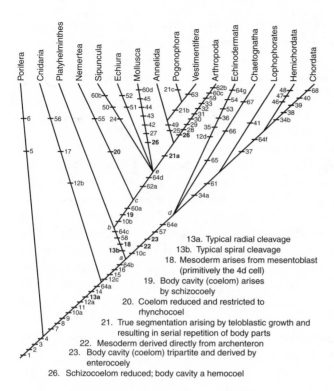

13a. Typical radial cleavage
13b. Typical spiral cleavage
18. Mesoderm arises from mesentoblast
    (primitively the 4d cell)
19. Body cavity (coelom) arises
    by schizocoely
20. Coelom reduced and restricted to
    rhynchocoel
21. True segmentation arising by teloblastic growth and
    resulting in serial repetition of body parts
22. Mesoderm derived directly from archenteron
23. Body cavity (coelom) tripartite and derived by
    enterocoely
26. Schizocoelom reduced; body cavity a hemocoel

FIG. 4.2 A cladogram indicating the distribution of selected characters—presumed synapomorphies and apomorphies—among sixteen of the major animal phyla. Schizocoels and enterocoels have been considered as distinct characters, metameric segmentation as a single character. The choanoflagellates were used as an outgroup. See text for discussion of some of the identified characters, and Brusca and Brusca 1990, from which this figure is adapted, for a key to all characters.

resemblance among phyla, a variety of schemes can result from such methods. It is clear that some of the resemblances must result from convergent homoplasies and are not based on homology.

The explicit distribution of characters such as are displayed on cladograms or phylogenetic trees is essential evidence in the reconstruction of pathways of descent. A number of carefully thought-out, morphologically based cladograms of metazoan phyla are available (e.g., Brusca and Brusca 1990; Schram 1991; Eernisse et al. 1992; Nielsen et al. 1996), but these disagree in many respects. A major problem is in the selection of homologous characters. For example, in the cladogram of Brusca and Brusca (fig. 4.2), in which morphological characters are evaluated by parsimony, the coelom is not regarded as a single character but rather has originated twice, once as a schizocoel (character 19 in fig. 4.2, present in primitive protostomes) and once as an enterocoel (character 23, present in primitive deuterostomes). Because of the distribution of other characters, it is then necessary to postulate reduction of

FIG. 4.1 Eight phylogenies proposed for twenty-six metazoan taxa believed to represent phyla, based on morphological (including developmental) characters. The variations in topology form a conservative indication of the variety of phylogenetic opinions that can be maintained, even though the workers have essentially the same facts at their disposal. Selected from a number of comparative phylogenies prepared by Eernisse et al. (1992).

the coelom (insofar as it represents an important body cavity or eucoelom) independently three times within the protostomes: in ancestral nemerteans (character 20), mollusks (26), and arthropods (26). For arthropods, the coelomic reduction is required partly because "true segmentation" is a character (21a) that arises in the common ancestor of annelids and arthropods and is homologous in those phyla, and thus a eucoelom, present in annelids, is assumed to lie in arthropod ancestry.

The topology of the cladogram could be quite different if the coelom were regarded as arising only once, and arthropod and annelid segmentation were regarded as separate characters. The topology would be quite different again if the coelom were permitted to have originated many times. In other words, a cladogram of phyla does not provide an unambiguous indication of the relationships of characters. The character distribution is uncertain because the status of the characters themselves is not known. A given theory that entails the specification of characters and thus speaks to their origins will produce a given cladogram. Of course, cladograms are most accurate when the characters involved are well understood and are reduced to elements whose possible homologies can be judged. This problem and many other pitfalls in the phylogenetic analysis of morphological data have been extensively critiqued by Jenner and Schram (1999) and by Jenner (2001, 2002), who produced a wealth of examples of the misspecification of characters, chiefly because of misjudged homologies or misdefined characters.

Criteria by which to judge morphological homologies across gaps in a succession of bodyplans remain problematic, despite the extensive literature on homology. An attempt to develop convincing criteria to identify homologues, by Remane (1952), has been widely cited, and I paraphrase a somewhat modified version of his ideas here (see Hanson 1977). Remane's major criteria are threefold. In homological relationships (1) features occupy similar positions in the system of which they are a part (positional criterion); (2) features have constituent parts that are similar in form, shape, or function, and in relation to other parts (compositional criterion); and (3) features can be sequentially arranged so that extreme members are connected by intermediate steps or intergradations (sequential criterion). There are three additional "minor" criteria, suggesting that homology is (4) probable when features similar in placement and composition appear in forms already known to possess homologies; (5) probable when features occur in large numbers of similar forms (so that independent origins become unlikely); and (6) possible when two or more different features have similar distributions in similar forms. These are logical criteria, thoughtfully formulated by an outstanding zoologist.

An example of a possible application of these criteria to phyla is furnished by the coelomic cavities (metacoels) present in the trunks of phoronids (chap. 9) and pterobranch hemichordates (chap. 11), groups that have somewhat similar bodyplans. The metacoels clearly meet criteria 1 and 2, and criterion 3 does not apply, as the features do not represent extremes in a sequence but are rather similar. Criterion 4 may apply; both of these phyla have coelomic regions anterior to the metacoel (at least in some taxa), and both have tentacular feeding crowns that have been suggested as homologues. Criterion 5 may also apply; two other groups (brachiopods and Bryozoa) have putative metacoels similar in position to phoronids, as do the enteropneust hemichordates. All of these nonphoronid taxa also possess coelomic compartments anterior to the postulated metacoel, associated with the feeding apparatus or at least with the oral region, probably fulfilling criterion 6. Thus the metacoels score highly as likely homologues. However, molecular evidence strongly suggests that the phoronids and hemichordates are only distantly related, and that their last common ancestor is unlikely to have had a metacoel.

Such failure of perfectly logical criteria to refute nonhomologies is common. For example, Hansen (1977), in studying the origins and early diversification of the more primitive metazoans, carefully considered homological arguments and observed logical guidelines wherever possible, which led him to conclude that turbellarian flatworms are descended from ciliate protozoans and represent a basal branch of the bilaterians, hypotheses now known to be incorrect. Much of the lack of consensus indicated by the trees in fig. 4.1 stems from similar failures.

Many of the Classic Phylogenetic Hypotheses Involve Assumptions as to the Phylogenetic History of the Coelom

Interpretations of coelomic spaces have played a particularly important role in the history of phylogenetic hypotheses. Some workers have assumed at least tacitly that "the coelom" originated only once, and thus that identification of its origin would provide a starting point from which the phylogenetic position of many of the phyla might be inferred. Schemes of relationship among phyla with different coelomic ontogenies could then be devised, and intermediate conditions could be postulated. Hypotheses based on the various inferred evolutionary histories of coelomic cavities provide a good sampling of historically important morphological approaches to metazoan phylogenies; some of the major schemes are reviewed below.

Dichotomous Coelom Theories Postulate an Early Branching between Protostomes and Deuterostomes from a Common Ancestor

A common class of phylogenetic hypotheses, traditionally well supported among British and American workers, implies that the distinction between enterocoely and schizocoely developed either independently or very early in the history of the coelomates, and that the split between protostomes and deuterostomes is associated with this divergence. This leaves a triploblastic paracoelomate as the logical ancestral form of the more complex metazoans in which intramesodermal spaces are found. Among the permissible phylogenies in this group are those that arrange the bodyplans more or less in the order of their complexity, both before

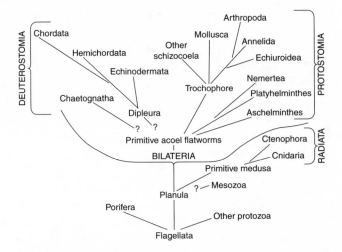

FIG. 4.3 A phylogenetic hypothesis by Hyman (1940) that involves the rise of the coelomates from flatworms, with a dichotomous division into protostome and deuterostome clades. Whether the coelom evolved once and then diverged developmentally, or evolved independently in each major clade, is moot in the figure. Although the branching pattern precludes a precise alignment of bodyplans according to their complexity, in general the phyla are arranged from less to more complex along any given branch.

the dichotomy and within each following branch. Fig. 4.3 shows one of the more widely cited phylogenies that is basically dichotomous in this sense, by (of all people) Hyman (1940). However, note that it is not clear in Hyman's figure whether schizocoely or enterocoely evolved independently, or whether one of those developmental processes evolved from the other. There are a great many proposals that follow this general dichotomous pattern but that differ in detail; many of those variations involve the very early metazoans and do not affect the implied history of the coelom.

Enterocoel Theories Postulate That Enterocoely Is a Primitive Feature of Bilaterians

Recall that in a number of invertebrate phyla the main coelom originates by enterocoely—by outpocketings from the gut or enteron. As the diploblastic cnidarians would seem to lie near the base of the metazoan tree, the origin of coelomic spaces was sought in their anatomy. Jellyfish, which do not have tissues that are generally recognized as true mesoderm, do have a tetraradiate gastrointestinal symmetry, with four gastric pouches (extended into radial canals in fours or multiples thereof; chap. 6). It was therefore suggested that the coelomic spaces of higher invertebrates had evolved from such enteric pouches, a hypothesis that was most widely held by European workers. This "enterocoel" theory was first given a thorough

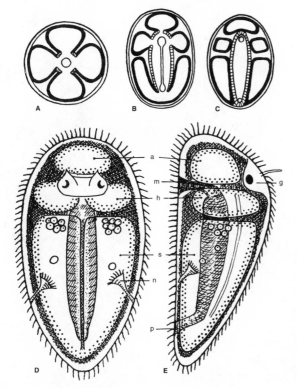

FIG. 4.4 The origin of the coelom as suggested by Remane. (*A*) Cnidarian with tetraradiate enteron. (*B*) As mouth and anus become separate openings, bilateral symmetry appears in the disposition of the enteric pouches. (*C*) Primitive bilaterian, with gut and three coelomic cavities (trimeric) as in some deuterostomes. (*D, E*) Dorsal and lateral views, respectively, of a hypothetical ancestral bilaterian. *a*, protocoel; *g*, brain; *h*, mesocoel; *m*, mouth; *n*, nephridium; *p*, anus; *s*, metacoel. From Remane 1954.

treatment by Sedgwick (1884), who suggested that the pouches were transformed into a segmented coelom. No primitive coelomates with segmented architectures are enterocoelous, however. As the simpler enterocoels have a triregionated architecture, it is sometimes postulated that tetraradiate cnidarian pouches were transformed into three coelomic regions as an evolving bodyplan was transformed from radial to bilateral (fig. 4.4; Remane 1954, 1963). Once this idea is accepted, the early part of the eumetazoan tree is occupied first by cnidarians and then by enterocoels, and thus other alliances of triploblastic phyla branch off from the deuterostomes, or at least from an enterocoelous deuterostome ancestor. The phylogenies with such a basis are often termed archicoelomate, for, after cnidarians, coelomates become the most primitive group. Some workers (e.g., Jägersten 1955) consider both cnidarians and enterocoels to be derived independently from a still more primitive hypothetical ancestor with enteric pouches. With a coelomate as a primitive form, the acoelomate and pseudocoelomate bodyplans must be regarded as secondarily simplified (see Remane 1954; Siewing 1980; Rieger 1985), and the schizocoels must arise from an enterocoel. An archicoelomate phylogeny is shown in fig. 4.5.

The archicoelomate hypotheses have been opposed by many workers on conventional morphological grounds (particularly effective comments are by Hyman

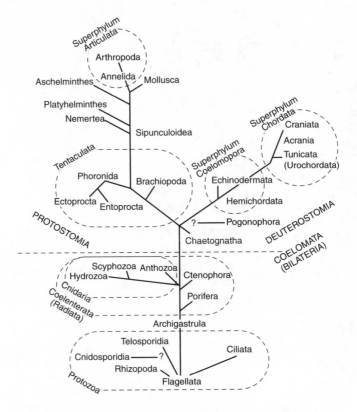

FIG. 4.5 An archicoelomate metazoan phylogeny. Coelomates arise directly from cnidarian ancestors, with enterocoely arising first and characterizing the deuterostomes; schizocoely is derived from an enterocoelous ancestor, from which lophophorates (tentaculates), then flatworms (among others), and, finally, mollusks, annelids, and arthropods descend. After Marcus 1958.

[1959]; Hartman [1963]; and Clark [1964]). Among the difficulties with these hypotheses are the lack of any functional justification for the transformation of the gut pouches into coeloms, and of any explanation for the change from endoderm-lined to mesoderm-lined pouches, which is geometrically implausible, at least in Remane's reconstruction (fig. 4.4).

Schizocoel Theories Derive "the Coelom" from a Spiralian Acoelomate

Another group of phylogenetic schemes proposes that "the coelom" first arose as a schizocoel, and thus that the first coelomate was a protostome. The resulting trees indicate that the spiral cleavage pattern found in flatworms was inherited by early protostome coelomates, with the characteristic deuterostome complex of developmental features, including radial cleavage and enterocoely, arising at some

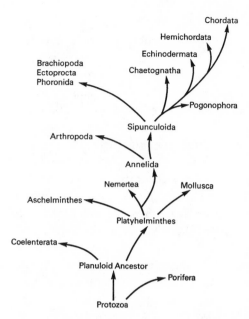

FIG. 4.6 Schizocoel phylogenies consider the early coelomates to be schizocoels with the deuterostomes arising significantly later from one or another protostome ancestor. Here, annelids are ancestral to other protostomes including "Sipunculoida," which gives rise to lophophorates and to the classic deuterostomes. From Nichols 1971.

later point. For example, Nichols (1971; fig. 4.6) places annelids at the base of the coelomates, giving rise to sipunculans, postulated to be ancestral to lophophorates on one hand and deuterostomes on the other. Hadži (1963) suggested that two branches arose from spiralian acoelomates, with arthropods on one branch and annelids on the other, the lophophorates arising early from the annelid branch and giving rise in turn to deuterostomes. Numbers of other trees with schizocoels preceding enterocoels have been proposed, most of which are quite contradictory in other features.

Morphological Phylogenies Can Use Some Help

Many systematists who study closely allied living species have turned to molecular methods. It has turned out that numbers of populations that had been thought to be conspecific on morphological criteria are in fact separate sibling species. When sibling species are discovered, it is then commonly possible to develop morphological criteria to separate them, but without prior knowledge that they are genetically distinct, it is not clear which features to trust as taxonomic indicators. Although the questions of relationships among phyla are on quite a different level—there is no shortage of distinctive morphological and developmental features to be found in the disparate bodyplans—there is certainly a problem with which features to trust as indicating relationships. There are two obvious places to turn for clues. One place is to molecular evidence, and the other is to the fossil record; neither is a panacea, but each can help.

Evolutionary Histories Affect Molecularly Based Estimates of the Timing, Branching Patterns, and Order of Origins of Phyla

The Dating of Deep Nodes by Molecular Clocks Is Problematic as Yet
The notion that protein structure should make a good taxonomic tool was suggested as early as 1958 by Francis Crick, and explored in some detail by Zuckerkandl and

Pauling (1965). The idea is that, once speciation occurs, the genes should tend to diverge, as should the amino acid sequences of their protein products, and the nucleotide sequences of their RNA products. Therefore the dissimilarity in homologous nucleotide or amino acid sequences between any two species should form an estimate of their relatedness and of the length of time since their last common ancestor lived. Zuckerkandl and Pauling used amino acid–sequence differences between alpha hemoglobin chains in different species (such as human and horse) to convert the estimated rate of alpha hemoglobin evolution to a molecular "clock." The clock was calibrated by fossil evidence as to when species may have first separated from a common ancestor. If the clock kept reasonably accurate time, the number of sequence differences between alpha hemoglobin chains in any two species would provide estimates of the time of their first divergence from their last common ancestor. Zuckerkandl and Pauling were quite aware of the major sources of error that could arise in such calculations, including the possibility that changes in protein sequences would accelerate during times of rapid morphological evolution. The fossil record certainly indicates that rates of morphological evolution have sped up and slowed down at different times in the history of different clades. Also, the morphologies of some clades seem to have changed little over hundreds of millions of years, while other clades have undergone numerous important changes. If rates of molecular evolution are correlated with rates of morphological change, then molecular changes should be quite variable among taxa, and episodic within taxa. Early branching orders of lineages (and therefore their cladistic relations) might be obscured or confused by such rate differences.

However, the plausibility of establishing an accurate molecular clock seemed to increase with the advent of the notion that most molecular evolution was neutral or nearly so with respect to selection (Kimura 1968, 1983; King and Jukes 1969). Thus if nearly all gene changes were fixed by drift, they would outweigh the far fewer changes that were effected by selection. In this event, molecular and morphological evolution would be effectively decoupled, and changes in molecular sequences in any two lineages could be calibrated to yield the ages of their last common ancestor. Furthermore, if rates of molecular evolution have been relatively constant, then the branching order should be well preserved in the pattern of sequence similarities, and the topologies of phylogenetic trees could be established by molecular evidence.

Unfortunately, rates of molecular evolution are not very clocklike. Different parts of genes evolve at different rates (Vawter and Brown 1993). Nontranslated regions evolve more rapidly than translated ones (Jeffreys 1982), and some translated portions of genes evolve more rapidly than others, presumably because the more conserved sequences code for critical protein domains wherein change is likely to destroy protein function. Even within codons, transversions—from pyrimidine to purine or vice versa—are generally more frequent than transitions—from one pyrimidine or purine to another (Brown et al. 1982; but see Vawter and Brown

1993). Different genes also have evolved at radically different rates (Ayala 1997; Rodrigues-Trelles et al. 2001). If the rates for each gene were constant, though, they could be calibrated separately. But rates of evolution of genes also vary among different groups of organisms. Rates of change in homologous genes in different lineages of protistans may vary by up to an order of magnitude (Sadler and Brunk 1992). Rates of change in a ribosomal gene differ by a factor of three in different clades of sea urchins (Smith et al. 1992) and a factor of five in different clades of sea stars (Lafay et al. 1995). Thus estimating molecular-clock dates for divergences in one group by genes that have been calibrated in another group can lead to enormous errors, especially for deep nodes. For example, if the divergence time of Fungi, Metaphyta, and Metazoa is estimated by evolutionary rates of three proteins (GPDH, SOD, and XDH) calibrated from *Drosophila*, it is 7,045 Ma for GPDH (long before Earth was formed), and 451 and 398 Ma for SOD and XDH, respectively (long after the Cambrian explosion; see Rodrigues-Trelles et al. 2001). To avoid this problem, the rate of evolution of a gene in a given taxon should be calibrated by the fossil record of members of its nearest possible relatives. There is still no guarantee that rates would be the same between the calibrated and estimated taxa, but the chances that rates would be similar should be much improved.

It seems possible that genes may evolve more rapidly during times of major morphological changes such as are associated with invasions of new adaptive zones or with extensive radiations. This relation was suggested for insects by Friedrich and Tautz (1997), and strongly indicated for regulatory genes in plants (and found in structural genes) by Barrier et al. (2001). Thus molecular evolution associated with the origination of the bodyplans of phyla, or of other morphological changes, may have speeded up, in which case divergences will appear to be farther in the past than they actually were (see Sanderson 1997; Valentine et al. 1999; Foote et al. 1999). Whether there may be genes that are insensitive to changes in rates of morphological evolution is not yet known.

There have been a number of attempts to correct for some of these problems and to produce molecular-clock dates for important early metazoan branchings on the phylogenetic tree. The results have not been encouraging. For example, the date of the protostome-deuterostome divergence that is implied by a number of molecular-clock studies, using different molecules and techniques, has been variously estimated as near 627 Ma (Lynch 1999), about 670 Ma (Ayala et al. 1998), 830 Ma (Gu 1998), near 1,200 Ma (Wray et al. 1996), and about 1,500 Ma (Bromham et al. 1998). Other clock dates have been scattered within that range. The time spanned by these molecular date estimates is longer than the known fossil record of metazoans. One feature in common to all of those dates is that they estimate the divergence to be significantly earlier than the appearance of the clades in the fossil record. Molecular-clock dates for branching events between other major clades have also been consistently earlier than suggested by the fossil record, for example, for orders of mammals and of birds (Hedges et al. 1996; Kumar and

Hedges 1998) and for early angiosperms (Goremykin et al. 1997). The plausibility of there being a long missing early record of mammalian lineages has been tested by Foote et al. (1999), who in effect evaluated the confidence to be placed in the nonoccurrence of their fossil records. Judging from the preservability of mammals and the density of their occurrences in their known fossil record, and using a variety of estimates of lineage diversification and species turnover rates, Foote et al. concluded that it is unlikely that mammalian orders arose much earlier than their earliest recorded fossil occurrences.

Similar tests cannot be performed on metazoan phyla whose ancestors or early members lacked durable skeletons, because of the obvious difference in preservability. If taxa whose bodyplans absolutely require durable skeletons can be identified, however, their order of stratigraphic appearance should carry some weight in evaluating their order of branching, or at least of bodyplan evolution. Unfortunately, the list of such taxa is quite short. Brachiopods are very likely candidates. Early arthropods probably required a fairly rigid skeleton also, but as it was largely organic, it was not necessarily very durable. However, the arthropod trace-fossil record should help in establishing their first appearance. Some other phyla include taxa that require durable skeletons—Bivalvia among the mollusks, for example—but that are probably not primitive within their phyla. Still, the geologically abrupt first appearances of many living phyla during the Early Cambrian includes the first brachiopod and bivalve skeletons, and arthropod traces. Thus, to the extent that some evidence of origins of bodyplans can be gleaned from the Cambrian fossil record, the geologic and molecular data seem to be at odds.

Dating Nodes Does Not Date the Origin of Bodyplans

Molecular clocks are under heavy study, and there is hope that appropriate molecules and/or corrective methods will be found that permit more accurate dating of ancient nodes. However, nodes do not necessarily (and will usually not) indicate major morphological changes. Morphologically based phyla do not originate until their defining bodyplan features have evolved (Valentine 1996; Runnegar 1996; Fortey et al. 1996). On the other hand, the fossil record will always indicate a later date for the origin of phyla than the actual historical date, because the record is imperfect (chap. 5). The molecular clock, such as it is, records the splitting of lineages, while fossils, such as they are, record the appearance of bodyplans. Lineages must diverge significantly before the evolution of a new bodyplan. Thus there will always be a lag between molecular and morphological evidence of divergence, a lag that could involve many tens of millions of years.

Genes That Are Phylogenetically Informative for Higher Taxa
Must Evolve Slowly but Not Too Slowly

If a gene evolves so slowly that no sequence changes have occurred within its homologues since they were isolated in separate lineages, then that gene will not provide any phylogenetic information about those lineages—the sequences will be plesiomorphic for the taxa involved. If the sequence changes have been quite frequent over a long time period, the gene will not be phylogenetically useful either. The changes will have been so extensive that the evidence of homology is lost—the sequences of homologues will have become "saturated" by change, and any sequence similarities will be coincidental, that is, will not be owing to common descent. A useful gene, then, will have undergone enough change so that its branching order in the taxa of interest may be inferred and their cladistic relatedness thereby estimated, but not so much change as to blur the relationships.

When the nodes being studied lie in the very remote past, it is clearly necessary to employ sequences that have evolved very slowly, so that they have not approached saturation over the years. With such highly conserved sequences, however, time resolution is rather coarse, because any small changes that may have occurred between a succession of branchings within a relatively short interval of time may be insufficient to distinguish the branching order. An ideal solution to this problem would be to find molecules that were evolving rapidly during early diversification among the lineages being studied, and that then slowed down to "freeze" the early branching pattern in the various sequence divergences, and did not change or at least speed up ever again. Possibly such molecules will be found for some radiations. At present, however, the branchings of phyla are best studied by comparisons of very slowly evolving molecular sequences, such as the small-subunit (18S-like) ribosomal RNAs (SSU rRNA) or their DNA genes (see Woese 1987). There is evidence that the average rate of change in these molecules has been about 1% per 50 million years (see Wilson et al. 1987), a rate that seems suitable for study of branchings in the range of 500 to 600 Ma. Some other molecules that may be useful in investigating branching events in this age range include elongation factor 1α (EF-1α), elongation factor 2 (EF-2), large subunit RNA polymerase II (POL 2) (Friedlander et al. 1994), heat-shock protein 70 (Hsp70) (Borchiellini et al. 1998), 28S rRNA (Medina et al. 2001), and most promising of all, myosin heavy chain (myosin II) (Ruiz-Trillo et al. 2002). (For other molecules see Giribet et al. 2001.)

Variations in the Rate of Gene-Sequence Change among Taxa Can Produce False Molecular Phylogenies

If a gene sequence evolves rapidly in one taxon and slowly in another, it will appear that the rapidly evolving taxon is more distantly related to their last common ancestor than is the slowly evolving one, which is of course incorrect. It is clearly best for estimates of relatedness if genes being compared have evolved at the same rates. Furthermore, it seems intuitive that if longer and longer molecular sequences are compared, the accuracy with which the relationships between their lineages would be estimated should be correspondingly improved (statistical "consistency"), but as shown by Felsenstein (1978), this is not necessarily the case. When some lineages' branches have changed much more than others, most methods of establishing a tree will indicate that the branches that have changed more (on average) are more closely related, regardless of the actual case. The longer the sequences that are compared, the more grievous this sort of error becomes. In other words, bringing more data to bear on the problem merely worsens it (statistical "inconsistency"). The conditions under which this sort of inconsistency appears in molecular comparisons have been termed the Felsenstein zone (Huelsenbeck and Hillis 1993). An erroneously close relation indicated between longer branches has been termed the unequal rate effect, or better, as it depends on the branch lengths rather than rates per se, has been epitomized as "long branches attract" (see Hendy and Penny 1989).

The reason that long branches attract is depicted in fig. 4.7 for one method of molecular-sequence comparison. Four lineages are related so that lineages 1 and 2 have recent common ancestors, as do lineages 3 and 4, while the last common ancestor of all lineages is more remote. Their relationships are assessed at five positions by parsimony, a method that selects a tree that requires the least number of nucleotide substitutions—is most parsimonious in this respect—to produce the known sequences from a common ancestral sequence. For example, at position 1 in fig. 4.7A, only a single nucleotide change is required to account for the change from U in lineage 1 and 2 to C in lineages 3 and 4, if the common ancestor had either U or C at that position. Indeed four of the five positions require only a single nucleotide change if lineage 1 is most closely related to 2 and lineage 3 is most closely related to 4. The pattern of change reflects the "true" tree because change has not been so extensive as to obliterate the original patterns, and because all of the lineages changed at the same rate.

If one member of each closely related pair has undergone more changes than the other since the time that they branched from their last common ancestor, however, that branch will appear longest on a plot that relates closeness to similarity (fig. 4.7B). When both long- and short-branch taxa are being compared, incorrect pairings can become the more parsimonious—and commonly will. This result occurs partly because the more extensive changes in the long branches swamp and conceal the original similarities. The signal from the sequence of the last common ancestor will be stronger (more parsimonious) in less related but slowly evolving lineages than in the pairs of lineages that are actually more closely related but have changed in unequal amounts. Furthermore, there are only four nucleotides, so a rapidly evolving pair of lineages will on average eventually display support for an incorrect tree at three out of every sixteen positions just by chance. Thus, if the branches began as more different than that, they will tend to become more similar as time passes. It can be seen that, although the long branches in the plot have by definition undergone more nucleotide changes than the short branches, they are "attracted" because of the assumptions that underlie estimates of the relatedness of lineages at the branch tips. Similar departures from true relationships probably arise as artifacts in all of the tree-building methods when branch lengths are unequal. The attraction of long branches is a general problem in phylogenetic reconstruction.

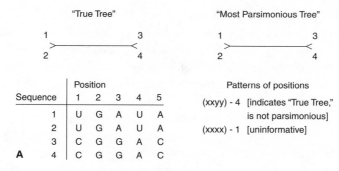

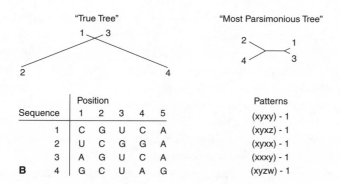

FIG. 4.7 Example of long branches attracting so as to lead to a parsimonious but false phylogeny. (*A*) Four taxa (sequences 1–4) are compared at five positions; pair 1 and 2 have identical nucleotides at all five, as do pair 3 and 4. Position 2 is uninformative because all taxa have the same nucleotide, but the pairs differ at positions 1, 3, 4, and 5, giving the most parsimonious phylogeny as illustrated. (*B*) Although taxa 1 and 2 are more closely related than either is to 3 or 4, taxa 2 and 4 have long branches, accumulating significantly more change than taxa 1 and 3 since they all diverged from their common node. Taxa 1 and 3 have not changed much, retaining a close similarity; taxa 2 and 4 have departed from the nodal condition at several sites (three out of four on average) and resemble each other at one out of four sites on average. Therefore the most parsimonious phylogeny allies taxa 1 and 3, and (less closely) taxa 2 and 4. From Lake 1989.

An example is provided by a study by Aguinaldo et al. (1997). In manipulating sequence data for nematodes, they found that sequences of both long- and short-branched species were similar enough to unite them into a nematode clade. The long-branched nematodes, when evaluated separately, appeared to be allied with flatworms, which are themselves long-branched. A shorter-branched species of nematode was found, however, and when evaluated alone, it nested within the ecdysozoan clade. As the short-branched form is less likely to be affected by branch-length artifacts than are the long-branched forms, it seems likely that it reflects the actual branching topology of the nematode clade. When all the nematodes were included, long- and short-branched species alike, they again appeared as flatworm allies; the long-branched forms dominated the topology. This behavior serves as a cautionary example of branch-length effects in phylogenetic evaluations. Because topologies can be extensively affected by numbers of complexly interacting artifacts, the selection of exemplars is important. A great many exemplars, especially if chosen to cover a broad taxonomic range, may provide a reasonable sample of a clade. If few exemplars are used, though, it should pay to search for as short-branched a form as can be found, even at the cost of reducing the number of exemplars.

Homologous Positions Must Be Aligned When Comparing Gene Sequences

Given that sequence changes have been large enough but not too large, and thus provide useful information, problems remain in deciding which sequences exactly to compare, what metric to use, and how to analyze the data. The sequences to be compared are, after all, different. Many changes may be accounted for by a simple nucleotide substitution, and if these have been the only changes in two sequences being compared, their lengths will be identical and their alignment will be obvious. The similarity (S) of the sequences may be measured as the percentage of nucleotide positions that display change, or $S = M/L$, when M is the number of nucleotide matches and L is the sequence length. If DNA has been inserted or deleted differentially, however, it will usually produce sequences of different lengths. Proper alignment then requires that gaps be inserted into the sequences to allow for the nonmatching segments created by insertions or deletions, so that the segments being compared are homologous. If gaps may be freely inserted at any position, any two segments may be brought into alignment whether they are homologous or not; there must be some constraint on placing gaps. In the analysis, gaps may be treated as single substitutions or they may be weighted as fractions or multiples of substitutions, lowering the similarity accordingly. The length of a sequence, L, now becomes $M + U + wG$, when M is the number of nucleotide matches, U is the number of nonmatches, G is the number of nucleotide positions at which gaps have been inserted, and w is the weight given to the gaps (see Swofford and Olsen 1995; Swofford et al. 1996). When parts of sequences cannot be aligned with any confidence (without including many gaps), they are often simply excluded from the comparison, which of course lowers the number of data points in use, but is preferable to including erroneously aligned segments.

Such a simple factor as the order in which pairs of segments are aligned can lead to an incorrect assessment of relationships (Lake 1991). If one is comparing four taxa, A, B, C, and D, and the pair A and B are aligned with each other, and C and D with each other, a bias will arise in favor of selecting the pairs AB and CD as closest relatives. Other more subtle alignment biases may occur as well. It is helpful in avoiding alignment bias to select a reference sequence to which each of the other sequences is then aligned.

Some Clades Are Characterized by a Natural Bias in Nucleotide Substitutions

Marshall (1992b) has considered the effect of substitution biases that may be particularly important in the study of very deep branches, such as those among phyla. When a particular change in bases is favored, as is true of some taxa, the favored nucleotide will be overrepresented relative to taxa with no such bias. Thus the sequences from taxa that happen to share the same substitution bias will appear to be more closely related than is the actual case, because they will happen to share the favored nucleotide at many positions. Marshall gives the example of birds and mammals, two clades in which there happens to have been a bias in favor of T-to-C substitutions in SSU rRNA sequences. Because this overrepresentation of C at many positions in both sequences increases the sequence identity, in effect producing convergent homoplasy at many sites, a tree-building algorithm that ignores this problem will consider birds and mammals to be too closely related.

Morphological and Molecular Homologies Are Decomposable

During Evolution, Morphological Hierarchies Are Dynamic Structures, and the Composition of Their Entities Is Somewhat Fluid

Hierarchies are beautiful, logical structures when static. But when processes occur that change the nature of hierarchical units, such as evolution does within morphological hierarchies, the relationships of units to levels can change, and the levels tend to become partially decomposed in rather elusive ways. Evolution is not required to construct higher levels from the whole units in preceding ones, but can select parts of those units to form newly combined entities. An evolving organ, for example, may not be made up of precisely the tissues present in the ancestor, but may recruit ancestral cell types that are not found together in any single ancestral tissue, and may eliminate some cell types from a tissue. Recruited cells may often be modified, and some may evolve so as to constitute novel cell types. If there is cell-type recruitment, cell-type origination, and cell-type elimination, the phylogenetic trees of cell types and tissue types become different. This situation can arise between any levels of the somatic hierarchy. Heterogeneous recruitment and retention do not present many problems at levels above the cell, at least for the reconstruction of relations among phyla, but become important problems at the molecular level.

Molecular Homologies Do Not Necessarily Map on Morphological Homologies

Morphological features need not evolve in concert with all of the genes that are responsible for their development (see Wagner 1989; Wagner 1994; Roth 1991; Hall 1994); as a result, the homologies of developmental genes are often dissociated from homologies of morphology. For example, morphological homology exists between

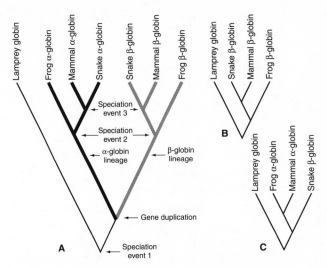

FIG. 4.8 Phylogeny of taxa can be confused by comparing mixtures of orthologs and paralogs. (*A*) Tree showing origins of globin paralogs, α and β globin, within some vertebrate taxa. (*B*) Correct phylogenetic relations indicated by a tree of β globins alone (the same tree would be recovered by using α globins alone, in each case using an ortholog). (*C*) Tree recovered when using a mixture of orthologs and paralogs, which reflects the phylogeny of the genes but which is incorrect for the taxa. From Hillis 1994 based on Goodman et al. 1987.

the archenterons (embryonic guts) in agnathans (lampreys), sharks, teleost fish, urodeles (salamanders), anurans (frogs), and lepidosirenid lungfishes, but different, nonhomologous genes are now involved the developmental pathways of the archenteron in each of these taxa (Streidter and Northcutt 1991). There must have been considerable switching among the genes that regulate different steps along those pathways, while the general function of their morphological product was conserved.

The phenomena of paralogy (chap. 3) introduces difficulties into the recovery of phylogenetic trees based on molecular sequences. Gene trees of orthologs will map on the phylogenetic tree of their taxa, barring losses. However, paralogs that derive ultimately from the same ancestral gene may originate along different branches within a clade, or may be sorted into different branches. Some branches contain paralogs of one paralog, while other branches contain paralogs of another paralog. In these cases a gene tree that includes both orthologs and paralogs will reflect the relationships among the genes but will not map on the relationships among their taxa (fig. 4.8). As genes within the same family, originating as paralogs at one time or another, have rather similar sequences, it is often difficult to untangle gene history and to tell paralogs from orthologs. Trees based on different protein-coding genes from the same set of species commonly imply different phylogenies. In these cases it is possible that one is dealing with complications from paralogs that have been duplicated and inherited in different patterns, and thus the differences in the trees are differences in gene trees, not taxonomic ones.

There Is a Large Variety of Ways to Form Trees from Molecular Sequences

Different methods of reconstructing trees commonly produce different tree topologies from the same data, and thus there is a problem in deciding which, if any, is correct, or more correct. There is an ongoing search for tree-building methods that are the most robust (i.e., that are more likely to estimate the true topology even when the evolutionary assumptions are violated), consistent (that converge on the true topology as more data are added), and efficient (that converge on the topology most quickly). These attempts to improve the accuracy of molecularly based phylogenies have resulted in the development of hundreds of methodological variations to take account of biases that result from different evolutionary processes. For example, various methods aim to correct for inconstancy of overall evolutionary rates, or for inequalities in transition and transversion rates. Reviews of many methods are given by Swofford and Olsen (1995) and Swofford et al. (1996). Phylogenetic methods that have been used to evaluate relationships among metazoan phyla include distance methods, parsimony methods, invariant methods, and likelihood methods. Other, more powerful techniques will surely be developed, but these methods provide examples of the ways in which molecular phylogenies can be constructed and evaluated.

Distance Methods Estimate the Mean Number of Changes between Species

Nucleotide sequence similarities (S) may be converted to sequence distances (d) in order to calculate the positions of the nodes of a cladogram. Common conversion formulas are $d = -\ln S$ and $d = 1/S - 1$. If mutations occur at random, there will be a certain number of positions at which, although a mutation has changed the nucleotide, a subsequent mutation has changed it back into the original state. Thus, although the sequence is evolving, a fraction of the evolutionary change will be hidden, and to employ a distance that is proportional to a sort of evolutionary distance, correction must be made for back mutations. As transitions are more common than transversions, the S calculated separately for these two classes of nucleotide substitutions will commonly be different. These inequalities may be corrected in the distance calculation or, if it is desired to reduce the effective branch lengths, distances may be calculated using transversions only. Other complexities, such as other inequalities in rates of change among positions, are at least partially taken into account in some distance measures. Swofford et al. (1996) present a clear discussion of basic distance calculations, with references for more complicated approaches. Distances are calculated for all pairs of sequences under study, thus creating a distance matrix.

A distance matrix may be transformed into a tree by a variety of methods. Neighbor joining (Saitou and Nei 1987) has been used in studies of phyla; it produces an additive structure, that is, one in which the distances between nodes are represented by the sums of the branch lengths that join them (fig. 4.9). The first step is to adjust the divergence of each sequence according to its average rate of change, producing a normalized distance matrix. Then the distance between the closest pair of nodes and their common ancestral node is calculated. The ancestral node then replaces this pair, and a new matrix is calculated with this node in a terminal

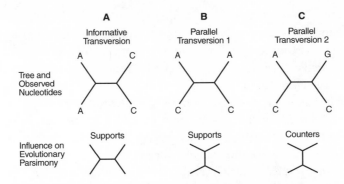

FIG. 4.9 Effects of transversions at one site on prediction of a correct four-taxon case. *Column A*, a transversion on the internal branch has produced descendant sisters that each differ from ancestral sisters; the tree that joins the terminal branches with A and those with C as sisters is correct, as shown. *Column B*, parallel transversions have occurred between each pair of sisters; the phylogeny that joins the terminal branches with A and those with C as sisters is incorrect, as shown, but cannot be distinguished from a phylogeny with a central branch transversion. *Column C*, transversions have occurred to produce different nucleotides; a phylogeny that unites the terminal branches with C is incorrect, as shown, and it is clear that this tree cannot have resulted from a central branch transversion. Lake's method (Lake 1987) uses such arrangements as in column C to indicate the frequency of peripheral branch substitutions relative to central branch substitutions, and subtracts them from support. After Swofford and Olsen 1995.

position. The process is repeated until only the least similar pair of nodes remains in the terminal positions, and the distance between them is calculated. The distances among the neighbors are then used to construct a phylogeny. More sophisticated distance metrics are available (Swofford et al. 1996).

Parsimony Methods Find Trees That Minimize the Amount of Evolutionary Change Required to Produce Observed Sequences

Cladograms that minimize the number of evolutionary steps required to produce differences observed among the sequences are "parsimonious," attempting to produce the simplest explanation for the observed sequence divergences (e.g., see Fitch 1971 and references in Swofford et al. 1996). The example in fig. 4.7 is a parsimony method. Given a phylogenetic arrangement of taxa, parsimony methods determine how many steps are required at a minimum to create the sequence divergences observed among the branches. A large number of arrangements may be created and examined to determine which among them is most parsimonious. The generation and assessment of cladograms form two separate steps in this process. As parsimony algorithms are quite sensitive to branch lengths, some parsimony methods are based on transversions only, which has the effect of making branches shorter and more nearly equal, but also reduces the number of informative sites that are used. One parsimony method (Williams and Fitch 1990) weights sites on a given cladogram according to the frequency of each kind of substitution required, and calculates a new parsimonious cladogram on this weighted basis, thus reducing substitution bias. This algorithm does not seem to have been used to evaluate relations among phyla.

Invariant Methods Concentrate on Reducing Long-Branch Attraction

An attempt to find a method that minimizes the attraction of long branches led to the development of "evolutionary parsimony" (Lake 1987), an invariant method. This method focuses on transversions, using a four-taxon case. For a given position, parallel transversions on two peripheral branches will produce one of two outcomes (fig. 4.9): (1) nucleotide frequencies that are indistinguishable from the original ones though indicating an incorrect phylogeny (column B), and (2) frequencies that are different from the original (column C). These two outcomes are assumed to be equally likely (which implies that a balance exists between types of transitions and of transversions, and that nucleotides change independently). The method identifies the changes that produce new nucleotide frequencies and removes them from support, thus neutralizing the effects of the presumably equal number of changes that produce similar frequencies but indicate a false tree. This method does greatly reduce the effects of long-branch attraction, but as the assumptions do not usually hold, it is less reliable than other methods under many circumstances (see below).

Likelihood Methods Estimate Actual Change under a Given Evolutionary Model

Likelihood methods require the specification of evolutionary hypotheses that contain such parameters as the equilibrium frequency of nucleotides, the relation between nucleotide abundance and rate of change, the independence of substitutions, and the relative rates of transition and transversion substitutions. Sequence segments that include gaps are discarded. For a given phylogenetic arrangement, the question is: what is the likelihood that evolution under the specified parameters will produce the observed nucleotide sequences? For a given evolutionary hypothesis, the likelihood of the observed change is computed for each position, and then the product of the likelihoods is expressed as a distance or branch length between the sequences. The parameters are then varied, and the combination with the highest likelihood is accepted. This procedure is then repeated for another arrangement, the two topologies compared, and the one with highest likelihood selected. The selective process is continued (via some algorithm) until an arrangement is found with the combined maximum likelihood of both an evolutionary hypothesis and a topology. This method is computationally demanding and can become prohibitively lengthy when large numbers of taxa are studied. The maximum-likelihood method may be the most powerful of any that is generally available for evaluating molecular phylogenies (Huelsenbeck 1995; Swofford et al. 1996).

Several Methods Help Evaluate the Quality of Support for Given Nodes
The preferred phylogenetic topology that is established by one or more of the available analytical methods may be very weakly supported, even though it is the best that the data will provide. In general, the quality of support varies among the nodes, available data supporting some relationships strongly while for others the data may be quite equivocal. A number of methods have been developed to provide some insight into which nodes are best supported in any given analysis. One technique that has been widely applied to the molecular phylogenies of phyla is bootstrapping (Felsenstein 1985; Efron et al. 1996). The basic procedure is conceptually simple (see Felsenstein 1988). A data set—of nucleotide sequences in this case—is randomly sampled repeatedly (often 1,000 or more times), with replacement. Each sample is made as large as the original data set. Any given nucleotide may fail to be included in a given sample, or may be included multiple times. Each sample is then evaluated by the method of choice. The percentage of times that a given node is supported in the replicate samples is taken as a measure, or at least an indication, of the reliability of that node with respect to the database. Thus strong bootstrap support reduces the possibility that random errors have influenced the topology, but does not certify that the phylogeny is correct.

Another estimate of support that has been used for phyla is the Bremer support index (Bremer 1988). In principle one begins with the most parsimonious cladogram for a given data set, and all cladograms that are one step longer are constructed, and a consensus tree is found. There will be some groups, present in the most parsimonious cladogram, that are not supported in this tree. This process is repeated for two steps longer, three steps longer, and so forth, until a tree is found in which there is no unambiguous support for any clade. Each node can now be assigned a Bremer support index—the number of steps at which the node disappears from the tree. Thus the lower Bremer indexes indicate higher levels of support for a given branch within the topology of the most parsimonious cladogram. The computer algorithms that produce Bremer support indexes do not proceed in this precise manner because of the number of calculations involved, but the logic of support is the same.

There Are Some Remedies for What Ails Molecular Trees
Huelsenbeck and Hillis (1993) provided a survey of conditions under which various methods of DNA sequence comparisons are likely to be most successful in identifying true phylogenies, for four-taxon cases. Simulation analyses confirm that many of the methods perform best when branches are short and relatively equal, and increasingly less well as these conditions are increasingly violated. The numbered regions in fig. 4.10 can be used to evaluate how different methods perform under increasingly poor conditions. Distance and parsimony methods performed well in region I. In region II, where branch lengths are long, large numbers of informative sites must be employed to achieve high performance. Many methods become

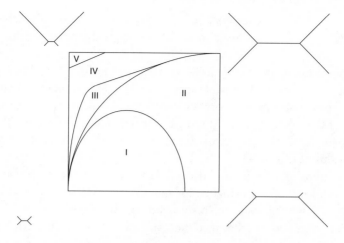

FIG. 4.10 A graph indicating regions of confidence in four-taxon arrangements of different branch lengths, as indicated by the figures at the corners. For example, at the bottom left corner, the taxa have short equal branches and a short internal branch; at the upper right, the taxa have long branches and a long internal branch; other corners are occupied by taxa with very different branch lengths (the three-branch length varies along the x-axis and the two-branch length along the y-axis). The regions are numbered in the order of the performance of phylogenetic methods with arrangements of the indicated pattern, I indicating the best and V the worst. Many methods enter the Felsenstein zone in areas III, IV, and V. From Huelsenbeck and Hillis 1993.

inconsistent in regions III, IV, and V, entering the Felsenstein zone. Performance was improved, however, by shortening the branches examined, for instance by using only transversions, or by using rate-insensitive distance methods. In region V most methods were inconsistent, and none performed very well; Lake's method of invariants performed best there, although it performed relatively poorly elsewhere. Although maximum-likelihood methods were not included in this study, they are generally regarded as performing well.

Another general problem is that topologies depend partly upon the mix of taxa used. In many cases removing or adding taxa will not only alter the position of branches closely involved with the taxa but may alter the topology in surrounding regions as well. Topologies may also be altered by substituting exemplar taxa— using a different species to represent a phylum may alter the position of the phylum, and of other phyla. This phenomenon can be best understood in the case where a substitute taxon has a very different branch length from the original, so that, for example, the attraction of long branches in the first case may be largely eliminated by the use of a short-branch taxon in the second case. Close relationships that were indicated because of long-branch attractions may be replaced by relationships more closely based on common ancestry. However, substitution of exemplars may also alter elements of sequence distances that are based on artifacts other than just branch length, placing taxa closer to or farther from their relationships based on common ancestry.

Adding multiple exemplars to a tree commonly improves its accuracy. Within a tree the branching pattern of multiple species within clades tends to break up longer branches and to ameliorate long-branch problems. There are other advantages; the monophyly of the clades is tested, and the species that are more basal within the clade can be identified, providing evidence of bodyplan features that were present early in clade history.

Finally, emphasis here is placed on SSU rRNA trees because they have provided the bulk of molecular evidence for metazoan phylogenies to date. However, trees based on proteins should have some advantages—for example, long-branch attraction should be lowered because there are twenty or so different amino acids, rather than four nucleotides, in the sequences. Paralogy remains a problem for protein-based trees, however.

In sum, perhaps the most reasonable approach to phylogenetic analysis using DNA is to use molecules that display enough but not too much change from the last common ancestral sequences (and thus that tend to maximize the number of informative positions), and to use methods that provide the shortest branches consistent with this requirement. Results should also be improved by the use of as many exemplars as possible for each taxon, although if few taxa are available and most are very long branched, this may not be the best strategy. It would also seem prudent to use as many methods as possible; results that are consistent among many methods are at least likely to be free from contamination by the idiosyncrasies of a particular algorithm (Kim 1993; and see the discussion in Avise and Nelson 1995). The use of more than one gene to establish topologies is clearly to be recommended, assuming that orthologs are used and that the individual gene properties—that is, different net rates of change—can be evaluated.

Although Molecular Phylogenies Produce Conflicting Topologies, They Have Also Produced a Growing Consensus on Major Alliances of Phyla

Early Studies Suggested Surprising Alliances of Phyla

The earliest major molecular phylogeny of metazoan phyla was that of Field et al. (1988), who produced RNA sequences for twenty-two classes in ten phyla, quite a remarkable achievement considering the state of the art at that time. The sequences were aligned by hand with gaps inserted where necessary, and nodes determined by a distance matrix method; the topologies they found are shown in figs. 4.11A, C, E, and G. Fig. 4.11A shows the major groupings, with slime molds as an outgroup, while figs. 4.11C, E, and G examine relations within each of three major groups. Note that some branches change order or position when associated with different groupings of taxa (e.g., the position of the earthworm *Lumbricus* between

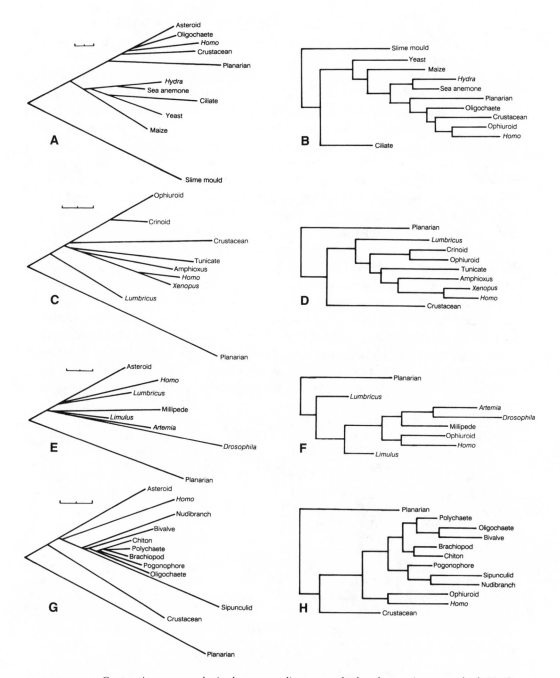

FIG. 4.11 Contrasting tree topologies between a distance method and a parsimony method. (A, C, E, G) Slightly modified phylogenies of Field et al. 1988, which were generated by a distance matrix method. (B, D, F, H) Corresponding phylogenies generated by a parsimony method; note that a few substitutions have been made (e.g., the ophiuroid in B is substituted for an asteroid in A). Scale bars represent a distance of 0.02 nucleotide substitutions per sequence position. From Patterson 1989.

figs. 4.11C and E). In this pioneering molecular phylogeny the kingdom Metazoa appears to be polyphyletic, with cnidarians arising independently of bilaterians. The flatworms are sisters to the remaining metazoans, which can be divided into four groups: echinoderms (in fig. 4.11C), chordates (in fig. 4.11C), arthropods (in fig. 4.11E), and a "eucoelomate protostome" group that included annelids, mollusks, and, surprisingly at the time, brachiopods (in fig. 4.11G).

Patterson (1989), using the data of Field et al., tried a variety of approaches to comparing sequences and employed parsimony rather than a distance method. Fig. 4.11 contrasts the Field et al. phylogeny with those produced by the parsimony method PAUP (fig. 4.11B, D, F, and H; note that a few taxa have been removed or substituted). Some of the differences are dramatic, as those of fig. 4.11G and H, between which every detail of the protostome relations differ. Some of the relations indicated by the parsimony analysis are surely erroneous (e.g., the positioning of a clade including *Limulus* as a sister to other arthropods plus deuterostomes in fig. 4.11F). Clearly, methodological artifacts greatly influenced the topologies; certainly the long branches seemed to attract each other. Patterson was nevertheless able to indicate that metazoans are probably monophyletic.

In attempting to formulate an analysis that could defeat the attraction of long branches, Lake (1987) developed the invariant method mentioned above, and applied it to the SSU rRNA sequences of metazoan phyla available from the work in Raff's laboratory (Lake 1989, 1990). The resulting tree (fig. 4.12) has some features that have characterized most subsequent molecular metazoan phylogenies.

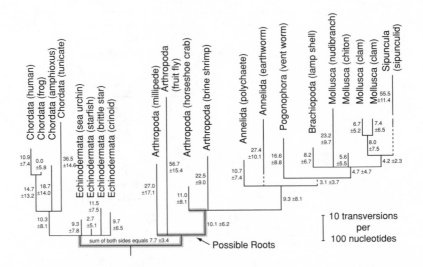

FIG. 4.12 Bilaterian data of Field et al. 1988, arranged by Lake's invariant method. Three major groups appear, the deuterostomes, the arthropods (which are paraphyletic in this arrangement), and the "eucoelomate protostomes." Distances in transversions per 1,000 nucleotides, followed by ± figure that represents 1.96 standard deviations. From Lake 1989.

When rooted conventionally so that the protostomes and deuterostomes are sepa-rate clades, three major bilaterian groups appear: the deuterostomes; an arthropod alliance; and a group of protostomes that includes the annelids, mollusks, and bra-chiopods, more or less the "eucoelomate protostome" group of Field et al. 1988. Cnidaria was used as an outgroup, and Platyhelminthes (represented by a long-branch exemplar) appeared to be a possible sister to the higher metazoans but not quite significantly so.

Subsequent Work Has Tended to Support the Existence of Several Major Metazoan Alliances

Additional studies have added longer sequences and sequences of many additional taxa to the database of SSU rRNAs of phyla, using a variety of methods to assess their branching patterns. As might be expected from the problems of determining "true" phylogenies from molecular data, briefly discussed above, the topologies have varied as new taxa are added, as different mixtures of taxa are studied, or as different algorithms are employed. Nevertheless, a general consensus as to the major groupings of phyla has been established by this work, particularly impor-tant findings emerging from the Lake lab. Four groupings have usually been found. One consists of the prebilaterian phyla: sponges, ctenophores, cnidarians, and pla-cozoans. Then there are three major groupings of bilaterian phyla: Deuterostomia (Grobben 1908), including hemichordates, echinoderms, urochordates, and chor-dates; Ecdysozoa (Aguinaldo et al. 1997), consisting of molting groups such as arthropods and their allies, including some paracoelomates; and Lophotrochozoa (Halanych et al. 1995), including lophophorate groups, classic spiralians, and some paracoelomates. Not all phyla have been confidently placed within one of these groupings, and in many cases the branching order within groupings is uncertain, but a core of phyla is consistently located within each group. The SSU rRNA data that bear on the relations among the phyla are discussed below in chapters 6–11, where basic features of individual phyla are reviewed.

Some common characteristics of the topology of metazoan phylogenetic trees are well-illustrated by the work of Giribet et al. (2000), who have evaluated the SSU rRNA sequences of 145 bilaterian species, including representatives of most of the phyla that are generally recognized, but concentrating on protostomes. Fig. 4.13 shows one of the topologies found by Giribet et al.; deuterostomes are collapsed into a single terminal branch, and terminal taxa belonging to each protostome phylum have been collapsed whenever possible (numbers in parentheses indicate the number of exemplar taxa involved in each terminal branch). Note that exemplars of several phyla are scattered through the tree. Molluscan exemplars occur in five different positions, with a different phylum or combination of phyla as sister(s) in each case. Some molluscan classes occur as terminal clusters, but for others, exemplars are scattered at different positions. Annelida occurs in four different positions within a group of eutrochozoans, to say nothing of a fifth occurrence

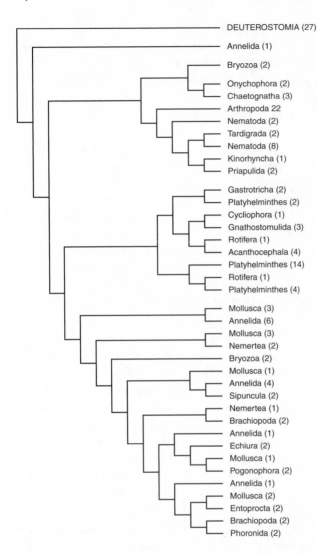

FIG. 4.13 Relations among bilaterian phyla evaluated by the SSU rRNA gene. Deuterostome phyla have been collapsed, and single nodes are used for all exemplars of the same phylum where possible; the number of exemplars at each terminal point is in parentheses. The parameters used in the analysis minimized incongruences; five hypervariable regions of the gene were excluded. See text for discussion. From topology of Giribet et al. 2000, fig. 2b.

at the base of the protostomes. It is essentially impossible that the mollusks and annelids are polyphyletic in the pattern suggested by this topology. However, most exemplars of the eutrochozoan phyla have been found to be associated in many molecular studies, and most share important developmental features. Thus there is support for the group but little for the branching pattern within it.

It has been suggested that the relatively slow rate of nucleotide substitution in the SSU rRNA gene, coupled with a rapid radiation of the eutrochozoan phyla, account for the confusing situation in their phylogeny. Not enough time had passed between the establishment of the crown ancestor of eutrochozoans and their principle radiations for the SSU rRNA molecules in the different clades to acquire highly distinctive sequences. Any small differences that did accumulate have been in large part swamped by random changes and perhaps by substitution bias in some cases. If this is the explanation, it implies not only that the clade ancestors of the phyla diverged in rapid succession, but that crown classes diverged very early in the history of their phyla, in order for their sequence similarities to be intermingled among phyla as they are. As we shall see (chaps. 5, 13), the fossil record indicates that major diversifications have occurred early in the history of many clades, a finding that is consistent with this interpretation of SSU rRNA data. Nevertheless, many protostomes show patterns of SSU rRNA similarities that suggest some special disruption of their sequences. For example, the annelid found at the base of protostomes is certainly out of place; evolutionary change within its SSU rRNA gene has been unusual in some way. That form may have evolved relatively rapidly over a long period of time so that sequence similarities to other eutrochozoans have been lost, or it may have undergone some major mutation(s), perhaps intercalating sequences that are not found in other eutrochozoans.

Combined Morphological/Molecular Phylogenies of Phyla May Require Improved Assessments of Homologies to Be Successful

Some workers have attempted to join evidence from molecular-sequence similarities with evidence from morphological character distributions in the hope of producing more accurate phylogenies than either can provide separately. For example, Wheeler et al. (1993) used a parsimony method in an attempt at a combined phylogeny. The morphological study was based on 100 characters and produced a cladogram with mollusks as sisters to annelids + onychophorans + arthropods, with annelids as sisters to onychophorans + arthropods, and onychophorans as sisters to arthropods (fig. 4.14A). SSU rDNA sequences of two to eighteen species belonging to those phyla produced a cladogram in which representatives of the phyla were not clearly separated (fig. 4.14B). Sequence data from the ubiquitin molecule were also used (fig. 4.14C). Finally, in a combined phylogeny, the SSU rRNA polytomy was broken up, and a topology of the phyla emerged that was like that indicated by morphology alone (fig. 4.14D). For the phyla involved, the combined topology is identical to that of Brusca and Brusca (1990), which was based on morphology alone and also evaluated by a parsimony algorithm (see fig. 4.2). It would appear that the identification or choice of putative homologues used in the scoring of morphological characters among the phyla was the overriding influence on the final topology. The topology differs fundamentally from those suggested by most analyses of SSU rRNA data alone (see below). There are significant homoplasies within one set of data or the other, or within both.

A combined morphological and molecular approach was also used by Zrzavy et al. (1998); by that time much additional molecular data had become available. Morphological and 18S rDNA molecular data were first evaluated separately

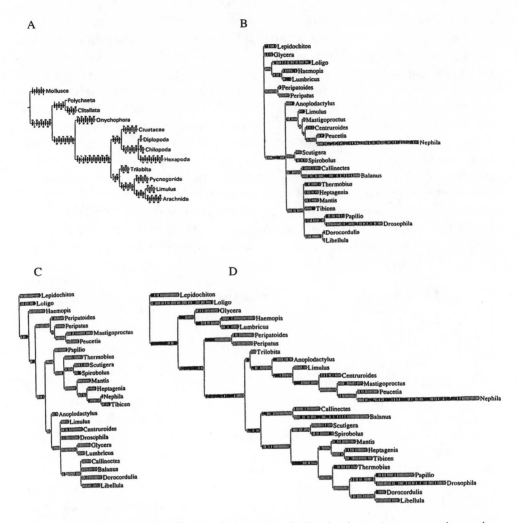

FIG. 4.14 Evaluating combined morphological and molecular data by parsimony to produce a phylogenetic hypothesis. (*A*) Topology found by evaluation of 100 morphological characters. (*B*) Topology found by evaluation of SSU rDNA sequences. (*C*) Topology found by evaluating Ubiquitin sequences. (*D*) Topology found by combining data in A, B, and C. Topologies from Wheeler et al. 1993.

by maximum parsimony methods. Morphological data were compiled from the literature and treated straightforwardly; molecular data were subjected to an elaborate system of screening and taxon selection; both data sets were subjected to various sensitivity analyses. The resulting molecular and morphological topologies are quite different, and in a combined tree, morphological data tended to affect the topology more strongly, as found by Wheeler et al. For example, in the morphological phylogeny, deuterostomes are sisters to brachiopods + phoronids; by contrast, in the molecular phylogeny they are sisters to the priapulids + kinorhynchs, and the lophophorates are associated with mollusks, annelids, and some other spiralian groups. In the combined phylogeny the deuterostomes are sisters to brachiopods + phoronids.

The study by Giribet et al. (2000), which presented the molecularly based tree summarized in fig. 4.13, combined that data with the morphological data set of Zrzavy et al. (1998) to produce a very large combined phylogeny. In general, using the combined data sets produced a tree in which members of the same phyla, though scattered in the molecular tree, were united at single terminations (the combined tree is summarized in fig. 4.15). The flatworms formed the only important exception, with one order, Nemertodermatida, appearing as sister to Chaetognatha

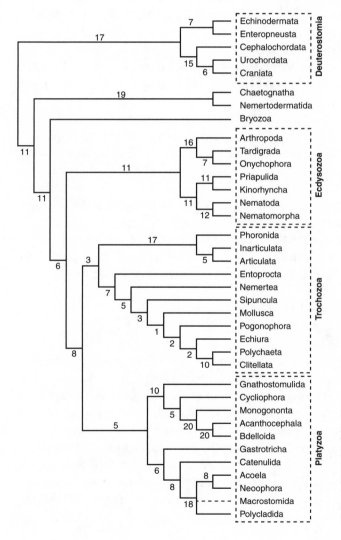

FIG. 4.15 Summary of the topology of bilaterian phyla found when combining the molecular data set of Giribet et al. (fig. 4.13) with the morphological data set of Zrzavy et al. 1998. Numbers on the branches are Bremer support values. From Giribet et al. 2000.

at the base of the protostomes. An interesting feature of this tree is that a clade of paracoelomates, termed Platyzoa (Cavalier-Smith 1998), emerged as a distinctive alliance.

It appears that morphological topologies are dominating or at least unduly influencing these combined topologies (Wheeler et al. 1993). Presumably most or all of the morphological characters are chosen because they have been found to characterize a phylum—they are taxonomically high-level characters—and they will appear in all of the exemplars of those phyla that are used in analyses, providing a strong signal. Characters that appear only within subtaxa of the phyla and that originated after the phyla were founded will not ordinarily be chosen by the workers. The nucleotide positions are used as found, however, whether or not they represent only a given phylum, and furthermore they have continued to change after the founding of the phyla, independently among subtaxa, right up until the most recent changes, which might have been just last year. Thus nucleotide positions will not necessarily be faithful to higher taxa. Given this situation, morphological homologies are not being evaluated by appropriate molecular-sequence data in combined studies. As the identification of morphological homologies is highly uncertain, combined results remain uncertain as well.

These are difficult problems to address. The problems may well be ameliorated as molecular data become more dense within phyla. At this time a more prudent approach may be to use the phylogeny from one source as a hypothesis, and to evaluate it with the aid of phylogenies from other sources. Marshall (1992a) has shown that such an approach can be useful, with examples from molecular and morphological trees of sand dollars, for which molecular data appear to be the more reliable, and of amniote classes, for which paleontologically supported morphological data appear to be the more correct. If one is quite certain of, say, morphological assessments of relatedness, then the morphological tree may be used to examine the reasons for any misleading patterns in molecular sequences (and vice versa).

Stratigraphic Data Can Add Useful Information to Phylogenetic Hypotheses

Fossils are an obvious source of useful evidence of the relationships among taxa that can be traced one into another through intermediates in the fossil record. For fossil taxa that are morphologically distinct and are separated by important stratigraphic gaps, however, phylogenetic relationships become more difficult to specify. Nevertheless, there is phylogenetic information simply in the stratigraphic record of taxa (see Gingerich 1979; Fisher 1994). For taxa that are quite common within their known stratigraphic ranges, those that have the earlier stratigraphic occurrences are more likely to have originated earlier, at least if the taxa have roughly similar preservation potentials. The branching orders of taxa within a tree may thus be suggested by their order of appearance.

One way to include stratigraphic evidence in a phylogeny is to consider it as a character like any morphological character, and to evaluate it accordingly in constructing a tree. Another approach is to construct a morphological tree and then to evaluate it in the light of stratigraphic data. Fisher (1992) has defined a concept termed stratigraphic parsimony debt, which is the difference between the order of stratigraphic appearance of taxa expected by a given phylogenetic hypothesis, and the order of appearance that is observed. The debt is owing either to an incorrect tree or to the perfidy of the stratigraphic record. The debt may be partially corrected for stratigraphic incompleteness by taking into account the probabilities of occurrence of the taxa when judged by the frequency of occurrence within their known stratigraphic ranges. Trees established by morphological criteria may be compared with trees established by stratigraphic criteria, evaluating them in part by whether the debt owing to possible homoplasies is greater than that owing to possible stratigraphic perfidy. Computational methods for such comparisons are discussed by Fisher (1992). In an important theoretical test of stratigraphic data in phylogeny reconstruction, Fox et al. (1999) contrasted the relative abilities of cladistic methods alone and cladistic methods supplemented with stratigraphic information, to recover a known (simulated) phylogeny. The addition of temporal data to cladistically evaluated characters significantly improved tree reconstruction, recovering the true tree in over twice as many cases as with cladistic data alone. Likelihood methods of testing phylogenies against the fossil record have also been introduced (Huelsenbeck and Rannala 1997; Wagner 1998). Stratigraphic information can be particularly useful when dealing with taxa that show high degrees of homoplasy, as it can sometimes sort out homoplastic features that originated at different times (Fox et al. 1999).

The chief circumstances under which such methods are unsatisfactory involve clades with very low preservation potentials or those that change their preservation potentials during their history. The preservability of metazoan clades varies widely. Small, soft-bodied taxa clearly have a low potential for preservation, and many phyla consist exclusively of such forms, so if they lack early records, it does not necessarily mean they did not appear early. For example, nematodes are known at present from single occurrences at only five horizons in the entire Phanerozoic, the earliest being of Early Carboniferous age. The relatively late appearance of these minute, soft-bodied organisms does not mean that they evolved after all of the durably skeletonized taxa that appear much earlier; the negative evidence of their nonappearance in earlier rocks counts for little. Moreover, the first occurrences of phyla with durable skeletons, whose subsequent fossil records are very dense temporally, are complicated by the fact that the phyla have evolved from soft-bodied ancestors, so that their first appearance as body fossils may record the acquisition of durable skeletons but not necessarily of bodyplans. Stratigraphic evidence (chap. 5) for these forms is not useful in interpreting the prelude to the Cambrian explosion. On the other hand, there should have been less difference in the preservation potential of trace fossils from the late Neoproterozoic through the Early Cambrian (actually better for some traces in older rocks when bioturbation was less important). Stratigraphic confidence intervals associated with trace-fossil taxa should be useful during that critical interval of time.

The Alliances of Phyla Indicated by Molecular Methods Provide Evidence for Evaluating the Origin and Early History of Phyla

In order to discuss the origin of phyla it is clearly necessary to have some idea of their general relationships. But instability in the position of phyla in phylogenetic trees, whether based on morphological, developmental, or molecular data, is commonplace. However, although we lack an incontrovertible, detailed phylogenetic tree, the SSU rRNA molecular data have identified some major branch points that appear to give rise to monophyletic assemblages of phyla. These alliances provide a general framework within which hypotheses of the origins and relationships of individual phyla can be discussed.

Bilaterian Alliances Can Be Identified in a Very Conservative SSU rRNA Tree

Fig. 4.16 indicates the relations among groups that are commonly considered to be phylum-level taxa, based on what can be most reliably asserted from SSU rRNA molecular sequences, compared by several methods and using as many exemplars as have been studied to date. The phyletic status of these groups, as indicated in the tree, is listed in table 4.1. Considering the difficulties outlined above, it is understood that some of even these rather minimalist relations may be incorrect, but they are the ones that are most likely to be correct on present evidence. In this tree the metazoans appear to be monophyletic, a finding also supported by sequences of the Hsp70 gene (Borchiellini et al. 1998). Sponges are basal among living metazoans, and are paraphyletic with respect to the Eumetazoa. The diploblastic phyla are basal with respect to bilaterians. A significant gap, represented by a long branch on the molecular trees and by the morphological difference in constructional grade, separates the diploblastic forms from living bilaterians. Groups that appear to lie at the base of the bilaterians tend to be quite long-branched and may owe their basal positions to this factor. The remaining bilaterians show several alliances—Deuterostomia, Ecdysozoa, and Lophotrochozoa stand out—but the relations among them are not resolved. A number of phyla (such as lophophorates) that have sometimes been placed in the deuterostomes are actually allied with the protostomes according to their SSU rRNA sequences. Furthermore, paracoelomate phyla for which data are available, which have often been depicted as branching earlier than the protostome-deuterostome branch, are evidently protostomes on molecular evidence. Protostomia has thus become quite a large group, with the Ecdysozoa, containing molting phyla, and the Lophotrochozoa, containing nonmolting phyla, each containing at least twice as many phyla as the entire Deuterostomia. The phyla placed between

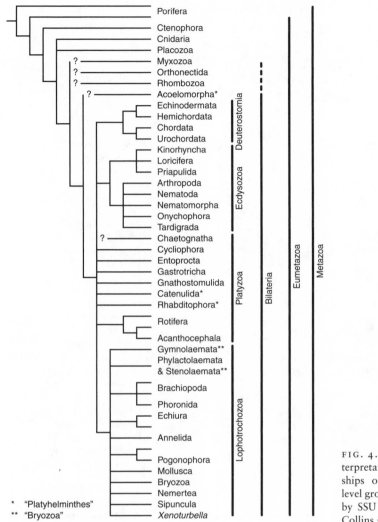

FIG. 4.16 A conservative interpretation of the relationships of metazoan phylum-level groups that are indicated by SSU rRNA studies. From Collins and Valentine 2001.

Ecdysozoa and Lophotrochozoa in fig. 4.16 seem to be protostomes (or at least not to be deuterostomes) but cannot reliably be assigned to a major alliance on present SSU rRNA evidence alone. According to these data, a few phyla (Platyhelminthes, Bryozoa, Annelida, perhaps Mollusca) may be polyphyletic, and some (Porifera, Rotifera, Brachiopoda, Annelida) may be paraphyletic.

The Conservative Tree May Be Modified by Other Criteria to Produce a More Liberal Hypothesis

The conservative tree lacks temporal structure between the major alliances, and leaves a number of phyla as virtual orphans. If a somewhat more liberal view

TABLE 4.1 Phyletic status of crown phyla indicated by SSU rRNA sequences. From Collins and Valentine 2001.

Number	Phylum	Species Known[a]	SSU rRNA Samples[b]	Phyletic Status and Notes
1	Acanthocephala	1,150	25	Likely holophyletic
2	Annelida	11,600	119	Possibly polyphyletic; likely paraphyletic with respect to Pogonophora
3	Arthropoda	1,000,000	990	Likely holophyletic
4	Brachiopoda	325	48	Possibly holophyletic; possibly paraphyletic with respect to Phoronida
5	Bryozoa	5,000	6	Possibly polyphyletic with Gymnolaemata separate
6	Chaetognatha	70	3	Possibly holophyletic
7	Chordata	45,000	105	Possibly holophyletic
8	Cnidaria	9,000	107	Likely holophyletic
9	Ctenophora	100	3	Likely holophyletic
10	Cycliophora	1	1	Uncertain
11	Echinodermata	6,000	55	Likely holophyletic
12	Echiura	140	2	Possibly holophyletic
13	Entoprocta	150	3	Possibly holophyletic
14	Gastrotricha	450	3	Possibly holophyletic
15	Gnathostomulida	80	3	Possibly holophyletic
16	Hemichordata	90	9	Likely holophyletic
17	Kinorhyncha	150	1	Uncertain
18	Loricifera	50	0	Unstudied
19	Mollusca	50,000	128	Possibly polyphyletic
20	Myxozoa	1,200	18	Likely holophyletic
21	Nematoda	12,000	89	Likely holophyletic
22	Nematomorpha	320	5	Possibly holophyletic
23	Nemertea	900	3	Possibly holophyletic
24	Onychophora	70	2	Possibly holophyletic
25	Orthonectida	10	2	Uncertain, one species sampled
26	Phoronida	14	8	Likely holophyletic
27	Placozoa	2	2	Possibly holophyletic
28	"Platyhelminthes"	3,000	331	Possibly polyphyletic with Acoela separate
29	Pogonophora	80	2	Possibly holophyletic
30	Porifera	5,000	24	Likely paraphyletic with respect to Eumetazoa
31	Priapulida	16	5	Possibly holophyletic
32	Rhombozoa	70	1	Possibly holophyletic
33	Rotifera	2,000	7	Likely paraphyletic with respect to Acanthocephala
34	Sipuncula	320	3	Possibly holophyletic
35	Tardigrada	600	7	Likely holophyletic
36	Urochordata	1,250	16	Likely holophyletic
37	*Xenoturbella*	1	1	Uncertain

[a]Rough estimate of known species; actual species richness is probably much greater for some phyla.
[b]Approximate number of complete or near complete (>1,700 base pairs) SSU rRNA sequences in GenBank.

is taken of SSU rRNA data, and other molecular information is also considered, branching sequences can be suggested that produce considerably more structure in the tree, although some of the data from other molecules are in conflict. The sorts of criteria that can best complement the SSU rRNA data at present are very early developmental patterns that seem most likely to reflect primitive conditions, such as the cleavage stages in indirect development. Adult bodyplan similarities are also useful when they involve similarly complex structures that are likely to be homologous. Of course, such criteria have been in common use, on their own or in combination with other developmental or morphological criteria, and have not been successful in defining a widely acceptable tree. But arguably, when in concert with molecular data that help to sort out some of the nonhomologous features, they become more useful. By combining molecular, morphological, and early developmental data in a more or less judicious way, a tentative phylogenetic framework can be erected that may then be examined critically in the light of what is known from the fossil record and what can be inferred of macroevolutionary processes. Such a provisional grouping of phyla, first framed by the more robust of the molecular data and then modified to take account of less robust molecular data, of patterns of development, and finally of similarities in bodyplan architectures, is presented in fig. 4.17. The four major groupings of phyla suggested by molecular data are taken here as phylogenetic alliances: pre-Bilateria (chap. 6), Ecdysozoa (chap. 7), Lophotrochozoa (chaps. 8–9), and Deuterostomia (chap. 11). Platyzoa (chap. 10) may represent a fifth grouping. The developmental and morphological features used to evaluate the position of phyla within the alliances are reviewed in chapters 6–11.

As new phylogenetic evidence has accumulated, clades of phyla that have never been postulated or that have been considered quite speculative have come to be well enough supported by new data that they have been named. In some cases these associations have resulted in reducing two or more phyla into one phylum (such as Rotifera and Acanthocephala into Syndermata; Haszprunar 1996b). More commonly, phyla have been assembled into newly named combinations (such as Neotrichozoa for Gnathostomulida + Gastrotricha, and Paracoelomata for Ecdysozoa + Syndermata + Neotrichozoa; Zrzavy et al. 1998). As phylogenies are not yet definitive and new evidence continues to suggest different relationships, the number of such named combinations tends to increase with each revision. I have tried to avoid using most of these new nomenclatures, and to stay with more traditional or at least familiar taxonomic names at the level of phylum and above. This is not meant to imply that these supraphyletic combinations may not have merit, but as it is, there is quite an array of taxonomic names to be dealt with, and clarity would not be served at this stage of our knowledge by attempting a history of these many usages. Some of the new combinations are suggested or reviewed in Zrzavy et al. 1998, Cavalier-Smith 1998, and Giribet et al. 2000.

The phylogenetic hypothesis in fig. 4.17 will be used hereafter as a framework for a discussion of metazoan bodyplans, which necessarily involves some evaluation

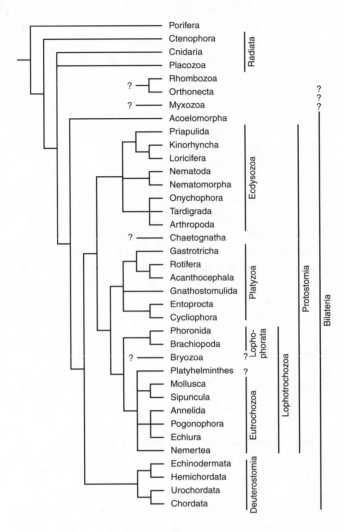

FIG. 4.17 A hypothesis of the relationships of metazoan phylum-level groups based on molecular studies but modified and interpreted in the light of additional information on development and morphology. The bases for choosing branching patterns within the alliances are discussed in chapters 6–11.

of the morphological divergences and homoplasies that are implied. The many decades of shuffling and reshuffling of the phylogenetic topology, both before and after the advent of molecular techniques, suggests that there is little chance that we now know the actual evolutionary history of all the phyla. Nevertheless, the molecularly based pattern of at least the major branchings provides fresh hypotheses that have not been available for serious evaluation until recently, and that merit careful study and cautious interpretation.

CHAPTER 5

The Fossil Record

Darwin had a lot of trouble with the fossil record, which did not seem to support a pattern of gradual and incessant morphological change such as he envisioned. In chapters 10 and 11 of *On the Origin of Species,* "On the Imperfection of the Geological Record" and "On the Geological Succession of Organic Beings," Darwin rationalized the lack of detailed support for his ideas, essentially by attributing it to the spottiness of the fossil record. Today the fossil record is far better known than in Darwin's time, and methods have been developed that help circumvent its imperfections. Analyses of evolutionary events that are registered by fossils are becoming available at increasingly finer resolutions. The fossil record is an indispensable sampling of the history of life that would otherwise be entirely lost. The record can inform us of the paces and styles of evolution that were associated with the origins of phyla, and that have been responsible for creating the present biosphere by exploiting potentials of those early bodyplans. Despite these wonderful properties, the record remains incomplete and biased, and poses as many fascinating questions as it provides answers.

The Stratigraphic Record Is Incomplete in a Spotty Way

Sedimentary Rocks Are Accumulated and Preserved Episodically
All of the early metazoan fossils, including those that form our record of the early metazoan diversification of phyla, are entombed in marine sedimentary rocks. A first step in understanding the nature of the fossil record is to understand the character of those rocks. Most of the marine fossils are from shallow-water sediments of epicontinental seas. Such seas are relatively restricted at present (the ocean now covers about 16% of the continents) but were quite extensive at times during the Phanerozoic, covering perhaps up to 50% of the continents at their greatest extents—more on some continents, less on others. Most sediment is supplied to the ocean from the land, after erosion and transport by rivers or by wind, and by coastal waves. However, important amounts of marine sediment are also derived from the sea itself through the biological production of mineralized algal, protistan, and metazoan skeletons—by potential fossils themselves. The sediment particles are denser than water and tend to sink to the seafloor. If the water is in motion, particles may be carried along with the current until they are deposited. If the water is highly turbulent,

as in waves or very strong currents, particles may be buoyed up in rising eddies and remain suspended in the water column for indefinite periods, and particles already deposited may be resuspended and transported. Sediments, then, tend to come to rest and to accumulate in quiet water, and to be eroded in turbulent regimes. Over geologic time, the accumulation and preservation of sediments at any given locality is episodic.

Sediments are least likely to accumulate in the more turbulent regions in the shallow sea, which are near shore where wave activity is greatest. Sediment supply to offshore sites may be great enough to build the seafloor up to the depth at which wave erosion will occur, and the sedimentary buildup is then halted. Thus, off highly exposed coasts, the depth at which sediments can accumulate will be greater than along quiet shores within embayments where waves are weaker. In some regions, as at the well-protected mouths of large rivers that carry heavy sediment loads, the supply of sediments may overwhelm the ability of waves and currents to erode or transport them even in shallow water. In these cases sediment may even build up above the sea surface, as in some river deltas.

The conditions of sediment accumulation and erosion vary through time. Relatively short-term variations include fluctuations in sediment supply owing to seasonal rainfall and erosion on land, or to seasonal peaks of productivity among the skeletonized marine biota. There are also numerous longer-term changes. For example, the climate may change to institute a new regime of weathering and erosion. Perhaps most important of all are changes in sea level, which have amounted to swings of several hundred meters as measured from continental platforms. In a given coastal region the sea may advance or retreat across the local topography owing to local warping of the continental margin or to a more general rise or fall in a continental platform, compared to surrounding ocean basins. On a global scale sea level varies with the volume of the ocean basins and with changes in the volume of ocean water. For example, the growth of ocean ridges reduces ocean-basin volume, while the growth of deep-sea trenches increases it. Such tectonic activity, though slow even in geologic terms, has been going on since long before the origin of metazoans. Obviously, when ocean-basin volume decreases enough, water climbs up the continental slope and spills onto continental platforms; when ocean-basin volume increases, water drains off the continents into the ocean basin—sea level rises or falls. Furthermore, during the last few million years, and during a number of other periods in geologic time, ocean water volume has varied significantly as continental glaciers and ice caps have waxed and waned, locking up or releasing water, and altering global sea level accordingly. From these and other causes, sea-level fluctuations have been common throughout geologic time.

Barrell (1917) invented a simple way to indicate the incompleteness of the sedimentary record. Consider a locality near shore at the beginning of a period of geologic time characterized by an overall rising sea level, as in curve AA′ of fig. 5.1. Perhaps the general rise results from a decrease in ocean-basin volume as seafloor spreading accelerates. As the water depth increases, the seafloor falls below the depth

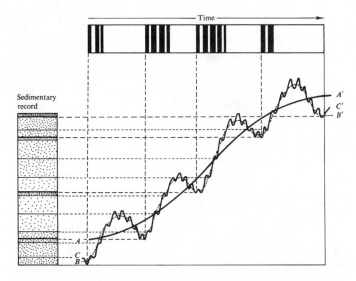

FIG. 5.1 The sedimentary record remaining (column at left) after deposition and erosion during base-level oscillations having three periods (indicated by curves AA′, BB′, and CC′). Time is from left to right; the time represented by sediments after this history is indicated by the black bars at the top, and the time represented by gaps is indicated by the white spaces between the bars. After Barrell 1917.

at which waves or currents erode sediment, permitting sediments to grow thicker as time passes. But many of the other factors that contribute to sea-level change may affect this general trend. Suppose, for example, that glacial-interglacial cycles produce shorter-term sea-level fluctuations during this rise; they would impose a set of oscillations on the curve, so that sea levels would sometimes fall (as in curve BB′, fig. 5.1) during this general rising trend. When sea level drops, previously deposited marine sediments are brought first into the zone of strong wave action, and then into the subaerial environment with its many erosional processes, so that, typically, the upper part of the accumulated sedimentary record would be eroded away in shallow-water or subaerial regions. When sea level resumes its rise and sediments again accumulate at those localities, there would be a gap in the local sedimentary record that includes both the time represented by any eroded sediment and time represented by any period of erosion or nondeposition. The surface representing the gap is termed an unconformity. A set of still shorter oscillations may be imagined, perhaps owing to local crustal warping, producing the local sea-level fluctuations traced by curve CC′ (fig. 5.1), and creating additional gaps in the record. The actual times represented by accumulated sediment at the end of such a sea-level history are indicated by the black bars at the top of the diagram; each packet of sediment would be separated by an unconformity from the preceding and succeeding packets. Thus even though the general history of the region has been favorable for sediment accumulation during this period of rising sea level, not all of geologic time is represented by sediments. During periods of generally falling sea levels, large portions of the continental platforms are brought above sea level and exposed to myriad erosional processes of the terrestrial environment. Older sediments are liable to be entirely stripped away except where fortuitously protected in local or regional basins or at other sites that favor preservation. Like extinction, the gaps in the geologic record, once they occur at some locality, are forever.

Fig. 5.2 shows curves for sea level over late Neoproterozoic and Phanerozoic time, estimated chiefly from geophysical evidence and calibrated to present sea level. The curve on the left shows the general sea-level trend, suggesting the effect of a long-term factor, while the curve on the right suggests that there are secondary (and perhaps tertiary) cycles superimposed on that trend. These curves are analogous to Barrell's curves in fig. 5.1. Sea-level changes based on increasingly smaller time increments tend to be increasingly more volatile, as the averaging effects of larger time bins are removed. Note that the early fossil record of metazoans accumulated during generally rising sea levels.

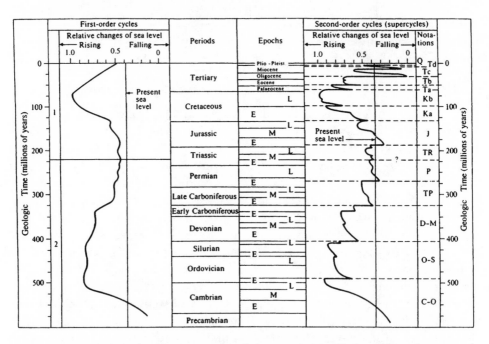

FIG. 5.2 Sea levels during late Neoproterozoic and Phanerozoic time. The first-order cycles suggested by the smoothed curve on the left have secondary and perhaps tertiary cycles superimposed on them to produce the curve on the right. Sea-level changes, even more variable when averaged over smaller time increments, have led to shifting environmental patterns and frequent interruption of sedimentation with erosion of sediments on regional or even global scales. Cycles: *C-O*, Cambrian-Ordovician; *O-S*, Ordovician-Silurian; *D-M*, Devonian–Lower Carboniferous; *TP*, Upper Carboniferous; *P*, Permian; *TR*, Triassic; *J*, Jurassic; *Ka, Kb*, Cretaceous a, b; *Ta, Tb, Tc, Td*, Tertiary a, b, c, d; *Q*, Quaternary. After Vail et al. 1977 modified by Hallam 1981.

Sedimentary "Completeness" Varies with the Resolution That Is Desired

Studies by Schindel (1980, 1982), Sadler (1981), Strauss and Sadler (1989b), Sadler and Strauss (1990), and others have led to an increasingly quantitative under-standing of the stratigraphic rock record. This work has clarified the concept of stratigraphic completeness by requiring the specification of a time period at which completeness is desired. If one has a 100-year sedimentary record, one may ask how complete that record is with respect to months (dry seasons might not be rep-resented) or years (most of which might be represented, barring extensive drought). The record might be 60% complete for months and 98% complete for years—there is no sediment representing 40% of the months, but there is at least some sediment for all but 2% of the years. The interval of interest, months or years in this case, is called the resolution interval, and the time to which the resolution interval is applied, 100 years in this case, is the time span. In general, the longer the resolution interval, the more complete the stratigraphic section will appear to be within a given span; the critical value is the resolution interval divided by the time span (see also Strauss and Sadler 1989b; Sadler and Strauss 1990).

It has also been determined empirically, both for Recent sediments (Schindel 1980) and for ancient sedimentary rocks (Sadler 1981), that for a given resolution interval, short spans are generally more complete than long spans. Fig. 5.3 depicts, for an average shelf sequence of the sort that has yielded early metazoan fossils,

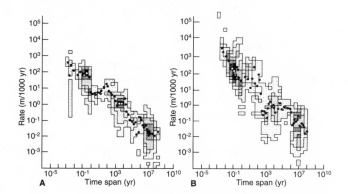

FIG. 5.3 Relation between the average rate of sediment accumu-lation and the time span of sediment accumulation displayed by (A) carbonates and (B) other sediment types, found in continental-shelf depths. The greater the number of observations of a given rate-span ratio, the darker the corresponding symbol. It is clear that, in general, the longer the time span, the lower the rate of sediment ac-cumulation observed (owing to the greater chances of encountering larger gaps in sequences of longer duration). The study from which this figure is taken was based on over 25,000 points, and the general relations of rate to span held for all depositional environments. After Sadler 1981.

the empirical relations between the net rate at which sediments accumulate on continental platforms, and time spans of accumulation. The relations are based on many thousands of sedimentary intervals. The longer the span, the lower is the rate of retention of sediment. The slope of the regression of rate on span is a measure of the rate at which sedimentary retention falls off with increasing span. The reasons for this relationship are simple: sedimentation is episodic and sediments are subject to erosion, much as in Barrell's diagram. The slope indicates that the longer the span that is examined, the more gaps, and the longer some of the gaps, that are encountered. Therefore, completeness at a given resolution interval diminishes as the span increases and as larger gaps are encountered. As an example, for successive geologic stages that averaged 7 million years in duration in a system that lasted 30 million years, the average carbonate shelf section is about 61% complete at a 7-million-year resolution interval. For a resolution interval of 1 million years, the same average section is only about 33% complete.

Such calculations involve single sections only. If a given time span of interest is found in many localities, then there is a chance that the time represented by a gap in one region will be represented by sediment in another. If gaps are randomly distributed in time among sections, the global completeness will rise with the number of sections available. At a 1-million-year resolution interval, a 30-million-year span rises from 33% complete at one section to 54% for two independent sites to 98% for ten (Valentine, Awramik, Signor, and Sadler 1991). As all of the possible 30-million-year spans of Phanerozoic time occur at more than ten sections, it would seem that the global stratigraphic record would be nearly complete at that resolution *if* the gaps are truly random.

On the other hand, not all of those sections (each being 33% complete) would contain sediments within the same 1-million-year intervals. The probability that, within a 30-million-year span, a given 1-million-year interval contains a record at each of two independent sections is 11%, and at each of five independent sites, only 0.4% (see Sadler in Valentine, Awramik, Signor, and Sadler 1991). Clearly, correlation should be very difficult if the rate of faunal turnover is at all high. But our experience is that faunal correlation, though difficult enough, is commonly possible among numerous sites scattered across the world, at 2-million-year and perhaps on occasion at 1-million-year intervals or even better. Thus the gaps are *not* independent, or not completely so. One obvious reason is that falling sea levels should produce widespread, contemporary gaps in shallow-water sedimentary sequences. By the same token, rising sea levels should produce packets of sediments that correlate around the globe. To be sure, local or regional events, such as crustal warping, commonly produce local or regional unconformities and sediment packets that do not correlate with global events, and overprint the more general patterns. Nevertheless, a quasi-hierarchical system of sedimentary packets, termed "sequences" in a seminal paper by Sloss (1963), are found to be roughly correlative on an intercontinental scale. Boundaries between the most inclusive sedimentary sequences correspond with the major sea-level drops indicated in fig. 5.1B, and at many localities sequences can be recognized at finer scales (see Vail et al. 1977).

If widespread unconformities are common, which seems to be the case, at least regionally, there are numerous gaps in the shallow marine fossil record of those regions that can never be filled. This makes it difficult to study shallow marine lineages through time, especially at the level of species or sometimes of genera or even higher taxa, because the lineages found within the regions disappear into the gaps. When the gap is wide and morphological change has been rapid, the ancestral taxon is not always evident. On the other hand, it is better for fossils to be preserved in packets rather than scattered at random through time. The times at which major sedimentary packets are deposited, between the widespread unconformities, have a higher-than-average chance to be represented by sediments over broad areas, preserving global or near-global faunal patterns during relatively narrow time intervals.

The Completeness of Sedimentary Sections Is Independent of Their Ages

One might expect that both the stratigraphic and the fossil records would deteriorate monotonically with the increasing age of rocks, but this is not correct. For example, Sadler (1981) has shown that the completeness of stratigraphic sections is independent of their ages as such; some systems may be more complete than others, but that is not a matter of their age but rather of their depositional and erosional histories during accumulation. Surviving sections from the Cambrian are not necessarily less complete than surviving sections from, say, the Cretaceous, environment for environment. There tend to be fewer Cambrian sections available for study than those of some later systems, because older rocks tend to be covered by younger rocks, and to be sure many Cambrian sections may have been destroyed by erosion. But the sections that we have should not be of less than average completeness, and we have enough of them that we should have a good record preserved at least between the more pervasive gaps. Furthermore, it happens that the Cambrian was a time of generally rising sea levels (fig. 5.2), so the sections that we have should be more complete than average. Also, the geography of deposition and the tectonic settings of Cambrian deposits, coupled with the subsequent geometry of their uplift and exposure via erosion, imply that we have a better record of the Cambrian than of some much later systems. It is unarguable that as time goes by more and more fossils are destroyed, yet more and more are exposed as well, and whether the *available* record of a particular time is improving or deteriorating does not depend upon its age. Indeed, the available Cambrian record is probably improving as I write, as more of it is being exposed by erosion, and perhaps in 10 million years or so it will be significantly better than it is now. The late Neoproterozoic, spanning the time that the earliest metazoans were evolving, is evidently less well exposed than the Cambrian but appears to have at least an average rock record in which fossil remains could be captured.

The Marine Fossil Record, while Incomplete, Yields Useful Samples of a Rather Consistent Fraction of the Fauna

Most marine sediments today contain living organisms, but, unfortunately, the presence of appropriate sedimentary rocks is no guarantee that they contain fossils. Most organisms are destroyed shortly after death or, if fossilized, are commonly destroyed later, so that many rocks are rendered barren of fossils. Nevertheless, some marine sediments are spectacularly fossiliferous. Given the *sedimentary* record that we have, then, how complete is the *fossil* record? For our purpose here, fossil completeness can be simply defined as the proportion of taxa, alive at a given time or place, that have left a fossil record.

Local Fossil Assemblages Are Largely Durably Skeletonized and Time-Averaged
Completeness is difficult to estimate for ancient fossil associations, because there are commonly few clues as to the original size of the biota from which the fossils have been assembled. The discipline that deals with the preservation of fossils is termed taphonomy (see papers in Allison and Briggs 1991; Kidwell and Behrensmeyer 1993; see also Kidwell and Flessa 1995). A number of taphonomic considerations are relevant to understanding the adequacy of the fossil record during the origin of phyla.

Other things being equal, soft-bodied forms have less chance of being preserved than forms with stiff organic skeletons or parts, and these have less chance than forms with mineralized skeletons or parts, and thin-shelled forms have less chance than robust-shelled forms. Thus organisms that are entirely soft-bodied, lacking mineralized skeletons or other parts, have little chance of fossilization. When soft-bodied forms do occur as fossils, it is almost always in "exceptionally preserved" faunas, also termed fossil lagerstätten, that have accumulated under highly unusual circumstances. About two-thirds of the marine invertebrate species living on continental shelves today can be fossilized only by rare events. The size of organisms and of their durable parts also affects the chances of preservation; in general, smaller skeletons are more easily destroyed.

Factors other than just the size and durability of organisms contribute heavily to their preservability. Abundance is certainly one of the more important of these factors. For a fossil taxon, the larger its average population, or the shorter its average generation time, or the longer its geologic duration, the more individuals it will produce and therefore the more chance it will have of appearing in the fossil record, other things being equal. Somewhat related to abundance is another important factor, breadth of distribution. If a taxon is geographically widespread, the chances of its being preserved in the fossil record are certainly greater than if it is highly localized. Similarly, the broader the environmental tolerance of a taxon, the better its chances of living at a locality that happens to contribute fossils to the record. And organisms that preferentially inhabit environments of deposition, rather than of erosion, have the better fossilization potential. It is thus likely that the taxa that we are least likely to know as fossils are those that were soft-bodied, rare, ecologically narrow, geographically restricted in erosional settings, and geologically short-lived. The fossil record is clearly not representative of the living faunas from which it is drawn, but is biased in favor of forms that are more easily preserved.

A phenomenon that is a problem for paleoecologists but that is probably a boon to macroevolutionists is time-averaging (Walker and Bambach 1971). Most marine invertebrate fossil associations represent the assemblage of skeletal elements from organisms that lived within a period ranging from a few years to a few thousand years (Kidwell and Bosence 1991; Kidwell 1993; Flessa and Kowalewski 1994). Most fossil faunas are thus not representative of any single living species associa- tion, or even of several contemporary associations, but reflect in some (unknown) way the succession of species that have inhabited the (uncertain) area from which the fossils were assembled. This means that, usually, many more species are pre- served in a given assemblage than would be present if a simple snapshot of a living fauna were preserved. Nature has kindly put together a large sample for the curious paleontologist, and enhanced the opportunity for locally rare species to be found. It is difficult to learn how representative these samples are, though Kidwell (2001) has shown that even the relative abundances of species in fossil assemblages commonly reflect their living abundances, which argues for reasonable fidelity between living and fossil associations.

One way to begin to assess fossil completeness is to examine the degree to which well-known living faunas contribute to the fossil record. With relatively re- cent fossil associations it can be assumed that the fossils have been assembled from faunas much like those of today, and actual fossils may be contrasted with the potential fossils in living communities. Such a comparison was made between the living bivalves and gastropods of an entire biotic shelf province, the Californian Province, and the species recorded as fossils in well-studied assemblages deposited in the Californian region within the last million years. About 77% of the living species had been found as fossils by 1988 (Valentine 1989b). The "missing" species are chiefly quite rare or poorly known or found in habitats not well-represented by sediments, and many of them are small; no common or ecologically important species is missing from the record, and additional fossil species are still being dis- covered. There is no reason to believe that species of other durably skeletonized taxa are not equally well represented. This example suggests that quite a large part of the durably skeletonized fraction of the shelf fauna, from which the marine fossil record is largely derived, is captured by the fossil record—perhaps 80% or more. Yet estimates of the percentage of durably skeletonized Phanerozoic species that are known to science range up to 19% (see Signor 1985). It is a reasonable guess that there are about the same number of species that have been preserved but that have not yet been discovered or described, so perhaps 38% or so of the durably skele- tonized species may eventually become known as fossils. Soft-bodied fossil species, though extremely important qualitatively, do not add so much as a percentage point to this total. Thus over 60% of Phanerozoic invertebrate species that should have been captured by the fossil record may be missing.

Where are these missing species? Presumably most of them were once fossilized, but have been destroyed. It is plausible that the great majority have been removed by the processes that created the stratigraphic gaps, though many have probably also succumbed to in situ destruction through diagenesis and metamorphism. Nevertheless, when rich fossil localities or locality clusters are found, it is likely that quite a large sample of the living, skeletonized biota of the time and region is preserved. Fossil-rich localities are not uncommon in the marine stratigraphic record across the Phanerozoic, and the general outlines of marine life are probably captured by the fossils. However, Neoproterozoic animals, with minor exceptions, lack durable skeletons, and thus their fossil occurrences fall under the category of lagerstätten.

In many Neoproterozoic rocks, the best fossils are the trails and burrows made by animals that have left no other remains. These are termed trace fossils, as opposed to body fossils (which include skeletal remains or just impressions of once-living organisms). Trace fossils, like body fossils, vary in their potential for preservation. The traces that are least likely to be preserved are entirely surficial. The surface sediment of a seafloor is easily disturbed by currents or by the activities of organisms, and markings, such as worm trails or the scratches left by limbs, are likely to be erased. Trails comprising deep grooves, or semi-infaunal burrows formed at or just beneath the surface are more easily preserved than shallow traces. Deeper sediment layers are even more likely to be preserved, so infaunal burrows are more likely to be found than are trails. Nevertheless, even rather delicate surface markings are preserved in some cases, just as are the impressions of bodies. Late Proterozoic and earliest Cambrian sediments were not much disturbed by burrowing animals (see below), and thus were firmer at the surface than later sediments, other things being equal, providing a relatively good record of surface trails (Droser et al. 2002). Once the sediment is lithified, trace fossils (and body fossils that are impressions) become structures within the rock and are far harder to destroy than are mineralized skeletons.

Many Local Faunas Are Required in order to Estimate Global Diversity at Times of High Environmental Heterogeneity

The completeness with which the fossil record has sampled the ancient global marine biota at any time must depend greatly upon the heterogeneity of the ancient biosphere. Recalling the hierarchy of ecological units (chap. 1), if the marine shelves of the entire world were ever under the same climatic regime, with no isolated regions and with a homogeneity of substrate and a narrow restriction in depth, then there would be little partitioning of the shelf biota into different communities, and the same associations would be found almost everywhere on the globe that shelves occurred. A few good fossil associations, accumulated almost anywhere, would produce a nearly complete fossil record of the durably skeletonized biota of that biotic realm. If, however, the shelf environment was partitioned into a number of distinctive habitats and into a number of climatic regions, and some shelves were separated beyond the normal dispersal abilities of species, then there would be many biotic provinces, with many communities within each province. This is in fact the situation in the marine biosphere at present. To represent the durable biota adequately, at least a few fossil assemblages of each community in each province are required. When species diversity is high within communities, experience suggests that disproportionately more of the species are rare, so more or larger samples will be required than in a community of low diversity, to capture a given fraction of the fauna. In general, times of high biotic heterogeneity and diversity will be less well sampled by the fossil record.

Cambrian fossil communities have fewer species on average than later ones (Bambach 1977). Cambrian shelf provinces contain fewer distinctive assemblages than do later shelf provinces, at least partially because of increased post-Cambrian depth zonation (Sepkoski 1988). Cenozoic shelves became increasingly provincialized as continents drifted apart and as the poles cooled (Valentine 1968). Thus it requires far more samples to represent the global fauna for a time in the Cenozoic than for a time in the late Paleozoic, and more samples to represent a time in the late Paleozoic than in the Cambrian. To paraphrase a remark of Signor's (1985), we must know durably skeletonized fractions of some of the earlier Paleozoic faunas, say those near 450 million years old, better than we do the global faunas that are only, say, 10 million years old—a striking comment on the lack of correlation of fossil completeness (or at least fossil discovery) with age.

Curves of generic and familial diversity, usually compiled for stages that average 6 to 8 million years, seem to reflect general trends in standing diversities, for which there is some independent confirmatory evidence. Times when diversity appears to have been particularly low follow extinction events when major clades disappear from the record, and therefore appear to be times of real faunal depauperization, rather than only representing some failure in fossil preservation. Similarly, times of greatest faunal diversity are associated with faunal radiations that can be confirmed taxonomically. Thus there appears to be a real signal from the fossils that is related to former patterns of diversity and evolutionary activity (see chap. 12). Of course, estimates of the completeness of the fossil record at any particular time or locality must be based on evidence from the sites involved; minor fluctuations in diversity trends may often be related to variations in fossil completeness.

Jumping Preservational Gaps Is Possible by Extrapolation between Rich Faunal Horizons

Although the fossil record is incomplete, it is, as mentioned above, incomplete in a very fortunate way. While it contains gaps that are commonly long enough to interrupt the temporal continuity of data on species lineages and species associations, the record between the gaps is often rather good. Richly fossiliferous horizons are sometimes widely developed regionally or even globally, representing the occurrence of conditions that favored preservation of marine fossils (e.g., high sea levels). For many purposes it is obviously better, if we have only, say, 33% of the rock record preserved, to have a good record every third time unit rather than to average 33% of the record randomly smeared over all time units. Although our actual record is by no means perfect, it contains a large and reasonably consistent fraction of the biota—durably skeletonized forms—and within that fraction even the rank order of species abundance is commonly approximated (Kidwell 2001). There is, to be sure, a sparse and relatively spotty record of the remaining soft-bodied forms. If we must make do with a record of only a fraction of the ancient biotas, however, having that fraction form a fairly consistent sample from one time horizon to the next is the best that we could ask for in studying macroevolutionary dynamics.

On the other hand the presence of many regional and even global gaps virtually ensures that very few lineages can be traced through long periods of time via closely spaced samples. As many species "fall" into the gaps, just which species found in an earlier fauna are direct ancestors of the species that are found later is commonly impossible to learn. It may be possible to decide from which species group—a genus perhaps, or a family—the descendants have originated, but all too often there are missing intermediates. The gaps interrupt the traces of phylogenetic branches irrespective of whether the lineages are abundant or rare.

The Known Geologic Ranges of Taxa Are Sensitive to Their Fossil Abundances

To extract the most information possible from the fossil record, we would like to know what is missing, so that we may correct for it. "Missing" fossils can be put into two categories: those missing in gaps for which there is no record at all, and those missing from preserved sediments in which they could have occurred but do not. The entire geologic range of a species is not expected to be represented in the fossil record, for the chances of discovering the founding members of a species, or the last individuals surviving before its extinction, are slim indeed. Essentially all of the actual geologic ranges of taxa must be longer than their observed fossil ranges. How well the geologic range recorded for a species is likely to represent its actual range can be estimated from the record itself. Shaw (1964) and Paul (1982) have shown that the number of stratigraphic occurrences of a species can provide an index to the likely extent of its true duration beyond its geologic record. That is, if a species occurs at every sampled horizon within a given range in a long fossiliferous sequence of sediments from comparable depositional environments that continues below and above the first and last occurrences of the species, then it is likely that the observed range end points are relatively close to the historical geologic range of the species. By contrast, if a species with a similar observed range is found at only a few widely spaced horizons, it may have lived well before and after the observed range—it fossilizes too sparingly to be sure that its absence from the record is informative.

The probable error associated with an observed range may thus be estimated from the gaps in occurrences within that range. A simple correction would be to add to each end of an observed fossil range an interval as long as the average gap within the range. This general approach has been refined by several workers (see Paul 1992; Wagner 1995a).

Strauss and Sadler (1989a) discuss methods of calculating the extent to which the observed stratigraphic range of a species with a given number of occurrences is likely to underestimate its actual range, for a given level of confidence. The methods apply to collections within a section, and assume that the depositional regimes were stochastically constant throughout the actual range of the species and did not change so as to truncate the record; that the fossil occurrences were independent and random stratigraphically; and that collecting intensity was uniform. Marshall (1990) examined the application of this approach to higher taxa and to composite (multiple) sections, which require care to ensure that the assumptions of the method are not significantly violated even though collections are made at different sites.

Other models for evaluating stratigraphic ranges have been developed by Foote and Raup (1996). Those approaches depend upon how relationships among the true durations, preservation probabilities, and observed ranges are distributed among taxa. Model faunas with log-linear range-frequency distributions were experimentally degraded under a variety of preservation probabilities, randomized in time and among the model taxa. The preservation probability was closely approximated as a function of ratios between taxa of a given duration and those of the immediately longer and shorter duration intervals. With these data the distribution of fossil completeness among the parent fauna can be estimated for the taxa involved, and the diversity of that fauna could therefore be reconstructed. Solow and Smith (1997) and Foote (1997) have applied various methods of estimation to databases of range distributions, and have shown that these estimates and reconstructions hold reasonably well for the sorts of data that are available to paleontologists. Interestingly, estimates of the fossil completeness of species records of trilobite, bivalve, and mammal faunas studied by Foote and Raup range from 60% to 90%, a range encompassing the completeness recorded for the Pleistocene mollusks of the Californian Province, discussed above.

Foote and Sepkoski (1999) devised a method of estimating the completeness of the record of various marine taxa globally over Phanerozoic time. They reasoned that the taxa with higher preservation probabilities should be present in more successive stratigraphic intervals, while those with low probabilities would be confined to single intervals. Accordingly they calculated a per-interval preservation probability, $f_2^2/(f_1 f_3)$, where f_1, f_2, and f_3 are the frequencies of genera with stratigraphic ranges of one, two, and three intervals. The calculations were repeated for two time scales. The resulting probabilities of generic preservation for a number of higher taxa are presented in fig. 5.4 (vertical scale). In order to test the plausibility of

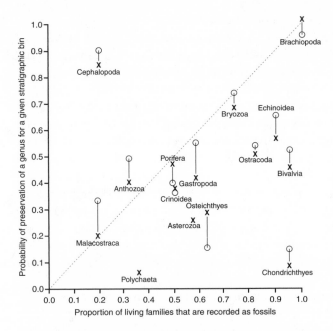

FIG. 5.4 Probabilities of preservation for some important marine clades during the Phanerozoic. The good correlation of these two independent estimates of preservation probabilities suggests they are generally correct; exceptions (Cephalopoda, Chondrichthyes) are discussed in the text. Xs are probabilities based on a time scale of 107 bins, Os on a time scale of 103 bins. From Foote and Sepkoski 1999.

the results, they also determined the percentage of the living families of those taxa that are present as fossils, as being another rough but completely independent assessment of preservability (horizontal scale). These two estimates correlate rather well, with two exceptions. Cephalopods have high generic preservability, but few of their living families are represented by fossils. This situation is easily explicable, for ancient cephalopods included heavily shelled ammonoid and nautiloid faunas, while most living groups are poorly skeletonized. The chondrichthyan fishes, by contrast, tend to be well represented by the fossil remains of living families in multiple intervals—mostly teeth—but their Paleozoic and Mesozoic fossil representatives are chiefly whole-body specimens preserved in lagerstätten, and therefore tend to be restricted to single occurrences. Otherwise, the fairly good correlation of these different preservability estimates suggests they are generally correct. Generic preservability averaged around 50%, with brachiopods (durably skeletonized) much better, and polychaetes (chiefly soft-bodied) much worse.

Soft-bodied fossils are preserved only very rarely, in such settings as peat bogs, amber-bearing deposits, lithographic limestones, and anaerobic basins into which sediments are introduced in episodic bursts, to create fossil lagerstätten. Such settings are rare enough that they do not provide anything like the relatively consistent record required to evaluate fossils by numerical methods. Not enough is known about the taphonomy of soft-bodied fossils to assert that the absence of a soft-bodied taxon from any given locality is very strong evidence that the deposit is beyond the geologic range of the taxon.

There Are Ways of Coping with Incomplete Records

Paleobiologists wish to recover from the fossil record those signals that result from biological processes and effects. While the aim is to accomplish this in the greatest possible detail, care must be taken to avoid interpreting preservational patterns or quirks as paleobiological signals. There are a number of cures for taphonomic (preservational) noise. The best cures involve detailed, case-by-case taphonomic studies, but there are many more cases than there are paleobiologists, and it will be some time before such labor-intensive remedies are available across spans of time and space in which macroevolutionary processes produce important evidence. Immediately accessible approaches involve striking a compromise between the degree of detail that we would like and the chances of erroneous overinterpretations.

Taxonomic Completeness Increases at Higher Levels of the Taxonomic Hierarchy

Most compromises aimed at overcoming incompleteness involve coarsening the databases. For fossil clades this means working, not at the level of species, but at the level of genera or of families. Taxa at those levels obviously contained more individuals, were more widespread, lasted longer than species (except for monospecific taxa), and have better chances of jumping stratigraphic gaps and generally of being found. The use of the generic level probably improves the invertebrate marine Phanerozoic record of durably skeletonized taxa from about 19% completeness to about 50%, and use of the family level improves it further to perhaps about 75% (see Raup 1975). For large-scale questions about species we shall have to extrapolate from such higher taxonomic levels. As it happens, there are interesting questions to be asked at those levels for their own sakes as well. At increasingly higher taxonomic levels, however, the volatility of changes in actual species diversity is progressively damped.

Taxonomic Completeness Rises as Larger Bins Are Used to Increase Time-Averaging

Most fossil localities are not closely dated as yet, and some that may be contemporaneous represent different communities or provinces with few taxa in common; correlation among them, chiefly accomplished by biostratigraphic methods using the fossils themselves, cannot be precise. It is therefore necessary to use relatively coarse time units. Geologic time units have been based on fossils, and as horizons appropriate for correlation do not occur at any regular temporal scale, geologic time units vary in length. In general, the longer the interval used to bin stratigraphic range data, the longer the estimated durations of the taxa will appear to be, as range end points of taxa will be extrapolated to increasingly longer bin intervals (see Foote and Raup 1996). The shorter the bin interval that can be achieved, the more accurate the range estimates will be. A finer time scale is possible for restricted regions than for the entire globe, as regional events reflected within large endemic faunas commonly permit relatively short subdivisions of geologic time that cannot be recognized globally. For global faunas across the Phanerozoic Eon, stages are most commonly used as time units. Radiometric age estimates indicate that an average Paleozoic stage is nearly 8 million years long, while an average Mesozoic or Cenozoic stage is nearly 6 million years. In order to calculate rates or standing diversities it is necessary to normalize the stage-level data (or data for whatever other geologic intervals are used), but this does not improve the error associated with binning.

Paleoecological and Biogeographic Completeness Increase at Higher Levels of the Ecological Hierarchy

For the ecological hierarchy we need not attempt to interpret data at the levels of the individual or of the population for macroevolutionary purposes. Some data analyzed at the community level yield consistent and plausible patterns (e.g., Sepkoski and Sheehan 1983; Jablonski and Bottjer 1990; and papers in Zhuravlev and Riding 2001), and some compilations of diversity data require analysis at the provincial or global levels. Most Phanerozoic diversity compilations have been made at the global level, although they can be analyzed in terms of contributions from communities (Sepkoski 1988) and provinces (Valentine et al. 1978).

Data from Coarser Units May Be Tested by Local Fine-Scale Studies

Although the strategy of using more inclusive units does improve accuracy at the level of the units employed, the resulting figures for these generalized units are accordingly distanced farther from the direct effects of ecological and evolutionary processes than is desirable. Once the patterns displayed by the coarse units have been established, it is certainly appropriate to attempt to infer the processes that must have occurred within finer units to provide the coarser effects. A useful technique in the case of hierarchical units is to compile data for several higher levels where accuracy is presumably high, and to extrapolate the trends that are found from the higher to the lower of those levels and then to the even lower levels for which data are rather spotty but where interesting processes occur. The most obvious assumption, that the observed trends hold when proceeding to lower levels, must be spot-tested against the results of finer-scaled, more local studies, for ratios of taxonomic richness vary through time and among regions.

The Neoproterozoic-Cambrian Fossil Record Provides the Only Direct Evidence of Early Metazoan Bodyplans

The Proterozoic Eon is usually defined as beginning about 2,500 Ma, and has been somewhat arbitrarily subdivided into Paleoproterozoic (2,500–1,600 Ma), Mesoproterozoic (1,600–1,000 Ma) and Neoproterozoic (1,000–543 Ma) times (see Schopf and Klein 1992). Only the Neoproterozoic has yielded metazoan fossils, in the "late Neoproterozoic," which is defined as beginning at 650 Ma. The Phanerozoic Eon and the Cambrian Period began at 543 Ma. Undoubted fossils of metazoans first appear in rocks whose ages are estimated at near 600 million years old, although this is obviously only a minimum age for the origin of the metazoan kingdom, which must have evolved at some undetermined earlier time. It is plausibly inferred that all crown phyla had appeared by about 520 Ma. The nature and sequence of fossils from rocks of this time interval yield basic data on the early history of Metazoa.

Satisfactory Definition and Dating of Late Neoproterozoic and Cambrian Rocks Have Been Accomplished Only Recently

The boundary at the base of the Phanerozoic Eon, a time of great importance in the origin of phyla, has always been a difficult horizon to study, to correlate, or even to define. The type section of the Cambrian, that is, the section of rock to which the name was first applied, is in Wales (for a brief history of the establishment of the Cambrian System see Rudwick 1976). The boundary there and elsewhere was long characterized by a lack of faunas known below the Cambrian, and simple lack of fossils is not a very useful criterion of age. Partly for this reason and its consequences in the history of late Neoproterozoic–Cambrian studies, the stratigraphic nomenclature of this general interval has been in a shambles. Fortunately, discovery and description of Neoproterozoic and Early Cambrian fossils in numbers of marine sequences around the world (fig. 5.5) has much improved our understanding of the stratigraphy of those ancient faunas. Light has been thrown on the sequence of events leading up to the Cambrian, though much still remains in shadow.

Criteria for Defining the Neoproterozoic-Cambrian Boundary Have Varied over the Years. As the base of the Cambrian type section in Wales lies unconformably upon highly metamorphosed "basement" rocks, an early strategy in correlating the boundary was to locate similar unconformities elsewhere below the lowest strata with Cambrian fossils. Indeed, the Cambrian was sometimes hypothesized to be preceded by an essentially universal unconformity, the "Lipalian interval." It was subsequently discovered that many thousands of meters of sediments were sometimes interposed between fossiliferous Early Cambrian strata and basement rocks, and that correlation using a hypothetical Lipalian unconformity was unsatisfactory. An

FIG. 5.5 Some areas where fossiliferous sediments of the late Neoproterozoic–Early Cambrian interval crop out today on continents (the paleogeographies at the times of deposition were quite different; see fig. 5.8). The continental outlines extend to the edge of the continental shelf, not the shore. From Valentine, Awramik, Signor, and Sadler 1991.

alternative suggestion placed the boundary at the first unconformity that happened to occur below the earliest appearance of such skeletonized Cambrian forms as trilobites or archaeocyathids. Other workers wished to divorce altogether the definition of the boundary from any unconformity; it seemed foolish to establish a boundary precisely where there was a gap in the record. This view has prevailed, and there has been a long search to locate the section with the *best* record, preferably one that was as fossiliferous and complete as possible, to serve as a stratotype (a reference section). The search has been under the aegis of international bodies, chiefly a working group of the International Commission on Stratigraphy of the International Union of Geological Sciences (see Cowie and Brasier 1989), and has produced important biostratigraphic results.

There have been a number of important contenders on several continents for the type locality. Pride of place served to stimulate intensive stratigraphic and paleontological research in candidate regions, and there has been extensive international cooperation among boundary workers as well. As this work proceeded, descriptions of some of the better sections, for example from Siberia in Russia, from Yunnan Province in China, and from Newfoundland in Canada, became available. Local stratigraphic names from these regions, particularly stages from the Siberian Platform, which had long been under study, came into wide though informal usage, filling the need for a stratigraphic terminology with which to express the sequence of late Neoproterozoic and Early Cambrian events (fig. 5.6). The Siberian work indicated that there were horizons with assemblages of metazoan skeletal fossils, mostly minute (the "small shelly fossils"), that were older than the earliest appearance of trilobite body fossils. These beds were termed the Tommotian

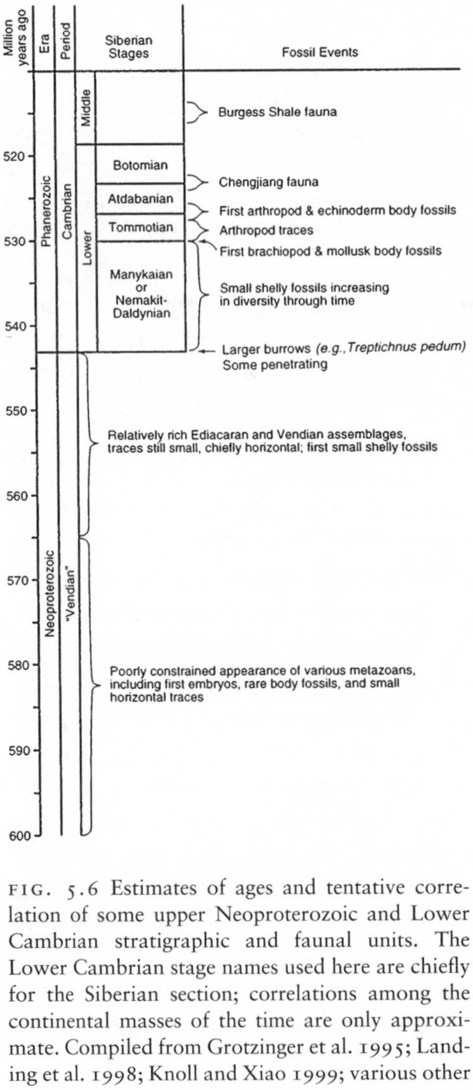

FIG. 5.6 Estimates of ages and tentative correlation of some upper Neoproterozoic and Lower Cambrian stratigraphic and faunal units. The Lower Cambrian stage names used here are chiefly for the Siberian section; correlations among the continental masses of the time are only approximate. Compiled from Grotzinger et al. 1995; Landing et al. 1998; Knoll and Xiao 1999; various other sources.

FIG. 5.7 *Treptichnus pedum*, the trace fossil chosen to mark the beginning of the Cambrian period. It is formed of series of straight to curving burrow segments that intersect to form projections along a sinuous path, occasionally looping. (*A*) A specimen from the Lower Cambrian of Sweden; scale bar 10 mm. (*B*) Sketch to illustrate the trace geometry inferred for a looping specimen. A from Jensen 1997; B from Crimes 1989.

Stage, and it was suggested that the base of the Tommotian be taken as the base of the Cambrian (see Rozanov et al. 1969; Rozanov 1984). However, the upshot of the attempt to attain international agreement on the Neoproterozoic-Cambrian boundary is the recommendation that it be taken at the first appearance of a characteristic trace fossil, *Treptichnus pedum* (fig. 5.7; formerly called *Phycodes pedum*; see Jensen 1997), in a section in southeastern Newfoundland that has become the stratotype (see Narbonne et al. 1987; Landing and Westrop 1998). This horizon is significantly older than the base of the Tommotian (see below); the Cambrian stage thus interposed beneath the Tommotian but above the Neoproterozoic boundary is usually termed either the Manykaian or the Nemakit-Daldynian, from stratigraphic terms used in different regions of Russia. Correlations of Neoproterozoic rocks have been based on such fossils as are available, including acritarchs (unicellular algal phytoplankton) and metazoans, on radiometric dates, on the distribution of glacial deposits (tillites), and most recently on such geochemical markers as carbon isotope ratios.

The Neoproterozoic-Cambrian boundary has clearly had a checkered past; only a bare outline is presented here. When evaluating discussions in the literature that relate paleobiological events to this boundary, it is necessary to determine where the boundary was located at the time of the discussion. The rather unusual metazoan fossils that occur near the boundary are under heavy study, and the metazoan data are being supplemented by a maturing stratigraphy. The sequence of events across the boundary is thus far better known today than even a few years ago.

The Age of the Late Neoproterozoic–Early Cambrian Sequence Has Been Established Chiefly by Precision Dating of Zircon Crystals. Beginning in the mid-1960s, the base of the Cambrian was generally taken to be about 570 million years old, although radiometric dates bearing on the age ranged widely. Many of the radiometric age estimates then available were based on potassium-argon or uranium-lead ratios from whole-rock samples. Sediments frequently include particles that are reworked from preexisting deposits and that can be radiometrically much older than the bed in which they are finally deposited. Dating whole-rock sediments averages older and younger particles and tends to overestimate the age of the beds themselves.

On the other hand, the isotopic ratios in many minerals can be reset by thermal or chemical activities to yield radiometric ages that are younger than the ages of the rocks in which they occur.

High-precision dating of single crystals of zircon, a mineral common in some volcanic rocks, has been made possible by the development of techniques involving analyses by ion microprobe and mass spectrometry using uranium-lead isotope ratios. The application of these techniques to Neoproterozoic and Cambrian stratigraphic ages (see Compston et al. 1992; Cooper et al. 1992; Bowring et al. 1993; Grotzinger et al. 1995; Martin et al. 2000) has shown that the Lower Cambrian boundary, even after the downward extension caused by its redefinition, is surprisingly young, near 543 million years old. Probable dates for boundaries of the Siberian stages are shown in fig. 5.6; although the dates are probably quite accurate for the rocks that were sampled, they have been extrapolated to the boundaries, and that correlation is less certain. These data do indicate that lowering the boundary from the base of the Tommotian to the horizon at which *Treptichnus pedum* first appeared has the effect of making the Cambrian about half again as long. With this definitional increase, the duration of the revised Cambrian period turns out to be a bit over 50 million years. The base of the Tommotian Stage probably dates from well before 520 Ma in its type region in Siberia, but it rests unconformably on earlier rocks, and as much as 10 million years may be missing from the stratigraphic record there (Landing et al. 1998). The most reliable dates of rocks harboring the earliest known metazoans, in the Doushantuo Formation of China, suggest an age near 600 Ma (Barfod et al. 2002).

Late Neoproterozoic and Early Cambrian Geographies Were Very Different from Today's

Reconstruction of the paleogeography of the Neoproterozoic is very much a work in progress. It has been thought that in the early part of that time interval most continental masses were in close proximity and perhaps formed a supercontinent (e.g., see Valentine and Moores 1972; Moores 1991; Dalziel 1997 and references therein) usually termed Rodinia (McMenamin and McMenamin 1990). Plate tectonic processes were continually rearranging continents, and there is evidence that a supercontinent was partially broken up and that a second assembly of most continental masses to form another supercontinent, termed Pannotia (Powell 1995), occurred later in Neoproterozoic time. Some authorities also use the name Rodinia for a late Neoproterozoic supercontinent. At any rate, when the Neoproterozoic supercontinent broke up, the larger continents during the latest Neoproterozoic and earliest Cambrian were Laurentia (most of North America and Greenland), Siberia, Baltica (most of Europe), Gondwana (a large continent composed essentially of Africa, South America, India, Australia, and Antarctica), and north China (which is independent in some reconstructions) and south China (which may have remained part of Gondwana during much of those times). Fig. 5.8 represents a

A   580 Ma (late Neoproterozoic)

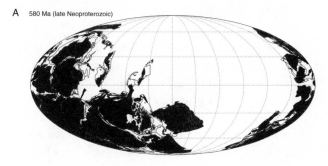

B   540 Ma (earliest Cambrian)

C   500 Ma (latest Cambrian)

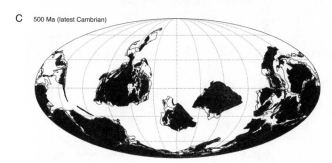

FIG. 5.8 Late Neoproterozoic and Early Cambrian paleogeographic reconstructions for major continental masses. (*A*) Late Neoproterozoic (Vendian). (*B*) Earliest Cambrian (Manykaian). (*C*) Latest Cambrian. After Smith 2001.

recent attempt to reconstruct continental positions from Pannotia in the late Neoproterozoic to more scattered continental masses of the latest Cambrian, based on paleomagnetic data supplemented by tectonic information (Smith 2001). These reconstructions are quite similar to some others that also include data from fossils and from climate-sensitive geologic indicators (e.g., Seslavinsky and Maidanskaya 2001), although there are other analyses of biodistributional patterns that do not fit very well with the reconstructed late Neoproterozoic continental positions (e.g., Waggoner 1999). A major point is that many of the important late Neoproterozoic and Early Cambrian fossil localities, as in Australia, Siberia, and Greenland and along the North American Cordillera, were in low latitudes. The main exceptions are Avalonia (which included Newfoundland) and Baltica, which were chiefly in midlatitudes. At least during the Cambrian, however, the climate in midlatitudes may have been quite mild (see Eerola 2001 and references therein).

Knowledge of Late Neoproterozoic and Cambrian Faunas Has Greatly Increased in Recent Decades

In Darwin's day the earliest metazoan fossils known were found in rocks that are now considered to be well up in the Atdabanian Stage. These faunas include trilobites, and Darwin was puzzled that such complex animals appeared so abruptly, without known ancestors. The fossil record has now yielded a sequence of faunas stretching back over 40 million years before the advent of the trilobites that Darwin knew about, permitting us to infer some of the broad outlines of early metazoan history. Despite this improved record, the early ancestors of Darwin's trilobites remain unknown.

Late Neoproterozoic Fossils Include Enigmatic Soft-Bodied Forms and Traces. The first invertebrate fossil association for which a Neoproterozoic ("Precambrian") age was argued persuasively and successfully is from the Ediacara beds of South Australia, discovered by R. C. Sprigg in 1946 and brought to light through collections in the late 1940s and 1950s. A general account of this fauna, dated but still informative, is by Glaessner (1984), who described and evaluated much of the fauna (see also Glaessner and Wade 1966; Gehling 1991). Other faunas sharing a Neoproterozoic aspect with the Ediacaran fossils have been discovered at many sites around the world. Neoproterozoic faunas from Russia are termed Vendian faunas, from the Vendian system, a stratigraphic unit below the Cambrian on the Russian and Siberian platforms (Sokolov 1952; Sokolov and Fedonkin 1984). Late Neoproterozoic fossils in general are sometimes referred to as either the Ediacaran or the Vendian fauna. The oldest well-dated assemblages assigned to this fauna are from Newfoundland and are about 565 million years old (Benus 1988). Neoproterozoic fossils are dated to about 600 Ma, but most appear to be between 543 and 550 million years old, late in the Neoproterozoic. The most diverse faunas have come from the Russian Vendian, reviewed by Fedonkin (1992 and references therein), who described and evaluated much of that fauna. Relatively rich Vendian assemblages from the White Sea are on the order of 555 million years old (Martin et al. 2000). The biological affinities of most of the Neoproterozoic body fossils are uncertain. Neoproterozoic metazoan fossils have been tabulated by Runnegar (1992a, 1992b) and Bengtson (1992b).

The earliest body fossils known at present from the late Neoproterozoic predate the Ediacaran and Vendian assemblages but are spectacular from the point of view of preservation. These are fossil tissues and embryos from the Doushantuo Formation, Guizhou, China, which is likely to have been deposited near 600 million years ago (Barfod et al. 2002). The fossils are phosphatized with such fidelity that individual cells can easily be discriminated (Li et al. 1998; Xiao et al. 1998; Xiao and Knoll 2000; Xiao et al. 2000; Xiao et al. 2002). This assemblage has yielded eggs and blastula-stage embryos of uncertain affinities (fig. 5.9), as well as cellular tissues of possible sponges, small tubes that just might be cnidarians, and varied algal

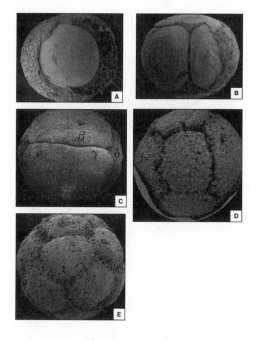

FIG. 5.9 Fossil egg and embryos from the Doushantuo Formation, South China, dated at about 600 Ma. (*A*) *Megasphaera*, an egg. (*B–E*) *Parapandorina*. (*B*) Two-cell embryo. (*C*) Four-cell embryo. (*D*) Eight-cell embryo. (*E*) Later cleavage stage. From Xiao and Knoll 2000.

remains. Some other minute structures from the Doushantuo Formation have been tentatively identified as bilaterian embryos (Chen et al. 2000), but further study has indicated that they are remains of egg cases or algal cysts encrusted by phosphatic material (Xiao et al. 2000). No definitively identified bilaterians or bilaterian traces are known from these rocks (see Xiao et al. 2002).

Fig. 5.10 sketches some of the characteristic body fossils from the late Neoproterozoic; most of them are impressions that somewhat resemble cnidarians. Their interpretation has been characterized by extremes. One extreme, exemplified by Glaessner (1984) and in part by Gehling (1991), is to assign these forms to the living phyla that seem most similar, a practice known as shoehorning. The other extreme, exemplified by Seilacher (1989, 1992), is to consider that these forms not only are not members of living phyla but may not be metazoans at all. Seilacher suggested that they represent the remains of a multicellular clade that originated independently of the metazoans and had its own peculiar architecture and lifestyle, and that became extinct during the Vendian or perhaps the Early Cambrian. Seilacher termed this putative clade the Vendozoa, later modified to Vendobionta. More recently, Buss and Seilacher (1994) suggested that some of these forms may represent diploblastic metazoans after all, but do not belong to living phyla. Many other interpretations have been proposed; some are reviewed in chapter 6. In fact the assemblage is probably taxonomically heterogeneous.

A few of the Neoproterozoic body impressions may be bilaterians. Some of these forms can be interpreted as possessing a jointed construction with cephalic lobes or shields anteriorly (e.g., *Spriggina* in fig. 5.11). One small form, *Parvancorina*, resembles probable stem arthropod fossils from the Early Cambrian (Zhang et al. 2003). These fossils have also been reconstructed as frondlike forms (e.g., *Spriggina* in fig. 5.10); their affinities remain in doubt. Another possible bilaterian is a sluglike form called *Kimberella* (Fedonkin and Waggoner 1997). The earliest mineralized skeletons appear in late Neoproterozoic rocks. These fossils include minute calcareous cones called *Cloudina* (fig. 5.12), several kinds of minute

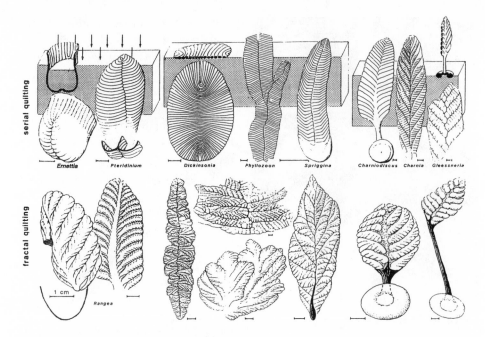

FIG. 5.10 A number of late Neoproterozoic forms from Russia, Australia, and southern Africa. The scale bars are 1 cm; some of the larger forms range in size to over 1 m. These sketches are drawn in support of an interpretation that all of these forms have a quilted construction, which is probably not the case (see chap. 6), but the seriation they display is quite real. Forms at lower right have not been formally described. From Seilacher 1992.

mineralized (commonly phosphatized) goblets and tubes (e.g., Xiao et al. 2000), and even a large (to 1 m) modular, perhaps spongelike or colonial, skeletal form (Wood et al. 2002).

In addition to body fossils, Neoproterozoic rocks have yielded numbers of trace fossils, markings such as trails and burrows made by animal activities (fig. 5.13; Glaessner 1969; Fedonkin 1985b; Crimes 1989). Undoubted, well-dated traces are continuously present from about 565 Ma, but some traces may be earlier. One difficulty in Neoproterozoic trace-fossil interpretation is that fossil algal debris is difficult to distinguish from metazoan traces (see Seilacher 1999). Some algae possess beadlike structures that have probably been interpreted as fecal pellets. Inorganic structures can also be confused with traces; for example, roll marks of tectonically displaced particles can be confused with trace fossils (Seilacher et al. 2000). Possible fossils older than 570 million years include structures described by Brasier and McIlroy (1998) in rocks of about 600 million years from Scotland, and forms thought to date from 1.1 billion years in India (Seilacher et al. 1998). The metazoan origin of those structures is, however, quite uncertain. There have been other structures from older rocks (some even billions of years old) that have been claimed to be biogenic, but they have been shown either not to be traces or not to be so old.

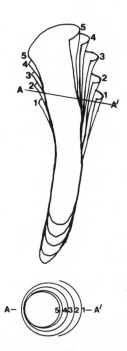

FIG. 5.11 *Spriggina*, latex cast of type specimen from Ediacaran beds of South Australia; scale bar 10 mm. Compare with sketch in fig. 5.10. From Glaessner 1984.

FIG. 5.12 *Cloudina*, evidently the earliest metazoan with an exoskeleton, dating from the late Neoproterozoic. Schematic (*upper*) longitudinal and (*lower*) transverse section of the fragile conical skeleton, based on specimens from Namibia, Africa; note successive deposition of tubular cones 1–5. Most specimens are between 1 and 3 mm in cross section. From Grant 1990.

The undoubted Neoproterozoic traces are generally minute, usually under 1 mm in width but, more rarely, ranging to 5 mm. Many of these early traces are simple curved structures, but there are complex types, some with transverse rugae, some with levees, and some bilobed in cross section. Most of these traces are horizontal, and even where studied most carefully, Neoproterozoic sediments contain little evidence of bioturbation (Droser et al. 2002), though a few traces do penetrate the sediment. These penetrating burrows are short, small in diameter, and rare. In some settings traces are associated with the remains of microbial mats. If the traces were made under the mats, as seems to be the case in some instances, they may have been somewhat protected from erosion and their chances for preservation significantly enhanced (see Gehling 1999). Although the preservation of minute surface trails would be quite a rare event on today's marine shelves, the relative lack of mixing of shallow sediment layers by animal activities during the Neoproterozoic would have

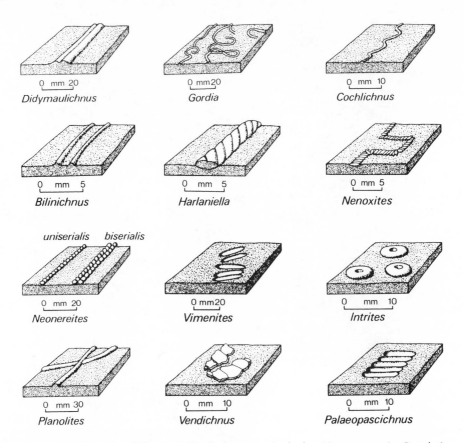

FIG. 5.13 Structures identified as trace fossils that occur in the late Neoproterozoic. Correlation of the late Neoproterozoic–Cambrian boundary is still uncertain in many sections, and it is possible that a couple of these traces first appear in the earliest Cambrian. *Neonereites* may be based on algal remains. Neoproterozoic traces are generally quite minute; note that the scales are in millimeters. After Crimes 1989.

increased the chances of trace preservation. Furthermore, without mixing, surface sediments would have had less water content and be firmer than today's on average, enhancing trace preservation (Droser et al. 2002). The bulk of Neoproterozoic traces are generally conceded to represent the activities of bilaterians, chiefly of worms. The small size of the traces suggests that the trace makers were at a paracoelomate grade of construction (chap. 2). However, even diploblastic organisms can leave similar traces (Collins et al. 2000), and it is possible that they have also contributed to the Neoproterozoic trace assemblages.

Earliest Cambrian Faunal Traces Indicate Increases in Body Size and in Biological Activities. The Manykaian or Nemakit-Daldynian Stage lies between the base of the Cambrian and the base of the Tommotian Stage, representing an interval that is likely between 13 and 20 million years long. Some of the Vendobionta are known from Manykaian rocks (Jensen et al. 1998), and a few may even have persisted into the Middle Cambrian (Conway Morris 1993a). The earliest mineralized fossils, including *Cloudina,* are joined by other minute mineralized forms in the latest Neoproterozoic and the earliest Manykaian; these include *Anabarites,* a triradiate cone, and hooklike forms termed conodontomorphs. This small shelly fauna gradually increases in diversity toward the top of the Manykaian Stage (Brasier et al. 1996). Most of these fossils are interpreted as sclerites—dissociated elements of mineralized skeletons, consisting chiefly of spicules, spines, plates, and cones.

The study of Early Cambrian trace fossils (and early metazoan traces in general) was pioneered by Seilacher (1955). Early Cambrian traces include significantly larger, more diverse, and more complex forms than their Vendian predecessors (fig. 5.14). The trace used as an index for the base of the Cambrian, *Treptichnus pedum* (fig. 5.7), is a series of branched burrows; the curved branches evidently represent feeding probes, sometimes penetrating the overlying sediments. During the Cambrian, vertically penetrating burrows become longer, larger in diameter, and more common, and bioturbation becomes deeper and more common, at progressively younger horizons (Crimes and Droser 1992; Droser and Bottjer 1993; McIlroy and Logan 1999).

Numbers of Crown Phyla Appear during the Cambrian Explosion. Aside from sponges, none of the Neoproterozoic fossils can be assigned to living metazoan phyla with assurance, although cnidarians are most likely to be present. From the base of the Tommotian Stage in its type region to the top of the Atdabanian Stage, numbers of living bilaterian phyla make their debuts. How long these groups had been around before they actually appeared is uncertain; they could have been waiting in the wings for a long time, or they could have evolved only a geologically short time before they are first found. Because the appearance of these forms is relatively abrupt, it has been dubbed the Cambrian explosion. The onset of the explosion corresponds with the evolution of durably mineralized skeletons within a variety of major animal groups. Therefore, some workers have supposed that the explosion is not real, but is an artifact of the preservation of fossils, representing an increase in preservability rather than representing the origins of new groups. There is considerable evidence that the explosion was real, however, though our appreciation of it is certainly affected by the difficulties in interpreting a rather incomplete record of life that has been overprinted by hundreds of millions of years of subsequent earth history (for reviews see Valentine et al. 1999; Knoll and Carroll 1999; Budd and Jensen 2000). For example, some of the groups that appear during the explosion require durable skeletons as part of their early bodyplans, so their origins can be no earlier than the origins of their durable skeletons. Trace fossils also increase in abundance, in variety, and especially in average size. This trend begins a bit before the explosion, and traces become particularly large and abundant during the explosion period, to

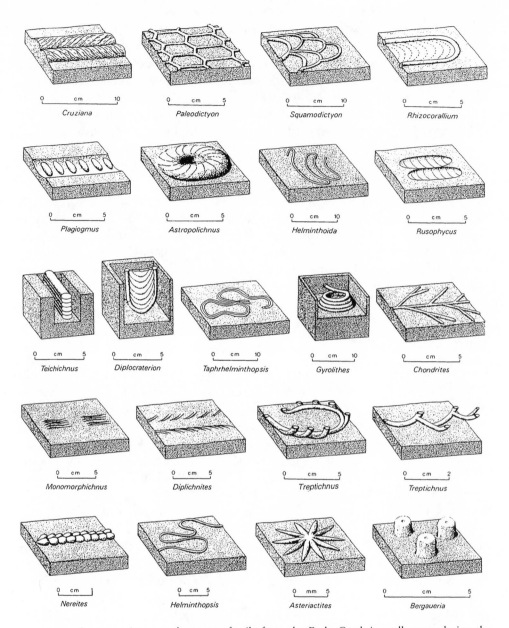

FIG. 5.14 Structures interpreted as trace fossils from the Early Cambrian; all occur during the Tommotian. The size, diversity, and abundance of these traces far outstrip Neoproterozoic traces (see fig. 5.13). Note that the scales are in centimeters. From Crimes 1989.

display a miniexplosion of their own. The appearance of many significantly larger animals implies that biomechanical and physiological problems associated with larger body sizes, and problems associated with ecological interactions within new communities, had been "solved." It appears that the interpretation of the explosion as an artifact of the evolution of durable skeletons has got it backward: the skeletons are artifacts, more or less literally, of the evolutionary explosion.

During the explosion the small shelly fauna became much more diverse (fig. 5.15). Sclerites are sometimes found in life associations (thus composing scleritomes), and provide indications of the bodyplans of the animals that bore them; rarely, sclerites are actually found in place within soft-bodied fossils. Tommotian fossils include mollusklike forms, both univalve and bivalve, and primitive brachiopods. Atdabanian assemblages have yielded echinoderm and arthropod skeletons. Lower Cambrian deposits in China have also yielded phosphatized embryos (Bengtson and Zhao 1997). While trace fossils increase significantly in abundance and diversity during the explosion, bioturbation is rare and shallow during the Tommotian, and although it becomes more marked in the Atdabanian, it is still

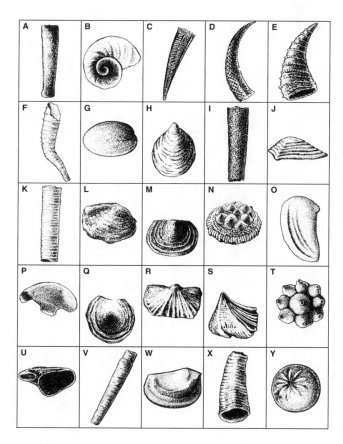

FIG. 5.15 Small shelly fossils of the Tommotian and Atdabanian of the Siberian Platform, from a facies characterized by patch reefs and bioherms. (A) Tiksitheca. (B) Aldanellid (a mollusk). (C) Hyolithomorphan. (D) Rhombocorniculum. (E) Lapworthellid. (F) Hyolithelid. (G) Rostroconchan (a mollusk). (H) Stenothecoidan (possibly a mollusk). (I) Tommotitubulus. (J) Sunnaginia. (K) Coleolid. (L) A bivalve (mollusk). (M) Kutorginid (a brachiopod). (N) Hadimopanella (probably a paracoelomate). (O) Halkierid. (P) Coreospirid. (Q) Aldanolina. (R) Orthid (a brachiopod). (S) Tommotiid. (T) Chancelloriid. (U) Pelagiellid (a mollusk). (V) Orthothecimorphan. (W) Bradoriid (an arthropod). (X) Kelanellid. (Y) Mobergella (possibly a mollusk). These fossils range from less than 1 mm to about 2 mm. After Rozanov and Zhuravlev 1992.

shallow by modern standards (Droser and Bottjer 1988a, 1988b). The increase in body size, intensity, and sophistication of activity indicated by the traces roughly parallels the increase of body-fossil sizes and types.

One of the more important fossil assemblages known, the Chengjiang fauna, occurs in Yunnan, China (Hou et al. 1991; Chen and Zhou 1997; Hou et al. 1999). It is evidently of latest Atdabanian age or slightly younger, from near the close of the explosion interval. This fossil assemblage is entombed in laminated muds and silts. The rocks have been interpreted as representing distal deltaic deposits of midshelf depths, and the preservation explained as a consequence of sudden burials (Chen and Lindström 1991; Chen and Zhou 1997). Another depositional model invokes tidal currents in presumably shallow-shelf depths that created fluctuating salinities near shore, protecting the remains of organisms, perhaps killed by salinity fluctuations, from burrowers or scavengers or decay organisms that could not tolerate the episodic low salinities (Babcock et al. 2001). These models are not mutually exclusive. Both indigenous benthic elements and pelagic forms are present, the latter suggesting more offshore conditions. The fauna consists largely of soft-bodied invertebrates and includes sponges, cnidarians, priapulids, extinct worms that may be paracoelomates, several extinct but rather onychophoran-like lobopods, possible phoronids, the earliest known cephalochordate, and a number of extinct forms that are not confidently placed within existing phyla (fig. 5.16). There is also an array of arthropods and arthropod allies, many of which cannot be assigned to living classes (fig. 5.17). And most spectacularly, there are forms that are plausibly interpreted as primitive craniate fish from essentially contemporaneous rocks at Haikou, near Chengjiang (fig. 5.18; Shu, Luo, et al. 1999).

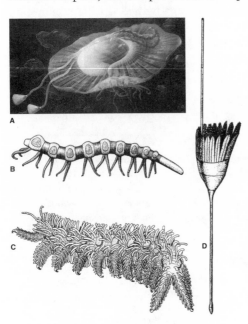

FIG. 5.16 Body fossils from the Chengjiang Formation, probably of latest Atdabanian age, from Yunnan, China. (A) Artist's conception of *Eldonia*, a presumably pelagic form of uncertain affinities, with commensal(?) lobopod. (B) Reconstruction of the lobopod *Microdictyon sinicum*, interpreted as an onychophoran ally. Note that each appendage is overlaid by a honeycomblike, phosphorite plaque. *Microdictyon* plaques were widely known as sclerites in Lower Cambrian "small shelly fossil" assemblages before the animal that bore them came to light. (C) Reconstruction of *Onychodictyon*, probably another onychophoran ally. (D) Reconstruction of *Dinomischus*, another enigmatic organism that cannot be confidently assigned to a crown phylum, first described from the Burgess Shale by Conway Morris. A from Chen and Zhou 1997; B, C from Hou et al. 1999; D from Chen et al. 1989a.

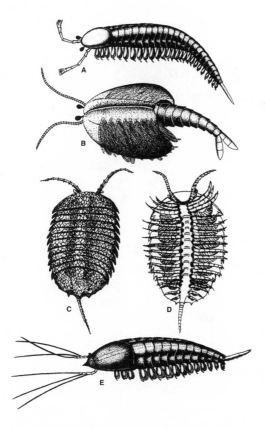

FIG. 5.17 Some arthropods and allied forms from the Chengjiang Formation. (*A*) *Jiangfengia*, not a crown-group arthropod; the large preoral "great appendages" and flaplike trunk appendages with setae are features held in common with a number of other Early Cambrian forms; jointed trunk appendages have not been found. (*B*) *Fuxianhuia*, which does not have great appendages (merely antennae) but has jointed trunk appendages (ventrally) and is considered to be quite basal among arthropods. (*C, D*) *Retifacies*, dorsal and ventral views, assigned by some workers to an extinct arthropod class, Artiopoda. (*E*) *Alalcomenaeus*, which has "great appendages" as has *Jiangfengia*; however, the flaplike trunk appendages shown are outer branches of limbs that have jointed inner branches. From Chen and Zhou 1997.

There are some less spectacular but important Early Cambrian faunas that have yielded soft-bodied specimens. A probable late Tommotian fauna from Guizhou, China, is under study (Zhao et al. 1999). A smaller but interesting soft-bodied fauna, which may be of latest Tommotian or earliest Atdabanian age, is known from the Baltic Shield (Dzik and Lendzion 1988). An important late Atdabanian soft-bodied assemblage roughly correlative with the Chengjiang fauna is known from Greenland (the Sirius Passet fauna; Conway Morris et al. 1987). The diversity of bodyplans indicated by combining all of these Early Cambrian remains is very great. Judging from the phylogenetic tree of life, all living phyla were probably present by the close of the explosion interval.

The Middle Cambrian Contains Spectacular Faunas, but No Crown Phyla Appear for the First Time. A host of unusual soft-bodied forms has been described from the Burgess Shale (fig. 5.19), a Middle Cambrian unit from British Columbia, and from other roughly correlative Middle Cambrian rocks chiefly in western North America (see Whittington 1985a; Conway Morris 1992a, 1998). Most of the benthic fauna of the Burgess Shale evidently lived along a submarine escarpment at shelf depths comparable to the Chengjiang assemblage. However, those forms were transported by slumping and consequent turbidity flows into deeper water and preserved as a death assemblage, most likely beneath an anoxic seafloor (Piper 1972; Fletcher and Collins 1998). Another large Middle Cambrian soft-bodied assemblage that may be slightly earlier than the Burgess Shale, the Kaili fauna, is found in Guizhou, China (Zhao

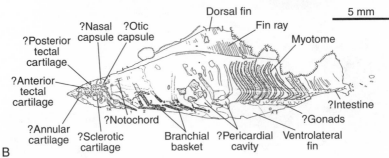

FIG. 5.18 A probable fossil craniate, *Haikouichthys ercaicunensis*, a likely stem vertebrate from the Quiongzhusi Formation, roughly contemporaneous with the Chengjiang fauna. It may be a representative of the lineages that branched during the rise of agnathans from cephalochordates. (*A*) Entire specimen. (*B*) Morphological interpretation. From Shu, Luo, Conway Morris, Zhang, Hu, Chen, Han, Zhu, and Chen 1999.

et al. 1994; Zhao et al. 1999; and other papers in those issues). Both of those Middle Cambrian faunas include many fossils allied to Early Cambrian forms, especially those from the Chengjiang assemblage. Evidently the main elements of the Cambrian fauna had appeared well before the Middle Cambrian and persisted for a significant period of time over a wide area, at least at inshore to midshelf depths. Trace fossils remain at about their latest Lower Cambrian level of richness throughout the Cambrian.

Phyla That First Appear after the Explosion Are Soft-Bodied with One Exception (Bryozoa). The Cambrian explosion had probably run its course by the close of the Early Cambrian, with the results registered whenever fossils, accumulated under appropriate environments and conditions of preservation, happen to come to light. No phyla are known to appear for the first time during the Late Cambrian. Phyla continue to appear later in the Phanerozoic, however, particularly in rare cases when soft-bodied forms are fossilized (fig. 5.20). The only durably skeletonized phylum to appear after the close of the Cambrian is Bryozoa, found early in the Ordovician. As nonmineralized bryozoans are not uncommon today, it seems plausible that bryozoans originated in the Cambrian, but first became mineralized in the Early Ordovician and entered the fossil record then. Linnean classes of

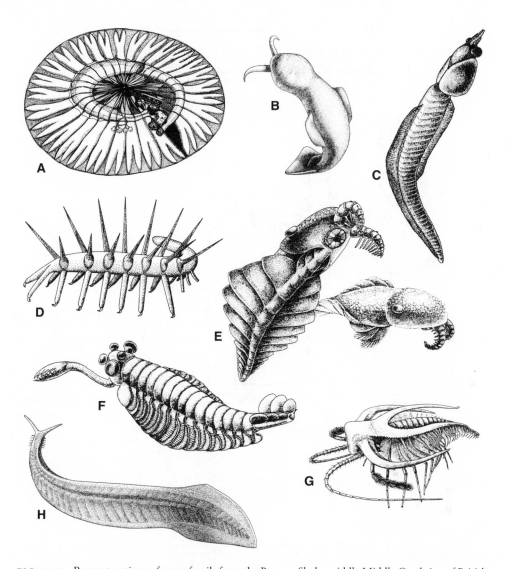

FIG. 5.19 Reconstructions of some fossils from the Burgess Shale, middle Middle Cambrian of British Columbia, Canada. Most of these forms are 2–4 cm long. (*A*) *Eldonia*, probably a planktonic form that reaches sizes over 10 cm, has not been definitively assigned to a crown phylum, but has been variously considered to be related to Echinodermata (as a holothurian), to Cnidaria (as a scyphozoan), and to Lophophorata; it is also found abundantly in the Chengjiang fauna. (*B*) *Amiskwia*, another enigmatic form, presumably planktonic, which may be a paracoelomate. (*C*) *Nectocaris*, which appears to be segmented and was presumably planktonic; its affinity is unknown. (*D*) *Hallucigenia*, the first of the fossil lobopods to be described, by Conway Morris; it also occurs in the Chengjiang fauna. (*E*) *Anomalocaris*, a relatively giant predator that may reach 40 cm in length; the body is segmented like an arthropod. Anomalocarids are known also from the Atdabanian of China, the Baltic Shield, and Greenland; several species have been described. (*F*) *Opabinia*, presumably segmented and with a single large anterior appendage; jointed limbs have not been found on the trunk. (*G*) *Marrella*, abundant in the Burgess Shale, is an arthropod ally that does not fall within crown groups. (*H*) *Pikaia*, an early cephalochordate. Reconstructions B, C, E–H by Marianne Collins; A by Duncan Friend; D by Pam Baldro. From Briggs et al. 1994.

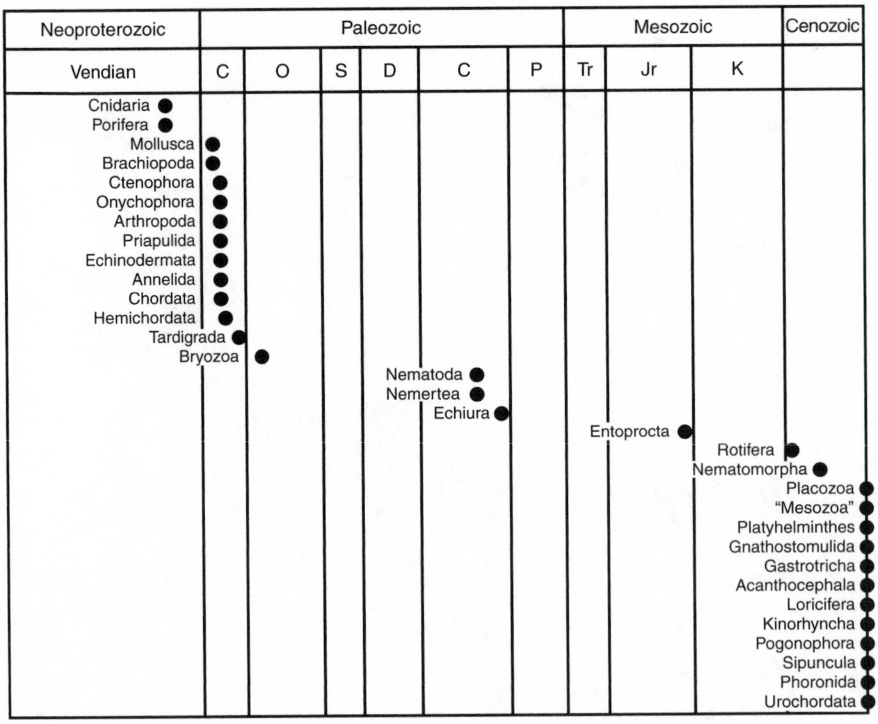

FIG. 5.20 Times of first recorded appearances of body fossils of living phyla and some other high-level taxonomic groups. *C*, Cambrian; *O*, Ordovician; *S*, Silurian; *D*, Devonian; *C*, Carboniferous; *P*, Permian; *Tr*, Triassic; *Jr*, Jurassic; *K*, Cretaceous. Sources are given in chapters 6–11. After Valentine 1995.

mineralized metazoan groups continued to appear throughout the Paleozoic, and orders into the Cenozoic. For a brief review of the subsequent history of phyla see chapter 12.

If All Phyla Were Present by the Close of the Explosion, Their Records Agree Well with Expectations Based on Their Preservabilities

Fig. 5.20 indicates the earliest well-established appearances of body fossils of living metazoan phyla as they are now known. The bimodal distribution of appearances is striking. If a phylum has a reasonable chance of being preserved, it is likely to be found during the Cambrian; thirteen phyla are known from that period, and one just afterward in the Early Ordovician. If a phylum is entirely soft-bodied, and particularly if the body size of its members is quite small, it may well escape fossilization (or at least escape detection as a fossil) and be known only among the living fauna; fourteen phylum-level taxa are unknown as fossils, all soft-bodied. Four phyla, all soft-bodied, make their first appearances at times scattered between these extremes, usually in situations where unusual fossilization processes were involved, such as in Carboniferous lagerstätten that preserve the Bear Gulch and Essex faunas (providing the first records of Nematoda and Nemertea) or in lignites (providing the first records of Nematomorpha). Some phyla are known from only a single fossil occurrence, failing which they would appear only among the living fauna along with the other soft-bodied groups unknown as fossils.

The distribution of first appearances of phyla thus agrees very well with expectations based on the quality of the fossil record, reviewed above. On one hand, the lack of a fossil record for many soft-bodied forms leaves open their times of origin. The lucky discoveries of some soft-bodied fossil forms are not very informative as to the times of origin of the phyla involved. On the other hand, many of the skeletonized phyla that appear in the Cambrian are common in later fossil faunas. The brachiopods, for example, are found in literally thousands of localities in the post-Cambrian Phanerozoic. This distribution implies that it is likely that the brachiopods first appear in the record quite near their time of origin, or more strictly, quite near the origin of mineralized brachiopod skeletons, for the average gap in their Phanerozoic record is quite small. Of course, it is certain that we do not possess the earliest members of any of the phyla. A phylum first appears in the record not when it originates but when sediments representing a locality that it inhabits, and containing organisms from such a locality, first happen to be preserved. However, the relatively large numbers of fossil localities available within the geologically narrow time range that encompasses the Cambrian explosion makes it quite unlikely that gaps in the stratigraphic record are large enough to seriously mislead us as to the general timing of the great radiations of the late Neoproterozoic and Early Cambrian. Radiation, in this context, refers to increasing disparity arising between and within many phyla, which would plot in a morphospace as a series of diverging branches—a radiation of metazoan morphology.

In sum, there is nothing in the fossil record to disprove the hypothesis that all phyla had appeared by the end of the Cambrian or, for that matter, the end of the Early Cambrian. However, there is little fossil evidence to preclude a longer early metazoan history than is recorded, if the organisms were small and soft-bodied.

The Lack of Neoproterozoic Fossil Ancestors of Living Phyla Is Not Inconsistent with the Quality of the Fossil Record

Considering the pattern of incompleteness of the fossil record, and the nature of the late Neoproterozoic–Cambrian record of the first appearance of phyla, it is possible to reach some tentative conclusions about the quality of the record of early metazoan evolution. Although the spotty nature of the fossil record dictates that some inferences cannot be based on direct fossil evidence, the situation is not nearly as bad as it could be, chiefly because it is possible to employ the sorts of remedies reviewed above. As the chief concern here is with the evolution of bodyplans such as characterize phyla and classes, we are working at very high taxonomic levels within the metazoans. The records of all of the species, genera, or families of the times—levels at which the record must be poor—are not required. Furthermore, the perspective is global as a first approximation, although the record does contain some hints at geographic and ecological patterns in the appearance and early diversification of phyla. While dating and correlation of many of the rocks containing fossils that bear on early metazoan history remain uncertain, several critical sections have been well-dated.

Very few fossils can be hypothesized as ancestral to phyla, despite the many strategies used to find ancestral lineages, and other suggestions remain rather speculative (see discussions in part 2). The fact remains that in the general case we have not unequivocally identified the taxa representing ancestral lineages leading to the bodyplans of phyla, or located the regions and environments where those evolutionary changes occurred, or dated them precisely, from fossil evidence. This situation is not unique to the Neoproterozoic-Cambrian transition faunas. Fossil data on lineages ancestral to higher taxa are most elusive, no matter what the age. For example, Ordovician fossils record an explosive radiation of echinoderms that produced numerous higher taxa commonly assigned to classes (chap. 11). Skeletonized echinoderms are present from the Early Cambrian. Yet we cannot trace the branches that led to these separate Ordovician classes, even though they were surely skeletonized, any more than we find lineages leading to the Cambrian phyla.

The paucity of records of ancestors may well be an important bit of evidence as to the mode of evolution of bodyplans (Valentine and Erwin 1987). What sort of evolutionary conditions would be least likely to produce a recognizable fossil record during the origin of a major morphological innovation? The single factor that would best hide such an event would be for it to occur entirely among soft-bodied forms. Most, probably all, of the phyla evolved from soft-bodied ancestors; insofar as can be told at present, each phylum that is durably skeletonized evolved its hard parts independently. For this reason alone the preskeletonized records of the phyla should be hard to find.

When extensive fossil assemblages such as the Vendian faunas include no skeletonized forms, it is very likely that they were absent, for if skeletons were present but have been destroyed everywhere, their former presence should be indicated at least by the sort of impressions that preserved the soft-bodied fossils. Cnidariomorphs *are* preserved in Vendian rocks, but body fossils of bilaterian worms are exceedingly rare (or possibly absent). This situation suggests that the cnidariomorphs were either lightly sclerotized (Seilacher 1989) and perhaps had unusually tough mesoglea (Valentine 1992a), or in some other way were more durable than the worms and other groups that we don't find. The cnidariomorphs are clearly not directly ancestral to bilaterians and do not help to solve the puzzle of their ancestries. The few possible Vendian bilaterians have not produced clues that help close the gaps between the bodyplans of living phyla.

However, it is not possible to dismiss the sudden appearance of novel bodyplans as resulting entirely from soft-bodied ancestral histories. The early radiation of skeletonized higher taxa among all of the phyla is equally obscure. It is common for several classes or orders to appear early in the record of phyla, but the ancestors of those branches are not found (chaps. 6–11). Either each of these classes and orders became skeletonized independently or there is some consistent bias against the preservation of their ancestors. As the details of the skeletal structures make it appear that skeletons are apomorphic for the phyla but in most cases plesiomorphic for their classes and orders, it is more likely that skeletonization evolved in the common ancestors of the higher taxa within phyla, and then diversified with their radiations. What, then, are the possible biases against discovery of the skeletonized intermediates leading to these descendant taxa?

Being small and fragile would seem like good possibilities, but in this case we happen to have a very good record of minute forms, some of which appear to be quite fragile—the small shelly faunas of the Early Cambrian. The widespread, repeated occurrence of small shellies suggests that preservation of small mineralized skeletons was at least no worse than average during that time. Good skeletonized faunas of varying richness are found throughout the remaining Cambrian. The answer to the lack of intermediates may lie among the other factors that lower the probabilities of preservation or discovery. Evolution in an environment that is poorly represented in the stratigraphic record can probably be eliminated; the relatively well represented habitats of shallow to moderate-depth shelf environments seem perfectly suited for the origin of many of the bodyplans that evolved.

It appears that bias against skeletonized intermediates may be explained as bias against lineages that do not produce many skeletons and thus have a poor chance of penetrating the many filters that destroy faunal remains. Lineages with rare, geographically and ecologically localized populations that evolved rapidly and were relatively undiversified, or at least that had several of those properties, would answer nicely. It follows that the clades that we do see are likely to have been the more abundant and widespread ones, although many of the small shelly fossils seem to have only regional distributions. It is uncertain how large a fraction of the fauna these represent. Some idea of the richness of the nonskeletonized fauna of the Middle Cambrian is given by the reconstructions of the Burgess Shale assemblage in fig. 5.21. It seems from those reconstructions that a smaller percentage of the fauna may have been skeletonized in the Middle Cambrian than is skeletonized today. The relatively high diversity of sclerites in the Early Cambrian, however, makes it appear that there was quite a variety of bodyplans at that time.

There Is a Vast Range of Hypotheses That Attempt to Explain the Cambrian Explosion

If the Cambrian explosion is not a taphonomic artifact, it must reflect some very special circumstances in life history. Lacking an obvious explanation, workers have not been shy about speculating on less obvious possibilities. Few of those various

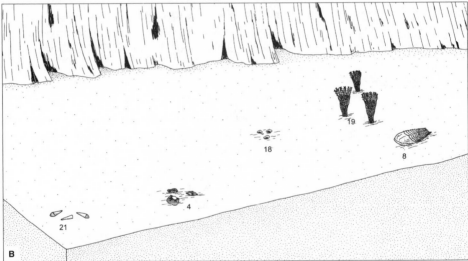

FIG. 5.21 Taphonomy and the fauna of the Burgess Shale. (*A*) Both durably skeletonized and soft-bodied components, reconstructed as at a fortuitous meeting on some small area of the seafloor during the middle Middle Cambrian. (*B*) Durably skeletonized components only; if the soft-bodied fauna were not exceptionally preserved, this would be the fossil record yielded by the Burgess Shale, with the addition of trace fossils, which have not been indicated. Sponges: *1, Vauxia; 2, Choia; 3, Pirania.* Brachiopod: *4, Nisusia.* Annelid: *5, Burgessochaeta.* Priapulids: *6, Ottoia; 7, Louisella.* Arthropods and relatives: *8, Olenoides; 9, Sidneyia; 10, Leanchoilia; 11, Marrella; 12, Canadaspis; 13, Molaria; 14, Burgessia; 15, Yohoia; 16, Waptia.* Lobopod: *17, Aysheaia.* Mollusk: *18, Scenella.* Echinoderm: *19, Echmatocrinus.* Chordate: *20, Pikaia.* Phylum uncertain: *21, Haplophrentis; 22, Opabinia; 23, Dinomischus; 24, Wiwaxia; 25, Anomalocaris.* From Briggs et al. 1994 after Briggs 1991.

ideas concerning the appearance of phyla were really intended to represent the whole Cambrian story. Without presenting an exhaustive review and critique, it is worth considering a few particularly popular or timely notions to give some idea of their range, which is remarkable. I have tried and failed to organize these hypotheses into a comprehensive framework, so instead here is a sort of laundry list of some representative ideas that can at least fall under a few general headings.

Perhaps There Was No Cambrian Explosion

The oldest fossils known to Darwin in 1859 included relatively complex forms, such as trilobites, that are of mid–Early Cambrian age in modern terms. Darwin postulated gradual evolutionary change, and therefore he believed that those forms must have been evolved over very long stretches of preceding geologic time that did not happen to be represented by fossiliferous rocks. Perhaps, he reasoned, those rocks are under the sea at present. Alas, we now know that subduction has quite removed all of those early rocks from the ocean floors. We now have the Neoproterozoic faunas from rocks deposited just before the Cambrian and preserved on the continents, but there are no convincing signs of animals that are just a little less advanced than the Cambrian forms, as Darwin would have expected. Instead, those fossils are difficult to relate to living phyla. Arguments much like Darwin's are nevertheless still raised; for example, Fortey et al. (1996) suggest that the morphological complexity and the widespread distribution of Early Cambrian trilobites imply a long prehistory, perhaps hidden because the ancestral forms were small and escaped fossilization. In the same vein, a stem crustacean has been described from the late Lower Cambrian (Siveter et al. 2001), and it is claimed that not enough time was available for evolution to produce a crustacean during the Lower Cambrian (Fortey 2001). These workers suggest that the explosion was not important as an evolutionary event.

There have also been a number of modern studies that attempt to date the times of origins of phyla based on extrapolations from the estimated evolutionary rates of molecules, used as molecular clocks (chap. 4). The molecular-clock studies have placed the origins long before the Cambrian, some nearly three times earlier than the age of the earliest body fossils known. However, molecular-clock dates for Neoproterozoic events that are based on different molecules or different protocols vary widely. Clearly, severe problems still plague the interpretation of rates of molecular evolution. The times that major metazoan clades were founded, and that the bodyplans of phyla were first represented in stem ancestors, remain uncertain.

The Explosion Was Due to Physical Changes in the Environment

One of the favorite explanations for the Cambrian explosion has been that free oxygen levels, which were clearly exceedingly low early in Earth history, rose as photosynthetic algae evolved and flourished. Eventually enough oxygen accumulated to change the atmosphere from a reducing to an oxidizing state, raising dissolved oxygen levels in the oceans high enough to permit the active lifestyles implied by Cambrian fossils. Berkner and Marshall (1964) were early champions of this hypothesis; they scaled the rise in oxygen levels over time to the minimum requirements of the succession of organisms then known from the fossil record. That evidence is of course circular, and to be on firm ground the hypothesis requires estimates of oxygen levels that are independent of the fossils themselves. That Cambrian oxygen levels must have been at least some significant fraction of today's is suggested by physiological studies of oxygen usage in low-oxygen environments at present (see Rhoads and Morse 1971). Furthermore, the biosynthesis of cuticles and of skeletons, features that are first known to be common during the explosion, have relatively high oxygen demands (Towe 1970). Unfortunately, it has proven difficult to trace the rise in oxygen very accurately, and while such considerations do set oxygen minima, it is not yet certain that metazoan complexity tracked the historical oxygen rise at all closely. In fact there is some evidence that oxygen levels were appropriate for active metazoans long before the explosion (Canfield and Teske 1996; Knoll 1996).

Various other changes and perturbations of the physical environment have been invoked to explain the Cambrian explosion. One idea is that a Cambrian transgression of the oceans created widespread epicontinental seas with many novel environmental settings (Brasier 1979). Such an event occurred but is not especially characteristic of or limited to the time of the explosion; extensive platform seas had been present earlier, and extensive transgressions occurred later. Neoproterozoic glaciations closely preceded, geologically speaking, the first indications of animals in the record, and a number of authors have suggested some tie between the global stress and release from glacial conditions and the early evolution of Metazoa. The glaciations were clearly extensive, and it has been suggested that in fact the entire surface of Earth froze (the "snowball Earth" hypothesis; Kirschvink 1992; Hoffman et al. 1998) and that the environmental changes attending this event paved the way for metazoan evolution. It is difficult to understand how the biota managed to survive such an extreme event, but many lineages of prokaryotes and protistans seem not to have suffered (Corsetti et al. 2003). Indeed, modeling has indicated that oceans are not likely to have frozen in low latitudes (Hyde et al. 2000). Furthermore, recent paleogeographic reconstructions also suggest that the glaciations may not have been global (Smith 2001). The glaciations may have lasted longer on some continents than others, perhaps overlapping the origin of metazoans (e.g., in North America; Barfod et al. 2002).

Geochemical anomalies detected in Cambrian sediments have led to suggestions of unusual ocean chemistries that suppressed earlier metazoan evolution, or of biotic extirpations that cleared the way for rebounding evolutionary radiations. One such study identified a "Strangelove Ocean," from which much life was thought to have been extirpated as indicated by a carbon-isotope shift, interpreted as caused by a decrease in oceanic fertility, that was believed to occur just before the explosion (Hsu et al. 1985). Evidence for this particular event turned out to occur in rocks deposited well after the explosion began. Still, there are other important geochemical excursions recorded in rocks deposited near the beginning of the explosion that may well speak to changes in the marine environment that might have encouraged or permitted metazoan radiations (see Knoll 1992a; Kaufman et al. 1997). The possibility that metazoan evolution was limited before late Neoproterozoic time by a scarcity of trace metals, required for biogeochemical cycles related to primary productivity, has also been suggested (Anbar and Knoll 2002). The trace-element limits are inferred from a reconstructed stratification of the Proterozoic oceans, with fairly well oxygenated surface layers but with sulfidic conditions in deeper waters, based partly on sulfur isotopic studies (Canfield 1998).

The Explosion Was Due to Biological Changes in the Environment

As major evolutionary changes were clearly under way before and during the Cambrian explosion, the changes in the biotic milieu itself must have given rise to new ecological and evolutionary requirements and opportunities, leading perhaps to biotic feedbacks that caused or aided the explosion. Expansion of the level of primary productivity, perhaps related to the rise of new planktonic forms, might support higher diversity levels. Simply stabilizing planktonic productivity might have a similar effect (Valentine 1973b). To go one step further, Butterfield (1997, 2001) suggested that the evolution of the mesozooplanktonic tier in the ecological hierarchy should have moderated fluctuations in the plankton, which could have contributed to the diversity rise in the benthic communities. Logan et al. (1995) explored the biogeochemical effects of the removal of carbon from the Neoproterozoic water column by sinking of the fecal pellets of newly evolved zooplankters, leading to marked changes on the seafloor that may have contributed to a Cambrian radiation.

Within the benthic communities, the evolution of effective predators might have resulted in widespread antipredator adaptations, including skeletons, thus creating the abrupt entrance of many lineages into the fossil record (Hutchinson 1961). New predators might also have placed new limits on the population sizes of prey species, freeing resources and permitting a rise in taxonomic diversity (Stanley 1973, 1976). The notion that light was introduced into metazoan behavioral systems for the first time during the Cambrian (an unlikely assumption) has led to conjectures about its effect on species interactions and on diversity and perhaps to promotion of the Cambrian explosion (Parker 1998). I should stress that the present characterizations of all these studies do not begin to do them justice, and the original papers should be consulted for their richer accounts. These ecological ideas chiefly speak to how diversity might be raised within ecosystems, and do not directly address the problem of the origin of bodyplans. However, many of these features may have accompanied the explosion.

The Explosion Reflects Intrinsic Evolutionary Change

Just as it can be supposed that ecosystems were changing during the origin of phyla, the characters of the genomes and of the developmental pathways of metazoans must have been changing as well, and it has been speculated that some such internal changes led to the Cambrian explosion. It has been suggested that the horizontal transfer between taxa, by virallike particles, of genes that promoted skeletonization might have produced the sudden appearance of Cambrian novelties (Mourant 1971), or that virallike transposons disrupted genomes to permit the evolution of novelties (Erwin and Valentine 1984). Another suggestion is that the explosion was enabled by more normal processes of growth and mutation within the regulatory portion of the genome (Britten and Davidson 1969, 1971; Valentine and Campbell 1976). Developmental processes might have reached a level of sophistication appropriate to the evolution of complex animals just before the Cambrian. A problem with this notion is that some unicellular taxa appear (such as foraminifera; Lipps 1992) or diversify extensively (such as planktonic algae; Vidal 1997) at the same time. Still another idea is that the evolutionary invention of set-aside cells, cells sequestered in simple ancestral lineages that could be used in the differentiation of complex bodies, permitted the rise of the higher metazoans that appear in the Cambrian (Davidson et al. 1995; Peterson et al. 1997; see chap. 13).

Clearly, there is no shortage of hypotheses to explain the Early Cambrian fossil record; many more have been suggested than are listed here, but this brief account gives some idea of their scope. It is no surprise that, in explaining Cambrian events, geologists invoke physical conditions, ecologists prefer environmental changes, microbiologists favor viral interventions, and developmental biologists look to the evolution of body-patterning systems. In fact it seems certain that many different ecological and evolutionary changes were indeed involved in Cambrian evolution.

In Sum, the Cambrian Fossils Imply an Explosion of Bodyplans, but the Underlying Causes Remain Uncertain

The stratigraphic record of the late Neoproterozoic and Early Cambrian interval is preserved as well or perhaps better than an average section of the geologic column. There are good marine sections of rocks deposited during this interval and exposed in several regions scattered around the world. Neoproterozoic rocks contain a body-fossil assemblage of cnidariomorphs, with some enigmatic seriated forms, some of which cannot be assigned to living groups, and some of which may be bilaterians; a few mineralized skeletal forms are known, but they are simple and not very informative. Many of the Neoproterozoic trace fossils, however, probably represent bilaterians. During the Early Cambrian, mineralized skeletons became diverse, chiefly during the Tommotian and Atdabanian. Although the bulk of those skeletons are sclerites of uncertain affinity, a number of them clearly represent primitive members of living phyla. From lagerstätten such as the Chengjiang fauna in the

Atdabanian and the Burgess Shale fauna in the Middle Cambrian, it can be inferred that the evolutionary explosion also produced a rich fauna that lacked mineralized skeletons. Both skeletonized and soft-bodied faunas included numbers of forms whose morphologies do not fit easily within crown phyla or classes, though many are probably stem groups. If the ratio of soft-bodied to durably skeletonized forms suggested by the Burgess Shale assemblage is characteristic of (or was exceeded by) the Early Cambrian fauna, the breadth of the Early Cambrian explosion, impressive as it appears, is actually significantly underestimated by the fossil record as it is now known. That there was a very broad early radiation of metazoans seems incontrovertible. However, the nature and timing of events during the prelude to the explosion, which made it possible, remain in doubt.

PART TWO

The Metazoan Phyla

Architecture is frozen music. FRIEDRICH VON SCHELLING

This part reviews the bodyplans and some developmental aspects of the living metazoan phyla. This is essentially an overview of what evolution has accomplished with the architecture of the Metazoa. The emphasis is on those features that will shed most light on the primitive bodyplan of each phylum, and on the place of each phylum in the tree of life. The various body subplans that have evolved within phyla, for example composing classes and orders, are certainly of interest insofar as they are of help in establishing synapomorphies and the polarities of morphological change, and thus in providing evidence as to the primitive condition. However, the arrangement of taxa into the classes or orders, and interrelations among them, are commonly disputed by workers who have systematic expertise within these groups. I have tried to use the taxonomic schemes that are best supported by recent work, especially if a scheme is bolstered by molecular evidence, and have tried to ignore opinions that are based only on polarities that have been imposed on classes and orders because of a commitment to an evolutionary scheme. Many early fossils represent extinct groups, chiefly stem forms (Budd and Jensen 2000), which of course can provide important information as to the ancestry of their crown groups, and to patterns of macroevolution among early metazoans. These extinct groups are not ordinarily covered in any systematic way in invertebrate texts that deal chiefly with the living fauna and which tend to underestimate the diversity and significance of fossils. On the other hand, derived forms such as insects, although they play large roles in the present biosphere and are well covered in invertebrate texts, are not included here, as they throw no special light on the origins of phyla.

In presenting a panorama of metazoan bodyplans it is usual to begin with sponges and end with chordates; in between, however, the ordering can vary significantly. The order in which the phyla are taken up here is based on the major alliances suggested by SSU rRNA evidence, but modified by developmental and morphological evidence as presented in chapter 4 (fig. 4.17). Taxa for which little or no molecular data are available are simply positioned within the framework of the molecular tree according to classical developmental and morphological evidence. The alliances are evaluated for general plausibility against classical and fossil evidence, and the possible branching patterns within the alliances are explored. It is worth repeating that the amount of molecular evidence that is available is still so small that the chances are very great that the tree will be modified as new data come to light; indeed, it is a virtual certainty.

CHAPTER 6

Prebilaterians and Earliest Crown Bilaterians

Evolutionary transitions from unicellular protistans to multicellular organisms have occurred many times; Buss (1987) estimated the minimum number as twenty-three. The cells of multicellular organisms should retain many of the features of their unicellular forebears. Protista display a vast array of cell structures and complexities, most of which differ in significant ways from those found in animal cells, so that those protistan groups are not likely to be animal ancestors. Animals share some common attributes that suggest that they have originated only once (see Müller 1998), most likely from an ancestor within or allied to the protistan phylum Choanoflagellata. The prebilaterian crown phyla are Porifera, Ctenophora, Cnidaria, and Placozoa. One other metazoan group, the Myxozoa, is composed of obligate parasites with complex life cycles and appears to have arisen from within Cnidaria on present evidence. The earliest branch of crown Bilateria may be represented by acoelomorph "flatworms."

Sponges and Spongiomorphs

Bodyplan of Porifera

General poriferan anatomy and developmental stages are well known. Two subphyla, Cellularia, with two classes, and Symplasma, comprising the single class Hexactinellida, are commonly recognized at present (table 6.1), although the placement of some orders is in dispute. The symplasmans are organized so differently from the cellularian classes that they are sometimes considered to be a separate phylum (Bergquist 1985). Skeletal features have traditionally been used to define the higher sponge taxa, although developmental, biochemical, and cytological features are coming to play important roles in morphological systematics, and molecular-sequence studies are beginning to clarify relations among the major groups.

Cellularia. At its simplest, the cellularian body wall consists of two epithelial layers, each one cell thick, separated by an extracellular matrix, the mesohyl (fig. 6.1). The external epithelium is a pavement of flattened cells, the pinacocytes. The inner epithelium includes globular to pear-shaped, flagellated collar cells, the choanocytes. Specialized cells, the porocytes, have central pores that pierce the entire body wall. Cell junctions are not commonly reported within sponge tissues, though

TABLE 6.1   Subphyla and classes of living Porifera. After Brusca and Brusca 1990; Reiswig and Mackie 1983.

Taxon	Remarks
Subphylum Cellularia	Tissues cellular
Class Calcarea	Spicules calcareous
Class Demospongia	Spicules siliceous, chiefly monaxons or tetraxons, rarely with basal solid calcareous skeletons
Subphylum Symplasma	Tissues largely syncytial
Class Hexactinellida	Spicules siliceous, chiefly hexactines

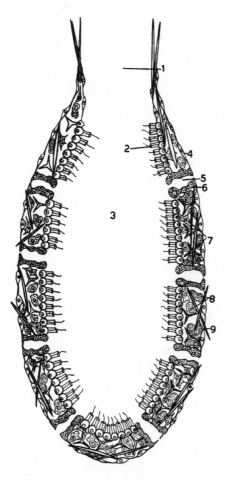

FIG. 6.1 Cross section of an ascon chamber. *1*, osculum; *2*, choanocyte; *3*, spongocoel; *4*, epidermis; *5*, *6*, porocyte; *7*, mesohyl with cells; *8*, amoebocyte; *9*, spicule. After Hyman 1940.

desmosomelike structures are known between pinacocytes (see Simpson 1984). These junctions seem to be transitory; some occlusion may be secured by a regular close spacing of dermal cells at intervals between 10 and 20 nm (see Green and Bergquist 1982). Within the mesohyl are archaeocytes, amoeboid cells that function in digestion and in other processes, and sclerocytes and other cell types that are responsible for secreting the skeletal elements. The basic functional unit of a sponge (fig. 6.2) is a flagellated chamber, usually a vase-shaped structure lined by choanocytes enclosing a spongocoel. Flagellar action creates a flow of water out of the opening (osculum) at the "neck" of the vase, and therefore water is drawn from the exterior (or from canals) through pores into the chamber. Food items suspended in the water can be captured by most cells, and especially by the "collars" of the choanocytes, which consist of numerous closely packed microvilli sometimes coated by a sticky substance. The flagellar waves force water distally along the flagellum and draw a current into the depression within the collar at the flagellar base; this current must pass between the microvilli, and food items are caught on their exterior surfaces and engulfed by pseudopodia. Thus feeding is largely localized at the collars.

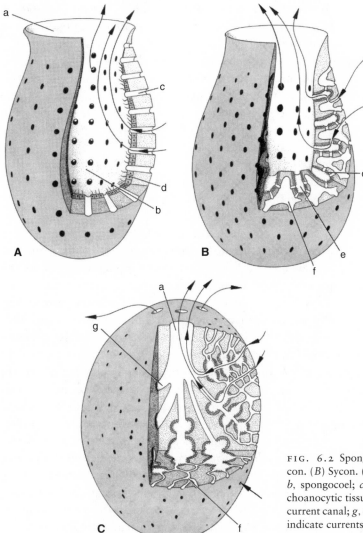

FIG. 6.2 Sponge body types. (*A*) Ascon. (*B*) Sycon. (*C*) Leucon. *a*, osculum; *b*, spongocoel; *c*, intracellular canal; *d*, choanocytic tissue; *e*, radial canal; *f*, incurrent canal; *g*, excurrent canal. Arrows indicate currents. From Bayer and Owre 1968.

The sponge bodyplan is built of modules of flagellated chambers (fig. 6.2); the simplest, the ascon type, is formed by a cluster of simple chambers. A more complex form is the sycon, which can be thought of as composed of compound ascons, their walls thickened to accommodate flagellate chambers that are supplied water via canals and that empty into a common excurrent chamber. The incurrent canals and the excurrent chamber alike are lined by pinacocytes. Ascon and sycon architectures are rare, ascons being known only in the class Calcarea. The most common body type is the leucon, which is still more complex and can be thought of as composed of compound sycons lying on their sides within the leucon wall, so that their excurrent chambers can empty into a common central chamber. A series of canals furnishes water to the pores of the flagellated chambers. Living sponges are generally not as regular as the drawings in fig. 6.2, although rather regular axial symmetries do occur in some members of all classes. The increasing complexity of chamber arrangement from ascon to leucon types is interpreted as a response to increasing body size. More choanocyte chambers are produced by the complex folding of the tissue sheets within the sponge, thus accommodating the increased nutritive requirements and providing a stronger body-wall structure, as body volume increases.

Cell types that are devoted to secreting distinctive skeletal elements are rather characteristic of different classes, which in fact have been largely distinguished by their skeletal types. Another notable type of cell is the contractile myocyte; these surround pores and vents in some demosponges. In some groups myocytes form networks within the mesohyl. The contraction of a myocyte can produce a progressive contraction of abutting cells along the network, so that a modest amount of coordinated, quasi-muscular control is achieved in the sponge wall despite the lack of nerve cells (Bergquist 1978).

Sponge skeletons may be organic or mineralized. Despite being termed keratose, organic sponge skeletons are collagenous. Two forms are present, fibrillar collagen and spongin. Fibrils occur in all sponges (indeed in all animals), secreted by most cell types but particularly by the collencytes and by lophocytes. Spongin fibers, secreted by spongocytes working in concert, are less rigid than fibrils. Spongin is a major skeletal component of demosponges. Mineralized skeletal elements of sponges are either siliceous or calcareous. Silica is secreted across a specialized membrane (silicalemma) which is evidently assembled just before spicule formation (Garrone et al. 1981; Simpson 1984). Siliceous spicules are secreted extracellularly in the mesohyl and thus are endoskeletal (though some protrude from the sponge body, for example as basal anchors); they usually, perhaps always, contain an axial filament of unknown organic composition upon which mineralization first occurs. In Calcarea the calcareous spicules are also secreted extracellularly but within organic sheaths, and do not contain organic filaments; usually, pairs of cells operate to produce each spicular ray (see Ledger and Jones 1977; Simpson 1984). These scleroblasts are usually derived from pinacocytes which, in common with other sponge cell types, localize calcium at several sites. A few living sponges have, in addition to either calcareous or siliceous spicules, solid calcareous exoskeletal elements, often as relatively thick basal sheets. These forms were once separated as a taxon of their own, the Sclerospongia, but are now believed to represent a polyphyletic assemblage of species allied variously to Calcarea and Demospongia. Solid skeletal elements were more common in Paleozoic and Mesozoic sponges than they are in living forms.

Archaeocytes are totipotent, capable of giving rise not only to all somatic cell types but to gametes (at least through choanocyte intermediates); collencytes can give rise to pinacocytes and myocytes; some pinacocytes form sclerocytes. Most of the more specialized cells are thought to be terminally differentiated. Just which cell types may dedifferentiate is not clear; perhaps dedifferentiation is not a normal in vivo activity, although even the reality of many of these cells as distinctive types has been questioned (see discussions in Bergquist 1978 and Simpson 1984). The simplest imaginable adult sponge would have perhaps five somatic cell types, and indeed some sponges may be this simple. Some sponges have about twelve cell types (or more if cell-type varieties are finely split).

Symplasma. Hexactinellid tissues are largely syncytial (see Mackie and Singla 1983). The mineralized skeleton is chiefly of six-rayed (hexactine) siliceous spicules, which may be free or welded into a rigid though rather brittle framework that is draped by a cavernous tissue network that has been likened to an irregular cobweb (Ijima 1901). This network consists chiefly of the trabecular syncytium, which substitutes for an outer tissue; it is a continuous cytoplasmic network with scattered nuclei and organelles. The trabeculum is supported by extracellular matrix that forms a sort of basal membrane, a sheet of collagen fibers evidently secreted by trabecular tissue, lying in the position of the mesohyl of other sponges. Flagellated chambers are lined by an enucleate choanosyncytium containing numbers of collar bodies—microvillar collars encircling flagella—recalling the collar structures in choanocytes (fig. 6.3). Archaeocytes are immersed below the epithelium; they are not very mobile but produce flagellated sperm and, presumably, eggs. Four or more other cell types are present (though myocytes are absent). In one hexactinellid the various syncytia and the individual cells are interconnected by cytoplasmic bridges that contain a unique type of pore-bearing plug (Mackie and Singla 1983). The flagellated chambers are irregularly arrayed, falling in complexity between the arrangements of sycon and leucon architectures.

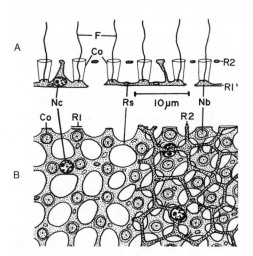

FIG. 6.3 Choanosyncytial tissue of a hexactinellid sponge, *Aphrocallistes*. (*A*) Vertical section, showing several collar structures per syncytium. (*B*) Diagrammatic view from the interior of a chamber. *Co*, collar body; *F*, flagellum; *Nb*, nodal body; *Nc*, choanosyncytium nucleus; *R1*, main reticulum; *R2*, secondary reticulum; *Rs*, reticular strands. From Reiswig 1979.

Development in Porifera

Lying as they do at the base of living metazoan phyla, sponges might well provide some clues as to the origins and ancestral state of metazoan development. However, sponge development is quite varied (Borojevic 1970; Fell 1974; Bergquist 1978). Asexual reproduction is common, either via gemmules (a sort of resting state for overwintering, consisting of masses of archaeocytes enclosed in a collagenous layer) or via budding, with either archaeocytes or representative suites of cells present in the buds. Septatelike junctions, usually absent in sponges, have been observed in the walls of gemmules.

Cellularia. In sexual reproduction cellularians produce both sperm and eggs but at different times. Both spermatogonia and oocytes are sometimes formed from choanocytes; in at least some cases the choanocyte first becomes archaeocyte-like. It is not yet clear whether other archaeocytes, or other cell types altogether, are sometimes involved in gametogenesis. In viviparous species, sperm first penetrate a choanocyte, which becomes amoeboid and carries the sperm to an oocyte in the mesohyl, where fertilization occurs. Ensuing cleavage is usually equal or nearly so and commonly produces a ball of cells known as a morula, from which a ciliated larva develops through extensive cell migrations (fig. 6.4).

In some demosponge embryos there are two larval cell forms, small rapidly dividing cells (micromeres) that become ciliated and form the larval epithelium, and larger cells (macromeres) that form the larval interior, resulting in a so-called parenchymella larva. In some calcareans and a few demosponges, cleavage consists at first of three meridional cleavages (producing macromeres) followed by a

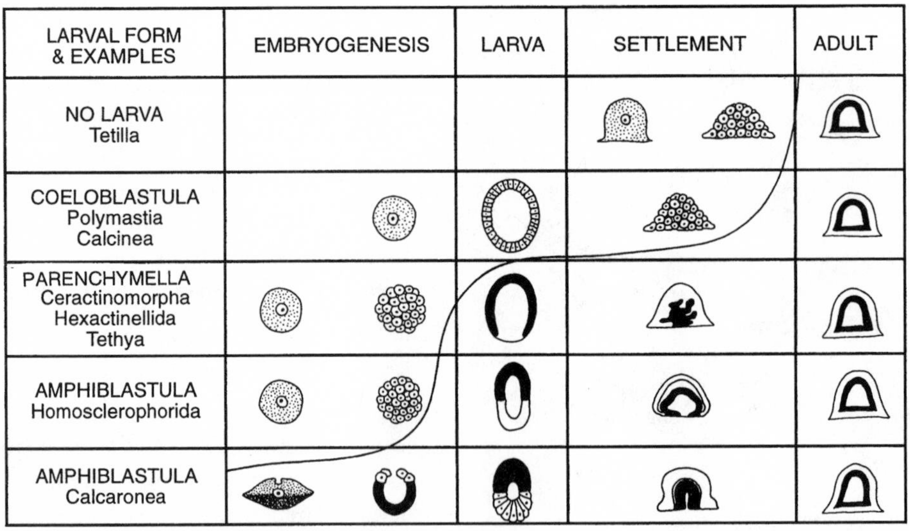

LARVAL FORM & EXAMPLES	EMBRYOGENESIS		LARVA	SETTLEMENT		ADULT
NO LARVA Tetilla						
COELOBLASTULA Polymastia Calcinea						
PARENCHYMELLA Ceractinomorpha Hexactinellida Tethya						
AMPHIBLASTULA Homosclerophorida						
AMPHIBLASTULA Calcaronea						

FIG. 6.4 Developmental sequences in a variety of cellularian sponges, from eggs, to morula or blastula (if present), to a settled adult form. Cell differentiation begins at the stage indicated by the solid curve, varying from (*below*) a well-differentiated egg to (*above*) a coeloblastula undifferentiated at settling. The black regions are micromeres or their derivatives. From Borojevic 1970.

latitudinal cleavage that produces a cycle of micromeres arranged in radial positions. These two cell forms become arranged in anterior (flagellated micromeres) and posterior (macromeres) positions, forming another larval type, the amphiblastula. Finally, a third larval type is the coeloblastula, with a blastocoel surrounded by a sphere of subequal flagellated cells, found among a number of major sponge taxa. In a few cases cells are connected by parallel, desmosomelike junctions. All the larvae are nonfeeding and swim from hours to days before settling to metamorphose. During metamorphosis, choanocytes arise through dedifferentiation of ciliated micromeres, which in at least some demosponges also produce another cell type. The general sequence of development is thus not too different from that found in many cnidarians (see below), although the cell types differ (see Leys and Degnan 2002).

The onset of cell differentiation varies widely; fig. 6.4 shows some of the developmental patterns. Unequal cleavages characterize some embryos, suggesting that pre-positioned or at least unequally distributed developmental factors are present in the egg, from which differential larval cell fates are determined. Simpson (1984) suggests that the coeloblastula larva is the more primitive, and if so, conditional specification via cell-cell signaling in a regulative embryo might be the more primitive method of sponge cell differentiation. In some sponges with very yolky eggs, cells cannot even be identified until 100 or so nuclei are found (Simpson 1984).

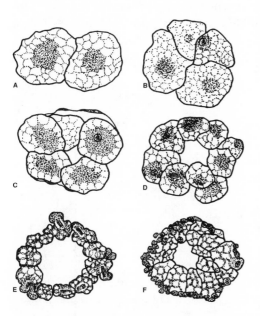

FIG. 6.5 Early embryogenesis in a hexactinellid, cross sections. (A) Two-cell stage. (B) Four-cell stage. (C) Eight-cell stage; cleavage is spiral; note blastocoel. (D) Sixteen-cell stage. (E) After sixth division, cleavage is tangential, with smaller cells given off to exterior. (F) Gastrula. From Boury-Esnault et al. 1999.

Symplasma. Developmental stages of a hexactinellid have been described from a population found within scuba depth in a Mediterranean cave (Boury-Esnault et al. 1999). Gametes derive from archaeocytes. Early cleavage (fig. 6.5) is equal and has a spiral geometry at the eight-cell stage, though cleavage is radial at the sixteen- and thirty-two-cell stages. A blastocoel is present at the eight-cell stage, and a classic coeloblastula is present at thirty-two cells. Micromeres are then given off toward the outside, and a gastrula-like embryo arises through delamination (see fig. 2.13C) and develops into a free-swimming larva, termed a trichimella (Boury-Esnault and Vacelet

TABLE 6.2   First appearances of major sponge and spongiomorph taxa in the fossil record.

Taxon	Age	Reference
Calcarea	Atdabanian	Bengtson et al. 1990
Demospongia	Tommotian or Atdabanian	Zhuravlev 1986; Chen et al. 1989b
Hexactinellida	Neoproterozoic	Steiner et al. 1993; Gehling and Rigby 1996
Archaeocyatha	Tommotian	Debrenne and Rozanov 1983
Radiocyatha	Tommotian	Rozanov and Zhuravlev 1992
Chancelloriida	Tommotian	Rozanov and Zhuravlev 1992

1994). This larva has seven or more cell types, including multiflagellate cells linked by parallel junctions resembling desmosomes, and choanoblasts, cells that bud the collar bodies of choanosyncytia. It is noteworthy that the early embryo is entirely cellular, syncytial tissue first appearing in the larva. Metamorphosis has not been described.

Fossil Record of Sponges and Spongiomorphs

Porifera. The early record of fossils assigned to living sponge classes is summarized in table 6.2. The earliest well-dated remains that may be sponges are in Neoproterozoic strata tentatively dated near 600 Ma from Guizhou Province, China (Li et al. 1998). These fossils consist of phosphatized tissues associated with siliceous monaxonic spiculelike structures. Sponge spicules usually dissociate after death, and therefore most early sponge records are based on spicule morphology alone. Inorganic shards or algal filaments can be mistaken for spicules, however, and uncertainties surround many possible early sponge fossils. Probable hexactinellid spicules are recorded from other Neoproterozoic rocks in China (Steiner et al. 1993). Additionally, there are records of probable sponge bodies from the Neoproterozoic of South Australia. Some fossils not only display impressions of spicular networks but have structures interpreted as oscula (fig. 6.6; Gehling and Rigby 1996). The Hexactinellida evidently radiated extensively during the Early Cambrian. Calcarea and Demospongia are first certainly identified in the Early Cambrian and radiated during the Middle Cambrian (see review in Reitner and Mehl 1995).

Other fossil problematica suggestive of sponge remains are found in rocks of Neoproterozoic or (mostly) Cambrian age; most of these are putative spicules or minute forms that are not well enough understood to serve in interpreting sponge history, and are not considered further here (see Rozanov and Zhuravlev 1992). Three living families of demosponges contain species that are known to bore, chiefly into calcareous skeletons. Early Cambrian borings similar to those caused by living sponges and associated with fossil spicules have been interpreted as the work of sponges (Kobluk 1981).

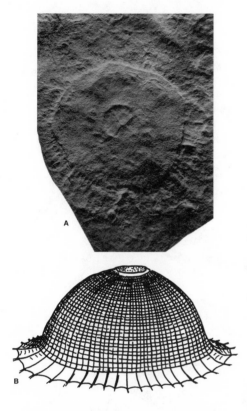

FIG. 6.6 The earliest body fossil that is probably a sponge, *Palaeophragmodictya reticulata*, from the Ediacaran of South Australia. (*A*) A paratype; spicular reticulations on the body are faint on this specimen, but the general form is well shown, with central "crater," network field, and peripheral frill; about 8 cm in diameter. (*B*) Reconstruction. From Gehling and Rigby 1996.

Archaeocyatha. Three extinct taxa that may have been at the same grade as sponges but that do not have spicules are also listed in table 6.2. The best known of these is Archaeocyatha (fig. 6.7; Hill 1972; Debrenne 1983; Debrenne and Vacelet 1983), a group with porous cup-like skeletons, chiefly solitary and anchored by spines or, more rarely, cemented to form compound structures. The skeleton is composed of a microgranular carbonate; there is no trace of spicules. It is likely that archaeocyathids had epithelia. The cups are single-walled or, more usually, double-walled with radial septa connecting the walls. Pores pierce the conical walls, smaller on outer and larger on inner walls, and also pierce the septa. A common cup height is about 8–10 cm, but some achieve more than 30 cm. Archaeocyathans form bioherms (reeflike buildups of skeletons) throughout most of their history, living in shallow and often turbulent water. Archaeocyathans appear in the Tommotian and radiate rapidly, displaying high rates of taxonomic turnover to provide a succession of characteristic forms that are excellent zone index fossils. Their last appearance in the record is probably in Late Cambrian (Debrenne et al. 1984).

Radiocyatha. Radiocyatha is a group of late Early Cambrian age, which has been compared with sponges, archaeocyathans, and algae (Zhuravlev 1986). The basic element of the radiocyathid skeleton is the nesaster, composed of two rosettes connected by an axis; as nesasters are of microgranular carbonate, they are not spicules. Radiocyathans are globular to ovoid in body shape, built so that the surface is a mosaic of nesaster rosettes (fig. 6.8).

Chancelloriidae. Chancelloriid skeletons are composed of sclerites (fig. 6.9A), which have two to twenty radiating rays, sometimes with a central boss. Chancelloriids

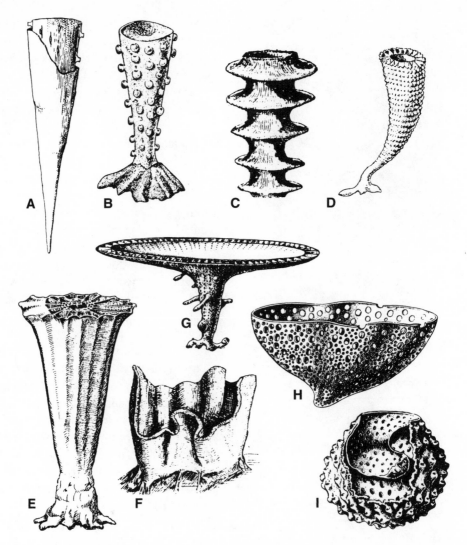

FIG. 6.7 Archaeocyathans. (A) *Dokidocyathus*. (B) *Tumuliolynthus*. (C) *Orbicyathus*. (D) *Kotuyi-cyathus*. (E) *Beltanacyathus*. (F) *Coscinoptycta*. (G) *Paranacyathus*. (H) *Cryptoporocyathus*. (I) *Fransuasaecyathus*. From Hill 1972.

have commonly been assigned to or allied with sponges, but as their rays are each hollow, they do not seem homologous with sponge spicules, and this relationship is doubtful. It has also been suggested that chancelloriids are related to a group of sclerite-bearing Cambrian forms called the coeloscleritophorans (Bengtson 1992a), which seem to have had creeping soles and to be molluscoid in gross body shape (chap. 8). Butterfield and Nicholas (1996) investigated the interiors of unusually preserved organic-walled chancelloriid rays, and have described an internal "pith" that is unlike the interior structures known for coeloscleritophorans.

FIG. 6.8 A radiocyathan. Reconstruction of *Girphanovella*. The outer skeletal wall is composed of the fused rosettes of nesasters; average radiocyathan height is about 5 cm. From Zhuravlev 1986.

FIG. 6.9 *Chancelloria eros*, a chancelloriid from the Chengjiang fauna. (*A*) An isolated sclerite. (*B*) A scleritome. (*C*) Reconstruction. From Janussen et al. 2002.

Further, chancelloriid sclerites have been found associated as scleritomes in a few cases (Rigby 1978; Butterfield and Nicholas 1996; Janussen et al. 2002) and reconstructions of articulated sclerites suggest that chancelloriids were more vase- or baglike, more spongiomorph than molluscoid (fig. 6.9B–C). Particularly well preserved chancelloriid scleritomes have been described from the Chengjiang fauna (Janussen et al. 2002), which indicate that the bag-shaped bodies were sessile and that the sclerites were imbedded in an epidermal covering. These findings suggest that chancelloriids were therefore at a different grade of construction than sponges (Janussen et al. 2002). Chancelloriids are common in archaeocyathid bioherms.

Poriferan Ancestry and Early Radiation of the Sponge Grade

At present there is no way to be sure how many times the sponge grade of organization was achieved by separate lineages. According to some SSU rRNA sequence comparisons, the Hexactinellida are the more basal, and the Calcarea are the most derived of living classes (West and Powers 1993; Wainright et al. 1993; Collins 1998), but support for this arrangement is not unequivocal. As sponges probably arose sometime in the Neoproterozoic from a protistan ancestor, the evolution of sponge development represents the first establishment of metazoan ontogenies. It is likely that choanocytes were primitive sponge cell types, along with collagen-secreting cells, lineages of which became specialized as collencytes and spongocytes. Although sponge cell types are said to be unusual in their ability to differentiate into each other, their existence as morphotypes indicates the presence of distinctive patterns of gene expression. Gene regulation of sponge development has not been studied at the molecular level, although a number of homeobox genes, which control so much of bodyplan patterning in higher metazoans, are known in sponges (Manuel and Le Parco 2000).

Among the important features that separate sponges from protistans is the synthesis and employment of collagen as the main structural protein, the extracellular component of connective tissue that gives animal bodies their structural integrity (Towe 1970, 1981). Collagen is unknown outside of the animal kingdom. Other extracellular molecules have also been identified in sponges (see Müller 1998), and these are presumably metazoan apomorphies (see Morris 1993). Collagen is a class of proteins that has a triple-stranded helical structure. Synthesis of collagen strands begins on ribosomes, and the strands then undergo chemical changes in the cytoplasm, where the hydroxylation of certain proline and lysine residues also occurs to produce the amino acids hydroxyproline and hydroxylysine, which are crucial to the bonding of strands to form collagen structures (see Towe 1981). Strands are then joined into triple helices, carried in vesicles to the plasma membrane, and exocytosed. Extracellularly, they are converted into collagen types that vary from site to site and among taxa. Of the four best-known collagen types, three (I–III) are found in vertebrate connective tissues and are fibrillar, while one (IV) is composed of a layered network sheet and is widely distributed in basement membranes throughout the metazoan phyla (Yurchenko and Ruben 1987). Type IV collagens may be similar to primitive collagens (Runnegar 1985). The fibrillar collagens are composed of similar modules and appear to have evolved from a similar common ancestor. As Towe (1981) has emphasized, hydroxyproline and hydroxylysine are nearly unknown in protistans. If it is judged that the complexities of synthesis of a useful collagen structure suggest a single evolutionary origin, then it is necessary to argue for a common collagenous ancestor for all animals. Perhaps cementation to a hard substrate for support in flowing water was among the early advantages of collagen secretion. As early sponges would have been quite small and probably had no mineralized parts, they would be nearly immune from fossilization and their absence from the fossil record means little; they may have originated tens of millions of years before they appear in the record.

If sponges arose from colonial choanoflagellates, the simplest pathway is from a cellularized and not a syncytial colony. A syncytial ancestor would require a reversal in the evolution of cell types, from cellular protists with a choanocyte-like cell phase to syncytial protists or syncytial sponges, and then back to cellular sponges in which choanocytes are developed again, and cells are capable of disaggregation and reaggregation: not impossible but unlikely. A stronger argument is that the origin of sponges occurred along a pathway of cell differentiation. An advantage of syncytial organization is that it promotes intracytoplasmic transport—the intake of materials and the products of biosynthesis may be distributed widely without barriers such as intercellular membranes or even special junctions or bridges (Reiswig and Mackie

1983). Cytoplasmic streaming has been identified within hexactinellid syncytia, and may play an important role in physiological homeostasis in this group (Leys and Mackie 1994). The apomorphic hexactinellid cytoplasmic bridge plugs seem designed to strike some balance between the ability to permit widespread transport within the cytoplasm of the entire organism, and the ability to maintain the individuality of several tissue or cell types (Mackie and Singla 1983). An advantage of cells, on the other hand, is precisely that the materials and products of biosynthesis are localized, and therefore that unique patterns of gene expression and of epistatic response may arise in different cell lines, producing that specialization of cell function that characterizes the Metazoa. Therefore the evolutionary trend of increasing tissue differentiation argues for a cellular condition during the origin of the sponge bodyplan. This pathway is consistent with the finding that early hexactinellid embryos are cellular. It follows that hexactinellids are likely to have had metazoan ancestors with cellular tissues.

Archaeocyathans probably evolved from early nonmineralized sponges. We do not know whether archaeocyathans had choanocyte-like cells or flagellated chambers, or whether they were syncytial. However, a strong case for a spongelike model that would require flagellar pumping of some sort has been made by Zhuravleva (1960, 1970b). The elaborate design of the skeleton certainly argues for localization of functions such as are associated with cellular tissues. If flagellated chambers were present, their arrangement must have been fairly complicated, at least in double-walled forms.

Experiments by Koehl (1982) have revealed that spicules serve to stiffen the sponge wall; the more tines per spicule, the stiffer the structure, and when spicules are separate and not fused into a rigid skeletal framework, the wall may be resilient as well as stiff. Spicules and other mineralized skeletal elements, including archaeocyathan cups, may have evolved partly for the properties they imparted to body walls as sponge bodies grew in size (and architectural complexity) and reached up out of the benthic boundary layer. Vogel (1977) has shown that the pressure differential created when a sponge osculum is higher in the water column than the incurrent openings can create a passive current flow through a sponge, which can thus supplement flagellar currents. Preliminary work suggested that a similar effect may have occurred in archaeocyathans (Balsam and Vogel 1973), and flow through a number of archaeocyathan models was subsequently studied by Savarese (1992), who confirmed the passive-flow phenomenon. Savarese also demonstrated that the presence of septa promotes the entrainment of passive-flow currents. Furthermore, modeling indicates that increases in the leakage of these currents at the higher flow velocities that accompany ontogenetic size increase can be nicely balanced by ontogenetic reduction in septal pore sizes, which decreases leakage. Such a relation is observed in archaeocyathans. These presumed adaptations for flow regulation and efficiency support a spongelike pumping model for the archaeocyathan bodyplan, but not necessarily a sponge grade of tissue organization.

The relation of chancelloriids to crown sponge taxa is also quite problematical; it is possible that they represent a clade of diploblastic organisms evolved independently of the radiate phyla; that they have sponges in their ancestry seems likely but is uncertain. Both archaeocyathans and chancelloriids may have achieved an epithelial grade (see below), or something closer to it than is found in living sponges. However, it is doubtful that either taxon represents an evolutionary intermediate between sponges and crown diploblasts.

Cnidarians and Cnidariomorphs

Bodyplan of Cnidaria

Living cnidarians are chiefly predators, although some forms harbor photosynthetic algal cells as symbionts within their tissues, receiving supplementary nourishment from the photosynthates. On the basis of cell-morphotype numbers, Cnidaria are more complex than Porifera, although the simplest cnidarian would overlap in cell-type numbers with the most complex sponge. Cnidarian body walls have two distinctive tissue layers separated by a layer of extracellular matrix, in this respect resembling sponges, but otherwise their bodies are organized differently (fig. 6.10). These organizational features are found in all more complex metazoans. Importantly, cells in tissue layers of cnidarians are united into epithelia by occluding septal junctions. The junctions space cells around 15 nm apart, and provide for the integrity of external and internal tissues, the ectoderm and endoderm, respectively. These tissues rest against the extracellular matrix, the mesoglea. The mesoglea consists of a gelatinous groundmass with few or no cells, in which collagen and other fibers are often common. Cnidarians have two body types, the chiefly benthic polyps and the chiefly pelagic medusae; these sometimes alternate from one generation to the next, although in many groups there is only one body stage during the life cycle (fig. 6.11). Two subphyla and four classes are recognized (table 6.3). Based on molecular sequence and mitochondrial genome structures, the earliest crown cnidarians appear to be anthozoans (Bridge et al.

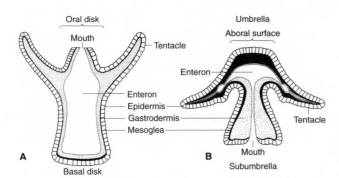

FIG. 6.10 Bodyplans of cnidarians; main body-wall features labeled. (A) Polyp. (B) Medusa. From Oliver and Coates 1987.

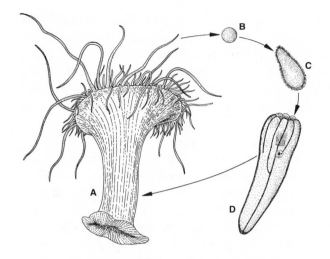

FIG. 6.11 Life cycle of an anthozoan cnidarian, stages not to scale. (*A*) An adult produces (*B*) eggs, which develop into (*C*) a planula larva and thence into (*D*) a juvenile polyp that grows into an adult. More complex life cycles involving alterations of body types between polyp and medusa may be derived. From Bayer and Owre 1968.

TABLE 6.3 Subphyla and classes of living Cnidaria. After Brusca and Brusca 1990; Collins 2002.

Taxon	Remarks
Subphylum Anthozoa	Polyps only
Subphylum Medusozoa	
Class Scyphozoa	Medusae budded or strobilated
Class Staurozoa	Medusae and polyps undivided
Class Cubozoa	Medusae metamorphose from polyps
Class Hydrozoa	Medusae budded from wall of polyps

1995; Collins 1998), polyps that lack medusae and have relatively simple life cycles.

Polyps may be solitary, presumably the primitive cnidarian state, or colonial. The body consists of a vaselike trunk region with a basal disk that may be free or attached, and an adbasal mouth with a circlet of tentacles. In the simplest polyp the trunk wall surrounds a digestive cavity, with the mouth as the single opening to the exterior. The ectoderm consists of a layer of epithelial-muscular (myoepithelial) cells; these are supporting cells, often columnar to cuboidal in shape and with contractile processes extending inward onto a basal matrix to form a longitudinally oriented, musclelike layer. The epithelium also contains sensory cells, gland cells of more than one type, and a network of nerve cells near the base. Finally, there is a scattering of interstitial cells that give rise to gametes or to other cell types, among them the nematocytes (or cnidocytes) that produce the various stinging and tangling cells of the cnidarian armamentum. There are about twenty-five or so types of nematocysts, complex intracellular structures that serve for prey capture, defense, and locomotion, and are found in characteristic associations that vary among taxa. Nematocysts tend to be concentrated on the tentacles and around the mouth. The outer epithelium rests upon the extracellular mesoglea, which is quite thin and in some cases contains amoeboid and other cells. The inner body wall is lined by a gastrodermal epithelium, the cells of which have contractile processes oriented so as to form a musclelike layer against the mesoglea. There are many gland cells in the gastrodermis, some of which (chiefly near the mouth) secrete mucus, and many of which secrete digestive enzymes. The major cnidarian cell types have been discussed by Chapman (1974), Thomas and Edwards (1991), Lesh-Laurie and Suchy (1991), and Fautin and Mariscal (1991). Simple cnidarians have perhaps ten broadly construed cell morphotypes, complex ones nearly twice that.

In the cnidarian gut or enteron, large prey items, captured by the tentacles and engulfed by the mouth, are broken down via extracellular digestion, and some of the resulting small particulate materials may be endocytosed and digested within cells. In small polyps the walls of the gastrovascular cavity may be essentially circular, but at larger body sizes they become more complex. In anthozoans such as sea anemones, the gastrovascular walls have folds that extend well into the central cavity to form mesenteries; each mesentery consists of two endodermal sheets (or folds) with a thin mesogleal layer between them. The area of the digestive tissue is thus greatly increased over the lining of a simple hollow cavity. At the oral margin, ectodermal tissue folds down to form a pharynx supported by the endodermal mesenteries near the top of the gastric cavity. The pharynx is elongated and bears one or more ciliated grooves termed siphonoglyphs, which drive water from the oral region down to the digestive cavity. Gonads, which arise from interstitial cells, form within the endoderm of polyps; in anthozoans gonads are found on the mesenteries.

Contractile fibers of the mesenteries form bands, the cells of which are separated from the endodermal epithelium and are partially sunk within the mesoglea to approach a muscular tissue. Just as the myofilaments within cells must be stretched by some external agent following contraction, so must the myoepithelial cells of cnidarians. To some extent the circular and longitudinal muscles antagonize each other, and the water in the digestive cavity can act as a hydrostatic skeleton. For example, if longitudinal muscles shorten the trunk of a polyp, the circular fibers are thereby stretched and the circumference expands; if circular contraction follows, the trunk will narrow and lengthen to become elongated, entailing a corresponding relaxation and stretching of the longitudinal fibers. By confining ambient seawater within the trunk through restriction of the oral opening, an anemone can cause concentric waves of contraction to pass down the body wall, the stretching of relaxed muscles being accomplished by the hydrostatic pressure within the gut. These peristaltic waves may be used to creep along or to form burrows within the substrate (fig. 6.12) or, rarely, to move through the water column—to "swim." Benthic creeping locomotion can also be accomplished by ciliary activity in some forms (Collins et al. 2000).

Medusae have bodyplans adapted to pelagic life (fig. 6.10). The body wall is expanded into a bell-like structure unique to the medusa body type. From the center of the bell hangs a columnar, often quadrangular structure (the manubrium),

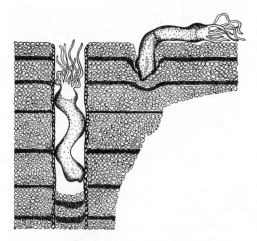

FIG. 6.12 Burrowing by the sea anemone *Cerianthus*. Antagonizing circular and longitudinal muscles expand and contract the body wall; the enteron, when filled with seawater and closed by contraction of the oral sphincter, can mediate shape changes. The animal on the right is just beginning to penetrate the sediment. From Schäfer 1972.

hollow and lined internally by endoderm, bearing the mouth distally. In some forms the margins of the mouth are four oral lobes or arms. Within the bell the manubrial space expands into (usually) a quadrilobate stomach. Endoderm-lined canals radiate from the stomach across the inner surface of the bell to its outer circumference, where they join a canal that encircles the bell margin. The medusae of several groups have four such radial canals, emphasizing the tetraradiate symmetry of the gastrointestinal system, while others have larger numbers of canals but in multiples of four; the canals may branch and thus multiply as they approach the bell margin. It is generally the larger medusae that have the larger number of radial canals. The canals furnish nourishment to the tissues of the bell. The mesoglea, so thin and inconspicuous in polyps, becomes the dominant layer within the bell, thickening toward the crown. Muscle fibers are oriented in circular bands beneath the ectoderm, lining the subumbrellar cavity, concentrated toward the bell margins; they provide for strong swimming contractions. In hydrozoan medusae the muscle tissues, which are surrounded by extracellular matrix, display both smooth and striated layers (Boero et al. 1998). Mesoglea acts to give the bell its body, shaping and rigidifying it, while permitting flexibility and providing an elastic and antagonistic response to the muscular pulsations involved in swimming. Tentacles, hollow or solid depending upon the taxon, fringe the bell margin; they are provided with rich concentrations of nematocysts for prey capture. Gonadal tissues develop on the manubrium, on radial canals, or on the floor of the stomach lobes. Hydrozoan medusae have a thin flangelike structure, the velum, which encircles the base of the bell, partly closing off the bell interior. The velum is formed of a fold of ectoderm with a thin mesogleal layer in the center, and contains abundant muscle fibers, which aid in swimming. Tracts of muscle fibers also run up and down the walls of the manubrium.

Both medusae and polyps form colonies in some taxa. The most spectacular colonies, which contain both medusae and polyps, are found among the pelagic hydrozoan siphonophores. As mentioned in chapter 1, some of these forms have developed bodyplans in which individuals (zooids) are so specialized that they function as organs—some as floats, some in securing prey, some in digestion, and so forth. The colony then becomes a virtual superorganism. The zooids do not display a particular abundance of novel cell types, however, so that despite their interesting evolutionary tangent, they remain essentially at the same structural grade as their solitary cousins.

A number of cnidarian taxa produce durable skeletons with significant fossilization potential, the scleractinian corals being the most notable among living groups. The coral polyps, organized like anemones, secrete ectodermal skeletons of aragonite. The skeletons are septate cups, with skeletal septa secreted within infolds of the base and walls of the polyps that lie in the interspaces between the endodermal mesenteries; the symmetry of the gastric partitions can therefore be inferred from the skeleton. Colonial corals further secrete a common skeleton (coenosteum) that binds and supports the individual polyps within a colonial architecture. It is clear from the success of anemones that the coral bodyplan does not require a durable skeleton, and from the success of corals that the ecological advantages accruing from the various colony architectures—branching, domical, sheetlike, and so forth—have provided the basis for radiations into adaptive spaces that would not otherwise have been accessible to anthozoans. Octocorals, another anthozoan group, may have horny organic or mineralized skeletons. The Hydrozoa also contain groups with fossilizable skeletons: the hydrocoralline skeleton is of massive aragonite; hydroid stalks and stolons are usually chitinous; and the solitary polypoid chondrophores, which are pelagic, secrete a float that includes a tough proteinaceous disk. Finally, spicules and other minor skeletal elements occur in a number of groups.

Development in Cnidaria

Cnidarians reproduce both sexually and asexually. Budding usually involves growth from an evagination of the body wall; medusae are produced in this manner in some hydrozoans. In scyphozoans, medusae may be produced by multiple transverse fissions of a polyp, a process termed strobilation. In sexual reproduction, gonadal tissues arise from interstitial cells, and either gametes are shed into the water or, in some cases, eggs are fertilized at the gonads and the embryos brooded. Indirect-developing hydrozoan embryos are highly regulative, cells commonly remaining unspecified up to the formation of the larva, although polarity is imposed in the egg and the normal blastomere fates may be mapped (Tessier 1931). Among cnidarians with direct development, however, some of the better-studied have only limited powers of regulation (Freeman 1983).

In indirect development, cleavage is radial, indeterminate, and somewhat unequal; in very yolky eggs, cleavage is irregular (Van de Vyver 1980). The blastulae may be either coeloblastulae or stereoblastulae; they become ciliated and motile while gastrulating. In coeloblastulae, gastrulation is usually achieved by introgression of cells to fill the blastocoel; less commonly, it involves an invagination of the blastula. The gastrulae elongate as they develop into larvae with external ectoblasts and internal endoblasts. Interstitial cells then form from endoderm (see Nyholm 1943; Kume and Dan 1968).

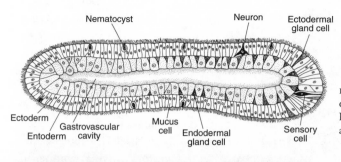

Nematocyst   Neuron   Ectodermal
gland cell

Ectoderm
Gastrovascular   Mucus
Entoderm   cavity   cell   Endodermal   Sensory
gland cell   cell

FIG. 6.13 The planula larva of the hydrozoan *Gonothyraea*. From Bayer and Owre 1968 after Wulfert 1902.

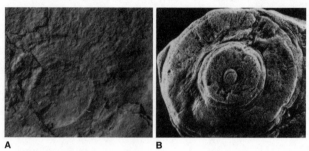

A   B

FIG. 6.14 Medusoids from the Russian Vendian. (*A*) Oral preservation of *Ediacaria;* poorly preserved marginal tentacles best shown at top. (*B*) Adoral preservation of *Ediacaria;* in life the base, with central boss, was sunk in the substrate. From Fedonkin 1992.

The most common cnidarian larval type is the planula (fig. 6.13). Planulae have an anteroposterior polarity, and most swim, to settle and attach to the substrate anteriorly and undergo a metamorphosis, the blastopore forming the opening to the enteron. A rarer larval type found among some hydrozoans, the actinula, is polyplike and not ciliated, and upon settling develops directly into an adult. Planulae are either lecithotrophic (nonfeeding) or planktotrophic (feeding), and there are many named stages within different taxa (review in Martin 1997). A few planulae are known that feed by ingesting material trapped on mucous strings (see Tranter et al. 1982 and references therein), which may be a rather common behavior.

Early Fossil Cnidaria and Cnidariomorphs

Interpretations of the early fossil record of the Cnidaria are difficult and disputed. Many of the body impressions found in Neoproterozoic rocks are of circular and frondlike forms that have been interpreted by many workers as medusoids and polypoid colonies, respectively. None of them can be placed within crown Cnidaria with certainty, however, and indeed the presence of Proterozoic cnidarians has not been confirmed beyond all doubt by fossil evidence. The roughly circular forms were first thought to be medusae, and some were assigned to living cnidarian classes and even orders after reconstruction (e.g., by Wade 1972b; Jenkins 1984, 1985). However, medusoids with central bosses or papillae, such as *Ediacaria* (fig. 6.14) and *Cyclomedusa*, have subsequently been reinterpreted as benthic forms attached to the substrate by a central adoral protrusion (Seilacher 1984; Fedonkin 1987;

FIG. 6.15 Three specimens of the Neoproterozoic "medusoid" *Cyclomedusa*, in positive hyporelief, from the Mackenzie Mountains, Northwest Territories, Canada. Note that, although the specimens are abutting, they do not overlap but appear to have interfered with each other's symmetrical growth—good evidence that they were benthic. From Narbonne and Aitken 1990.

Jenkins 1989; Narbonne and Aitken 1990; Farmer et al. 1992). This interpretation is supported by the complete lack of overlap among specimens even when they are densely aggregated on bedding planes (fig. 6.15). As with *Ediacaria,* some specimens of *Cyclomedusa* and those of some of its likely allies, such as *Hiemolora,* display numerous tentacles around their outer margins (see Fedonkin 1992), which suggests a cnidarian affinity; perhaps the Neoproterozoic "medusoids" are stem anthozoans. Another group of unusual Neoproterozoic fossils has a triradiate rather than a tetraradiate symmetry (e.g., *Skinnera, Albumares,* and *Tribrachidium*) and thus differs from all known living cnidarians (fig. 6.16). Those forms have been assigned to a variety of higher taxa (e.g., *Tribrachidium* has been postulated to be a primitive echinoderm by some workers), and Fedonkin (1985c) has united them as an extinct class of cnidarians, Trilobozoa.

A number of workers have experimented with modern cnidarians in order to determine how they might be preserved as fossils (Schäfer 1972; Norris 1989; Bruton 1991). Schäfer emphasized that modern cnidarians have about the same density as seawater and should not leave impressions on submerged seafloor sediments. However, many cnidarians, including pelagic medusae, normally alter their densities by passive osmotic accommodation to ambient seawater (Mills 1984). Thus, if placed in fresher water, medusae sink at first but eventually accommodate, and if placed in saltier water, they float at first but eventually accommodate. There are limits to this ability, however; in experiments with water of 75% normal salinity, medusae sank and were unable to recover, evidently because of the density of their mesoglea. Thus local freshening of ambient water could send a population of medusae to the seafloor (a ctenophore was also studied by Mills and reacted similarly). Once on the seafloor, the impression and relief of a medusa may be represented in fossil structures just like any other carcass. When stranded on a beach, medusae can imprint a variety of structures on underlying sediments, including bell outlines, tentacles, muscles, gonadal cavities, and the gastrovascular cavities and canal systems; not all of these structures are found in single specimens. In addition, gastrovascular systems and gonadal pouches are commonly filled with sand as jellyfish become stranded, and thus those structures may form sand molds, which can be preserved as casts in the overlying stratum.

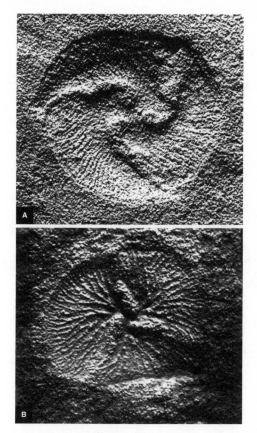

FIG. 6.16 Two Neoproterozoic forms with tri-radiate symmetry, from the Vendian of the White Sea, Russia; both ×6. (A) *Tribrachidium*. (B) *Albumares*. From Fedonkin 1994.

Norris performed experiments in which medusoids and frondose pennatulids were buried in sand, and then frozen either immediately or after decaying from 1 day to 2 weeks; the molds formed by this procedure were cast in plaster (fig. 6.17). The hydroid medusa *Aequorea* and experimental impressions (molds) of its subumbrellar and umbrellar surfaces are shown in fig. 6.18. *Aequorea* has an unusually large number of radial canals for a hydromedusan, and these appear clearly in both impressions. The thick mesogleal layer in the center of the bell is especially well reflected as a boss in negative hyporelief in the exumbrellar impression. However, there is no trace of muscle bands. Central concentric rings such as are found in so many Vendian "medusoids" are not present in any of the experimental impressions. Circular grooves are present on the bells of some living medusae (as in the coronates), and radial grooves and striae of several kinds mark the bells of others. These experiments do not strongly support interpretations of circular Neoproterozoic fossils as medusae.

Possible Neoproterozoic polypoids include impressions that may be the floats of the pelagic Chondrophora (Wade 1972b; Stanley 1986). Frondlike impressions that Seilacher has regarded as vendobionts are commonly interpreted as colonial cnidarians (fig. 5.10); some of these have been assembled into an extinct order Rangeomorpha and assigned to the Octocorallia by Jenkins (1984), and suggested as being ancestral to the living pennatulaceans. Experimental burials of modern pennatulids produce impressions that share some features with those Vendian forms (Norris 1989), although the living forms tend to become contorted and seem fragile by comparison with the fossils. The fossil fronds are evidently solidly bound together within an embracing framework. Conway Morris (1992b) has noted that a frondlike animal from the Middle Cambrian Burgess Shale resembles the Ediacaran *Charniodiscus* and displays evidence of possible zooids and other features that could support comparisons with living pennatulaceans. However, Williams (1997) has

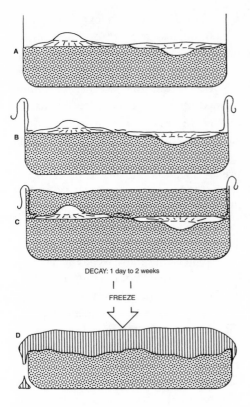

DECAY: 1 day to 2 weeks

FREEZE

FIG. 6.17 Experimental fossilization to form sand molds of cnidarians. (*A*) Specimens placed on water-saturated sand. (*B*) Layer of plastic wrap (not covering specimens) placed on sand surface. (*C*) Specimens buried with water-saturated sand, container sealed. (*D*) Container frozen and split along plastic layer to reveal sand molds, which could then be cast in plaster. From Norris 1989.

evaluated the morphology of the frond-like fossils in the light of the phylogeny and functional morphology of living pennatulaceans. Frondlike pennatulaceans have separate polyp leaves that permit the flow of water through the fronds, reducing resistance to currents and providing feeding currents for the polyps. There is no evidence that the frondlike fossils could function in this fashion. Williams suggests that the fossil fronds lay flat on the seafloor, or were perhaps buried, much like the highly derived pennatulacean sea pansy *Renilla* (and see Buss and Seilacher 1994). Williams concludes that the generalized resemblance of some pennatulaceans to the fossil fronds indicates neither phylogenetic relationship nor functional convergence between the groups.

One of the most famous of the forms that have sometimes been interpreted as cnidarians or cnidarian relatives is the fossil *Dickinsonia* (fig. 6.19); the largest are known from partial specimens that may have measured over 1 m in length when whole, but they are only a few millimeters in thickness (Runnegar 1982b). This odd form with its curious seriation has been particularly difficult to assign to a phylum. *Dickinsonia* has usually been placed in Cnidaria (as a medusoid, as by Harrington and Moore 1955), but has also been assigned to Platyhelminthes (as by Termier and Termier 1968), to Annelida (as by Wade 1972a), to an extinct phylum (Proarticulata Fedonkin 1985a), and to a separate kingdom from Metazoa by Seilacher (see below). Other workers, unsure of a phylum, have nevertheless suggested that dickinsoniids belonged to a polypoid (Valentine 1992a) or an annelid grade (Conway Morris 1979b). None of these assignments is without difficulties. Runnegar (1982b) argued that the dickinsoniids are not flatworms (he believed that dickinsoniids were too large to be viable under the constructional principles used by flatworms), and Seilacher (1989) suggested reasons that dickinsoniids are probably not annelids (e.g., there is no sign of an annelid-style gut). It is harder to argue that their structure is not consistent with a cnidarian

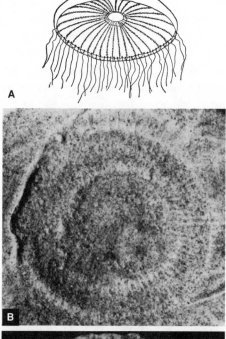

FIG. 6.18 Experimental fossilization of a hydroid medusa, *Aequorea*. (*A*) Sketch of *Aequorea* in life. (*B*) Subumbrellar impression, sand mold, after 2 weeks of burial, positive hyporelief. (*C*) Umbrellar impression, sand mold, frozen immediately after burial, negative epirelief. A after Hyman 1940; B and C after Norris 1989.

grade, but they are unlike any living cnidarian. The serial modules do seem to continue across the animal, and although a median furrow has been interpreted as an oral slit (Valentine 1992a), this is probably incorrect; indeed, their method of feeding is a major problem in their interpretation. McMenamin (1986) suggested that some Neoproterozoic forms harbored photosynthetic algae as symbionts, and this remains a possibility for *Dickinsonia*.

Schopf and Baumiller (1998) made models of *Dickinsonia* in materials of varying density. The models were then subjected to experimental and analytical procedures to determine whether they would have been stable in various living positions under flow regimes imputed to their paleoenvironments at collecting sites in Neoproterozoic rocks in Australia. The models were unstable at densities comparable to those of modern worms. The conclusion was that fossil forms of this type could not have lived unsecured on the surface of the ancient current-swept seafloor where they are found, and various alternatives (they lived during quiet intervals, or they were shallow burrowers or were fixed in place by algal mats, etc.) were considered. An additional alternative is that they were secured by mucus, which is produced in significant amounts by living cnidarians. Biomechanical approaches of this sort are clearly promising.

Some enigmatic Neoproterozoic fossils, such as *Pteridinium* and *Erni-etta*, best known from southern Africa (figs. 5.10, 6.20), do not have even distantly comparable counterparts among living cnidarian bodyplans; it is possible that they are also of diploblastic grade, however. These forms have been placed in an

FIG. 6.19 A small specimen of the enigmatic organism *Dickinsonia,* ×1; some specimens reach 1 m in length. Although this genus is best known from Ediacaran beds in Australia, this specimen is from the White Sea, Russia. From Fedonkin 1992.

extinct phylum, named Petalonamae by Pflug (1970), and are given a comparative review by Jenkins (1992).

The preservation of the Neoproterozoic fossils continues to be a puzzle. Most authors that have worked with the preservation of living cnidarians have remarked on their fragility and tendency to fold or deform, relative to the organisms that created the Neoproterozoic impressions. This evident robustness has been interpreted as indicating that the Neoproterozoic animals were sclerotized (Seilacher 1984) or at least stiffer than living cnidarians (Norris 1989). Yet there is conflicting evidence on this point. For example, some specimens of *Dickinsonia* are torn or folded very much as if they had a rather tough cuticle (see Seilacher 1984). Yet Wade (1972a) and Runnegar (1982b) have shown that *Dickinsonia* could shrink by up to 67% without noticeably deforming (though whether they could manage this in life is uncertain). Perhaps this anomaly may be explained by postulating that dickinsoniids, and probably many other cnidarian-like forms, had a body layer that displayed far more postmortem stiffness than does the mesoglea of living cnidarians (Valentine 1992a). The properties of gelatinous substances that may or may not contain much water and that are infused with such fibers as collagen or fibrin vary enormously with the conditions and with the characteristics—type, density, orientation, packaging, or tangling of the fibers (Schaffer 1930; Viidik 1972; Person 1983). Koehl (1976, 1977) has found that the elastic modulus of the body walls of two species of sea anemones differs by a factor of three owing to just such differences in constructional materials. The stiffer form lives at sites with higher current velocities.

Early trace fossils that are likely to be referable to cnidarians are probable burrows of sea anemones, called *Bergaueria,* present at least from early in the Cambrian. This form suggests that crown Anthozoa might have appeared, though what these anemones might have been like is unknown. Some modern anemones that can creep by ciliary activity leave mucous trails that form traces that much resemble the earliest, simple, meandering metazoan traces that appeared near 570 Ma (Collins et al. 2000). Remarkable embryos of probable scyphozoan affinities have been reported from the Lower Cambrian (evidently the Upper Tommotian or Lower Atdabanian equivalent) of China (Bengtson and Zhao 1997). Body fossils from the Lower Cambrian Chengjiang fauna (Upper Atdabanian or slightly later) include

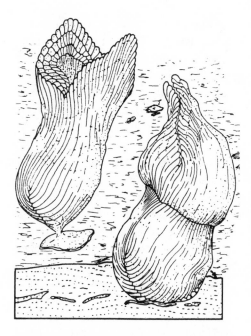

FIG. 6.20 Reconstruction of *Ernietta*, representative of a group of Neoproterozoic fossils known from Namibia, southern Africa, that have been referred to an extinct phylum, Petalonamae (Pflug 1970). From Glaessner 1984 after Jenkins et al. 1981.

columnar, tentaculate forms that appear to be sea anemones (*Xianguangia*; Chen and Erdtmann 1991).

The Vendobiont Hypothesis

Puzzling over the affinities of Neoproterozoic body fossils, Seilacher (1984, 1989) provided an ingenious alternative model to explain the fossils' distinctive morphologies and preservations. He suggested that most of those organisms had a quiltlike construction of thin, lightly sclerotized but rather flexible outer skins separated into compartments, termed pneus, by rather rigid internal struts (which in fossil impressions appear to be sutures). Such constructional modules may be shaped and combined into many of the soft-bodied forms that are known from Neoproterozoic rocks (fig. 5.10). As such a construction creates organisms without guts, feeding must have been by the absorption of dissolved organic substances in seawater, or through possessing bacterial symbionts; thus, these forms would not have possessed cnidae. This distinctive alliance, Seilacher suggested, might be descended independently of the Metazoa, and he accordingly dubbed it the Vendozoa. In a later paper Seilacher (1992) renamed the putatively pneu-bearing forms as a phylum, Vendobionta. Still later, Buss and Seilacher (1994) suggested that Vendobionta was sister to Eumetazoa (metazoans above the sponge grade). No convincing new fossil evidence that bears on this hypothesis has appeared, but advances in understanding the evolutionary history of bodyplan patterning provide grounds for further speculation about the placement of vendobionts in the metazoan tree.

As mentioned in chapter 3, both Hox and ParaHox clusters, which are paralogs, are found in eumetazoans. At least one ParaHox gene has been found in cnidarians and placozoans (Kourakis and Martindale 2000), implying the presence of both Hox and ParaHox clusters in their last common ancestor. By analogy with the functions of their homologous clusters wherever they have been studied in triploblastic forms, Hox and ParaHox clusters are presumably involved in the patterning of diploblastic bodyplans. As the diploblastic bodyplans are not bilateral, it has been unclear which of their axes, if any, are homologous with those of triploblasts,

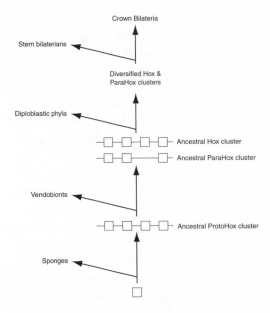

FIG. 6.21 A hypothesis of the placement of Vendobionta, based on reconstructions of the evolutionary history of the Hox gene cluster (see Finnerty and Martindale 1999). A ProtoHox gene in sponges is duplicated to form a ProtoHox cluster, and this in turn forms Hox and ParaHox clusters. Vendobiont bodies may have been patterned by ProtoHox clusters as shown. The early history of the Hox genes is not yet known; some sort of ProtoHox cluster may have been present in sponges, for example, and the Hox and ParaHox clusters may have originated before vendobionts arose. From Valentine 2001.

and how the patterning effects of gene clusters relate to their quasi-radial symmetries. Homeobox genes of the *empty spiracles* class (Mokady et al. 1998) and other regulatory genes (see Galliot and Miller 2000) are expressed in head structures in cnidarians as well as bilaterians, which suggests that the oral-aboral cnidarian axis is homologous with anteroposterior bilaterian axes.

It is possible to speculate that the modular construction of many Neoproterozoic forms noted by Seilacher was mediated by Hox-type gene clusters (see Fedonkin 1998; Valentine 2001); the patterning of modules into a variety of arrangements is just the sort of function provided by bilaterian Hox clusters, for example in arthropod tagmata (chap. 7). Vendobiont patterning could have been mediated by a ProtoHox cluster, or by Hox and ParaHox clusters if their duplication preceded the stem ancestor of vendobionts. In either case the most likely position for a vendobiont branch (the composition of which is not certain) would be between sponges and crown diploblasts, as suggested by Buss and Seilacher (fig. 6.21). A search for Hox-type genes and gene clusters in sponges and choanoflagellates should throw some light on the assembly and functions of Hox-type genes in early metazoans.

Cnidarian Relationships and Early Radiation of Cnidariomorphs

Which of the crown cnidarian groups is most primitive, and therefore most likely to indicate the nature of the founding cnidarians, has been a controversial question, partly because of differences among grand schemes of metazoan evolution and of relatedness among the major branches of the metazoan kingdom (chap. 4). Different founding types are preferred (though not always absolutely required) by the theoretical underpinnings of different systems. Anatomical evidence has been equivocal. It has been contended by many workers that the primitive cnidarian was a medusa (e.g., Hyman 1959; Hand 1963; Rees 1966), usually regarded as tetraradiate. However, this view rests chiefly on the conjecture that cnidarians are primitively pelagic and thus were radially symmetrical; polyps (such as anthozoans) that are biradial are thus considered to be derived. A host of other characters have been invoked to support this notion: that the smaller and simpler medusae should be more primitive; or that the greater variation in form in the Hydrozoa, and their larger numbers of nematocyst types, suggest that they have the longer history; and so forth. One problem with these points is that the morphological features in question are parts of coadapted bodyplans, and while they speak to differences in adaptations, they don't necessarily bear on the timing of origin of these groups. Furthermore, there is no particular reason to believe that the first Cnidaria were tetraradial. Few of the putative early medusoids had well-defined tetraradial symmetry (Wade 1969). Even living populations of medusae can show significant levels of nontetramery (Gershwin 1999), although whether these variations have a genetic basis is unknown.

It is also difficult to use morphological evidence to certify the primitiveness of the polyp stage. There are simple aspects to the anthozoans—they have simpler life cycles and fewer types of nematocysts than other groups (see Willmer 1990)—which could make them seem less derived than medusae, but these are not conclusive. However, two lines of molecular evidence do suggest that Anthozoa is the earliest extant cnidarian class. First, SSU rRNA sequence comparisons place anthozoans as basal cnidarians (Bridge et al. 1995; Collins 1998; Kim et al. 1999). And second, the mitochondrial DNA (mtDNA) molecules of anthozoans are circular, as are those of all other known metazoans except scyphozoans, cubozoans, and hydrozoans, in which mtDNA molecules are linear. This relation implies strongly that those taxa with linear molecules are more derived than anthozoans (Bridge et al. 1992).

The anthozoan groups with mineralized skeletons are usually separated at a high taxonomic level from their soft-bodied allies. Thus the living Scleractinia, mineralized corals, are considered to be ordinally distinct from the living sea anemones, the Actinaria, although they have essentially identical mesenterial development and symmetries. The septal patterns of the major extinct coral groups, the Rugosa, the Tabulata, and the Heterocorallia, imply that the mesenterial plans and developmental patterns in those groups were distinct from each other and from the Scleractinia. All of these groups appear abruptly, and cannot be traced one into another with fossil evidence. It is not known whether there were groups of sea anemones with the mesenterial plans of the extinct coral orders. For a short time during the Middle Ordovician, however, a coral group (the Kilbuchophyllida) had precisely the same pattern of septal insertion as the Scleractinia, which did not appear until the Middle Triassic, over 200 million years later (Scrutton and Clarkson 1991). Therefore, it is possible that there is a clade of anemones, now known as Actinaria, that dates at least from early in the Phanerozoic, and that has given rise both to those Ordovician corals and to Scleractinia, via the independent evolution of mineralized skeletons. There are still other coral groups, short-lived and poorly known but distinctive, recorded from the Cambrian (Jell 1984). It is easiest to account for the abrupt appearance of these corals, and of the extinct Paleozoic coral orders, as indicating times when various sea anemone clades, which we do not find as fossils, gave rise to mineralized branches. Those hypothetical sea anemones, however, must have had mesenterial patterns quite distinct from each other, suggesting an important early radiation of anthozoan polypoids.

In sum, the earliest cnidarians were probably benthic polypoids. It is tempting to hypothesize that the radial, baglike, and frondose vendobionts represented a larger diploblastic radiation, which produced the successful cnidarian bodyplan among a variety of body types. Recognizing cnidarian ancestors or even delineating the cnidarians themselves within this plexus of forms is not presently possible. Diploblastic architectures could be based on the wide array of mechanical properties available to mesoglea when utilizing structural fibers such as collagen. It is possible that some of those early forms possessed tissues so stiff as to be essentially cartilaginous, had stiff collagenous pellicles, and pursued modes of life unlike those of living diploblasts. An early radiation of stem diploblastic forms, exploiting the new metazoan architecture, was then succeeded by an early radiation of cnidarians. It is not inconsistent with the fossil record that all of the crown cnidarian classes originated during the late Neoproterozoic and Cambrian radiations.

Ctenophora

Bodyplan of Ctenophora

Ctenophores are chiefly pelagic, globose to ovoid diploblasts; they are biradially symmetrical (fig. 6.22). Living ctenophores possess eight comb rows. The combs are composed of several thousand cilia each, fused basally, with the cilia projecting like comb teeth; rows of these combs run longitudinally from near the adoral (dorsal) apex to near the ventral mouth. The coordinated beating of the combs propels the organism. The centrally located mouth leads through a pharynx to a stomach from which eight blind gastrovascular canals radiate beneath the comb rows; in some forms lateral diverticula proceed from the canals. The canals are sunk into the mesogleal mass and distribute digestive products and oxygen to the tissues. There is an adoral anal pore. A nerve net underlies the ectodermal layer much as in cnidarians, but contains eight radial nerve strands, one beneath each comb row. A pair of ectodermal tentacles that may bear side branches is present in most orders; some tentacles emerge adorally from internal tentacle sheaths. Fluid exchange with the exterior occurs through ciliated organs—rosettes—that may be the earliest excretory system in metazoans (Franc 1972). Although most ctenophores are pelagic, one order contains benthic forms, both sessile and creeping. Most ctenophores feed on pelagic organisms ranging from eggs and larvae and small crustaceans to larger cnidarians and other ctenophores.

Development in Ctenophora

Ctenophores, which have adultlike direct-developing larvae, foreshadow the construction of their bodyplan even during early cleavages (fig. 6.23; Reverberi 1971;

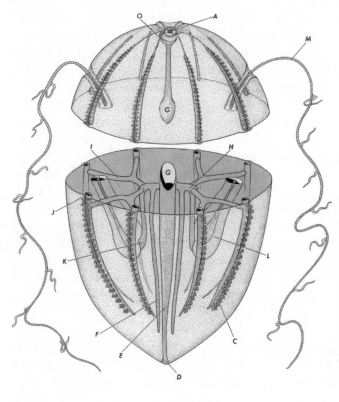

FIG. 6.22 A cydippid ctenophore, diagrammatic anatomy. A, statocyst; C, comb row of ctenes; D, mouth; E, pharynx; F, pharyngeal canal; G, digestive tract; H, transverse canal; I, interradial canal; J, meridional canal; K, tentacular canal; L, tentacle sheath; M, tentacle; O, anal canal. From Bayer and Owre 1968.

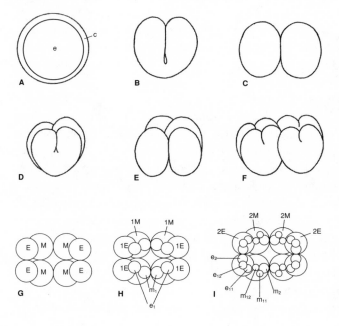

FIG. 6.23 Early cleavage in ctenophores. (A) A layered egg: e, endoplasm; c, cortex. (B) Onset of cleavage polarizes the embryo. (C) The first cleavage produces two macromeres. (D) Second cleavage plane forms. (E) Second cleavage produces four macromeres. (F) Third cleavage plane forms, slightly oblique. (G) Polar view of eight-macromere stage; E and M macromeres are differentiated. (H) Fourth cleavage produces eight micromeres. (I) Thirty-two-cell stage; micromere cleavages generate presumptive ectoderm. A–F after Reverberi 1971; G–I after Martindale and Henry 1997b.

Freeman 1983). Ctenophores are highly but not completely mosaic (Martindale and Henry 1997a, 1997b). The egg has a central yolky granular region (endoplasm) surrounded by a clearer cortical layer; endoplasm is required for endoderm formation, so its partitioning during cleavage contributes to the mosaic developmental pattern. The first cleavage establishes embryonic polarity. The first two cleavages are slightly oblique but nearly meridional (Hernandez-Nicaise 1991) and form four nearly equal cells. These cleavage planes establish, successively, the sagittal and tentacular axes of the larva, so each of the body quadrants is generated from one of these cells, though the cells are not identical but are differentiated diagonally (Martindale and Henry 1995). The third cleavage, also slightly oblique, divides these cells into eight macromeres arranged in a slightly concave plane, four inner (M) and four outer (E) cells, having different developmental potentials (fig. 6.23). The fourth cleavage divides the macromeres obliquely and unequally to produce eight micromeres. As cleavage proceeds, a stereoblastula is formed that gastrulates by epiboly. The derivatives of each of the micromeres from the E cells (e_1 micromeres) contribute to two comb rows. The micromeres from the M cells (m_1 and its derivatives) are induced by cells of the e_1 lineage to also contribute to the development of the comb rows; some endodermal cells are also induced by e_1 derivatives (Martindale and Henry 1997b).

Cells in the mesoglea are evidently derived from ectoderm, and they give rise in turn to a number of cell types, including myocytes. These muscle cells are particularly notable, for some are bundled into tissues. As these cells originate entirely from amoeboid cells that become situated in the mesoglea (Hyman 1940), the muscles are in a sense ectomesodermal (Harbison 1985). Adhesive cells termed colloblasts, which aid in prey capture, occur on side tentacles. Reproductive cells originate in gonads in the gastrovascular canals. A unique early stage is produced that resembles a miniature ctenophore (cydippid larva). The major cell types in ctenophores have been described by Hernandez-Nicaise (1991).

Fossil Record of Ctenophora

The ctenophores are quite delicate, and their fossil record reflects their poor preservation potential. Nevertheless, a number of striking fossils of Cambrian age, which vary greatly in the numbers of comb rows, have been assigned to Ctenophora. The earliest fossils are from the Early Cambrian Chengjiang fauna (Chen and Zhou 1997). One form, *Maotianoascus*, has eight lobes, each bearing two structures interpreted as comb rows. *Sinoascus* is a four-lobed form assigned to the Ctenophora on the basis of general shape, but it does not display comb rows. Other fossils assigned to the Ctenophora (fig. 6.24) are described from the Burgess Shale (Collins et al. 1983; Briggs et al. 1994) and other Middle Cambrian beds in British Columbia (Conway Morris and Collins 1996). These fossils show large series of up to eighty striated bands, which are interpreted as comb rows, and appear to represent variant bodyplans within a primitive clade from which living ctenophorans, with exclusively

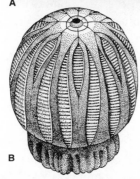

FIG. 6.24 Ctenophores from the Middle Cambrian; the numerous striated bands are interpreted as comb rows. (A) *Fasciculus,* which may have eighty comb rows; the specimen is badly flattened and distorted. (B) *Ctenorhabdotus,* reconstructed with twenty-four comb rows. A from Briggs et al. 1994; B from Conway Morris and Collins 1996.

eight comb rows, have descended. This pattern of quite disparate early morphologies being succeeded by a far narrower range of body types is analogous to that inferred for cnidarians, and for phyla in general.

Exceptionally preserved ctenophores from the Lower Devonian of Germany have been studied by X-ray techniques (Stanley and Sturmer 1983, 1987). These forms have eight comb rows, are tentaculate, and are compared to the cydippids. The fossil record of the approximately 400 million years since that time has failed to yield any ctenophores. Considering this spotty fossil record, the antiquity of the Cydippida relative to other living ctenophore orders cannot be judged from these two specimens, though as all ctenophore larvae are cydippidlike, it has been suggested that this group is primitive among living ctenophores (see below).

Ctenophoran Relationships

SSU rRNA sequence comparisons suggest that Porifera may be the living phylum most directly ancestral to the Ctenophora (Wainright et al. 1993; Smothers et al. 1994; Bridge et al. 1995; Collins 1998; Kim et al. 1999). Because of the obvious similarity of the gelatinous, radial or biradial ctenophores to medusae, it has often been postulated that ctenophores are descended from, or at least are closely related to, cnidarians. However, it is difficult to identify synapomorphies linking these phyla. Cnidarians are characterized by cnidoblasts and have myoepithelial cells; ctenophores have neither, but are characterized by colloblasts and, of course, comb rows. Therefore the last common ancestor of these phyla is likely to have lacked the features that now form characteristic elements in their bodyplans (see Harbison 1985). Primarily because of ctenophores' mesenchymal muscles and their multiciliation, some workers have considered them more closely allied with the flatworms than with cnidarians (Harbison 1985; Willmer 1990), although it is now known that multiciliated cells occur among sponges (Boury-Esnault and Vacelet 1994). If the molecular data cited proves to be correct, such similarities between flatworms and ctenophores are homoplasies.

As for evolution of major groups within the Ctenophora, there has been a divergence of opinion on which order is the more primitive and consequently on the polarity of evolution among these higher taxa. A cydippid morphology appears in the ontogeny of most orders. Harbison (1985) has constructed cladograms of major ctenophoran taxa in which the Cydippida appear polyphyletic. In one cladogram, the cydippid family Mertensiidae is the more basal, but in the other, Beroida is basal. Beroids lack tentacles and feed on gelatinous macroplankton, chiefly other ctenophores, by engulfing prey through a capacious mouth and pharynx; tentacles and colloblasts would have evolved in later forms. However, a study using the SSU rRNA molecule indicates cydippid polyphyly and suggests that Mertensiidae is indeed basal (Podar et al. 2001), with beroids being quite derived. The stem groups of the phylum, Cambrian fossils with their varied ctene rows, imply a complex history, while living species groups are relatively young (Podar et al. 2001). The nature of the stem ancestor cannot be reconstructed at present.

Placozoa

Placozoans (fig. 6.25) are minute (to 3 mm or so) flattened organisms consisting of two cell layers, a dorsal epithelium of flat flagellated cover cells, and a ventral

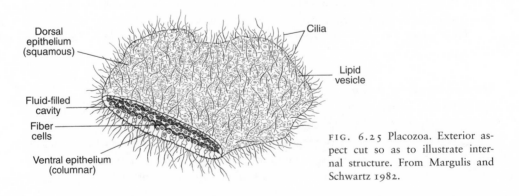

FIG. 6.25 Placozoa. Exterior aspect cut so as to illustrate internal structure. From Margulis and Schwartz 1982.

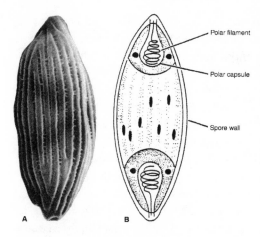

Polar filament

Polar capsule

Spore wall

A   B

FIG. 6.26 Myxozoa. (*A*) Exterior of spore of *Myxidium*; approximately ×7,800. (*B*) Diagrammatic interior of myxozoan spore. A from Hine 1980; B from Russell-Hunter 1979.

epithelium of crowded flagellated cylindrical cells interspersed with sparser nonflagellated gland cells (Grell and Ruthmann 1991). The epithelial cells are connected by belt desmosomes, but there are no basal laminae, and indeed, collagen is not confirmed in placozoans. Between the epithelia lies a narrow space crossed by a network of interconnected fiber cells, essentially a syncytium, bringing the somatic cell types to four (Ruthmann 1977). Locomotion is by gliding on densely packed ventral cilia; body-shape changes, described as amoeboid, are effected by contractions of the fiber cells (Grell 1982). Digestion may be chiefly extracellular despite the lack of a gastric cavity. Placozoans evidently position themselves over food items (such as algae) on the substrate and create a temporary pocket into which they secrete digestive enzymes. Fission and budding produce new individuals asexually, and while eggs, presumably arising from epithelial cells, have been observed to undergo early cleavage, further development has not been traced. Placozoans have no known fossil record of any sort.

SSU rRNA studies have placed placozoans among the prebilaterians. Wainright et al. (1993) first placed them as sisters to the Cnidaria, a position consistent with some later trees containing different taxa (Smothers et al. 1994; Philippe et al. 1994; Cavalier-Smith et al. 1996), while Bridge et al. (1995) found them to lie within the Cnidaria. Work by Collins (1998) suggests the possibility that they are more derived than Cnidaria, forming a sister group to the Bilateria, although they are certainly more similar in overall sequence to the prebilaterians than to the bilaterians. A study based on analysis of 23S-like rDNA placed Placozoa in still another position, as a sister to Ctenophora (Odorico and Miller 1997).

Myxozoa

Myxozoans (fig. 6.26) are an obligate parasite group, infesting cold-blooded vertebrates, chiefly fish, and platyhelminths, annelids, and sipunculids (Perkins 1991). Opinions have been mixed on whether the myxozoans are protistans or metazoans. Two major groups of myxozoans have been recognized, Myxosporidea and Actinosporidea. Work by Wolf and Markiw (1984) suggested that actinosporideans may be stages in the complex life cycles of myxosporideans, and this has been confirmed for a number of forms (Kent et al. 1994), though it is not yet certain that all actinosporideans have myxosporidean phases. One infective myxozoan stage is a multicellular spore with gametic cells and somatic valve cells that have desmosomelike connections (Siddall et al. 1995). This multicellular condition has long suggested to some workers that myxozoans are metazoans (e.g., Stolc 1899; Weill 1938). Myxozoans attach to host tissue via coiled filaments encased within "polar capsules" that resemble the nematocysts of cnidarians. Amoeboid cells are released from the spores and proliferate within the host, usually as plasmodia in a feeding (trophozoite) stage. Analyses of SSU rRNA sequences now indicate that the Myxozoa are indeed metazoans (Smothers et al. 1994; Hanelt et al. 1996; Schlegel et al. 1996; Kent et al. 2001). Using SSU rRNA sequences and both morphological and developmental data, Siddall et al. (1995) concluded that myxozoans are derived cnidarians related to Narcomedusae (as suggested by Weill 1938). Subsequent work has tended to support this conclusion (review in Kent et al. 2001), though the possibility that myxozoans are parasitic derivatives of early bilaterians (Cavalier-Smith et al. 1996; Kim et al. 1999) cannot be definitively ruled out as of now.

Diversification of Prebilaterian Metazoa

Nearly every evolutionary pathway that can connect the prebilaterian phyla has been proposed at one time or another. Full accounts of the many phylogenetic hypotheses, with their varied criteria and assumptions, would require exceptionally lengthy discussion. Instead, a well-supported SSU rRNA tree is adopted as a working hypothesis and evaluated in the light of evidence from development, morphology, and the fossil record. The tree in fig. 4.17 reflects a topology commonly though not exclusively recovered by studies using this molecule.

Fungi, though branching near the metazoans (Baldauf and Palmer 1993), have proven to be more distantly related to them than are some protistans. The sister to metazoans that is most often identified by SSU rRNA data is Choanoflagellata (Wainright et al. 1993; Cavalier-Smith et al. 1996; Collins 1998). Mesomycetozoa (Herr et al. 1999), a group of protistans that are chiefly parasitic on aquatic (and terrestrial) organisms, also branch between Fungi and Metazoa on molecular evidence (Kerk et al. 1995; Ragan et al. 1996; Collins 1998). In some molecular trees mesomycetozoans branch between choanoflagellates and metazoans (e.g., Medina et al. 2001); their exact position is still uncertain. The sponges lie at the metazoan root of the SSU rRNA tree, however, and the likely sponge and choanoflagellate synapomorphies associated with collar cells suggest that the earliest metazoans were spongelike and descended from choanoflagellates or their allies. If early sponges lacked mineralized skeletons, it will be difficult to identify them as fossils, though the phosphatized Neoproterozoic tissues from China hold out hope. Sponge skeletons may have evolved independently in each class (perhaps more than once in some cases), and there are few if any morphological synapomorphies among the classes to indicate their branching sequences, although it is simplest to postulate that the earliest sponges were cellular. In molecular-sequence trees for sponges the syncytial hexactinellids commonly branch most deeply, or in some cases are sisters to demosponges. As syncytia appear to be derived with respect to cellular tissues, there must have been cellular sponges, perhaps as yet unknown, which formed the common ancestors of living classes. The SSU rRNA data suggest that calcareans branched last, and that it is from that branch that ctenophore ancestors (and therefore all subsequent metazoans) descend. It is possible that the extinct archaeocyathans are members of or sisters to the demosponges, as nonspicular skeletons appear in both groups, although an independent origin of this skeletal type is quite possible.

The suggested evolutionary pattern for organisms of the sponge grade, then, is a major early radiation, beginning before sponge skeletons appear (and as noted from 28S rRNA results as well; Lafay et al. 1992), producing many distinctive branches. As many branches of the early sponge radiations may not have left any record at all, extensive blanks are likely in the record of sponge-grade organisms, perhaps never to be filled in by fossils.

A number of workers have suggested that diploblastic metazoans arose directly from protistan ancestors, independently of the sponge groups (e.g., Nursall 1962; Hanson 1977; Willmer 1990; some of these workers postulate independent origins of other animal phyla or combinations of phyla as well). Although early SSU rRNA results suggested possible metazoan polyphyly (Field et al. 1988), subsequent analyses (e.g., Patterson 1989; Hendricks et al. 1990; Wainright et al. 1993; Müller 1995; Collins 1998; Peterson and Eernisse 2001) have indicated monophyly. At present there is no evidence that requires a polyphyletic origin of animals. The arguments for polyphyly make it clear that it is possible, but metazoan monophyly can be supported by a number of likely morphological and molecular synapomorphies, which tend to be associated with the acquisition of multicellularity.

The SSU rRNA data suggest that the sponge lineage that gave rise to Calcarea branched to give rise to Ctenophora (Wainright et al. 1993; Smothers et al. 1994; Collins 1998). The diploblastic grade represents a distinct jump in organization, for here we find many functions performed on the level of tissues integrated by junctions and coordinated by nerves. There have been a number of attempts to find a nonsponge ancestor for the diploblastic phyla. Hanson (1977) suggested independent descent from a unicellular form other than choanoflagellates. The derivation of diploblastic forms from flatworms was advocated by Hadži (1953, 1963) but the features used to support such a phylogeny, leading from flatworms to cnidarians, are not synapomorphies (Carter 1954; Pantin 1966; see also the discussion in Hanson 1977). The molecular evidence seems to falsify both of these ideas, indicating that sponge-grade multicellular organisms were in fact interposed between the unicellular and diploblastic grades. There are no obvious groups that are intermediate in grade between sponges and diploblasts, with the possible exceptions of the extinct archaeocyathans and chancelloriids. Neither of these groups make very good candidates for cnidarian or ctenophoran ancestors, and are likely to have evolved independently of those diploblasts. Evidently, diploblastic and bilaterian forms were derived from sponges along a pathway involving the introduction of tissue-style cell junctions, the shifting of a primary digestive role to a gastrodermis, and the loss of the collar cell as a differentiated type. There are sponges in the ancestry of all eumetazoans.

Both cnidarians and placozoans have significantly less complex bodies, at least as measured by cell-type numbers, than ctenophores. It is possible that their last common ancestor was less complex than either ctenophores or cnidarians, and that ctenophores, though their ancestors branched more deeply, happened to be the group that developed a more complex bodyplan. Gastrulation is similar enough between ctenophores and cnidarians to strongly suggest that it is homologous and preceded their ancestral branching. This process implies an adult with an internalized digestive cavity, something other than a planula.

On the basis of SSU rRNA data, Placozoa has been placed as sister to Cnidaria (Wainright et al. 1993; Smothers et al. 1994), sister to Cubozoa within Cnidaria (Bridge et al. 1995), sister to Metazoa (Littlewood et al. 1998), sister to Bilateria (Collins 1998), and sister to Cnidaria + Bilateria (Peterson and Eernisse 2001). If Placozoa arose from a cnidarian ancestor, it has presumably been "reduced." Placozoans somewhat resemble a creeping larva such as a planula, although they evidently lack an anteroposterior axis. Living planula larvae have seven or eight cell types (Thomas and Edwards 1991), one of which is precursory to nematocysts and may be discounted, but even so some simplification would be required to produce a placozoan from such an ancestor.

Acoelomorpha: Earliest Crown Bilaterians?

The Acoelomorpha (Ehlers 1985) are paracoelomate worms that have been traditionally placed among the flatworms. There are two orders, Acoela and Nemertodermata, though it is not certain whether the orders are as closely allied as combining them within Acoelomorpha suggests. Cladistic morphological analyses have indicated that acoelomorphs are monophyletic, but have not been able to ally them unambiguously with other flatworms (e.g., Rieger et al. 1991). Most acoels for which SSU rRNA data are available have been too long-branched to be placed with confidence in the molecular metazoan tree. However, Ruiz-Trillo et al. (1999) found a somewhat shorter-branched acoel, *Paratomella*, considered to be basal among acoels (e.g., Ax 1987; Ehlers 1992; Raikova et al. 1997), that appears as a sister to the other Bilateria by SSU rRNA sequence analysis. This finding has been disputed, for example by Berney et al. (2000), using sequences of the elongation factor 1α gene. That molecule has not always recovered credible trees for some early metazoan branchings, however, and there have been strong rejoinders to the disputed points (Littlewood et al. 2001; Jondelius et al. 2002). That acoelomorphs fall outside the major clade of flatworms is supported by the finding that unique changes in the mitochondrial genes are shared-derived features among members of that clade but are absent in acoelomorphs (Telford et al. 2000). Finally, additional molecular studies support the placement of acoelomorphs as basal bilaterians (Baguñà, Ruiz-Trillo, Paps, Loukota, Ribera, Jondelius, and Riutort 2001; Ruiz-Trillo et al. 2002, using the myosin heavy chain type II molecule), and recent morphological and developmental studies tend to support this notion as well (see below). It seems likely that acoelomorphs are the only living bilaterians that originated before the last common protostome/deuterostome ancestor.

Bodyplan of Acoelomorpha

Acoela (fig. 6.27) and Nemertodermata are perhaps the simplest free-living bilaterians (Haszprunar 1996b). Acoels are small, usually 2 mm or less in length, and lack well-defined guts, with the mouth opening on the ventral surface, often subcentrally, sometimes through a single-walled, tubular pharynx, onto a mass of endodermal cells that may be amorphous or somewhat epithelized; there is no separate anus. The digestive tissues consist of a "central parenchyma" that is difficult to demarcate from the "peripheral parenchyma," the connective tissue that underlies the body wall or body-wall muscles. In some acoels the central portion of the central parenchyma is syncytial. The syncytium forms from surrounding parenchymal cells.

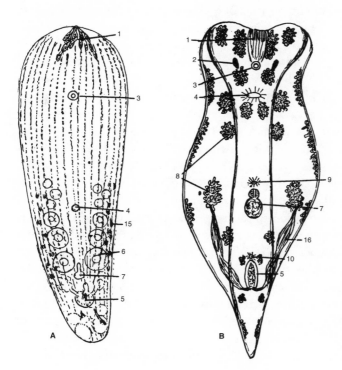

FIG. 6.27 Acoel flatworms. (A) Unidentified species from California. (B) *Convoluta*. *1*, frontal glands; *2*, eyes; *3*, statocyst; *4*, mouth; *5*, penis; *6*, sperm ducts; *7*, seminal bursa; *8*, pigment cells; *9*, female pore; *10*, male pore; *15*, sperm strand; *16*, sperm duct. From Hyman 1951a.

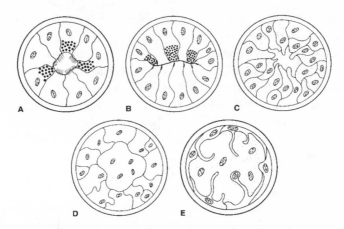

FIG. 6.28 Various states of the digestive tract of some acoelomorphs. (A) Ciliated lumen found in species belonging to several orders; gland cells stippled. (B) Occluded lumen as in nemertodermatids; gland cells stippled. (C) Irregular but open lumen as in some acoels and nemertodermatids; no gland cells. (D) Parenchymal cells surround a central syncytium, as in some acoels. (E) Small parenchymal cells surround a large syncytial digestive tract, as in many acoels. From Smith and Tyler 1985.

Some species—diatom feeders, for some reason—appear to have a permanent syncytial region; in others the syncytium forms only during feeding; and in still others it is never present (see reviews by Smith and Tyler 1985; Rieger et al. 1991). Those syncytial tissues are found in the more derived taxa, suggesting that the ancestral condition is cellular. In some nemertodermatids there is a digestive lumen (Karling 1974) although it is obscured by the projecting, interdigitating processes of endodermal cells in some species. Diagrammatic cross sections of some acoelomorph digestive tracts are shown in fig. 6.28.

In acoels the ectodermal cells are not underlaid by a basal membrane (Bedini and Papi 1974). While intercellular septate junctions are known in some forms, they are sometimes lacking or indistinct (Rieger 1981). Ectodermal cells are multiciliate. A ciliary rootlet system within cells interconnects the cilia, and intracellular fibers form a terminal cell webbing system, so that the outer cell surfaces, though lacking cuticles, are provided with intrinsic supporting structures (Rieger 1981). Muscle tissues line parts of the inner body wall, with circular outer and longitudinal inner layers. Within acoel parenchyma, extracellular matrix evidently occurs only around the statocyst, a sensory organ, though some is associated with muscle cells in nemertodermatids (see Rieger 1985). The acoel reproductive system is somewhat variable, generally consisting of ventral ovaries that lack oviducts, and dorsal collections of sperm cells—far simpler than other flatworms (chap. 8). Eggs are produced with yolk, and expelled through body-wall tissue or through the mouth. Acoel sperm have two flagella, which is the rule among flatworms, but nemertodermatid sperm possess a single flagellum (Tyler and Rieger 1975). Acoel nervous systems range from a subepidermal plexus, denser anteriorly, to a number of paired longitudinal cords, some of which are connected by lateral commissures, and there is a commissural rather than ganglionic "brain" (Raikova et al. 1998). The lack of a ganglionic brain in acoels thus suggests a gap between them and other flatworms. Nemertodermatid "brains" differ both from acoels and from other flatworms (Raikova et al. 2000). There are no nephridia in acoelomorphs. Locomotion in acoels is primarily by ciliary swimming, though ciliary crawling is evidently common.

Development in Acoelomorpha

Acoels have an unusual cleavage pattern, with micromeres appearing at the second cleavage and cleaving in duets rather than quartets of cells as in classic spiralians (fig. 6.29). Acoel duet cleavage planes are quite different, in some respects opposite, to those in classic spiralian quartets (Boyer et al. 1996b), and cleavage is somewhat regulative (Boyer 1971). Duets are produced by counterclockwise cleavage offsets from both the macromeres and the micromeres. Henry et al. (2000) comment that this cleavage pattern is essentially bilateral and that, if acoels are primitive bilaterians, the cleavage might be derived from a radially cleaving diploblastic ancestor. Nemertodermatid cleavage has not been described but is reported to be in duets and to resemble acoels (Israelsson in Lundin and Sterrer 2001), an observation that requires confirmation. There is no 4d cell in the acoel blastula. All mesoderm arises from blastomeres 3A and 3B and is endomesoderm. Gastrulation is by epiboly, the mouth lying at the site of the blastopore. Acoel development is direct. Three Hox-type genes have been detected in acoels (Baguñà, Ruiz-Trillo, Paps, Loukota, Ribera, Jondelius, and Riutort 2001; Saló et al. 2001), but there is as yet no information on their developmental roles.

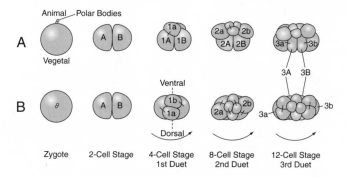

FIG. 6.29 Duet cleavage in the acoel *Neochildia* through the twelve-cell stage. (*A*) Lateral view. (*B*) Animal-pole view. Note that the spiral offsets of the micromeres are all in the same direction, in contrast to the alternating directions in classic spiral cleavage (see fig. 8.3). Endomesoderm arises from 3A and 3B, and ectomesoderm is lacking. From Henry et al. 2000.

Fossil Record of Acoelomorpha
No acoelomorphs are known as fossils.

Acoelomorphs and the Early Bilateria
The morphological simplicity of Acoelomorpha is quite in keeping with a position at the base of crown Bilateria. Their tissues and organs are simple, they have no secondary body cavities, and their diffuse nervous systems are reminiscent of Cnidaria. Their vermiform bodyplan is in keeping with the early trace-fossil record of small-bodied creeping and horizontally burrowing forms. Acoelomorph development is distinctive, and in particular the duet cleavage pattern is unique and does not serve as a very good ancestor to classic spiral cleavage (Henry et al. 2000). If acoelomorphs are at the base of the crown bilaterian tree, they are the sole known survivors of the bilaterians that preceded the protostome/deuterostome ancestor, and provide evidence that the early bilaterians were paracoelomates. Their position within the postulated early paracoelomate fauna is of course unknown, though their simplicity suggests that they may not be highly derived. The protostome/deuterostome ancestor may have been nearly as simple as the acoelomorphs, or may have been more complex in some features, as argued for example by Dewel (2000) and Dewel et al. (2001).

CHAPTER 7

Protostomes: The Ecdysozoa

The SSU rRNA tree of the Protostomia (fig. 4.17) displays three major branches. One branch, the Ecdysozoa, consists of seven phyla assigned there on the basis of molecular phylogenetics—Priapulida, Kinorhyncha, Nematomorpha, Nematoda, Onychophora, Tardigrada, and Arthropoda—and to these, Loricifera may be tentatively added on the basis of morphological evidence. Possible synapomorphies for these phyla are that they molt (giving rise to the name for this clade, the Ecdysozoa; Aguinaldo et al. 1997) and, consequently, lack locomotory or feeding cilia. The chiefly small-bodied ecdysozoan phyla are paracoelomates, and in the past many or all of them have been grouped with some other paracoelomates as Aschelminthes (see below). Ecdysozoans are direct developers or have larvae that appear to be modified juvenile instars; some living ecdysozoan taxa (insects, spiders, onychophorans) are entirely terrestrial and display highly derived patterns of early embryogenesis (see Anderson 1969, 1973; Gilbert and Raunio 1997).

Two phyla that are not reported to molt, the Chaetognatha and Gnathostomulida, have been tentatively allied with Ecdysozoa in some SSU rRNA sequence comparisons (e.g., Littlewood et al. 1998). The species that have been studied are long-branched, however, and thus doubt remains as to their true phylogenetic positions. Those phyla are treated in chapter 10.

Priapulida

Living priapulids are among the least diverse of phyla, with fewer than two dozen described species assigned to three families, though they are widely distributed in modern seas and have a fossil record dating from the Early Cambrian. Living forms are well discussed by Hyman (1951b), van der Land (1970), and Storch (1991). A cladistic morphological analysis of priapulids that included extinct forms found unambiguous support for their monophyly (Wills 1998).

Bodyplan of Priapulida

Adult priapulids range from a couple of millimeters to over 20 cm in length. The body (fig. 7.1) is divided into an anterior introvert bearing longitudinal rows of scalids—cuticular structures that are sensory or locomotory—and an unsegmented trunk with an annulated cuticle; the cuticle is molted with growth. Beneath the

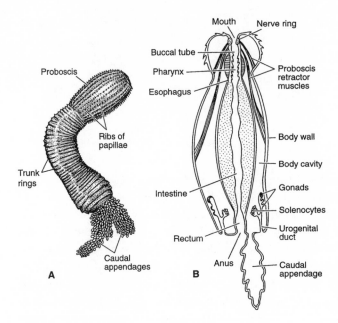

FIG. 7.1 *Priapulus*, illustrating priapulid bodyplan. (*A*) Sketch of animal. (*B*) Longitudinal section. From Brusca and Brusca 1990.

cuticle is an epidermis of flattened cells followed by a circular muscle layer. Longitudinal muscles act as retractors for the introvert, forming a hydraulic system with the trunk-wall musculature and the fluid of the body cavity. The body cavity also acts as a hydrostatic skeleton during burrowing (Elder and Hunter 1980). This cavity is rather capacious in large forms, up to half the body volume, but much less important in smaller species. Although splanchnic muscles (along the gut) are present, neither they nor the somatic muscles (along the body wall) have peritoneal linings, and thus the body cavity is technically a pseudocoel, lined by extracellular matrix. Because the body cavity is surrounded by both somatic and splanchnic muscles, however, it could be interpreted as a coelom that has lost its mesothelial lining, perhaps during body-size reductions associated with the evolution of minute ancestral forms (a possible analogy is with some meiofaunal annelids; see chap. 8). Alternatively, it could be a pseudocoel that is found in an intramesodermal position simply because of the development of an extensive splanchnic musculature. If the priapulid body cavity is pseudocoelomic, it conforms to all the other ecdysozoan phyla, which do not use a coelom as a principal body cavity, and this seems the most likely interpretation.

The central mouth opens atop a mouth cone, and is armed with spines that continue into the muscular pharynx. In at least one species the muscles that surround the posterior half of the mouth cone contain, between layers, small irregularly shaped chambers that are lined by epithelia and thus are technically coelomic, although their development is not known (Storch et al. 1989). These spaces are inferred to function hydrostatically in support of the mouth cone. The digestive tract is straight and the anus terminal. A nerve ring circles the pharynx, and a major longitudinal nerve cord is ventral, with a series of lateral extensions; there is a nerve plexus at the base of the epidermis. Gonads are paired and supported on mesenteries posterodorsally; paired protonephridia share common urogenital ducts with the gonads, with pores opening near the anus. There is no blood vascular system; evidently fluid in the body cavity provides adequate circulation.

Development in Priapulida

Priapulid development is not well known (for reviews see Storch 1991; Malakhov 1994). Cleavage is said to be radial and to produce a coeloblastula; gastrulation is by ingrowth of cells from the vegetal pole. At or following gastrulation the blastocoel fills with mesoderm from two cell strings; the development of the adult body cavity has not been reported. The mouth forms away from the blastopore as in deuterostomes. The larvae are benthic with introverts, and have thickened ventral cuticular areas termed loricas. The range of cell morphotypes in living priapulids is about eighteen to twenty or so (see Storch 1991).

Fossil Record of Priapulida

The fossil history of the priapulids, though spotty, is remarkable considering their lack of mineralized skeletons. Priapulids are reported from the Early Cambrian Chengjiang fauna of China (Sun and Hou 1987; Chen and Zhou 1997; Hou et al.

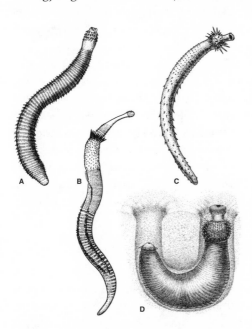

FIG. 7.2 Fossil priapulids from the Burgess Shale, Middle Cambrian. (A) *Ancalagon*. (B) *Louisella*. (C) *Fieldia*. (D) *Ottoia*, reconstructed in burrow. These forms range from less than 2 cm to over 15 cm. After Briggs et al. 1994.

1999) and are well represented in soft-bodied Middle Cambrian faunas, particularly in the Burgess Shale (fig. 7.2; Conway Morris 1977a; Conway Morris and Robison 1986, 1988). In both the Chengjiang and Burgess Shale faunas, priapulids are more commonly preserved than are annelids, which today are so much more abundant. The Cambrian forms, while different from living priapulids, are comparable to them in structure—that is, they are not notably simpler or more primitive in general appearance. In an analysis of morphospace occupation within the priapulids, Wills (1998) has demonstrated that the living clade tends not to overlap morphologically with fossil taxa. The living clade is more morphologically disparate than Cambrian fossil priapulids. Most living taxa are in deeper-water environments, but the living family that most resembles Cambrian

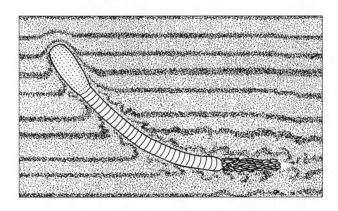

FIG. 7.3 Soft sedimentary layers deformed by burrowing of *Priapulus*. From Bromley 1990.

priapulids, the Tubiluchidae, inhabits shallow waters, as did the Cambrian forms so far as is known. A later fossil priapulid record, from the Upper Carboniferous of Mazon Creek, Illinois (Schram 1973), is also in shallow-water deposits.

Priapulids chiefly live infaunally, with the smallest species being elements in the meiofauna (benthic animals so small that they are not regularly retained by 1-mm sieves). Larger species burrow using peristalsis with alternating anterior and posterior anchors, commonly forming arcuate paths in the softest upper sediment layers where they do not ordinarily leave open burrows, but nevertheless disturb any sedimentary layering (fig. 7.3). Perhaps some early arcuate trace fossils were formed by priapulids.

Kinorhyncha

Kinorhynchs consist of about 150 species placed in two orders, the Cyclorhagida and the Homalorhagida, and have been associated with other paracoelomates in the complex of phyla termed aschelminths. Opinions as to phylogenetic relationships within this complex have been quite varied, but as these phyla have become better known, kinorhynchs have been linked by some workers with the Loricifera and the Priapulida (e.g., Lorenzen 1985; see Kristensen and Higgins 1991; Neuhaus 1994). SSU rRNA data also suggest that kinorhynchs are closely related to the priapulids (Aguinaldo et al. 1997).

Bodyplan of Kinorhyncha

Although they can be regarded as segmented, the kinorhynchs actually are jointed in the same sense as are the arthropods; that is, their integumentary structure is coadapted with their locomotory system. Subdivision of the chitinous exoskeleton gives a highly individualistic stamp to these organisms. Kinorhynchs are minute, chiefly less than 1 mm long, and live interstitially. For general accounts of this phylum see Hyman 1951b; Kozloff 1972; Kristensen and Higgins 1991; and Neuhaus 1994.

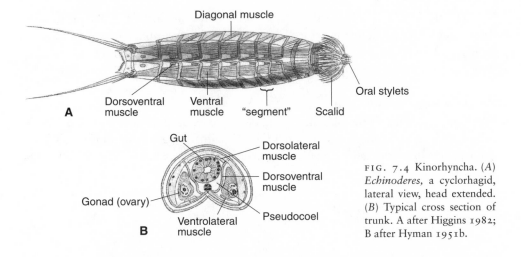

FIG. 7.4 Kinorhyncha. (A) *Echinoderes*, a cyclorhagid, lateral view, head extended. (B) Typical cross section of trunk. A after Higgins 1982; B after Hyman 1951b.

There are thirteen segments along the body (fig. 7.4), the first being a head (an introvert) ringed by several rows of sensory spines, the scalids; the second segment is a neck segment characterized by a ring of plates, the placids, that help to close off the head when withdrawn. Cyclorhagids have fourteen to sixteen placids, and homalorhagids have four to eight placids and generally fewer cuticular structures. The remaining segments are usually spinose and constitute the trunk. The exoskeletal segments are divided, usually from about segment 4, into tergal (dorsal) and sternal (ventral) sclerites, with the sternals usually subdivided midventrally. The mouth is situated on an oral cone and opens into a buccal cavity, following which there is a muscular pharynx and a rather straight gut leading to an anus on the posterior segment. Circular muscles are associated with the oral cone and introvert, and two sphincters encircle the pharynx. The body wall is a cellular, glandular epidermis resting on a basal lamina; there is some interdigitation with mesodermal muscle cells beneath. In the trunk, segmented longitudinal muscles insert (through epidermal cells) onto apodemes (commonly termed pachycycli in kinorhynchs) on anterior sclerite margins. The pseudocoel is compact and in addition to muscles and other organs may be packed by amoebocytes. The fluid-filled spaces within the body cavity are commonly lacunae; that is, they are small cavities among the cells. Dorsoventral muscles cross each trunk segment; when they contract, body fluid pressure increases and the introvert is extruded, the scalids press backward against sediment particles, and locomotion ensues. The head is retracted by longitudinal muscles that run from the introvert to insert on various body segments. There is a ten-lobed circumpharyngeal brain, and the segmental muscles of the trunk are served by ventral, lateral, and dorsal longitudinal nerve cords that bear ganglia in each segment. Female gonads are associated with paired gonopores. Excretion is via one pair of protonephridia. There is no blood vascular system.

The thickened cuticles of the cyclorhagids and homalorhagids are rather distinct, suggesting to Neuhaus (1995) that they are likely to have evolved independently, and that their common ancestors, and presumably the ancestral kinorhynchs, may have been only weakly sclerotized.

Development in Kinorhyncha

Cleavage is unknown; the only described developmental pattern is direct, the juveniles hatching with nearly adult bodyplans (Kozloff 1972). Postembryonic development has been studied by Neuhaus (1995). There are seventeen or so cell types (see Kristensen and Higgins 1991).

Fossil Record of Kinorhyncha

There are no records of fossil kinorhynchs.

Loricifera

There are no molecular data available on the phylogenetic position of loriciferans, but they are treated here because, on morphological grounds, they seem to be related to priapulids and kinorhynchs (Kristensen 1983; Higgins and Kristensen 1986; Nielsen et al. 1996). According to Nielsen (1995), over 100 living species are known, though few are described as yet.

Bodyplan of Loricifera

Loricifera is a phylum of exceedingly minute forms, chiefly less than 0.25 mm long. Although they are thus as small as many unicellular organisms, loriciferans possess over 10,000 cells, differentiated into about eighteen major cell types. Kristensen (1983, 1991) originally described the phylum and has reviewed recent work. The body (fig. 7.5) is bilaterally symmetrical but with an anterior feeding apparatus that has both biradial and triradial elements. The body is regionated into a mouth cone; a head that is an introvert; a neck, or anterior thorax; a posterior thorax; and an abdomen that is the dominant body section, covered by a sclerotized lorica composed of six layered cuticular plates that are molted during growth. The head bears rings of spiny scalids which vary from ring to ring, from club-shaped to jointed leglike appendages. The mouth cone is hexagonal and opens into a buccal region with triradiate teeth, succeeded by a pharynx. The pharyngeal bulb may be muscular with supporting rods (placoids), the whole forming a hexagonal armature around a triradiate lumen. There is some variation in the character of the mouth and anterior digestive tract among taxa. The gut is nearly straight and ends at an anus either just dorsal or ventral to the posterior body terminus. The gut lumen is irregularly oval to Y-shaped. Excretion is via protonephridia; a gonadal pore opens with the anus, or midventrally. A complexly ganglionated brain mass encircles

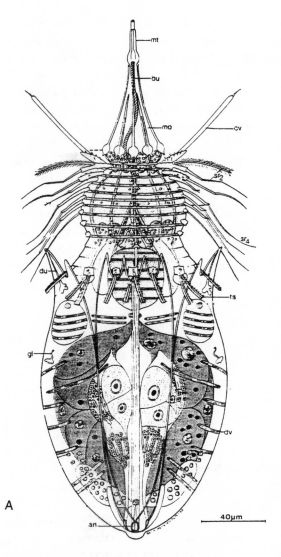

A

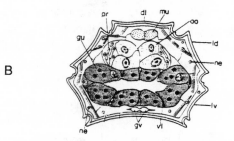

B

40µm

FIG. 7.5 Loricifera; *Nanaloricus*. (*A*) Anatomy. (*B*) Cross section. *an,* anus; *bu,* buccal tube; *cv,* clavoscalid; *dl,* dorsal lorical plate; *du,* glandular duct; *dv,* dorsoventral muscle; *gl,* epidermal gland; *gu,* midgut; *gv,* ventral ganglion; *ld,* lateral dorsal lorical plate; *lv,* lateroventral lorical plate; *mo,* mouth cone; *mt,* mouth tube; *mu,* muscle; *ne,* nerve cord; *oo,* oocyte; *pr,* protonephridium; *sr2, sr4,* scalid rings; *ts,* trochoscalid; *vl,* ventral lorical plate. From Kristensen 1991.

the digestive tract within the head; there are up to ten longitudinal nerve cords, with the two most ventral ones sometimes fused.

Beneath the epidermal cell layer of the body wall lie muscle bundles that attach to the cuticle; the muscles are longitudinal and diagonal, and in some species circular as well. There is no circular muscle coat on the abdomen but longitudinal, oblique, and dorsoventral muscle fibers occur. The body cavity, lined by an extracellular matrix, is a pseudocoel. In one family the body cavity is spacious and permits all of the anterior divisions of the animal, together with the most posterior section of the abdomen itself, to retract within the loricated abdomen; in other forms the body cavity is reduced or even absent. There is no blood vascular system. Loriciferans are known to live meiofaunally in sands of the continental slope and are infaunal in deep-sea muds; the full range of their habitats and distributions is probably not yet known. The meiofaunal forms adhere firmly to particles in their substrates, and this habit presumably delayed the detection of this group until the 1980s, when a meiofaunal sample that happened to include loriciferans was immersed in fresh water, which destroyed their adhesion, leading to their discovery.

Development in Loricifera
Early development is unknown; the larval stage, the Higgins larva, is benthic, moving about by means of the introvert. The adult has about eighteen cell types (see Kristensen 1991).

Fossil Record of Loricifera
There is no record of fossil loriciferans.

Nematomorpha

The early stages of the life cycles of nematomorphs are parasitic. While the adults are free-living, they evidently do not feed, subsisting on resources acquired during their parasitic stages. There are two classes. One class (Gordioidea) is terrestrial and largely aquatic, although some species live in nonaquatic but damp environments; this class contains over 200 species. The other class (Nectonematoidea) consists of only four nominal species that are pelagic marine forms; their young stages are found chiefly in crustaceans but are also reported from other invertebrate groups, including mollusks and annelids. The morphology of the marine forms is reviewed in Hyman 1951b, and important ultrastructural information is given by Bresciani (1991). Bleidorn et al. (2002) found Nematomorpha to be monophyletic and to be sister to Nematoda by both SSU rRNA and morphologic criteria (Bleidorn, Schmidt-Rhaesa, and Garey 2002).

Bodyplan of Nematomorpha
Nematomorphs (fig. 7.6) are usually exceedingly slender, with diameters ranging to 1–2 mm and lengths to tens of centimeters. The marine forms have rounded

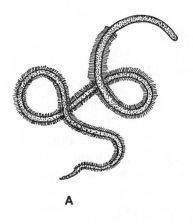

A

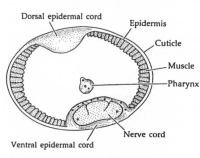

Dorsal epidermal cord

Epidermis

Cuticle

Muscle

Pharynx

Nerve cord

Ventral epidermal cord

B

FIG. 7.6 Nematomorpha. (A) *Necto-nema*, a marine form. (B) Cross section of a nectonematid through the pharyngeal region, illustrating the large body cavity. From Brusca and Brusca 1990; B after Hyman 1951b.

ends (females are sometimes blunt posteriorly) and double rows of locomotory (swimming) bristles dorsally and ventrally. The mouth is anterior but is often sealed closed, as nutrition is derived entirely by absorption from a host during juvenile stages. As the adults do not feed, their digestive tract has usually been considered degenerate, but recent work indicates that it is employed in adult uptake and storage of the previously absorbed nutrients, thus providing a food supply for the adult to draw upon while living free (and reproducing). The body wall is surrounded by a bilayered fibrillar cuticle, which at least in some species is molted. The epidermis is thin but forms dorsal and ventral longitudinal cords that somewhat interrupt the body-wall muscles; the epidermis rests on a basal lamina that is continuous with the extracellular matrix coating of the muscle cells. The muscles form a cylinder of longitudinal fibers. The fluid-filled body cavity is surrounded by the basal lamina of these muscles, and although some mesodermal muscles are found on the gut wall, the cavity can be regarded as a pseudocoel. A septum crosses the body cavity above the pharynx. An anterior nerve mass includes a circumpharyngeal ring, and a midventral longitudinal nerve cord lies above the ventral epidermal cord.

In males the testis is attached to the dorsal epidermal cord, while in females, eggs lie in the body cavity. There is no blood vascular system or any excretory organ. In the terrestrial gordioids the body cavity is more or less occluded by mesenchyme.

Development in Nematomorpha

Development has been described only for terrestrial forms, and of course the larvae are parasitic; cleavage is said to be modified radial, but in fact seems to be unique. In the gordiacean *Chordodes japonensis* the second cleavage is perpendicular to the first in at least one blastomere, and may be in a different plane for each blastomere (Inoue 1958). The adults have about seven to ten cell morphotypes (see Bresciani 1991).

Fossil Record of Nematomorpha

The only recorded occurrence of a fossil nematomorph is from the Eocene Braunkohle, a lignite, near Halle, Germany (Voigt 1938, 1988); it appears to be gordioid. Hou and Bergström (1994) have suggested that the Lower Paleozoic Paleoscolecidae (see below) may be nematomorphs.

Nematoda

Nematodes are a highly diverse group of generally minute worms that include both free-living species and parasites on marine animals and on terrestrial animals and plants. There are over 12,000 described species, and workers such as Hyman (1951b) have held that the living fauna is significantly greater, perhaps by several times. There are two classes, the Adenophorea, which contains nearly all of the marine forms, and the Secernentea, which is almost entirely terrestrial. Each class contains many parasites. Here only free-living marine forms are considered. Good references include Hyman 1951b; Chitwood and Chitwood 1974; Wood 1988; and Wright 1991. Most early evaluations of SSU rRNA data suggested that nematodes branched more deeply than the protostome-deuterostome node and were therefore quite primitive among living bilaterians. This placement probably resulted from a long-branch effect, however, for when a shorter-branching form was studied, it fell within the ecdysozoan branch of the molecular tree (chap. 4; Aguinaldo et al. 1997). A nematode phylogeny based on the SSU rRNA molecule has been produced by Blaxter et al. (1998).

Bodyplan of Nematoda

Nematodes (figs. 7.7, 7.8, 7.9) are nonsegmented cylindrical worms. Most of the free-living marine forms are under 1 mm long, although they range to about 50 mm at the largest. The cuticle is differentiated and is molted at least once. Cuticular spines and scales are common and presumably serve for purchase and stability within substrates. The digestive tract runs from a mouth, centered anteriorly, to a subterminal anus (fig. 7.8). Posterior to the anus is a terminal pore, sometimes opening from a cuticular tube that is supplied with adhesive material by a caudal gland. The mouth opens into a buccal capsule, which is followed by and may be incorporated within a muscular pharynx (termed an esophagus in much of the nematode literature). The pharynx is triangular and the pharyngeal lumen Y-shaped (fig. 7.9). Typically, six lobes or labia surround the mouth. Each labium bears two sensory papillae to form inner and outer rings of six labial papillae each; the papillae lie in dorsolateral, lateral, and ventrolateral positions. Four (sometimes six) cephalic papillae lie outside of the labial papillar rings in dorsolateral and ventrolateral positions; variations in these arrangements occur. The papillae function as anterior sense organs; they are served by six (dorsolateral, lateral, and ventrolateral) nerve strands that split so as to supply two or three papillae each; these strands connect

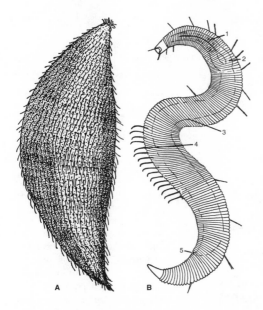

FIG. 7.7 Two free-living marine nematodes. (*A*) *Greeffiella,* a desmoscolecid. (*B*) An unidentified species of desmodorid from Puget Sound. *1*, pharynx; *2*, pharyngeal bulb; *3*, intestine; *4*, natatory bristles; *5*, anus. After Hyman 1951b.

with a nerve ring that encircles the pharynx. The anterior end of a nematode thus has a triradial or hexaradial appearance, the radial elements consisting of the feeding apparatus and its supporting sensilla.

Associated with the nerve ring around the pharynx (circumenteric ring) are ganglia (the lateral ganglia being the more important), and from these, dorsal, lateral, and ventral cords run to the posterior region to end in connectives and sensory organs. A pair of anterior sensory organs, termed amphids, lie in lateral positions, usually in or just outside of the cephalic ring of papillae, and connect by nerve strands to the lateral ganglia. There are also nerve systems in the pharynx and the rectum. Excretion involves unique renette cells that connect to a ventral pore that lies near the pharyngeal-midgut border. Body-wall muscles form a longitudinal cylinder. Small nematodes are compact and essentially acoelomate. Larger forms have perivisceral spaces, but the gut is often lined only by the basal laminae of the epithelia; although mesodermal cells sometimes occur on the gut, there is not a complete mesodermal tissue sheath. The perivisceral cavity is thus regarded as a pseudocoel.

Most studies of nematode locomotion have been on terrestrial forms (see Wallace 1968 and references therein). Free-living marine nematodes are benthic, although a few can swim weakly by serpentine motions in which the body bends in a dorsoventral direction (rather than from side to side), a motion produced by

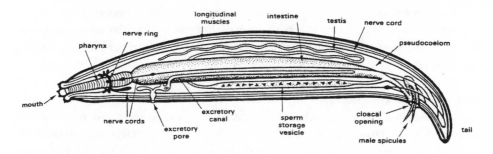

FIG. 7.8 Generalized nematode anatomy, lateral view of a male. From Pearse et al. 1987.

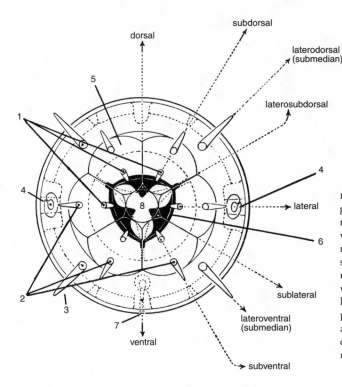

FIG. 7.9 Arrangement of organs around the nematode mouth, generalized anterior view. Note the triradiate symmetry of the jaws and associated structures, a common but not universal feature of organisms with pharyngeal pumps. *1*, inner labial papillae; *2*, outer labial papillae; *3*, cephalic papilla; *4*, amphids; *5*, labium; *6*, buccal capsule; *7*, excretory pore; *8*, mouth. After Hyman 1951b.

alternate contractions of dorsal and ventral longitudinal muscles. Nematodes may also crawl in this fashion by lying on one side, the alternate bendings of the body then producing a wavy trail, though evidently terrestrial nematodes are so light that they cannot easily make trails while submerged in water. The wavy trails are usually formed when locomotion occurs subaerially on wet substrates. Some nematodes move by waves that alternately shorten and lengthen the body; such locomotion is employed by a few nematodes that burrow (Stauffer 1924; Clark 1964). Cuticular features such as bristles and annulations are presumably used as accessory gripping structures in locomotory activity; in a parasitic form, cuticular longitudinal ridges have been interpreted as aiding in locomotion (Lee and Biggs 1990). The nature of nematode burrows themselves seems not to have been investigated. Many nematologists consider the marine forms the more primitive (Maggenti 1970), although there is evidence supporting primitive states for some characters found exclusively in terrestrial taxa (Riemann 1977).

Development in Nematoda

Very thorough studies of the relationship between cleavage and cell-fate inductions have been on the terrestrial nematode *Caenorhabditis elegans*. As luck would have it, the cleavage pattern shown by *C. elegans* is unique among invertebrates. However, the pattern does suggest the sort of distinctive developmental integration that

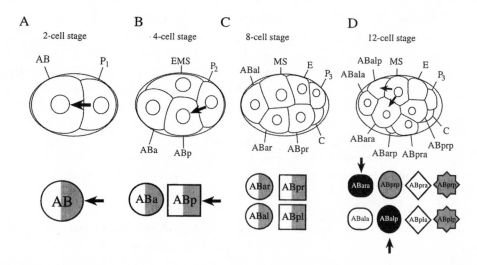

FIG. 7.10 A model of inductions in early cleavages in a nematode (*Caenorhabditis elegans*). (*A*) The first cleavage produces a large anterior cell, AB, and a smaller posterior cell, P_1. The spindle axis is oriented in an anteroposterior direction, and the P_1 cell induces polarization of the AB cell in an antero-posterior direction. (*B, C*) The AB cell divides. In order to preserve the polarization in both daughters, cleavage must be anteroposterior so as to parallel polarization, so the spindle in AB is orthogonal to an-teroposterior. P_1 divides in what would be the dorsoventral direction, except the blastomeres are shifting within the egg; spindles are oriented in the original P_1 anteroposterior but are skewed with the shift. Thus in the two blastomeres that participate in the second cleavage, the spindle orientations are more or less orthogonal. The EMS and P_2 cells are daughters of P_1; P_2 induces EMS so that one daughter, E, produces intestine. (*D*) Further inductions of the cells ABpra and ABplp complete the production of distinctive fates for each of the eight products of the AB cell. P_1 products not shown. The nema-tode cleavage pattern, orthogonal in different sister blastomeres, is termed rotational cleavage. Arrows indicate inductions. After Hutter and Schnabel 1995.

may be associated with any given cleavage type. In *C. elegans*, fertilization begins a series of movements of materials within the egg. A centrosome introduced with the male chromosomes moves toward the center of the zygote, while granules scat-tered in the egg cytoplasm migrate to the pole opposite that of sperm entry; this pole becomes the morphological posterior. In preparation for the first cleavage, the centrosome divides, and the daughters take up positions along what is to be the anteroposterior axis. The centrosomes contain microtubule-organizing centers and produce a spindle between them, so that the first cleavage plane is orthogonal to the anteroposterior axis. The cleavage plane is somewhat offset toward the pos-terior (fig. 7.10), and therefore the first two blastomeres, called AB and P_1, are slightly unequal; P_1, with the granules, is smaller. These blastomeres thus lie along the anteroposterior axis of the embryo.

A model that explains the subsequent series of early cleavages and blastomere fates has been worked out by Hutter and Schnabel (1995). In this model P_1 induces polarity in AB, producing a gradient along the anteroposterior axis of AB (fig. 7.10A). In order for each of the daughters of the AB blastomere (ABa and ABp) to preserve this gradient, the AB cleavage plane must lie along the anteroposterior axis. The spindle must be oriented with centrosomes aligned orthogonally to the gradient (otherwise one daughter would have the lower portion of the gradient and the other, the higher). Therefore, when the AB centrosome divides, the daughter centrosomes rotate 90° from their original position to lie in a plane parallel to the first cleavage and produce a spindle for the AB cleavage along the anteroposterior axis (fig. 7.10B). Slightly later, the granules have migrated to the posterior of the P_I blastomere, which divides orthogonally to the anteroposterior axis. The duplicated P_I centrosomes have rotated 90° into a plane parallel to the first cleavage, just as the AB centrosomes did, but before P_I cleavage they rotate a further 90°—180° in all—so that the P_I cleavage plane is not parallel to the anteroposterior axis but orthogonal to it. This plane is now in the same direction as the first cleavage plane but, in a sense, has rotated upside down from it (Rhyu and Knoblich 1995). The daughters of P_I are called P_2 (now with the granules) and EMS.

A most interesting event accompanies these early nematode cleavages. The nematode eggshell is oval and rigid, the long axis being the future anteroposterior axis. The dividing AB and P_I cells skew from their anteroposterior alignment to the asymmetric positions indicated in fig. 7.10B (see Hutter and Schnabel 1995). Evidently this motion is entirely mechanical, based on the unstable cell arrangement arising from the confinement of the larger, cleaving anterior blastomere within the oval egg, and the P_I blastomere slides so that, following P_I cleavage, EMS lies developmentally beneath the daughters of AB. P_2 contacts the most posterior of the AB daughters and induces this daughter, which is then ABp, while the other, non-induced AB daughter is ABa. These events, which depend on the packing geometry of the cells inside the oval eggshell, establish a dorsoventral axis.

As cleavage proceeds, EMS is polarized by an induction from P_2 and divides along a dorsoventral plane, orthogonal to the induced polarity, into E and MS cells, which have separate fates. The E line produces the gut, while MS gives rise to portions of the pharynx and other tissues (fig. 7.10C). P_2 divides, the granules segregating into one of the daughters, which divides in turn, with a daughter again receiving the P granules; this daughter (P_4) is then sequestered as the germ-line founder. In this sequence the cleavage planes have been so oriented (nearly dorsoventrally) as to deliver the posteriorly migrating granules to the germ line (see White and Strome 1996). Meanwhile, the ABa and ABp cells have divided (in a dorsoventral plane) to produce four blastomeres which then divide (along anteroposterior planes) to contribute eight of the cells at the twelve-cell cleavage stage (fig. 7.10D). The rotations of the centrosomes that are required to produce these cleavages are effected by the cytoskeleton. Each of the eight AB descendants has a different fate, owing to a complex series of inductions from cells with which they are in contact; for example, MS induction of some cells is required for proper establishment of the proximodistal axis (Hutter and Schnabel 1995).

The roles of many of the molecules involved in nematode inductions have been studied. For example, the gene *lit-1* appears to act as a binary switch to transform the identities of otherwise similar blastomeres, acting from the third to at least the eighth cleavage. Action of *lit-1* produces a posterior fate for one of the daughter blastomeres of EMS, then a posterior fate for one of its daughters, and so on in a series of transformational steps (Kaletta et al. 1997). *Lit-1* is also active in determining blastomere fates in other cell lineages.

The *C. elegans* cleavage is clearly choreographed to produce a developmental pattern in which the cleavage planes establish induction vectors, evidently involving a series of binary genetic switches, that through cell-cell interactions produce a pattern of nearly invariant blastomere fates. These processes eventually lead to the differentiation of adult cell morphotypes (see Schnabel 1996). The cleavage is an integral part of the developmental system. An unusual feature of this system is the mechanical skewing of blastomere positions, a phenomenon that the system requires, even though it is not controlled per se by the genome. It is possible that the very unusual sequence of cell divisions that characterizes this unique cleavage pattern is accounted for by the history of exploitation of the positional skewing as part of an induction pattern.

The nematode clade to which *C. elegans* belongs, the Rhabditida, is highly derived (Blaxter et al. 1998). Nematodes from other groups, while similar in overall development, do not share with *C. elegans* all the blastomere fates and the highly stereotypical features of very early development, and indeed some variation has been noted in *C. elegans* as well (see Voronov et al. 1998). For example, in the more basal marine nematode *Enoplus* (Enoplida), the only determined blastomere up to the eight-cell stage is an endodermal precursor, although gastrulation is similar to that in *C. elegans* (Voronov and Panchin 1998).

Gastrulation in nematodes is partly by epiboly as ectodermal cells spread over the endodermal precursors, while some cells invaginate around a blastopore. Mesoderm arises from some cells around the blastopore, derived from AB, EMS, and P_2 lineages, rather than from a specific mesentoblast. Toward the end of gastrulation the blastopore is elongate, and there is some variation in subsequent events, but it commonly closes in the center, and the anterior portion becomes the mouth. The remaining blastopore closes, and the anus arises later at the site of the posterior end (see reviews in Nielsen 1995 and Schierenberg 1997). Nematodes have about sixteen adult somatic cell morphotypes (see Wright 1991).

Fossil Record of Nematoda

The earliest body fossils that might represent marine nematodes are of Early Carboniferous age; there are records from Scotland (Størmer 1963) and from the Bear Gulch Limestone of Montana (Schram 1979). There is also a possible fossil nematode from the Late Carboniferous Mazon Creek fauna, Illinois (Schram 1973), and

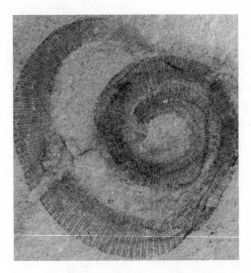

FIG. 7.11 *Palaeoscolex ratcliffei*, a vermiform fossil from the Middle Cambrian of Utah belonging to the Paleoscolecidae, a group that may represent an extinct paracoelomate phylum. There are about 400 highly papillate annuli. Specimens range from 2 to 5 mm wide. From Robison 1969.

a fairly convincing nematode is described from the Jurassic (Sinemurian) of Osteno, Italy (Arduini et al. 1983). Undoubted terrestrial nematodes have been described from lignites or ambers of Eocene and Oligocene age (reviews in Poinar 1977; Poinar 1981) and juvenile nematodes are identified in Pleistocene hyena coprolites from Italy (Ferreira et al. 1993). Wavy trails that closely resemble trails of nematodes are known from nonmarine rocks from the Triassic (Metz 1998) and Eocene (Moussa 1970). The trace fossil *Cochlichnus* is a similar wavy trail sometimes suggested as formed by nematodes, and fossils from the Neoproterozoic to the Tertiary have been assigned to that genus (Hantzschel 1975), but most are far larger than expected for nematodes. Some other minute Neoproterozoic trails have been suspected of representing nematodes, but they do not display definitive features.

Paleoscolecidae

The vermiform paleoscolecids are found from the Lower Cambrian (Atdabanian) to the Upper Silurian, and were important members of the benthos during much of that time (Müller and Hinz-Schallreuter 1993; Conway Morris 1997). These worms (fig. 7.11) are elongate and closely annulated, with the annuli encircled by rows of minute papillae or of sclerites that are now phosphatic; they may have been phosphatized posthumously. The sclerites had been known as disassociated problematica under the generic names of *Hadimopanella*, *Kaimanella*, *Milaculum*, and possibly *Utahphospha* (Hinz et al. 1990). The presumed anterior end of the body seems to be preserved on one specimen (Conway Morris and Robison 1986); it bears sclerites and is slightly expanded but lacks appendages. There are no parapodia, nor have setae been observed. The gut appears to be straight.

Paleoscolecids somewhat resemble oligochaetes (earthworms), and as there have been speculations that oligochaetes are closest in morphology to the earliest annelids, workers have proposed that paleoscolecids might be primitive annelids (Conway Morris 1977a; Runnegar 1982b). Presumably the sclerites, like the setae of many annelids, are gripping structures employed in locomotion (Conway Morris and Robison 1986). Later, Hou and Bergström (1994) suggested a nematomorph affinity for paleoscolecids. Conway Morris (1997) studied the ornamentation pattern and microstructure of sclerites of the type species of *Palaeoscolex,* and suggested that paleoscolecids are allied to or perhaps belong with the Priapulida, some of which have a similar ornament. These later findings indicate that the paleoscolecids are most likely to be paracoelomates and are possibly ecdysozoans.

Relationships of Paracoelomate Ecdysozoans

At one time taxa of pseudocoelomate grade were considered by many workers to represent a clade, based on their common lack of a "true" coelom. The entoprocts and tardigrades were often set aside as outliers. Most of the minute vermiform groups have been lumped together in various combinations—usually including Nematoda, Nematomorpha, Gastrotricha, Rotifera, Kinorhyncha, and Priapulida— as classes in a phylum Aschelminthes. In recent years the aschelminths have usually been considered polyphyletic (Lorenzen 1985; Clément 1985; Ruppert 1991b). They are treated as separate phyla here and in most recent accounts (though see Kristensen 1983 for an opposing view; and see Hyman 1951b and Ruppert 1991b for longer accounts and references). There has been a strong suspicion that at least some of these groups have descended from larger, coelomate ancestors, with the coelomic spaces being reduced and mesothelial linings lost as small body sizes evolved. Certainly the body cavities of these forms are rather heterogeneous. SSU rRNA data support the view that paracoelomate taxa have disparate kinships (Raff et al. 1994; Winnepenninckx, Backeljau, Mackey, Brooks, DeWachter, Kumar, and Garey 1995; Aguinaldo et al. 1997). However, if some paracoelomates are basal within their major clades, they could represent an early bilaterian paraclade from which the distinctive crown clades have descended. In any case it may be that some of the resemblances among paracoelomates are owing to parallel or convergent adaptations to small body size.

Onychophora

The onychophoran bodyplan was once considered transitional between those of the arthropods and annelids, though molecular data now place onychophores firmly within the Ecdysozoa (Aguinaldo et al. 1997 for SSU rRNA evidence; Boore et al. 1995 for mitochondrial evidence), while annelids are lophotrochozoans (chap. 8). Onychophora has sometimes been included in Arthropoda and united with insects and myriapods as Uniramia, a taxon that excluded crustaceans and other putatively biramous forms (Manton and Anderson 1979). However, data from SSU rRNA sequences (Giribet and Ribera 1998; Giribet and Wheeler 1999) and mitochondrial gene arrangements (Boore et al. 1995) indicate that Onychophora branches outside of crown Arthropoda (and see the morphological analyses by Wills et al. [1998]).

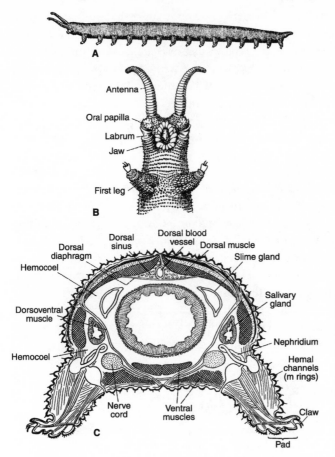

FIG. 7.12 Onychophora. (*A*) External appearance. (*B*) Ventral view of head region of *Peripatus.* (*C*) Partial transverse section of *Peripatopsis;* note spacious body cavity (white space surrounding gut and invading leg), which is a hemocoel, and the blood vascular system represented by hemal channels in leg wall. From Snodgrass 1938.

Bodyplan of Onychophora

The onychophoran body (fig. 7.12A) is covered by a cuticle that is molted during growth. While chitinous and presenting a tuberculated, rather annulated appearance, the cuticle is very thin, highly flexible, and unjointed. The paired onychophoran appendages are serially repeated and correlated with series of muscles, nephridia, and coelomoducts to form a segmented architecture. The elongate dorsal heart has ostea that also correlate with the segments, and the paired ventral nerve cords exhibit slight swellings within each segment. The body cavity, a hemocoel, is not septate. Although onychophoran bodies are neither jointed like arthropods nor compartmentalized like annelids, their uniquely segmented bodyplan has caused many workers to regard them as "intermediates" between those phyla. All living onychophorans are grouped into two families, united in a single order and class. There are about ninety described species. Evolutionary aspects of the Onychophora have been reviewed by Monge-Najera (1995).

Tagmosis is minimal; the head is not much demarcated from the body but evidently includes three segments and bears three pairs of differentiated appendages: antennae, jaws, and, between them, glandular protuberances termed oral or slime papillae. The innervated antennae are sensory. The slime papillae are connected to slime glands that run along most of the body and form organ coeloms; slime is used for entrapping prey. The jaws (fig. 7.12B) are modified appendages and bear claws similar to the legs, mincing food items with slashing motions that alternate from claw to claw, unlike any arthropod jaws. The body wall contains thin oblique and circular muscle layers, and is underlaid by thick longitudinal muscle bands. The hemocoel acts as a hydrostatic skeleton (fig. 7.12C). Pairs of nephridia occur in leg-bearing segments (except the segment that bears the genital opening). There are three nephridial types, ranging from a simple type anteriorly to complex metanephridia posteriorly, which arise in coelomic sacs and compose the so-called segmental organs (Storch and Ruhberg 1993). The limbs are lobopods, usually stumpy ventrolateral appendages with a cuticle and with muscles that are continuations of the general body-wall musculature, surrounded by ridges containing vascular channels. The lobopodial musculature, which is entirely extrinsic, is partly antagonized by extensions of the hemocoel (see Manton 1950, 1973, 1977). An anterior dorsal brain is joined to two strong ventrolateral longitudinal cords that are connected by numerous transverse commissures, with motor and sensory nerves to each segment.

Development in Onychophora

Onychophorans do not display spiral cleavage, though they have often been considered members of a spiralian clade. Some forms are viviparous, but Anderson (1973) considers that oviparous forms are less derived with respect to early development. Although the eggs are yolky, Anderson was able to form from them an interpretation of the basic onychophoran developmental pattern. Cleavage produces an envelope of small cells surrounding the yolk. Gastrulation proceeds largely by invasion of the interior by proliferating surface cells. Development is direct. A voluminous embryonic hemocoel surrounds the gut, to become somewhat divided by muscle sheets. Presumptive mesoderm is located posteroventrally, and as development proceeds, paired bands of mesoderm migrate anteriorly, eventually breaking into segments. The organ coeloms appear as ventrolateral splits in the segments in an anteroposterior sequence that persist to form the coelomoducts, excretory except at the anterior (where they form the slime glands) and posterior (where they form gonads).

Onychophora and Fossil Lobopods

Within the last few years a number of forms that somewhat resemble living onychophorans have been recovered from Cambrian lagerstätten, chiefly in the Chengjiang fauna of China, but also from the Burgess Shale of Canada and from Greenland and the Baltic Shield (Dzik and Krumbiegel 1989; Ramsköld and

Hou 1991; Hou and Bergström 1995; Ramsköld and Chen 1998). Modern ony-chophorans, all terrestrial, form a morphologically compact group. Each of the Cambrian forms is more different from the others and from living onychophorans than living onychophorans are from each other, and all were marine. The living and fossil forms are united by possessing a series of lobopods supporting generally cylindrical bodies. The fossil forms that are well preserved display body and limb annulations, presumably indicating hemal channels as in living onychophorans. The lobopods end in claws, but they do not have differentiated "foot" pads as do living onychophorans. It appears that these groups may form a clade, a phylum in Linnean terms, once quite diverse but now restricted to the surviving lineages of onychophorans. The term Lobopodia is sometimes used for these fossil forms plus the living onychophorans, but as Lobopodia originally included arthropods, and as lobopods may have been present in other, distinctive bodyplans, simply expanding the concept of Onychophora to embrace these fossils as stem groups seems less confusing.

Aysheaia, Luolishania, *and* Xenusion. The Cambrian fossil most closely resembling living onychophorans is *Aysheaia,* known from the Middle Cambrian Burgess Shale and correlatives (fig. 7.13; see Whittington 1978). It has a homonomously annulated, lightly tuberculated trunk and clawed, annulated limbs shaped much like those of living onychophorans, but with a pair of unique anterior limbs that project sideways, perhaps for grasping prey. The trunk is not ornamented. *Luolishania,* from the Chengjiang fauna, is not as well known but so far as can be told seems similar to *Aysheaia,* although it has a narrower body and limbs, and the trunk annulations at the site of limbs are wider than the other annulations and bear paired nodes (fig. 7.14A; see Hou and Chen 1989a). *Xenusion,* from presumably Lower Cambrian float (reworked rocks) of the Baltic Shield, is stouter than either of the preceding forms and bears even stronger paired nodes on limb annuli (fig. 7.14B; see Dzik and Krumbiegel 1989). None of these three fossil lobopods are known to bear antennae, unlike living ony-chophorans. Still other onychophoran-like genera are recorded from the Chengjiang fauna (e.g., *Paucipodia;* Chen, Zhou, and Ramsköld 1995b).

Armored Lobopods. Beginning with the description of a remarkable specimen of *Microdictyon* from the Chengjiang fauna (Chen et al. 1989a), an assemblage of "armored" onychophorans has been brought to light (fig. 7.15).

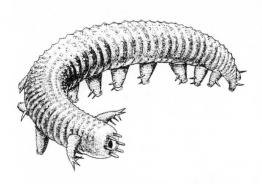

FIG. 7.13 *Aysheaia,* a lobopod resembling living onychophorans, from the Middle Cambrian Burgess Shale. Reconstruction by Marianne Collins. From Briggs et al. 1994.

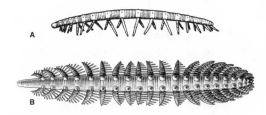

FIG. 7.14 Reconstructions of two Early Cambrian lobopods. (A) *Luolishania*, a lobopod with differentiated (heteronomous) segments, from the Chengjiang fauna. (B) *Xenusion*, a lobopod with differentiated and strongly noded segments, from the Baltic Shield. A from Hou and Bergström 1995; B from Dzik and Krumbiegel 1989.

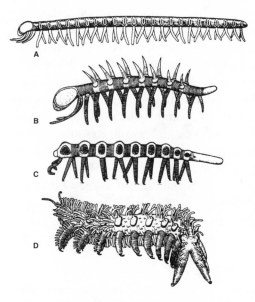

FIG. 7.15 Reconstructions of four "armored lobopods" from the Chengjiang fauna. (A) *Cardiodictyon*. (B) *Hallucigenia*. (C) *Microdictyon*. (D) *Onychodictyon* (some papillae removed to show spiny plates). These forms range around a few centimeters in length. From Hou and Bergström 1995.

These forms bear plates or spines, some of which occur in isolation as fossil sclerites, on limb annuli. Four genera are now known in the Chengjiang fauna, one of which (*Hallucigenia*) was previously known from a different species in the younger Burgess Shale but was not interpreted as a lobopod (Conway Morris 1977b). Still other Cambrian forms are known. Ramsköld (1992) has compared and contrasted these fossils in a search for homologies, and has arranged them in a morphocline (not intended to represent an evolutionary series), based on annulation, node strength, presence and character of plates or spines, and other features (see also Budd 1996). The times of appearance of these forms do not correspond well to the morphocline; the simplest (*Aysheaia*) is known from the Middle Cambrian, while *Microdictyon,* farther along on the morphocline, has the earliest record, being known from phosphatized sclerites in the Tommotian and ranging into the early Middle Cambrian. Considering the nature of the fossil record, however, the detailed stratigraphic order of such rare fossils is unlikely to reflect faithfully a temporal branching series.

Later Forms. Post-Cambrian Paleozoic records of Onychophora are from the Essex fauna, Carboniferous (Westphalian D) of Illinois (Thompson and Jones 1980) and from the Stephanian B of the Autun Basin, France (Rolfe et al. 1982). The preserved morphology of this fossil species, *Ilyodes inopinata,* resembles living onychophorans closely, although the "feet" are undifferentiated; unfortunately, the figured fossil is incomplete anteriorly. The associated fauna includes both marine and terrestrial (including aquatic) elements, so the life habitat of *Ilyodes* is uncertain.

Finally, fossils from Tertiary amber from the Baltic (c. 40 Ma) and the Dominican Republic (c. 20–40 Ma) closely resemble living onychophorans (Poinar 1996). Slime glands (and indeed fossil slime) can be identified in these forms, but they too do not have differentiated "feet," though they were certainly terrestrial.

Relationships among Living and Fossil Onychophorans

As emphasized by Ramsköld and Hou (1991), the disparate Cambrian fossils imply that a clade of organisms with an onychophoran-style bodyplan underwent a significant early radiation to produce the forms discussed above. Perhaps, then, a broad radiation of stem lobopod-bearing animals preceded or accompanied the arthropod radiation, and within this disparate clade the living onychophorans represent only one branch, the crown group that has survived to the present day. This evolutionary pattern is much like the early arthropod radiation, though more muted. Very different groupings of the fossils can be made, depending upon which characters are thought to be important. For example, antennae seem to be absent from the Cambrian fossil forms but are present in living onychophorans. This feature, together with the small terminal mouth of Cambrian forms, has served to unite those Early Cambrian lobopodians in a proposed new phylum, Tardipolypoda (Chen and Zhou 1997), believed to be most closely related to Tardigrada, which also lacks antennae. Grouping on the presence of platelike sclerites, of posterior body extensions, or of posterior limb pairs that lack corresponding nodes or plates produces different associations. The bodyplan of Onychophora (*sensu lato,* as the phylum containing the Cambrian lobopodians) originated at least by the Tommotian Stage, at about the same time that trails that are probably of arthropods, and burrows that are possibly of annelids, appear in the record. When onychophorans invaded the land, or when they became extinct in the sea, is unknown.

Tardigrada

Tardigrades (fig. 7.16) are minute, with body lengths from 0.05 mm to just over 1 mm, and are represented today by about 550 described species. They have commonly been placed in three orders (Ramazzotti 1972; Morgan and King 1976; Morgan 1982), but Kinchin (1994) has employed four. All tardigrades are aquatic, and most are terrestrial (many live in water films), but essentially all members of one order, the Arthrotardigrada, are marine, as is one family of the order Echiniscoidea. In general the marine forms are morphologically more diverse than the terrestrial ones, often having flamboyant cuticular extensions (fig. 7.16B–C). Tardigrades are firmly established as ecdysozoans related to Arthropoda by SSU rRNA data (Giribet et al. 1996; Garey, Krotec, Nelson, and Brooks 1996; Aguinaldo et al. 1997; Giribet and Ribera 1998) and were placed as sisters to the Arthropoda in a study by Wheeler (1998). However, the relationships among Arthropoda, Onychophora, and Tardigrada have not been completely resolved.

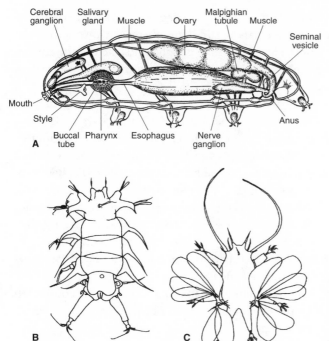

FIG. 7.16 Tardigrada. (*A*) Generalized anatomy of a terrestrial tardigrade. (*B, C*) Two marine arthrotardigrades, *Parastygarctus* and *Tanarctus*, respectively. A from Cuénot 1949; B, C from Kinchin 1994.

Bodyplan of Tardigrada

Tardigrades have an indistinct head region that lacks antennae and a trunk that bears four pairs of appendages—lobopodlike legs—terminating in claws or adhesive disks (fig. 7.16A). There is a layered but flexible chitinous cuticle that is molted frequently. The cuticle is usually broken into rings by transverse furrows. Usually, every other ring (cuticular "segment") bears a leg, but there is much diversity in this feature; in marine forms especially, the cuticle may be smooth or crossed by weak transverse creases. There is no circular muscle coating on the body wall, which is latticed by a network of muscle cells. In most marine forms, which feed chiefly on algae and detritus, the mouth is subterminal and leads into a buccal tube that contains stylets and is succeeded in turn by a muscular pharyngeal bulb that serves as a pump to ingest food items. The midgut is large but simple and is unsupported; it leads to a posteroventral cloaca and anus. Many of the terrestrial forms have glands, believed to be excretory, that are termed Malpighian tubules (after the insect organs), but these are absent in marine species. There is no blood vascular system. A single gonad, a mesodermal sac, lies in the dorsal body cavity. The nervous system consists of a large dorsal anterior lobate brain that gives off a doubled ventral ganglion from which a pair of ventral nerve cords run longitudinally. The entire brain of the tardigrades has been homologized to the arthropod protocerebrum, the anteriormost arthropod brain region, by Dewel and Dewel (1996). Four more doubled ganglia occur along the ventral cords, each producing nerves to innervate the legs; other minor ganglia and nerve cords occur. Thus there is a distinctive correlation between the seriation of the legs and nerve ganglia and often of the cuticle, with a head, three trunk segments with legs, and a posterior segment with legs usually being distinguishable.

Development in Tardigrada

Tardigrade cleavage is total and equal. During development, a pair of anterior ectomesodermal pouches and a following series of four pairs of endomesodermal pouches, all with schizocoel compartments, are elaborated (Marcus 1929; Bertolani 1990). The walls of the anterior pouch and of three of the following pouches give rise to the body musculature, while the posterior pouch walls fuse to form the gonad. None of the adult body cavity derives from these coelomic compartments; it forms on the site of the blastocoel. Therefore, although the tardigrades are sometimes called coelomates (e.g., Morgan 1982), their perivisceral body space is a pseudocoel (Dewel et al. 1993), and some workers decline to concede a coelomic status to any of their organs (e.g., Clarke 1979), though the gonadal spaces have been considered coeloms by others. Tardigrades have about eighteen cell morphotypes (Dewel et al. 1993).

Fossil Record of Tardigrada

The earliest known fossil tardigrades are marine, of Late Cambrian age from Siberia (Walossek, Müller, and Kristensen 1994). Terrestrial tardigrades quite similar to living species are described from Cretaceous amber (Cooper 1964). These occurrences appear to be the only recorded tardigrade fossils.

Arthropoda

Among living marine invertebrates, the arthropods (fig. 7.17) may equal or surpass the mollusks in species numbers and exhibit about the same disparity in body types; the major taxa are listed in table 7.1. Terrestrial arthropod diversity is legendary but involves derived forms (such as spiders and insects), which are not considered here. The position of arthropods as a member of a clade distinct from the classic spiralian protostomes was suggested in early SSU rRNA trees (e.g., Lake 1990) and has been supported in most subsequent studies. According to molecular phylogenies, arthropods are ecdysozoans (Aguinaldo et al. 1997).

Bodyplan of Arthropoda

Major Clades. As the name implies, arthropods have jointed appendages. The arthropod exoskeleton (fig. 7.18) is a complex cuticular structure of chitin and protein, sometimes thick, stiff, and elaborately layered, sometimes thin and flexible, that encloses the entire body, including the appendages, and extends into

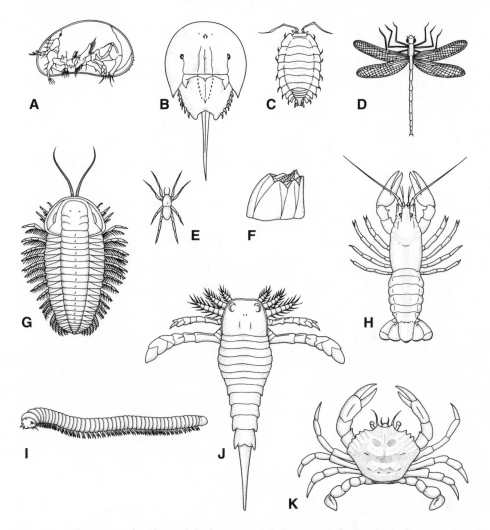

FIG. 7.17 An array of arthropod body types. (A) Crustacea: Ostracoda (valve transparent). (B) Chelicerata: Xyphosura (horseshoe crab). (C) Crustacea: Isopoda (sow bug). (D) Hexapoda (dragonfly). (E) Chelicerata: Arachnida (spider). (F) Crustacea: Cirripedia (barnacle). (G) Trilobita: Polymerida (extinct). (H) Crustacea: Decapoda (crayfish). (I) Myriapoda (millipede). (J) Chelicerata: Eurypterida (extinct). (K) Crustacea: Decapoda (crab). From Robison and Kaesler 1987.

the fore- and hindgut. The exoskeleton is composed of a series of segments between anterior and posterior body divisions. In most marine forms, each segment is more or less well sclerotized, but separated by narrow, lightly sclerotized sections that provide for intersegmental flexibility. The well-sclerotized sections, or sclerites, are commonly divided into dorsal tergites and ventral sternites, separated laterally by pleurites (fig. 7.19); these elements may become separated upon death of the

TABLE 7.1 Major taxa of living Arthropoda. Arrangement of subphyla based on gene order and molecular sequences. After Brusca and Brusca 1990.

Taxon	Remarks
Subphylum Chelicerata	
Class Merostomata	Horseshoe crabs; marine
Class Arachnida	Spiders, scorpions, etc.; terrestrial
Class Pycnogonida	Sea spiders; marine
Subphylum Myriapoda	
Subclass Chilopoda	Centipedes; terrestrial
Subclass Diplopoda	Millipedes; terrestrial
Subclass Symphyla	Terrestrial
Subclass Pauropoda	Terrestrial
Subphylum Crustacea	
Class Remipedia	Marine caves
Class Cephalocarida	Marine benthos
Class Branchiopoda	Brine shrimps, etc.; a few marine
Class Maxillopoda	
Subclass Ostracoda	Many marine, some freshwater
Subclass Mystacocarida	Benthic marine, interstitial
Subclass Copepoda	Chiefly marine, important zooplankters
Subclass Branchiura	Fish lice; marine to freshwater
Subclass Cirripedia	Barnacles; marine
Subclass Tantulocarida	Marine parasites
Class Malacostraca	
Subclass Phyllocarida	Marine
Subclass Eumalacostraca	Crabs, shrimps, lobsters, etc.
Subphylum Hexapoda	Insects; terrestrial

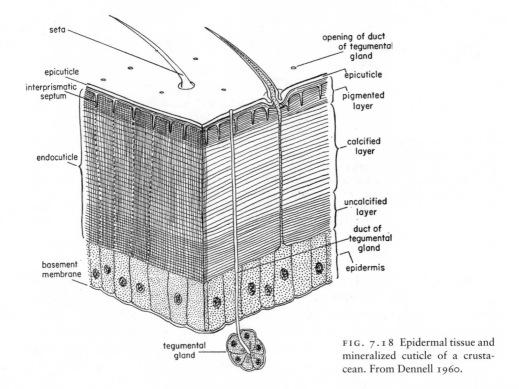

FIG. 7.18 Epidermal tissue and mineralized cuticle of a crustacean. From Dennell 1960.

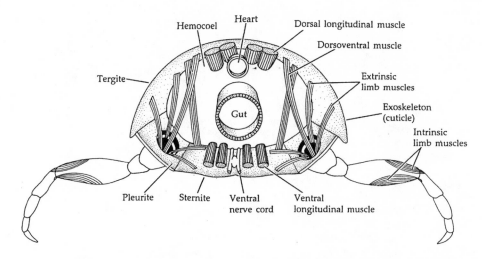

FIG. 7.19 Transverse section of a generalized arthropod to show major skeletal elements and arrangement of gut, blood vessels, and nerve cord. From Brusca and Brusca 1990.

animal. Small-bodied forms require only light sclerotization, and many are practically soft-bodied. In some marine forms the exoskeleton is partially mineralized, which greatly improves fossilizability. Once secreted, the exoskeleton cannot be enlarged to any significant extent, so that growth is accommodated by a series of molts. The hydrostatic skeleton of arthropods is hemocoelic; technically coelomic spaces are small and in adults surround some glands or such paired organs as metanephridia. Most arthropod body segments are characterized internally by sets of muscles, paired nerve ganglia, and blood vessels. Many of the muscles serve the appendages and do not necessarily insert on the segment in which their appendage is contained, while others help to move the segments themselves, and again they may straddle segmental boundaries; there is also a complex system of intersegmental tendons. Usually, neither gonads nor excretory organs are serially arranged. The nervous system consists of anterior dorsal ganglia that form a bi- or tripartite brain, consisting of the protocerebrum, deuterocerebrum, and tritocerebrum; the deuterocerebrum is absent in chelicerates. From these ganglia a ventral longitudinal cord, sometimes paired, extends back along the body, with a series of ganglia that usually correlates with segmentation but can be quite variable. Nerves from the cord serve appendages and body-wall muscles, though in the head, appendages and eyes are innervated from the cerebral ganglia, as is usually the anterior part of the alimentary canal. Arthropods are known for their abundance of sensory neurons and the relative paucity of motor neurons (presumably because they are efficiently organized). Clearly, the segmented nature of the arthropod body is related to the mechanics of body movement, particularly to locomotion, with nerve and blood supplies in support.

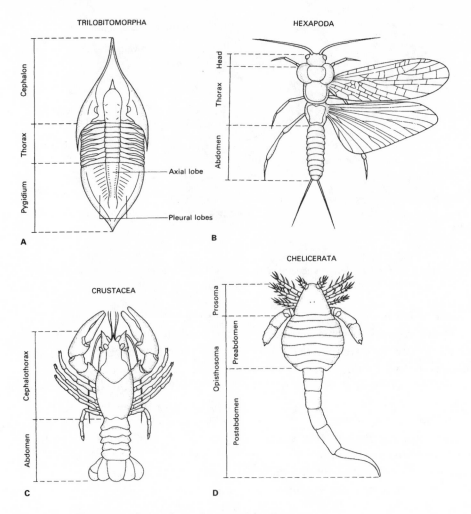

FIG. 7.20 Tagmosis in four arthropod groups. After Størmer 1959.

The arthropod body is often markedly regionated (fig. 7.20). The jointed appendages, encased within their sclerotized cuticles, have systems of muscles that operate chiefly through leverlike arrangements. Between joints the appendage sections, termed podomeres, are often heavily sclerotized or even mineralized, but at the joints themselves the cuticle is a flexible membrane. Appendages are commonly specialized, varying in function between tagmata and also within a tagma in many cases (fig. 7.21), or are lacking altogether in some segments and tagmata. The nature and position of the various appendages are among the characters used to define the higher taxa of living arthropods. Two basic appendage types are unbranched uniramous, the only type occurring in myriapods and hexapods (which have sometimes

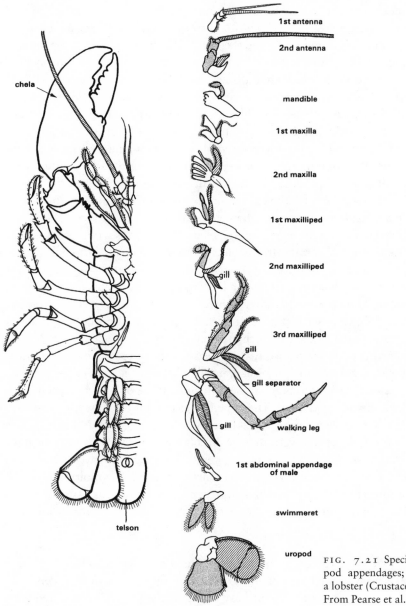

1st antenna

2nd antenna

mandible

1st maxilla

2nd maxilla

1st maxilliped

2nd maxilliped

gill

3rd maxilliped

gill

gill separator

gill

walking leg

1st abdominal appendage of male

swimmeret

uropod

chela

telson

FIG. 7.21 Specialized arthropod appendages; these are on a lobster (Crustacea: Decapoda). From Pearse et al. 1987.

been united as the taxon Uniramia), and biramous, with two branches, an inner leg branch and an outer gill branch, as found in most crustaceans (fig. 7.21). Living chelicerates are uniramous but were biramous primitively, having lost the outer branch on the appendages of the anterior tagma (prosoma) and the inner branch on the appendages of the posterior tagma (opisthosoma).

In crustacean and chelicerate limbs, the inner surface of the coxa supports a cusped projection, the gnathobase. Food items are moved into the space between gnathobases by a variety of methods, and then carried anteriorly toward the mouth, either mechanically or via forward-directed currents generated by limb movements. When moved toward each other, gnathobases produce a crushing effect, while if moved by each other, they cut and slash, and they may grind food between them in either case. Crustaceans possess mandibles, specialized appendages evolved from legs; these mouthparts move in essentially an anteroposterior direction, using muscles that are serially homologous with those employed in swinging the legs; opposing mandibles, essentially modified gnathobases, roll past one another during the swing to produce grinding and shredding effects. Marine chelicerates, on the other hand, lack mandibles but employ several pairs of gnathobases on anterior legs to prepare food. The primary movement between these gnathobases is transverse and uses the muscles employed in flexing the legs. Thus the feeding apparatuses of these two taxa have evolved separately, diverging from some common ancestor. Although myriapod jaws have mandibles, they are not multisegmented; the processing part is not a gnathobase but rather has evolved from a limb tip; the movements are transverse as in chelicerates. Thus myriapods have evolved a third feeding system. Manton (1977) and others have described a number of modified food-processing systems within each of these taxa, some of which are convergent.

Although arthropods are plausibly described as segmented, it could be argued that the bodyplan is better thought of as jointed. The point is that jointing of the elongate, exoskeletally armored body of the arthropod is the only obvious system to provide for flexibility in movement. The appendages exhibit the same problem and solution. The musculature is serially repetitive in order to support and manipulate the serial sclerites and appendages, and the circulatory and nervous systems have serial elements that support the muscular system. The incredible diversity of form found within the arthropod bodyplan was chiefly driven by exploitation of the system of jointing. There is no hint that the arthropod body was ever separated into isolated compartments such as characterize the segments of many annelids (chap. 8).

Pentastomids. The pentastomids or tongue worms have a bodyplan that is distinctive enough to have been classed as a phylum of its own by many workers. Pentastomids are obligate parasites and do not have many of the features associated with the arthropod bodyplan—they are vermiform and have only two pairs of hooklike appendages born anteriorly; trunk segmentation is obsolete. Nevertheless, some workers have developed evidence to suggest that pentastomids are aberrant crustaceans modified for a parasitic lifestyle (e.g., Wingstrand 1972), a hypothesis now supported by SSU rRNA evidence (Abele et al. 1989). Pentastomid nervous systems are thought to be reduced, but there are tripartite brains and a ventral cord with ganglia in the hook segments. Remarkable fossil pentastomid material from the Upper

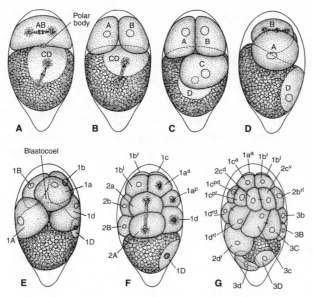

FIG. 7.22 Cleavage stages of a cirripede barnacle, *Tetraclita* (Crustacea: Arthropoda), worked out by Anderson (1969), who has labeled the blastomeres as if they were the result of classic spiral cleavage, albeit modified. (*A*) Two-cell stage, dorsal view. (*B*) Three-cell stage, dorsal view. (*C*) Four-cell stage, dorsal view. (*D*) Four-cell stage, left lateral view. (*E*) Eight-cell stage, left-lateral view. (*F*) About sixteen-cell stage, left-lateral view. (*G*) Twenty-eight-cell stage, left-lateral view. Note that cleavage planes do not pass through the yolk. From Anderson 1973 after Anderson 1969.

Cambrian of Sweden is interpreted as indicating a more basal position within Arthropoda, predating euarthropod branchings (Walossek, Repetski, and Müller 1994; Walossek and Müller 1998).

Development in Arthropoda

Although the terrestrial arthropods have highly modified cleavage patterns, some of the marine groups display cleavages that may represent a less derived state (Anderson 1973, 1979, 1982; Williamson 1982; Gilbert 1987). The cirripede barnacles are among the best studied, and it has been suggested that they indicate the basic mode of crustacean cleavage (Anderson 1969). The eggs are oval, with the smaller end being the embryonic posterior, and are rather yolky. The first cleavage creates cells that in the terminology of spiral cleavage are AB and CD; AB is yolk-free (fig. 7.22A–B). AB cleaves in the long axis of the egg, and then CD cleaves at an angle to the long axis so that one yolk-free cell (C) is the more anterior, and D, the larger, retains the yolk; as this occurs, cell A slips posteriorly. This leaves the cells packed in a sort of deformed tetrahedral arrangement (fig. 7.22C–D), like cells in many terrestrial annelids (clitellates) with yolky eggs, which led Anderson to conclude that, as in the clitellates, cirripedes display a modified spiral cleavage. It is clear that the presence of abundant yolk (and perhaps the shape of the egg) influences the cell packing.

Proceeding to the eight-cell stage, the cleavage planes are at right angles to the offset planes of the four-cell positions, so their associated spindles are not orthogonal to the embryonic axis (fig. 7.22E). In each of the next two cleavage stages the spindle axes also rotate 90° from the previous stage, much as in classic spiral cleavage. The axes continue tilting with respect to the embryonic axis, although the cells are not positioned in alternating circlets as in the classic spiralian case (fig. 7.22F–G). At the end of the fifth cleavage, there are usually twenty-eight cells—not all cells participate in all divisions—the yolk remaining in the most posterior D-quadrant cell, which then cleaves again to produce 4D, which retains the yolk, and 4d, a terminally posterior, yolk-free cell. Blastomeres arising in other quadrants have spread to nearly cover the yolk. The cell fates are distinct from those in classic spiralian development; the 4d cell is not the mesentoblast, but joins a variety of cells derived from all other quadrants to form ectoderm. According to the spiralian terminology adopted by Anderson, cirripede mesoderm arises from the 3A, 3B, and 3C blastomeres (which form endoderm in classic spiralians) to form paired internal bands. Costello and Henley (1976) have criticized Anderson's identification and terminology of the mesoderm and endoderm precursors, but at any rate it seems that none of the blastomeres involved in mesoderm formation is homologous with the mesentoblast of classic spiralians.

Whether or not crustacean cleavage is truly spiral has been long debated; Anderson established that it is possible to interpret it as a spiralian pattern modified to accommodate yolk, but the differences between classic spiralian and cirripede cleavages are striking even using his terminology (fig. 7.23). Other arthropods, including some cladocerans, copepods, euphasiids, and even decapods, have radial cleavage patterns (Anderson 1973). The radial patterns have been interpreted by some as secondary modifications of a primitive spiralian cleavage. If the relationships suggested by SSU rRNA sequences are generally correct, however, there may be radialian ecdysozoan clades that branch more deeply than the arthropods. It is thus possible that the spiral cleavage found in some arthropods (some cladocerans and ostracods in addition to cirripedes) is derived and results from a developmental system, involving axial tilting of the spindles, that evolved independently of classic spiralians. Such an innovation can produce early embryos whose blastomeres may

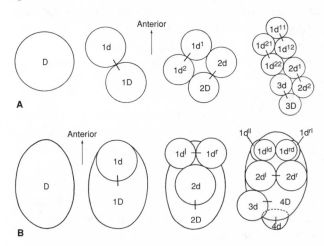

FIG. 7.23 Classic spiral cleavage contrasted with cirripede cleavage; only D quadrants illustrated, third to fifth divisions, viewed from dorsal surface, anteroposterior axis vertical. (A) Classic spiralian cleavage. (B) Cirripede cleavage. From Anderson 1973.

be numbered according to the Wilsonian system (as they start with a quartet of cells after the second cleavage) yet which are homoplastic with respect to the classic spiralians. This possibility is strengthened by the nonspiralian derivation of mesoderm in crustaceans, further calling into question the homology of arthropod and classic spiralian blastomeres and cleavages. Scholz (1998) reviewed arthropod cleavages and concluded that the ancestral cleavage was radial, but with some irregularities in spindle orientation.

In many direct-developing forms, gastrulation is accomplished by the proliferation of micromeres, while in others the presumptive mesodermal cells migrate to the interior. Chelicerates, with yolky eggs that cleave superficially, display no spirallike cleavage at all. Yolky cells come to occupy the interior of a stereoblastula, but they give rise to the midgut and its diverticula. Mesoderm, together with additional endoderm, arises from cells along the midline of the ventral surface of this blastula, the proliferating cells migrating into the interior. Myriapods are terrestrial today (with the minor exception of a few species that have readapted to marine conditions), but may be represented by marine fossils. The developmental patterns of living terrestrial uniramians may well differ significantly from marine forms from which they have descended, but they are all that we have. In chilopods (centipedes) eggs are quite yolky and do not display spiral cleavage (see Anderson 1973). The blastula develops from a superficial cleavage that produces a layer of cells surrounding a central yolk. Mesoderm arises from the posteroventral portion of the blastula, cells migrating into the interior to produce paired bands. The midgut is derived from ventral cells anterior to those that produce mesoderm. Other myriapods, and the insects as well, display a variety of patterns.

Thus late embryogenesis is exceedingly varied among arthropod groups, as are the succeeding stages of development. Some marine forms show a series of larval types, others have direct development. In general, the blastopore, when one exists, closes, and both the mouth and anus open at different sites—the embryo is not literally protostomous. The mesodermal bands tend to pinch off as serial tissue sheets, and as the circulatory system and musculature develop, the segmental architecture of the arthropod becomes evident. Thus a few segments differentiate within the embryonic body, and commonly the embryo hatches as a larva, the nauplius, at a stage with three appendage-bearing segments (fig. 7.24). This larva, like all arthropod larvae, lacks motile ectodermal cilia. As marine arthropods have durable exoskeletons and molt to accommodate growth, changes of form leading from the larva to an adult occur in steps associated with the skeletons secreted after a molt. Molting stages vary greatly in number and appearance among groups, but among crustaceans three successive larval types are recognized, first the nauplius (swims with cephalic appendages), then the zoea (swims with thoracic appendages), and finally the megalopa (swims with abdominal appendages). Postnaupliar segments are usually budded off from a generative zone just in front of the most posterior body unit, the telson. Peterson et al. (1997) considered

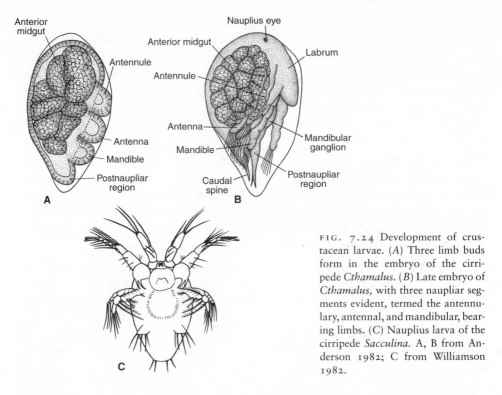

FIG. 7.24 Development of crustacean larvae. (A) Three limb buds form in the embryo of the cirripede *Cthamalus*. (B) Late embryo of *Cthamalus*, with three naupliar segments evident, termed the antennulary, antennal, and mandibular, bearing limbs. (C) Nauplius larva of the cirripede *Sacculina*. A, B from Anderson 1982; C from Williamson 1982.

arthropods to be direct developers because their adult bodyplans are organized in the nauplius.

The nature of body cavities has played a large role in arguments about arthropod phylogenetic relationships. Within the segments, cavities arise between the gut and the sheets of mesoderm, the site of the blastocoel if one has existed, coalescing to form the hemocoel. Usually the hemocoel is divided into two compartments by a horizontal septum that lies just above the gut, to form a dorsal pericardial sinus and a ventral perivisceral sinus. Never is there a vertical mesentery present to support the gut (see Clarke 1979). In some taxa other cavities arise as splits within the paired mesodermal segments of the larva, thus qualifying as schizocoels. The larval function of these cavities is uncertain, and their fate varies (Anderson 1973). In many forms the spaces are occluded as development proceeds. In others at least some of the spaces remain as coelomoducts, serving glands or excretory or reproductive organs. And in still others coelomic spaces become confluent with the hemocoel (Anderson 1973; Clarke 1979). In such cases the hemocoel has contributions from both blastocoelic and coelomic sources (a "myxocoel"). The arthropod circulatory systems, in all their ramifications and variations, never display evidence of having evolved from anything other than a primary body cavity (Clarke 1979). According to Bonner (1965), an arthropod bodyplan may be constructed with about fifty-five cell morphotypes. Some living marine arthropods are considerably less complex; judging from the account of branchiopods, which branch early among crown-group arthropods, a simple arthropod bodyplan might employ thirty-seven or so cell types (Martin 1992). Some living arthropods are considerably more complex; the fly *Drosophila* is estimated to have about ninety cell types (Sneath 1964).

Early Fossil Record of Arthropoda

Although many early arthropods had nonmineralized cuticles, a marvelous diversity of early arthropod body types has come to light, so many and so distinctive as to pose important problems in applying the principles of systematics. The most diverse of the extinct arthropod groups is the Trilobita, which were chiefly mineralized and are usually treated as equivalent in Linnean rank to the Crustacea and Chelicerata. However, a large number of nontrilobite fossils, with jointed bodies and appendages, display great disparity in just those features that form defining characteristics of the living higher arthropod taxa—tagmosis, including segment numbers, and the number, type, and placement of appendages (see Wills et al. 1997 for review of tagmosis and appendage types). Most of the Early and Middle Cambrian forms have such unique assemblages of those characters that they cannot be included in any of the living higher taxa as they are defined within crown groups, and many of the fossil taxa are quite distinct from each other as well. These disparate arthropod types are phylogenetically puzzling; some can be included as stem taxa in clades that include living groups, while others are commonly considered separate taxa of very high rank. Some of these taxa appear to be arthropod stem groups, but others seem even more distantly allied to basal arthropods. This evidently sudden burst of evolution of arthropodlike body types is outstanding even among the Cambrian explosion taxa (Jacobs 1990).

The earliest fossils assigned to arthropods are trace fossils left by organisms that had limbs that could scratch grooves in sediment and excavate hollows by raking or digging, or that made other trails indicating the activity of legs. *Cruziana* (fig. 5.14) is among the earliest of such traces to be identified with confidence. Probable arthropod traces appear at least by the Lower Cambrian Tommotian Stage, and are usually attributed to trilobites, even though trilobite body fossils do not appear until the next stage, the Atdabanian. Whittington (1980) analyzed the possible limb motions of the trilobite *Olenoides* and concluded that *Cruziana* traces were probably not made by trilobites at all, although perhaps by another sort of arthropod. However, Fortey and Seilacher (1977) presented strong evidence that some *Cruziana* are indeed trilobite traces. Interconnected two-dimensional burrows are reported from shallow-water carbonate facies, evidently of Atdabanian age, in the western United States (Droser and Bottjer 1988a, 1988b), and it is possible that these were formed by burrowing arthropods.

Arthropodan body fossils first appear in the Atdabanian; they include forms assigned to stem-group arthropod taxa as well as chelicerate-like and crustaceanlike

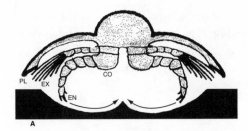

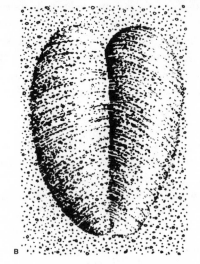

FIG. 7.25 Trace fossils attributed to arthropod activity. (*A*) Cross section of a trilobite and hypothetical feeding activity. *CO*, coxa; *EN*, endite; *EX*, exite; *PL*, pleural lobe. (*B*) The trace termed *Rusophycus*, interpreted as an arthropod resting burrow, in which trilobites have been found (not to scale). A after Seilacher 1970; B from Frey 1975.

fossils. The branching order of the taxa is unclear, and they are assigned to either crown or stem groups by different workers; Budd and Jensen (2000) have reviewed this problem and question whether any crown-group members of living classes are known before the Late Cambrian.

Among crustaceanlike forms, taxa from the Chengjiang fauna that include some soft-part preservation have been referred to Branchiopoda (Hou and Bergström 1997; Hou 1999), and fossils somewhat resembling Phyllocarida also occur in the Chengjiang (Hou 1987a), although appendages are not preserved. These taxonomic assignments are not secure. Small bivalved fossils from the Burgess Shale, with appendages, have also been tentatively referred to the Phyllocarida; the best known of these is *Canadaspis* (fig. 7.26; Briggs 1978, 1983). The phyllocarid assignment of canadaspids has been strongly disputed (Dahl 1984). An extinct group of bivalved forms, the Bradoriida, is widespread and locally abundant in Lower Cambrian rocks in Australia, China, and elsewhere (Öpik 1968; Jones and McKenzie 1980; Bengtson et al. 1990; Hou and Bergström 1997; Bergström and Hou 1998; Shu, Vannier, Luo, Chen, Zhang, and Hu 1999; and references therein). Bradoriid soft parts are preserved well enough in material from the Chengjiang fauna to permit a reconstruction (fig. 7.27). Although the bradoriids fall outside the limits of crown-group crustaceans, lacking, for example, head appendages differentiated in the crustacean pattern, they are undoubted arthropods and may represent stem-group crustaceans (Shu, Vannier, Luo, Chen, Zhang, and Hu 1999). A number of other early fossils known from bivalved carapaces have been referred to the Crustacea, but bivalved carapaces associated with soft parts from the Chengjiang fauna demonstrate that carapace morphology alone is not enough to support systematic placement within the arthropods (Hou 1999). *Isoxys* is a possibly crustacean-like bivalved form that is widespread in the Early Cambrian and persisted into

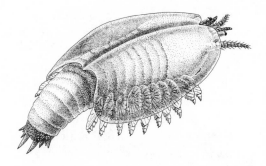

FIG. 7.26 *Canadaspis,* an early bivalved arthropod from the Burgess Shale, possibly a stem-group crustacean. Specimens range from about 1 to 5 cm in carapace length. Reconstruction by Marianne Collins. From Briggs et al. 1994. See Briggs 1978, 1992.

Middle Cambrian times (see Williams, Siveter, and Peel 1996). This form has several morphological features (thin carapace with projecting cardinal spine, paddlelike exopods fringed with setae) and a wide biogeographic distribution that suggest a pelagic existence. Accordingly, a strong case has been made that species of *Isoxys* inhabited primarily midwater niches of the open ocean and outer shelf (Vannier and Chen 2000). As those workers note, *Isoxys,* taken together with other possible pelagic organisms in Early Cambrian faunas, suggests an early invasion of pelagic environments and implies the establishment of fairly complex pelagic communities.

A stem of Chelicerata may be represented earliest by *Sanctacaris,* described from a Middle Cambrian locality closely correlative with the classic Burgess Shale (fig. 7.28; Briggs and Collins 1988). This interesting fossil has two tagmata that can be interpreted as prosoma and opisthosoma, with at least six pairs of appendages on the head, and an anus terminal on the last segment, a combination of features shared with chelicerates. The first five prosomal appendages are raptorial, while

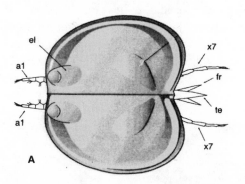

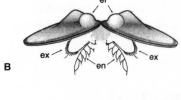

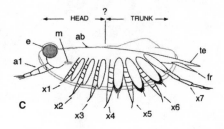

FIG. 7.27 Reconstruction of a bradoriid, *Kunmingella,* a possible stem crustacean. (*A*) Dorsal view. (*B*) Frontal view. (*C*) Lateral view from left, carapace removed. *a1,* first antenna; *ab,* attachment of body to carapace; *e,* eye; *el,* eye lobe; *en,* endopod; *ex,* exopod; *fr,* furcalike ramus; *m,* mouth; *te,* end of trunk; *x1-x7,* postantennular appendages. From Shu, Vannier, Luo, Chen, Zhang, and Hu 1999.

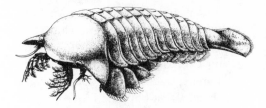

FIG. 7.28 *Sanctacaris,* a possible chelicerate stem taxon from the Middle Cambrian Burgess Shale, Stephen Formation, Canada, shown with lamellate trunk limbs. Specimens range from under 4 cm to over 9 cm long. Reconstruction by Marianne Collins. From Briggs et al. 1994.

the outer rami (branches) on the opisthosomal appendages are lobate or lamellate; *Sanctacaris* was probably a swimmer and certainly a carnivore (Briggs and Collins 1988). However, *Sanctacaris* lacks chelicerae, of all things, is biramous, and shares some features with nonchelicerate Cambrian arthropods (see Hou and Bergström 1997). Nevertheless, much about the morphology of *Sanctacaris* suggests that it may be a stem-group chelicerate.

A possible early record of Myriapoda is Middle Cambrian, from the Wheeler Formation of Utah (Robison 1990). The fossil, *Cambropodus,* is incomplete, but the anterior portion is well enough preserved to reveal a head with a pair of antenniform appendages and two slender head limbs, and the beginnings of a presumably elongate body with several segments that bear uniramian legs. It is not clear that the legs are jointed, and they appear to be undifferentiated. The assignment of *Cambropodus* to Myriapoda or even to Arthropoda has been questioned (Shear 1997) and remains uncertain. A likely myriapod is reported from the earliest of known terrestrial faunas, of Silurian age (Jeram et al. 1990), and there are other terrestrial myriapod occurrences from the mid-Paleozoic (Almond 1985).

Trilobites (fig. 7.29) have the jointed exoskeleton and appendages typical of arthropods. As most trilobite skeletons were mineralized, they have produced a good fossil record, second among arthropods only to the class Ostracoda in terms of numbers of fossil taxa. The trilobites were restricted to the Paleozoic and were diverse only in the early periods of that era. Trilobites dominate the fossil assemblages of Cambrian time, making up over 70% of the known Cambrian species of metazoans.

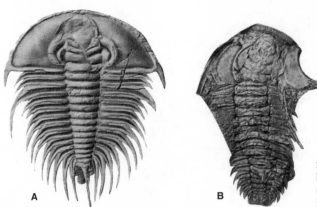

A   B

FIG. 7.29 Lower Cambrian trilobites. (A) *Olenellus,* Vermont. (B) *Redlichia,* Australia. From Boardman et al. 1987.

The famous trilobite skeletons have a cephalon of fused segments and then a vermiform axial lobe, beginning on the cephalon and continuing through the thorax onto the pygidium. The segmented skeleton is extended laterally into pleural lobes that protect a series of limbs. The limbs are biramous, walking legs on inner and gills on outer branches, and are all rather similar but usually grade to smaller sizes posteriorly. The anterior tagma, the cephalon, bears a pair of uniramous preoral antennae and three pair of biramous postoral limbs. This combination of appendages on the first tagma is unique among arthropod taxa and sets the trilobites apart from other clades. Most trilobite limbs display gnathobases, some of them impressively serrate, but there are no mandibles, although there is sometimes some specialization (more robust gnathobases) of the immediately postoral limbs, presumably for food handling. In general the limb articulations appear to be weak, the associated muscle scars small, and the space between opposing gnathobases relatively wide, all suggesting that feeding did not involve much crushing or tearing of prey (Manton 1977). For many trilobites, small soft-bodied organisms that could be passed forward between gnathobases seem the likeliest food items, though some trilobites may have been scavengers, and a few do appear to have fairly formidable apparatus for macerating food (see Bergström 1973; Cisne 1975; Whittington 1977, 1980). Several nonmineralized forms that resemble trilobites have been described from the Cambrian; *Naraoia* is the best known, from the Lower and Middle Cambrian (Chengjiang and Burgess Shale faunas). These forms may lie outside the trilobites proper, perhaps as a sister group (see Ramsköld et al. 1996 and references therein).

Other kinds of very distinctive early arthropods are known, chiefly from the Chengjiang fauna (table 7.2; Hou 1987a, 1987b, 1987c, 1999; Hou et al. 1989;

TABLE 7.2 A classification proposed for the arthropods that occur in the Early Cambrian Chengjiang fauna. Cambrian arthropods display significant disparity in just those aspects of bodyplans—tagmata and appendage types—that are used as defining characters in higher taxa of living arthropods; the application of similar criteria to the fossils therefore results in the erection of many higher fossil taxa. All these taxa except Crustacea and Chelicerata are extinct. From Hou and Bergström 1997.

Taxon	Remarks
Superclass Proschizoraia	
Class Yunnanata	Eye segment anterior, segments multilimbed
Class Paracrustacea	Includes *Canadaspis* (fig. 7.26)
Class Megacheira	Large second limb
Class Sanctacarida	Includes *Sanctacaris* (fig. 7.28)
Classes uncertain	Six ordinal-level taxa
Superclass Crustaceomorphia	
Class Pseudocrustacea	Agnostid trilobites, *Waptia*, etc.
Class Crustacea	Extant
Superclass Lamellipedia	
Class Marrellomorphia	*Marrella* (fig. 5.19)
Class Artiopoda	Trilobites
Superclass Chelicerata	Extant

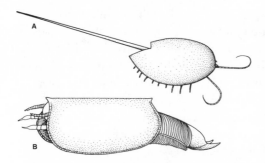

FIG. 7.30 Two arthropods from the Burgess Shale that do not fit into any living classes. (A) *Burgessia,* believed to be allied to the forms that gave rise to chelicerates and trilobites. (B) *Branchiocaris,* from perhaps a still more primitive stock that preceded the branching that led to modern groups. Reconstruction A by Isham; reconstruction B by Baldaro. From Briggs et al. 1994.

Chen and Zhou 1997) and the Burgess Shale (Whittington 1971, 1974, 1985a, 1985b; Briggs and Whittington 1985; see also Gould 1989; Conway Morris 1998). The most abundant Burgess Shale arthropod is *Marrella* (fig. 5.19G), and it is also among the most distinctive (Briggs and Fortey 1989). There are two tagmata; the anterior tagma, spectacular with its two pairs of spines, bears two pairs of uniramous antennae preorally but no other appendages. The posterior tagma consists of twenty-four to twenty-six segments diminishing in size posteriorly and bearing biramous appendages that have one or two fewer podomeres than do trilobite limbs, which they superficially resemble. Neither mouthparts nor gnathobases are known. *Marrella* stands in no clear relationship with living arthropod clades or with trilobites.

Many other Burgess Shale arthropods are also rather isolated morphologically from the major clades. For example, *Burgessia* (fig. 7.30A; see Hughes 1975) has a large ovate shieldlike carapace. The body is divided into two tagmata, the anterior bearing a pair of preoral antennae and three postoral biramous appendages with the outer branches flagelliform, but there are no mandibles. The posterior tagma bears seven pairs of biramous appendages with outer gill branches, and a final pair of uniramous appendages. An elongate spikelike telson completes the body units. As with *Marrella* there are no gnathobases on the legs or other obvious means of processing food. *Branchiocaris* (fig. 7.30B; see Briggs 1976) is another form with a bivalved carapace, in this respect resembling many crustaceans, but has two preoral antennae and no other appendages on the anterior tagma. The posterior tagma bears many appendages with lobate or bladelike outer branches and evidently with short inner branches. The posterior segments narrow gradually to a conspicuous bifid telson. The second antenna bears a short terminal process that may be a pincer or a spike, and was presumably used to secure food items. Carapaces resembling *Branchiocaris* but lacking appendage remains are reported from the Lower Cambrian of China (Hou 1987a). *Burgessia* may be allied to the Chelicerata-Trilobita alliance, while *Branchiocaris* cannot be confidently allied with the major arthropod clades and may well represent a stem lineage that antedates their origins.

There are more than fourteen other arthropodlike body types from the Cambrian that have appendages and tagmoses unknown among living groups (see for example Briggs and Fortey 1989). The remains of other jointed skeletons of arthropodan aspect that appear to represent additional types, but that are too poorly preserved to be characterized, are also known from Lower Cambrian rocks, chiefly from the Chengjiang (e.g., Hou and Bergström 1997); their existence testifies further to the breadth of the early arthropod radiation.

Early Fossil Relatives (or Perhaps Basal Stem Groups) of Arthropoda

In addition to the numerous early arthropod forms that cannot be assigned to living classes, there are a few organisms that have many arthropodan features but that may not be arthropod stem groups or even sisters. The most spectacular of these are the anomalocarids (fig. 5.19E), large predators that attained lengths of nearly half a meter (see Collins 1996). The pair of spiny jointed appendages on the head were presumably used in prey capture; the large jaws were furnished with arrays of "teeth," perhaps used to crush lightly sclerotized arthropods or other prey. A pair of large eyes flank the rear of the head. The trunk appears to be segmented and possesses lateral body lobes thought to have been used in swimming. The presence of limbs on trunk segments has been demonstrated in some Chengjiang taxa (Hou et al. 1995), but limbs have not been found on other forms. The trunk was not as well sclerotized as the head appendages or mouthparts, and interpretation of the poorly preserved structures there is difficult. Several species are known, ranging from the Atdabanian to the Middle Cambrian from such widespread localities as Canada, China, and Greenland.

Opabinia (fig. 5.19F), a presumably segmented organism from the Burgess Shale, was monographed by Whittington (1975). It has a number of compound eyes and lateral lobate extensions along the trunk, which carry lanceolate gill-like features. The mouth faces backward, but is in range of a flexible, trunklike anterior tube with a beaklike termination. Because of the likely segmentation and the compound eyes, Opabinia might be thought of as an unusual sort of arthropod. Whittington was unable to discover jointed appendages on specimens in which they seemed most likely to have been preserved, however, and the affinities of this organism were deemed problematic. Another fossil organism that may be related to Opabinia is Myoscolex, from the late Lower Cambrian of South Australia (Briggs and Neddin 1997). This form, roughly correlative with the Chengjiang fauna, is represented chiefly by phosphatized muscle tissues that suggest axial segmentation, and there are indications of lobate lateral flaps and of multiple eyes that recall those of Opabinia. Unfortunately, the preservation is too poor for a conclusive interpretation.

It has been suggested (Bergström 1986) that Opabinia is a relative of Anomalocaris, partly on the basis of the common presence of a segmented trunk, lateral lobes, anterior grasping structures (albeit quite different), and large eyes (of which Opabinia has a plethora). The more recent morphological cladograms

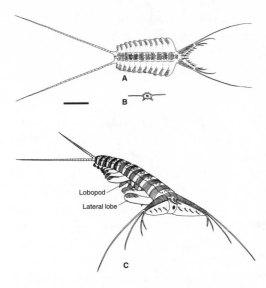

FIG. 7.31 Reconstruction of *Kerygmachela*, from the Early Cambrian of northern Greenland. (*A*) Dorsal view. (*B*) Transverse section. (C) Conception of *Kerygmachela* in life, with several lateral lobes omitted to show inferred lobopods. Scale bar is 20 mm. From Budd 1999.

and cladistically based trees do indeed identify these taxa as being allied and either as representing a clade basal to Arthropoda (Budd 1998) or basal to Onychophora (Wills et al. 1995; Wills et al. 1998).

Kerygmachela (fig. 7.31; Budd 1993, 1999), from the Sirius Passet fauna, Lower Cambrian of Greenland, has a vermiform body with segmental markings resembling those of the lobopod *Luolishania* (see fig. 7.14A). Anteriorly there is a pair of unusual appendages, unlike any known antennae, and the posterior bears lengthy paired spines. The thoracic region has seven segments, each of which has paired lateral lobate structures that have been interpreted as gill-bearing. *Kerygmachela* is associated with an anomalocaridlike form, *Pambdelurion,* in which limbs are preserved in some specimens (Budd 1997). Like many arthropods in Cambrian lagerstätten, *Kerygmachela* specimens have an axial structure interpreted as a gut. Along this axial structure is a series of paired, segmentally arranged, subtriangular extensions that were interpreted by Budd as lobopods (fig. 7.31C). Budd (1996) restudied *Opabinia* and found similar structures. The limbs of *Pambdelurion* were also interpreted as lobopods, and thus there seemed to be an alliance of arthropod-like forms that bore lobopods and that might be considered transitional between the onychophoran-like lobopodians and stem arthropods with jointed appendages. These forms would thus represent an intermediate fossil group standing between two crown phyla, a rare if not unique case.

Butterfield (2002) has studied the axial features in a number of Cambrian arthropods, however, and has been able to interpret the paired, axially segmented three-dimensional structures associated with the gut trace as representing midgut digestive glands. Such glands are present in many living arthropod groups, being particularly well developed among carnivorous forms. The three-dimensional preservations, unusual for Cambrian arthropod structures, are plausibly explained as resulting from early phosphatic permineralization of richly glandular tissues with reactive contents and internal phosphorus sources. This beautiful taphonomic study produces an alternative interpretation for imputed lobopodial structures in *Kerygmachela, Opabinia,* and other rather arthropodlike forms; the structures may

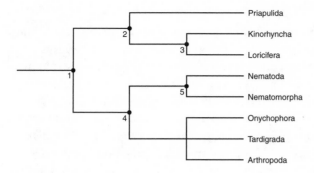

FIG. 7.32 A hypothesis of the phylogeny of living ecdysozoan phyla based chiefly on SSU rRNA sequence comparisons.

well be digestive diverticula and, if so, do not indicate lobopodian affinities. An important confirmatory test is that similar diverticula-like features are associated with the axial structures in many Cambrian fossils with undoubted arthropod (i.e., nononychophoran) affinities (e.g., *Leanchoilia, Sidneyia, Canadaspis*) as well as anomalocarids (Butterfield 2002). Furthermore, as paired digestive midgut glands are associated with carnivory in living forms, many early arthropods now appear to have been predators or scavengers, in contrast to some early notions that they were mostly deposit feeders (e.g., Valentine 1973b).

Some Branch Points within the Ecdysozoa

Fig. 7.32 depicts the topology of an ecdysozoan tree that is based chiefly on SSU rRNA data, supplemented by the developmental and morphological information discussed above. Like other phylogenetic trees in these chapters, it is only a tentative hypothesis, but provides a basis for discussion of possible branch points in ecdysozoan evolution.

Nodes 1, 2, and 3

Both kinorhynchs and loriciferans somewhat resemble priapulids on the basis of their paracoelomate grade, small sizes, molted cuticles, and elaboration of cuticular elements into spines and similar structures. The sorting out of homologous from analogous features within the paracoelomates has proven to be extremely difficult. Having said that, kinorhynch and priapulid introverts may be homologous, and possible synapomorphies of kinorhynchs and loriciferans include the structure of their mouth cones and first scalid rows (see discussions in Kristensen 1983; Nielsen 1995) and a terminal circumcentric brain (nerve ring in the mouth cone of kinorhynchs; Nebelsick 1993). However, it is difficult to choose among the many possible permutations in the placement of the kinorhynchs and loriciferans along the branches associated with the priapulids, their putative sister, on the evidence at hand. Whether the serial architecture of kinorhynchs is homologous with the seriation in tardigrades, onychophorans, and arthropods is an open question.

Nodes 4 and 5

To reconstruct the common ancestor of the nematode + nematomorph clade and the segmented organisms of its sister clade containing arthropods, it is necessary to evaluate the lack of any sign of segmentation or seriation in the former clade. The ancestor at the node in question need only have been a rather simple worm. However, the presence of the segmented bodyplan among the arthropods and their allies suggests the possibility that the ancestor they share with nematodes + nematomorphs may have been at least seriated or incipiently segmented, perhaps something like a kinorhynch, and that this architecture was lost during evolution of the nematode + nematomorph clade, possibly in association with the acquisition of parasitic habits. This scenario suggests the possibility that free-living nematodes are descended from parasites. Still other possibilities involve the postulation of even more gratuitous evolutionary complexities. There is no evidence at hand to decide whether seriation evolved before or after the branch point between the two clades in question.

Onychophora + Tardigrada + Arthropoda Clade

As for the last common onychophoran + tardigrade + arthropod ancestor, the segmented architecture held in common among these phyla was surely present and must have involved limbs.

Onychophora. In evaluations of SSU rRNA sequence data the Onychophora, Tardigrada, and Arthropoda tend to form an unresolved clade. Onychophorans have a series of paired nephridia associated with their segmental pattern, like annelids and unlike arthropods. Whether early lobopodians had such seriated nephridia is unknown. This organ pattern could have evolved in primitive lobopod stocks and then have been lost somewhere along the lineage leading to arthropods, or it could have evolved along the lineage leading to crown onychophorans after the arthropod clade ancestors had branched off—the simplest case. In either event, there is no evidence that onychophorans, or their ancestors, had perivisceral coeloms or hydrostatic coeloms of any sort, and nephridial seriation must have arisen independently in onychophorans and annelids.

Tardigrada. The tardigrades and onychophorans may share a lobopodial ancestor. Which of these phyla branched more deeply, and which most closely resembles the ancestral bodyplan, is not certain. As the tardigrades lack circulatory systems, it might be argued that they should be primitive with respect to onychophorans, but this condition may be related to the minute size of tardigrades and represent an apomorphy of that clade. A developmental feature for which tardigrades may be considered more advanced than onychophorans has been described by Dewel and Dewel (1996). On the basis of an analysis of the segmental origin of the ganglia that have coalesced to form the tardigrade brain, the mouth of tardigrades may be assigned a position between the first and second segments, presumably homologous to that in onychophorans. In this case the mouth apparatus of tardigrades and the jaws of onychophorans are both innervated from the second segment. During development, onychophoran jaws appear to arise from a single polarizing zone posterior to the mouth, growing forward on both sides to encircle the mouth region. The oral apparatus of tardigrades, however, appears to originate from two polarizing zones, one anterior and the other posterior to the mouth. Dewel and Dewel suggest that the tardigrade state is the more derived. It is certainly possible that a tardigrade/arthropod ancestor evolved from a lobopodian after the Onychophora had branched, and that a reduction in nephridia occurred in that stock, perhaps owing to a phyletic size reduction. Such a relationship is indicated by Wheeler (1998), using combined molecular and morphological evidence.

Arthropoda. For many decades the dominating hypothesis of arthropod origins involved descent from the annelids in a process termed arthropodization. This scenario includes the following steps: the loss of the extensive annelid coelom and its replacement with a hemocoel (or myxocoel); the appearance of the sclerotized, jointed exoskeleton; the evolution of jointed limbs from flexible coelomate parapodia; the loss of intersegmental septa; the loss, or rather consolidation, of segmentally arranged excretory and reproductive organs; and a host of other changes that, while not speaking directly to the fundamental body architectures of these phyla, are required to transform the annelids that we know into the arthropods that we find. Most textbooks have continued to hold to this notion (see Pearse et al. 1987 and Boardman et al. 1987 for neontological and paleontological examples). However, there is really nothing in the arthropod bodyplan that requires that it be derived from an annelid ancestor (see Eernisse et al. 1992; Valentine 1994).

Deriving arthropods from a lobopodian ancestor with roots in a paracoelomate clade is not only consistent with the SSU rRNA data but is considerably more plausible morphologically than imagining an annelid ancestor. Lobopodians provide all of the necessary plesiomorphies—a hemocoelic, serial bodyplan with limbs, cuticular molting, and so forth—and obviate the necessity of accounting for the loss of a coelomic architecture, a loss that is simply unexplained in the annelid-as-ancestor scenario. It does seem that the segmented exoskeletal arrangement of arthropods must have arisen as a modification of a preceding serial arrangement of limbs, enhancing locomotory efficiency as skeletonization proceeded (see Budd 1998). The sister clade to the arthropods, presumably either the tardigrades or onychophorans or both, must have shared with arthropods an ancestor that had paired unjointed locomotory appendages with the requisite supporting systems, and a molted cuticle.

Ecdysozoan Cleavages

The cleavage patterns and cell fates of tardigrades, onychophorans, and arthropods are not classically spiralian, and as the cleavage pattern of priapulids is radial, of nematomorphs is possibly modified radial, and of nematodes is unique, there appears to be no classic spiral cleavage among crown ecdysozoans. Perhaps the developmental changes that produced the varied early embryo types occurred within lineages that evolved very yolky eggs and therefore highly modified cleavage patterns (Weygoldt 1979). Branching among such lineages could have produced the distinctive early ontogenies of many of the ecdysozoan taxa.

Early History of the Lobopodian and Arthropodan Clades

Although the range of arthropod and lobopod-bearing morphologies that were present in the Early Cambrian must be significantly underrepresented by fossils, the number and disparity of body types that are known is remarkable. Trace fossils certainly indicate the presence of arthropods in the Tommotian, near 530 Ma. As arthropod disparity is known to be great by the Atdabanian, arthropods probably originated at least as early as the Manykaian. This situation in turn implies that lobopodians, the earliest fossils of which are Tommotian, were present in even older

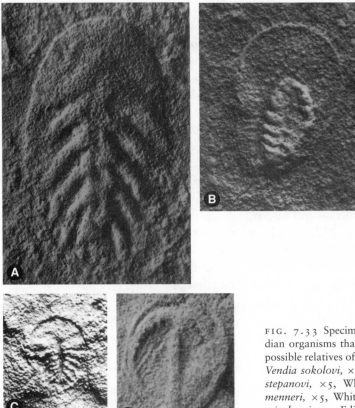

FIG. 7.33 Specimens of some of the Vendian organisms that have been suggested as possible relatives of arthropod ancestors. (A) *Vendia sokolovi*, ×5, White Sea. (B) *Onega stepanovi*, ×5, White Sea. (C) *Vendomia menneri*, ×5, White Sea. (D) *Parvancorina minchami*, ×1, Ediacara. A, B, C from Fedonkin 1992; D from Glaessner 1984.

stages. It would seem that while lobopodian origins probably lie, at the outside, between the onset of bilaterian traces and the close of the Vendian, between about 570 and 543 Ma, they are more likely to be in the later part of this interval rather than the earlier.

Several pre-Tommotian fossils may be related to the lobopodian-arthropodan clades. These are chiefly flattened bilateral forms with some sort of presumably anterior lobe or shield followed by serial units that can be interpreted as segments. These fossils include *Spriggina* and *Marywadea* (sprigginids), *Praecambridium, Vendia,* and *Vendomia* (vendomiids), and a few other less well known forms (fig. 7.33). Many of the frondlike cnidariomorphs from the Vendian have serial units arrayed along a central axis (chap. 5). Some workers have suggested that the putatively segmented Vendian bilaterians in fact belong to a frondlike alliance (Seilacher 1989; Bergström 1989); the putative anterior structure then becomes basal and might be interpreted as a holdfast of some kind, with a vertical axis in the center (see fig. 5.10). *Parvancorina* is another Vendian fossil that has been allied to arthropods (fig. 7.33D). If interpreted as arthropodlike, it has a shield-shaped carapace with an anchor-shaped ridge and a series of corrugations that have been interpreted as appendages (Glaessner 1979c; but see Bergström 1991). None of these fossils is entirely convincing as an arthropod ancestor, though some may be stem ecdysozoans. For a more sanguine view see Waggoner 1996.

CHAPTER 8

Protostomes: Lophotrochozoa 1: Eutrochozoans

A number of nonmolting protostome phyla are placed in a sort of superclade, Lophotrochozoa, on molecular evidence (fig. 4.17; Halanych et al. 1995). One group of phyla within Lophotrochozoa shows the classic spiralian pattern of early development; these include Mollusca, Sipuncula, Annelida, Pogonophora, Echiura, and Nemertea. These groups share a characterizing set of developmental features, such as spiral cleavage in quartets, a similarity in cell fates of early blastomeres, mesoderm usually derived from a 4d mesentoblast, schizocoely when a coelom is present, and, except for Nemertea, a trochophore-type larva in indirect developers. The term Eutrochozoa can be used for these forms (Ghiselin 1988; Eernisse et al. 1992). While SSU rRNA evidence indicates that eutrochozoan phyla are closely allied, their taxa are commonly interspersed in molecular trees in patterns that are so strongly contradicted by morphological evidence as to indicate that those branching patterns are unrealistic. Presumably this problem was caused at least in part by a rapid eutrochozoan radiation that the slowly evolving RNA molecular sequences are unable to resolve.

Rhabditophoran flatworms lack body cavities but otherwise display all of the eutrochozoan developmental features except for the possession of a trochophore larva, and have commonly been placed with the classic spiralian phyla, often at their base. When evaluated by SSU rRNA sequences, however, rhabditophorans (and catenulids) sometimes join other noncoelomate phyla in the paracoelomate grouping termed Platyzoa (Cavalier-Smith 1998; Giribet et al. 2000). Nevertheless, the similarities in cleavage and cell fates between these flatworms and the eutrochozoan phyla are striking, and considering the difficulties in resolving eutrochozoan relationships by molecular data, I treat them together here for comparative purposes. Molecular data suggest that the lophophorate phyla (Brachiopoda, Phoronida, and Bryozoa) are closely related to Eutrochozoa; they are treated in chapter 9.

Platyhelminthes: Rhabditophora and Catenulida

Morphologically based invertebrate phylogenies have traditionally placed the flatworms as the earliest truly triploblastic living phylum, the Platyhelminthes. Hyman (1951a) provides an outstanding though dated introduction to the flatworms, and papers in Littlewood and Bray (2001) review much subsequent work.

Morphologically based cladistic analyses of flatworms have been conducted by Ehlers (1985), Littlewood, Rohde, and Clough (1999), Littlewood, Rohde, Bray, and Herniou (1999), and a number of authors in Littlewood and Bray 2001. Campos et al. (1998), Carranza et al. (1997), Carranza et al. (1998), Ruiz-Trillo et al. (1999), Littlewood, Rohde, and Clough (1999), Baguñà, Ruiz-Trillo, Paps, Loukota, Ribera, Jondelius, and Riutort (2001), and Baguñà, Carranza, Paps, Ruiz-Trillo, and Riutort (2001) have published SSU rRNA evidence of relationships among free-living flatworms. New morphological and molecular evidence suggests that Platyhelminthes as usually defined is polyphyletic, with two of the orders, Acoela and Nemertodermatida, representing basal bilaterians, a position traditionally assigned to all Platyhelminthes. Furthermore, SSU rRNA sequence comparisons have placed the major flatworm clade Rhabditophora (Ehlers 1985), together with the order Catenulida, within the nonecdysozoan protostomes. Thus Rhabditophora appears to have arisen, not only after the last protostome/deuterostome ancestor, but also after the last ecdysozoan/eutrochozoan common ancestor, and is far from being a basal bilaterian group. Catenulida, which lacks derived gene code changes that are found in Rhabditophora (Telford et al. 2000), may be its sister group. There are about 3,000 described flatworm species, but many more are believed to be undescribed. Major flatworm taxa are listed in table 8.1.

TABLE 8.1 Some flatworm higher taxa. The parasitic subclasses contain numerous orders, the delimitation and arrangement of which are in dispute; some rhabditophoran orders are also unsettled. Chiefly after Karling 1974 and Rieger et al. 1991, modified slightly to accord with molecular evidence.

Taxon	Remarks
Phylum? Acoelomorpha	
Order Acoela	Shallow marine sediments, small
Order Nemertodermatida	Subtidal marine sediments, small
(Phylum?)	
Order Catenulida	Marine, freshwater
Phylum? Rhabditophora	
Class Turbellaria	
Order Macrostomida	Marine, freshwater meiofaunal
Order Haplopharyngida	Marine, freshwater meiofaunal
Order Gnosonesimida	Marine sediments
Order Prorhynchida	Freshwater, terrestrial
Order Rhabdocoela	Marine, freshwater, some symbiotic
Order Prolecithophora	Marine, freshwater, small
Order Polycladida	Marine, commonly large
Order Proseriata	Marine
Order Tricladida	Marine, freshwater, a few terrestrial
Order Temnocephalida	Freshwater, symbiotic, small
Class Monogenea	Chiefly ectoparasitic on fish
Class Trematoda	Chiefly endoparasitic
Class Cestoda	Tapeworms; endoparasitic

Bodyplan of Platyhelminthes

Free-Living Rhabditophora. The rhabditophoran body is bilaterally symmetrical, with a triploblastic construction of ectoderm, endoderm, and a more or less well-defined cellular mesoderm that is generally the most massive of the three layers. Muscles, reproductive systems, much of the excretory system, and connective tissues are developed from the mesoderm. The muscle fibers lie along the body wall, usually oriented and bundled so as to produce an outer "circular" layer, a thin intermediate diagonal layer, and an inner longitudinal layer. A cellular mesodermal connective tissue, the parenchyma, lies between the muscles and endodermal digestive tissue. Flatworms lack a circulatory system, respiring through the integument. As the mesoderm in flatworms is always cut off from the environment by a layer of ectoderm or endoderm (with partial exceptions provided by ducts), interior mesodermal cells would lack oxygen if the tissue layers were very thick. A flattened body shape permits diffusion of oxygen to inner tissues, and perhaps for the same reason, flatworms lack surface cuticles. Flatworms are chiefly hermaphroditic, usually with separate ducts for egg and sperm and with both a genital pore and a penis; their reproductive systems are sometimes quite complex. Multiciliated cells are present, and benthic locomotion is commonly mucociliary; many marine flatworms also swim readily.

The triclad turbellarians include the largest of free-living rhabditophorans (over 50 cm long in terrestrial forms, although most marine species are only a few millimeters long) and have an architecture that at one time was thought to be ancestral to segmented higher invertebrate groups. Triclads can be interpreted as more derived than the acoels in nearly every respect. There is a well-defined digestive cavity, consisting of a single anterior gut that forks to produce two posterior branches (fig. 8.1). A series of diverticula emerge from these branches to serve lateral body tissues. The digestive tissue, believed to be partly syncytial from light microscope studies, turns out to be entirely cellular in triclads according to electron microscopy (see Rieger et al. 1991). In the body wall the multiciliate epidermal cells display septate junctions. There is no interconnected ciliary rootlet system, as in acoels, and apical webs are weakly developed. Unlike acoels, however, the cells are supported by a well-developed basement membrane containing a marked fibrillar layer, probably collagenous, that invests the circular muscle bundles beneath (Bedini and Papi 1974; Rieger 1981; fig. 8.2). The lateral gut diverticula are quite regular in some forms, such as the figured marine species of *Procerodes,* so that a regular series of spaces for the placement of organs within the mesoderm occurs between them. It is in these spaces that series of gonads and excretory ducts, when multiple, are found (fig. 8.1). The nervous system has various patterns, but there is a ganglionic brain, and many marine forms have three pairs of nerve cords connected by a series of commissures, sometimes in a ladderlike fashion (Hyman 1951a). Thus many triclads have series of more or less paired organs, resembling in this respect the seriated architectures of arthropods and annelids.

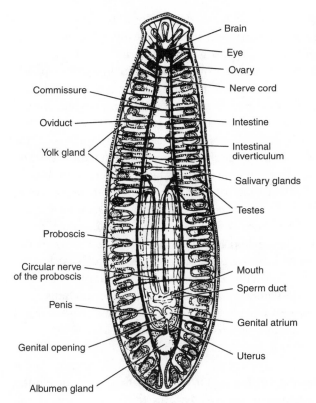

Brain

Eye

Ovary

Nerve cord

Commissure

Oviduct

Yolk gland

Intestine

Intestinal
diverticulum

Salivary glands

Testes

Proboscis

Circular nerve
of the proboscis

Mouth

Sperm duct

Penis

Genital atrium

Genital opening

Uterus

Albumen gland

FIG. 8.1 A rhabditophoran body-plan. Some organ systems in the triclad *Procerodes;* note that seriated systems, such as testes and nerve commissures, do not necessarily correlate. From Vagvolgyi 1967 after Lang 1882.

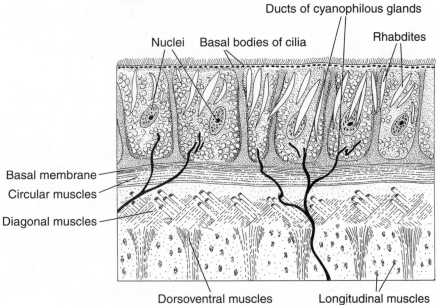

Ducts of cyanophilous glands

Nuclei   Basal bodies of cilia

Rhabdites

Basal membrane

Circular muscles

Diagonal muscles

Dorsoventral muscles

Longitudinal muscles

FIG. 8.2 Epidermal histology of the triclad *Geoplana.* The epidermis is cellular; the rhabdites, chiefly secreted by epidermal cells, break down to produce mucus. After Bayer and Owre 1968.

Marine triclads swim and creep via cilia on the seafloor. Terrestrial triclads are even larger, and although they are very flat, thus possessing a relatively large surface area that can be packed with cilia, they have enough mass that body-wall muscles must often become involved in locomotory activity, providing more power than ciliary action alone (Clark 1964). On land the muscular locomotion involves "pedal waves," so named because of their resemblance to gastropod locomotory waves. In both turbellarians and gastropods, waves of contraction may sweep forward along the body length, progressively raising the ventral body surface off the substrate. Contractions of longitudinal muscle strands then cause the raised portion to advance, and when the wave passes and the body surface contacts the substrate again, it has moved forward. Forward motion can also result from retrogressive pedal waves that move backward along the body; the key point is that muscular contractions advance the raised body portion (for data and discussions see Crozier 1918; Lissmann 1945a, 1945b; Clark 1964; Miller 1974). As wave follows wave, the worm glides along. Marine polyclads also creep on the seafloor, and some but not all display progressive pedal waves. Some marine polyclads, even large ones, simply glide on mucous trails via cilia without body-wave involvement (Collins et al. 2000), and many swim frequently. Cell-type numbers in individual turbellarians seem to range from about twenty to thirty, or more in some cases when many specialized gland cells are present (see Rieger et al. 1991).

Catenulida. The small catenulid flatworms do not belong within Rhabditophora. They have a simple bodyplan and reproductive features that have caused them commonly to be allied with acoels. Yet unlike acoels, their nervous system includes lobes that may function as a brain. The relation of catenulids with either acoelomorphs or to rhabditophorans is disputed, and they may not be closely allied with either.

Xenoturbella. A centimeter-sized vermiform invertebrate, *Xenoturbella,* is known from three free-living species found in the North Sea and northeastern Atlantic; it has been tentatively placed in several different phyla by different workers. The adult is uniformly ciliated and has a thick epidermis with a basal nerve layer, underlaid by circular and longitudinal muscles and parenchyma. Israelsson (1997) has suggested a molluscan affinity, but there are interesting similarities with catenulids and other flatworms (see Tyler 2001). Partial sequences of SSU rDNA and cytochrome *c* oxidase subunit I genes suggest that *Xenoturbella* lies within Eutrochozoa (Norén and Jondelius 1997). *Xenoturbella* may be a flatworm, but its precise phylogenetic position remains in doubt.

Some Flatworm Bodyplan Variations. Aside from the range of body types mentioned above, there are some free-living flatworms that, while not necessarily standing in any phylogenetic line that led to the origin of other phyla, nevertheless suggest some of the sorts of bodyplan variations that may have been important in the radiation of bilaterians. Numbers of species in several orders are cylindrical, for example. These are usually minute species, under 1 or 2 mm in length and only a fraction of a millimeter in diameter. Their small size has evidently led to a relaxation of body-shape restrictions, despite the lack of a circulatory system; they are round flatworms.

One group of rhabdocoels, the kalyptorhynchs, has a protrusible proboscis at the anterior tip, used to capture prey which are then presented to a protrusible pharynx, located more posteriorly, for ingestion (Meixner 1925; Hyman 1951a). The proboscis is circular, and is squeezed (and partly pulled) out of its sheath by muscular action. Rieger (1974) has discussed possible evolutionary pathways for this proboscis.

Development in Marine Rhabditophora

Development varies considerably among turbellarian orders. The classic form of spiral cleavage in quartets (fig. 8.3) is well-displayed among the polyclads. Developmental quadrants of most spiralians are founded by the first two cleavages that produce cells A–D (chap. 3). Although the pattern of blastomere contact is symmetrical at the animal pole, there is a cross furrow at the vegetal pole between cells B and D that isolates cells A and C from contact at the vegetal surface. Normally, these four cells form quadrants that give rise to the left (A), right (C), ventral (B), and dorsal (D) regions of the larva and adult (Henry et al. 1995). During gastrulation, which forms a coelogastrula by epiboly, the 4d cell divides to produce endoderm from one daughter and endomesoderm from the other; ectomesoderm arises from micromeres of the second and third quartets. The turbellarians have commonly been placed as basal to the spirally cleaving protostomes in which the 4d cell also acts as a mesentoblast. However, in many turbellarian taxa, subsequent development is unique, suggesting a radiation of development to produce numbers

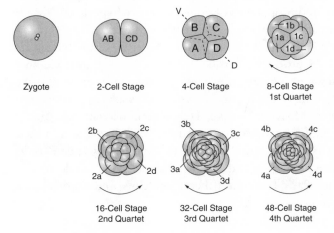

Zygote	2-Cell Stage	4-Cell Stage	8-Cell Stage 1st Quartet

16-Cell Stage 2nd Quartet	32-Cell Stage 3rd Quartet	48-Cell Stage 4th Quartet

FIG. 8.3 Spiral cleavage in quartets as found in the rhabditophoran polyclad *Hoploplana*, animal-pole views. Spiral offsets of the micromeres alternate in direction. Endomesoderm arises from blastomere 4d, while ectomesoderm arises from micromeres in the second and third quartets. *V*, ventral; *D*, dorsal. From Henry et al. 2000.

of highly derived states. Development may be either direct or indirect. Two types of planktonic larvae are found in forms with indirect development, Götte's larva (with four lobes) and Müller's larva (with eight lobes), both highly ciliated and not clearly related to other spiralian larvae.

Fossil Record of Flatworms

There are no verified records of flatworm body fossils, although several fossil remains have been assigned to the flatworms for one reason or another (see review in Conway Morris 1985b). Quaternary turbellarian eggs are reported from lake sediments in Venezuela (Binford 1982), and Ruiz and Lindberg (1989) identify internal pitting in marine molluscan shells of Late Miocene to Pleistocene age as traces formed by parasitic trematodes. Living polyclads can form mucociliary trails in marine sediments (Collins et al. 2000). Sedimentary particles entrained in mucous pads form ridges composed of mixtures of mucus and sediment that are flanked by depressions from which the sediment is recruited. Thus relief on these trails is associated with the placement of mucous strings along the body; for example, when there are two mucous strings (as from paired mucous glands), such a trail has two ridges in cross section. It is possible that some of the trails of small creeping animals found in late Neoproterozoic rocks are from the activities of flatworms or of organisms of the same grade. Nonmarine trails from the Early Permian of northern Italy have been assigned to planarians (Alessandrello et al. 1988).

Flatworm Relationships

Flatworms have been major actors in metazoan phylogenies, appearing in many roles. Flatworms have been considered to be holophyletic sisters to all other bilaterians (Salvini-Plawen 1978); to be basal to bilaterians but massively paraphyletic (Hyman 1940; Willmer 1990; Barnes et al. 1993); to be holophyletic sisters to spiralians (Brusca and Brusca 1990); and to be simplified descendants of coelomate ancestors (Rieger 1985 and proponents of archicoelomate hypotheses). Many other phylogenetic positions have been suggested. Although early SSU rRNA evidence was not inconsistent with the rhabditophorans being basal to bilaterians (see Turbeville et al. 1992; Riutort et al. 1993; Philippe et al. 1994; Campos et al. 1998), studies that include longer sequences and sequences from shorter-branched species have placed the rhabditophorans as nonecdysozoan protostomes (Winnepenninckx, Backeljau, Mackey, Brooks, DeWachter, Kumar, and Garey, 1995; Balavoine 1997; Aguinaldo et al. 1997; Carranza et al. 1997). Such a position is also supported by Hox gene assemblage data (Balavoine 1997; de Rosa et al. 1999), mitochondrial gene sequences, and the myosin heavy chain molecule (reported in Baguñà, Ruiz-Trillo, Paps, Loukota, Ribera, Jondelius, and Riutort 2001, data not shown).

Combined morphological–SSU rRNA analyses by Giribet et al. (2000) have suggested that nonacoelomorph flatworms are allied with numbers of paracoelomates in a clade (Platyzoa) that is sister to the Eutrochozoa (chap. 10). Those data suggest that rhabditophorans are not basal within Platyzoa. In light of the difficulty in resolving eutrochozoan branchings, and the tendency of flatworms to be long-branched, the concept of Platyzoa should be tested with additional data.

Mollusca and Mollusklike Forms

The phylum Mollusca is composed of a number of classes with such distinctive organizations that it is difficult to decide just what the molluscan bodyplan actually is, and which if any of the living groups resembles the most primitive mollusks. Partly this is because few of the major molluscan features are present in every group, making them extreme examples of a polythetic taxon, and partly because it is hard to polarize even important molluscan characters. Today the mollusks are among the most diverse of the skeletonized phyla in the sea; of over 50,000 described molluscan species, perhaps 35,000 or more are marine.

Bodyplan of Mollusca

Mollusks are triploblastic bilaterians. It is usual to describe molluscan organization as involving a ventral foot, dorsal visceral mass, and anterior head, although only the visceral mass is universally present. The body wall is elaborated dorsally into a mantle that more or less cloaks the body and that is responsible for the secretion of an exoskeleton, if present. The basic architectures of the major living molluscan groups are depicted in figs. 8.4 and 8.5, and the major taxa, including extinct forms, are listed in table 8.2. There is a nearly anterior mouth and a nearly posterior anus; sometimes the body is compacted by being bent (Cephalopoda) or torted and bent (Gastropoda) into a U-shape. In a number of classes, feeding is promoted by a radula, an organ, sometimes eversible, that contains teeth that can be used in scraping or tearing. The morphology of the radular teeth varies among taxa, correlating well with the nature of the food items. Taxa that lack a radular apparatus generally feed on minute particulate matter, either plankton or other items filtered from suspension, or detritus derived from sediments.

Most living mollusks have open circulatory systems (the exception is Cephalopoda, which has a closed system); blood is pumped into arteries from a chambered heart, but flows into a hemocoelic cavity. As the heart and arteries are submerged in mesodermal tissues, there is a pericardial space to permit heart pulsations, technically a coelomic space. The heart usually lies dorsally and toward the posterior. Gonads are commonly situated just in front of the heart; gonoducts pass through the mesoderm to penetrate the body wall as gonopores, and are technically coelomic. The qualification of these and other coelomic spaces in mollusks as

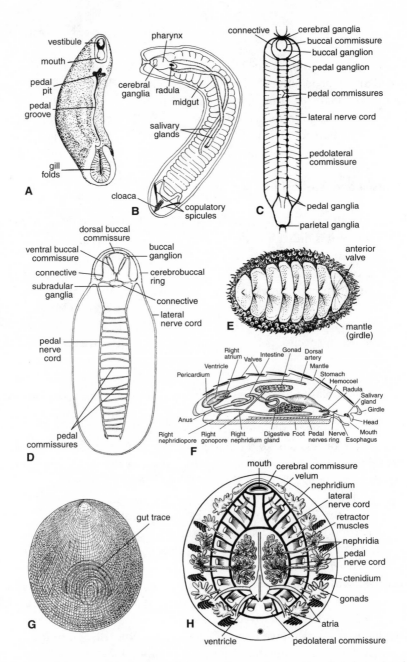

FIG. 8.4 General features of bodyplans of those molluscan classes most often deemed to be primitive. (*A–C*) Aplacophorans. (*A*) Exterior of a solenogaster (*Neomenia*). (*B*) Anatomy of a caudo-
foveate (*Limifossor*). (*C*) Nervous system in a solenogaster (*Proneomenia*). (*D–F*) Amphineura (chitons).
(*D*) Nervous system in *Acanthochitona*. (*E*) External dorsal view of *Mopalia*; the skeleton has eight
valves. (*F*) Generalized anatomy, lateral view. (*G, H*) Tergomyan monoplacophorans. (*G*) External dorsal view of *Neopilina*. (*H*) Anatomy of *Neopilina*. A from Hyman 1967 after Hansen; B from Hyman
1967 after Thele; C from Hyman 1967 after Pruvot; D from Hyman 1967 after Pelseneer; E from Light
et al. 1970; F from Brusca and Brusca 1990; G from Warén and Gofas 1996; H from Hyman 1967 after
Lemche and Wingstrand.

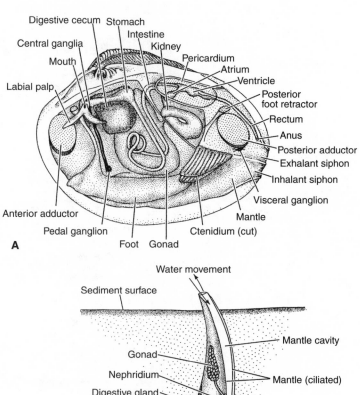

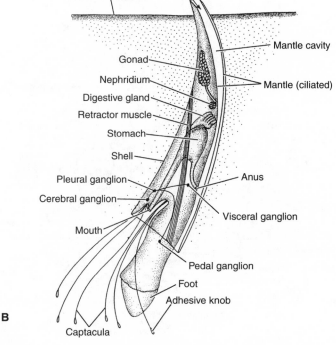

FIG. 8.5 General features of bodyplans of those molluscan classes most often deemed to be derived. (A) Bivalvia, sometimes suggested to be descended from the extinct class Rostroconcha. (B) Scaphopoda, sometimes suggested to be descended from the extinct class Rostroconcha but probably sister to Cephalopoda. (C) Gastropoda, sometimes suggested to be descended from the class Tergomya (monoplacophorans). (D) Cephalopoda, sometimes suggested to be descended from Gastropoda. After Brusca and Brusca 1990.

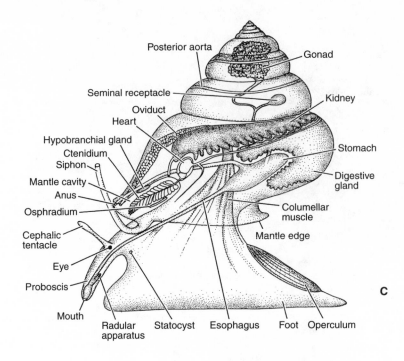

Posterior aorta

Gonad

Seminal receptacle

Kidney

Oviduct

Heart

Hypobranchial gland

Ctenidium

Siphon

Stomach

Mantle cavity

Anus

Digestive gland

Osphradium

Columellar muscle

Cephalic tentacle

Mantle edge

Eye

Proboscis

Mouth

Radular apparatus   Statocyst   Esophagus   Foot   Operculum

C

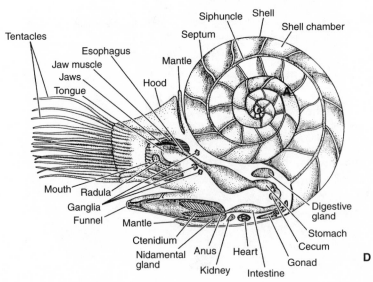

Tentacles

Siphuncle   Shell

Esophagus

Septum

Shell chamber

Jaw muscle

Mantle

Jaws

Hood

Tongue

Mouth   Radula

Digestive gland

Ganglia

Funnel   Mantle   Stomach   Cecum

Ctenidium   Gonad

Nidamental gland   Anus   Heart

Kidney   Intestine

D

FIG. 8.5 (*continued*)

TABLE 8.2   Some major taxa of Mollusca and mollusklike forms, living and extinct.

Taxon	Fossil Record	Remarks
Coeloscleritophora	Lower-Middle Cambrian	Molluscan affinities uncertain
Caudofoveata	Recent	Aplacophoran
Solenogastres	Recent	Aplacophoran
Polyplacophora	Upper Cambrian–Recent	
Tergomya	Lower Cambrian–Recent	Monoplacophoran
Helcionelloida	Lower-Middle Cambrian	Monoplacophoran
Rostroconchia	Lower Cambrian?–Upper Permian	
Scaphopoda	Middle Ordovician–Recent	
Bivalvia	Lower Cambrian–Recent	
Gastropoda	Lower Cambrian? Upper Cambrian–Recent	
Paragastropoda	Lower Cambrian–Devonian	Polyphyletic; some possibly gastropods
Cephalopoda	Upper Cambrian–Recent	
Stenothecoida	Lower-Middle Cambrian	Molluscan affinities uncertain
Hyolithida	Lower Cambrian–Upper Permian	Molluscan affinities uncertain

"technical" calls attention to the possibility that they may not be homologous with coelomic spaces in other eutrochozoans. They do appear to be homologous within the Mollusca, however (Morse and Reynolds 1996), despite some earlier skepticism on this point (e.g., Stasek 1972). In general, coelomic spaces do not serve among mollusks as hydrostatic or hydraulic skeletons, hemocoels of the blood vascular system being extensively used for those purposes. The cephalopods, however, with the most derived and elaborate of the molluscan body types, possess fairly roomy coelomic spaces dorsally, associated with the gonads and excretory system. Perhaps as a consequence of respiratory efficiencies stemming from a closed circulatory system, cephalopods have relatively small blood volumes for mollusks (Martin et al. 1958). It may well be that the enlargement of coelomic space is compensated by the reduction in volume of the hemocoel and the loss of its hydrostatic properties.

Between the cloaking molluscan mantle (and, if present, its shell) and the foot or body wall, a cavity is created through which, in aquatic forms, water may circulate. Within this mantle cavity, usually situated posteriorly or posterolaterally, are vascularized gills termed ctenidia that are usually paired. The anus and excretory pores open into the mantle cavity also. Ctenidial cilia create a circulating current; mantle-cavity geometry is such that fresh incurrent water bathes the gills first and then flushes away metabolic wastes in an excurrent stream. In some forms one or more gills in a pair have been lost. In polyplacophorans (chitons), the ctenidia are arranged in series within a mantle cavity that runs along the sides of the foot; there may be fewer than ten to more than ninety gills on each side. The lateral cavities join posteriorly, where a median anus empties. Gonoducts and excretory ducts open into the mantle cavity near the rear on each side. Chitons have a ladderlike nervous system, with anterior ganglia, paired ventral nerve cords, and series of transverse connectives. Aplacophorans, possibly basal within crown group mollusks, have a similar nervous system. Tergomyan monoplacophorans, with only about twenty-two living species known, are characterized by an even greater seriation of organ systems. The taxon Monoplacophora was first established on the basis of fossils that had a bilaterally symmetrical series of paired muscle scars, suggesting that they were untorted, though they resembled gastropods otherwise (Knight 1952). Subsequently, living untorted monoplacophorans were discovered (Lemche 1957). These mollusks have seriated pairs of ctenidia (three to six), excretory organs (three to seven), gonads (two), and vertical muscle bundles (usually seven or eight), and a series of nerve connectives (see Haszprunar and Schaefer 1997). The ctenidia and excretory organs are placed laterally in the mantle cavity, with the anus opening posteriorly. Solenogasters and caudofoveates have posterior mantle cavities containing a single pair of ctenidia, and aside from their ladderlike nervous systems they display no particular evidence of seriation.

Locomotion in most chitons, gastropods, and some solenogasters is by ciliary creeping in small forms and by pedal locomotion in large forms. In either case a flattened ventral "sole" is in contact with a ventral mucous trail or directly with the substrate, and the animal is propelled forward by either the beating of cilia or the passage of waves of contraction ("pedal waves" described above under Platyhelminthes; see also Jones and Trueman 1970). In some gastropods the ventral surface rests on a mucous layer that acts alternately as a glide plane beneath the moving portion, where it is liquefied, and as an anchor beneath the stationary portion, where it is solid (Denny 1984). Most caudofoveates, bivalves, and scaphopods burrow, while cephalopods can employ jet propulsion, truly an unusual locomotory system (also found in scallops and sea hares; Trueman 1975) and clearly highly derived. Early mollusks were quite small, and a hypothetical ancestral mollusk is often reconstructed with a broad, flatworm-style foot (see below); if this is correct, primitive molluscan locomotion was probably mucociliary.

The living mollusk classes are sometimes divided into two groups on the basis of their shells. One group, the Conchifera, includes the Bivalvia, Scaphopoda, Gastropoda, Tergomya (crown Monoplacophora), and Cephalopoda. Their exoskeletons are shells that are secreted by the mantle as incremental additions to the margins of a cone, usually coiled logarithmically, with various modifications associated with the requirements of distinctive life modes. Shell deposition is mediated by the surface of mantle epithelia at and near the growing shell edge, and occurs in at least some cases from an extrapallial fluid entrapped between the mantle and shell, the composition of which is regulated by the mantle (Wilbur 1964; Lowenstam and Weiner 1989). The other group, the Aculifera, is sometimes regarded, with respect to Conchifera, as sister or as paraphyletic. Aculifera includes the Caudofoveata and Solenogastres (aplacophorans) and Polyplacophora (chitons). The skeletons of these taxa include spicules (and sometimes scales in chitons) embedded in the mantle. Each spicule or scale is first formed in a vesicle within an individual cell, then protruded from the cell, which enfolds the growing spicule base. The cell and spicule come to lie within an epithelial invagination, and spicular or scalar growth may then devolve upon the epithelial tissue in some cases (Haas 1981; Lowenstam and Weiner 1989). Spicular elements are found in the girdle of chitons (fig. 8.4E) and form the only skeletons in aplacophorans. The eight shell plates of chitons are not spicular but are secreted by an area of mantle epithelium, but somewhat differently than conchiferan shells (see Scheltema 1996), and the shell structure is unique.

Development in Mollusca

Mollusks develop as classic spiralians. In the molluscan oocyte, maternal products are sequestered at the vegetal pole, defining the embryonic anteroposterior axis. There are two chief patterns of early cell specification among the mollusks, associated with different patterns of dorsoventral axis determination. In one case (van den Biggelaar 1977; van den Biggelaar and Guerrier 1979) the first two cleavages are longitudinal and equal, and the four resulting blastomeres are essentially the same size and have been shown experimentally to be developmentally equivalent (fig. 8.6B) . They display a vegetal cross furrow as in flatworms (fig. 8.6C),

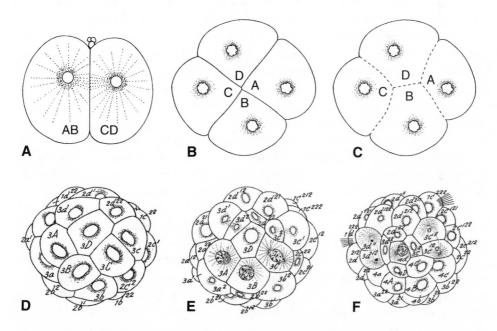

FIG. 8.6 Cleavage in the gastropod mollusk *Patella vulgata*, a member of a primitive clade. (A) Two-cell stage. (B) Four-cell stage, animal-pole view; note that macromere sutures meet at the center; the embryo is radially symmetrical in this view. (C) Four-cell stage, vegetal-pole view; note the cross furrow between macromeres B and D that keeps blastomeres A and C from meeting vegetally; the embryo is bilaterally symmetrical. (D) Thirty-two-cell stage, vegetal view. (E) Sixty-cell stage, vegetal view. (F) Eighty-eight-cell stage, vegetal view; note the 4d cell, the mesentoblast. After van den Biggelaar 1977.

however, which identifies developmental quadrants (chap. 3). Subsequent cleavages are classically spiral, with the centrosomes so located that the spindles incline to the polar axis, alternating in direction between cleavages (figs. 2.10 and 2.11, and see fig. 8.3). With daughter cells lying in the cleavage furrows, each of the blastomeres is in contact with two cells of each of the lower and higher tiers. The four macromeres remain equivalent until between the thirty-two- and sixty-four-cell stages, when one of the cross-furrow macromeres comes into contact with animal-pole micromeres and is induced to define a D quadrant, becoming blastomere 3D. The D quadrant is dorsoposterior, and the induction establishes dorsoventral polarity. This pattern is found most often in primitive groups (Freeman and Lundelius 1992) but is also known in some rather derived forms (such as some opisthobranchs; Boring 1989).

The other molluscan pattern involves autonomous specification of cell fates and of the dorsoventral axis at an early stage. The D blastomere is specified either through unequal cleavages that assort cytoplasmic factors differentially, or through the cytoplasm in a polar body that extrudes from dividing BD cells and contributes its contents to only one of the daughters. In either of these cases, a large D blastomere is produced at the four-cell stage (Freeman and Lundelius 1992), establishing dorsoventral polarity and identifying the cell line that will produce the mesentoblast. These types of cell specifications are found in more derived molluscan taxa.

Mesoderm is derived from cell 4d, and the mouth forms at the site of the blastopore. The origins of the various coelomic spaces during development are incompletely described for many molluscan groups, but it seems that, while these spaces may arise as schizocoellike partings within the mesoderm, there is never a schizocoelic perivisceral cavity (review in Haszprunar 1996a). In mollusks with indirect development the larva is commonly a trochophore, a larval type found also within the annelids, echiuroids, and sipunculans and assumed to indicate a close relationship among those groups. The trochophore larva is characterized by having a double ring of cilia, termed the prototroch, that encircles the body just anterior to the mouth (fig. 8.7A) and is used both for swimming and feeding. Additionally, a ciliary band may be present below the mouth (the metatroch), and a ciliary band may encircle the anus (the telotroch). In many mollusks, especially bivalves and gastropods, the pelagic trochophore develops further into another larval type, the veliger, unique to mollusks (fig. 8.7B); in many forms, larvae hatch as veligers. The veligers are equipped with some organs that characterize the adults—the bodyplans of the classes are foreshadowed in veligers—and are sustained by lobate expansions of the prototroch to form the so-called velum, used both in swimming and feeding.

Early Fossil Record of Mollusca

Many early fossil groups of mollusks are now extinct. Some of those groups are now generally ranked as classes, while the placement of others is disputed. The earliest mollusklike forms are small to minute (Runnegar 1982a); a variety of them, from the Siberian Platform and Mongolia, are illustrated in fig. 8.8. Earliest Cambrian fossils

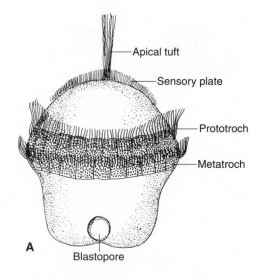

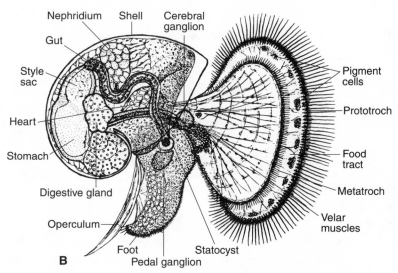

FIG. 8.7 Molluscan larval types. (A) Trochophore larva (of *Patella*). (B) Gastropod veliger larva (of *Crepidula*). A after Patten, B after Werner 1955, modified by Hyman 1967.

are not usually well preserved, commonly being represented only by phosphatic molds, by casts, or by coarsely recrystallized calcitic fossils that were originally aragonitic (see Runnegar 1985; Runnegar in Bengtson et al. 1990). Some of those fossils display characters that are unknown among living molluscan types, and they may represent stem groups.

Concerted efforts to classify these groups and to establish their relationships have been made by Runnegar and Pojeta, beginning in 1974 and continuing in a

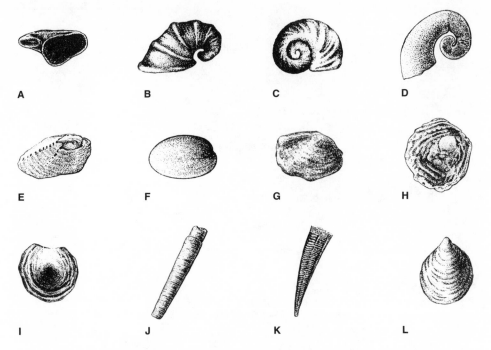

FIG. 8.8 Small shelly fossils from the Lower Cambrian of the Siberian Platform and western Mongolia that are probably mollusks or have molluscan affinities. They are very difficult indeed to interpret; discussion of these groups is scattered through the text. (*A*) A pelagiellid, a paragastropod. Though spirally coiled, it is unlikely that these forms were gastropods. (*B*) A helcionelloid monoplacophoran. (*C*) An aldanellid, possibly either a gastropod or a paragastropod. (*D*) A sinuitid, possibly a tergomyan monoplacophoran. (*E*) *Rozanoviella,* possibly either a tergomyan or a gastropod. (*F*) A rostroconch. (*G*) A very early representative of the Bivalvia. (*H*) *Maikhanella,* probably a coeloscleritophoran. (*I*) *Aldanolina,* a cap-shaped fossil of possible molluscan affinities. (*J*) An orthothecimorph hyolithid. (*K*) A hyolithomorph hyolithid. (*L*) A probivalve. From Rozanov and Zhuravlev 1992.

long series of contributions. The resulting classification (Runnegar and Pojeta 1974, 1985; Pojeta 1980 and references therein; Runnegar 1983) visualized the Monoplacophora as the earliest molluscan class represented by body fossils, and as being paraphyletic, with living classes descending in two great alliances from one or another monoplacophoran branch. One alliance is of forms with U-shaped guts (e.g., gastropods and cephalopods), and the other is of forms with relatively straight or gently flexed guts (included are bivalves and scaphopods). Many of the early mollusklike groups with unusual morphologies were included within the Monoplacophora, which requires a very broad interpretation of the monoplacophoran skeleton and, by implication, the bodyplan.

Coiled univalve shells have been particularly difficult to understand. They display a great range of coiling patterns; the basic nomenclature associated with coiling types in these forms is illustrated in fig. 8.9. The presence of torsion may be inferred from shell symmetries, from the presence of sinuses around the aperture or other structures that can be related to respiratory currents, and from the patterns of muscle

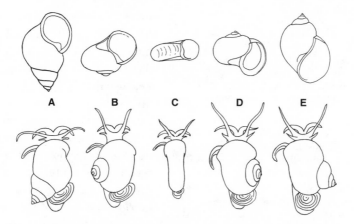

FIG. 8.9 Nomenclature and appearance of coiled shells with different coiling patterns. Two views each of a series of dextrally organized snails (in apertural view, the aperture is anatomically right-side up; in dorsal view, siphons are on the left). In similarly oriented but sinistrally organized snails, the apertures would be on the left. (*A, B*) Coiling is "downward"; this is hyperstrophic coiling. (*C*) Coiling is planispiral; this is discoidal coiling. (*D, E*) Coiling is "upward," as in most snails; this is orthostrophic coiling. From Cox 1960.

scars that indicate the attachment of the body to the shell. Some of the coiled Early Cambrian mollusks were probably torted and were presumably gastropods. Other early coiled shells seem to have been untorted and display series of paired muscle scars, and were presumably monoplacophorans. Still other helically coiled shells were evidently untorted but may have belonged to neither of those taxa.

One of the helically coiled groups, the Pelagiellidae, consists of minute forms that appear as early as any mollusk, in the Early Cambrian (Tommotian; fig. 8.8A). They are interpreted as being dextral and orthostrophic. Their apertures are elongated at less than 90° to the axis of coiling, unlike gastropods, and they possess sinuses that are easiest to reconcile with respiratory currents if the animal was not torted. Bischoff (1990) has reported phosphatized soft parts in two specimens of Middle Cambrian pelagiellids that are interpreted as digestive glands. Unlike such glands in living monoplacophorans they are offset to the left; this asymmetry may be owing to their helical coiling, however, rather than to torsion. Another group of early Paleozoic hyperstrophically coiled mollusks, the Onychochilidae, is also reconstructed as untorted on the basis of apertural features. Linsley and Kier (1984) proposed a new, extinct class, the Paragastropoda, to include these putatively untorted, coiled mollusks, but as onychochilids are quite different from pelagiellids, the class is overtly polyphyletic and should be abandoned.

The monoplacophorans have been formally divided into two classes by Peel (1991a, 1991b), the living Tergomya and the extinct Helcionelloida, both of which appear in the Early Cambrian. Tergomya includes several living genera and forms a

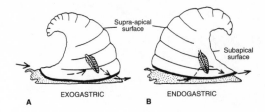

FIG. 8.10 Molluscan coiling directions relative to the anteroposterior axis, illustrated by alternative reconstructions of a helcionelloid (such as *Latouchella*). (*A*) In exogastric coiling the shell apex faces over the head, and the mantle cavity lies opposite the apex. (*B*) In endogastric coiling the shell apex faces away from the head, and the mantle cavity lies beneath the apex. From Peel 1991a.

fairly compact group (see Warén and Gofas 1996). The adult shell is planispiral and has a high rate of whorl expansion so as to be limpet-like, usually coiling through about half a whorl, with the apex pointing toward the anterior. Respiratory currents enter the mantle cavity laterally and exit posteriorly. For coiled shells there are two alternate orientations between the coiling direction and the anterior; the apex of the shell may be coiled over the head (exogastric), or it may be coiled away from the head (endogastric; fig. 8.10). Living tergomyans are exogastric, but whether the extinct helcionelloids are endogastric or exogastric is a controversial question. Helcionelloids are commonly rather high-peaked and flattened laterally, and some possess a "snorkel" (e.g., *Yochelcionella*, fig. 8.11). The function of snorkels, and whether they faced anteriorly or posteriorly, is uncertain, though as David Lindberg has pointed out (personal communication), the diameter of snorkels is so small that it is unlikely that incurrents could have been made to flow in them. Peel has interpreted snorkels as excurrent features (fig. 8.11). Both tergomyans and helcionelloids appear to be untorted and seriated, although the anatomy of the extinct helcionelloids is inferred only from shell form and muscle scars, which can be misleading.

Another class of extinct mollusks is Rostroconchia (fig. 8.12). Although many of the species now assigned to the rostroconchs have been described for a long time, only in 1972 was this class identified and the species assembled into coherent

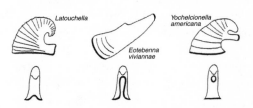

FIG. 8.11 Some helcionelloids, supporting a case for their being endogastric. Views of three forms (*Latouchella, Eotebenna,* and *Yochelcionella*) from (*above*) the side and (*below*) the apical end to illustrate the postulated excurrent functions of structures there: *Latouchella* has a raised sinus in the lower shell margin; *Eotebenna* has a deep slit-like sinus; *Yochelcionella* has a snorkel. These features could represent an evolutionary morphocline. From Peel 1991a.

associations by Pojeta et al. Rostroconch skeletons have two mirror-image sides called valves that are connected across the dorsal margin by one or more shell layers, and that are connected internally by a plate (or rarely two plates) known as the pegma (fig. 8.12). Rostroconchs are present at least as early as the Late Cambrian and persisted until the Late Permian; over 400 species are described. There are two main groups, Ribeirioida (fig. 8.12A, B, D–F) probably the more primitive, and Conocardioida (fig. 8.12C). Except along the dorsum, the margins of the valves of

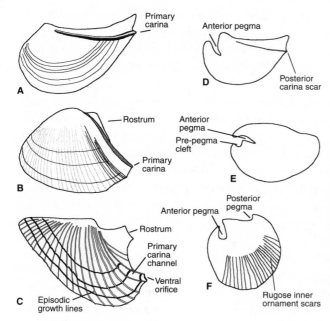

FIG. 8.12 Rostroconchs. (A–C) External views of left side. (D–F) Steinkerns (internal molds) viewed from the left side. Note that the steinkerns show evidence of pegmas, internal plates that connected the valves dorsally, represented by grooves in the molds. (A) *Technophorus* (compare with D). (B) *Ischyrinia*. (C) *Pseudoconocardium*. (D) *Technophorus*. (E) *Reibeiria*. (F) *Wanwoidella*. From Wagner 1997.

primitive ribeirioids are separated, but in more advanced forms they are joined except for one or two restricted apertures. In some species the valves are produced into a presumably posterior funnellike structure around an aperture.

Both the helcionelloids and rostroconchs have been interpreted as exogastric by some workers (e.g., Runnegar and Pojeta 1974) and endogastric by others (Peel 1991a, 1991b). The Bivalvia somewhat resemble bivalved rostroconchs and are themselves exogastric, their beaks facing forward. Runnegar and Pojeta have postulated that Rostroconchia descended from the helcionelloids and that Bivalvia descended from rostroconchs, partly on the basis of their inferred orientations. Scaphopods have also been suggested as being derived from the rostroconchs (Runnegar and Pojeta 1974), but they are not connected to the rostroconchs by fossil intermediates and their anatomy is highly derived, and they may be closer to Cephalopoda than to other living classes. Gastropods are exogastric and are usually considered to have originated from tergomyans with the onset of torsion. Cephalopoda, exogastric also, is often considered to be a sister to Gastropoda.

It is still uncertain what weight such features as exo- or endogastricity should be given in constructing molluscan phylogenies. In the arthropods, with their elongate, segmented bodyplans, the evolution of new body types was achieved in great part by variation in tagmosis and in variation in the relations between segments and appendages (chap. 7). In the unsegmented mollusks, however, new body types have involved the differential rotation, bending, suppression, joining, and compression of organs and other body parts. For example, bivalves and scaphopods are decephalized, gastropods are torted, mantle cavities are lateral (or ventrolateral) in bivalves and polyplacophorans while posterior in solenogasters, gastropods, and cephalopods, and ventral in scaphopods. In cephalopods the head and foot are combined. Siphons are present in bivalves (though not all), gastropods, and cephalopods. Additionally, Lindberg and Ponder (1996) point out that compressions, rotations, and flexures occur within the viscera of many subclades of molluscan classes, characterizing their anatomies. These sorts of evolutionary changes make it particularly difficult to interpret the morphology of extinct groups—which are likely to owe their distinctiveness to precisely these kinds of alterations of soft body-part positions and interrelations—on the basis of their skeletons alone. The arrangements of soft body parts are not always interpretable from fossil material, while analogies with the morphology of living classes are especially dangerous considering the style of molluscan bodyplan evolution.

Early Fossil Record of Mollusklike Forms

Acaenoplax. An exceptionally preserved aplacophoran-like fossil mollusk, *Acaenoplax,* described from Silurian rocks in England (Sutton et al. 2001), is notable as the only fossil plausibly referred to that group; it has been interpreted as representing a sister group to the Caudofoveata. *Acaenoplax* is unique in combining a narrow body that lacks an external foot with a series of seven mineralized dorsal valves. There is a series of eighteen spiny ridges dorsally, coincident with fleshy lobes ventrally that carry finer spines. The valves recall the chitons, and some of the isolated Lower Paleozoic sclerites that have been assigned to chitons may instead represent this multivalved, aplacophoran-like group. On the other hand, the ventral spines do not suggest the sort of ciliary locomotion found in aplacophorans, and a polychaete annelid relationship has been postulated as an alternative (Steiner and Salvini-Plawen 2001). The bearing of this highly seriated form on eutrochozoan phylogeny is uncertain, but its existence suggests the possibility that there was a Paleozoic radiation of small-bodied vermiform groups, now extinct.

Probivalvia or Stenothecoida. A number of curious bivalved fossils that resemble mollusks are known from Lower (Atdabanian) to Middle Cambrian rocks (fig. 8.13A–C; also fig. 8.8L). Although these forms do not show definitive molluscan characters, their calcareous bivalved shell and growth pattern (concentric, quasi-logarithmic) are suggestive of mollusks. These fossils have been thought to represent an extinct molluscan class by Aksarina (1968; Probivalvia) and by Yochelson (1969; Stenothecoida). Some workers have considered these forms to be monoplacophorans, although that group is otherwise not known to possess bivalved shells, and Rozov (1984) suggested that they might represent a phylum of their own. Yochelson has provided an imaginative reconstruction of the soft parts (fig. 8.13D–F). Whether or not this reconstruction is accepted, stenothecoid shells certainly do resemble mollusks, but the soft parts seem unlikely to have been organized anatomically like either the tergomyans or helcionelloids.

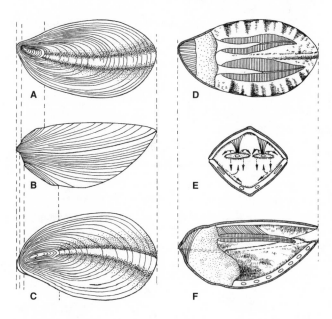

FIG. 8.13 Probivalvia (=Stenothecoida). The curious bivalved fossil *Stenothecoides*; this specimen is from the Lower Cambrian of Alaska. (*A–C*) Reconstruction of shell; dorsal, right lateral, and ventral views. (*D–F*) Interpretation of anatomy; soft parts are hypothetical. (*D*) Dorsal view with dorsal shell and mantle removed, showing ctenidia in mantle cavity. (*E*) Transverse view with suggested feeding currents. (*F*) Longitudinal section; visceral mass reconstructed toward narrow end. From Yochelson 1969.

Hyolitha. Among the small shelly fossils of the Early Cambrian are bilateral and roughly conical skeletons known as hyoliths (fig. 8.14), which are common only during the Cambrian and Ordovician but persist until the Permian, a span of some 280 million years. Hyoliths range in size from a few millimeters to 40 cm in length; the early ones tend to be small. This group has been ranked as a phylum by some workers (Runnegar et al. 1975; Pojeta 1987), while others believe them to represent a class of mollusks (e.g., Marek 1967; Marek and Yochelson 1976). Hyolith cones (conchs) are elongate with only one aperture, which is closed by another skeletal element, the operculum. There are two major taxa, Orthothecida, with planar aperture and operculum, and Hyolithida, with angled aperture and operculum and two additional skeletal elements, termed helens, crescent-shaped appendages extending from between the operculum and conch (fig. 8.14). Hyolith life orientation and habits are speculative; hyolith anatomy is hardly known although looping gut fillings have been identified in several specimens, suggesting that they were sediment feeders, and a pyritized Devonian hyolithid preserves some obscure body features, including an anterior "tentacular mass" (Houbrick et al. 1988). The gut presumably turned back upon itself to provide apertural access for an anus. Many, probably most, hyoliths lay on their flatter side in life, and Kruse (1996), reasoning from a gut trace, has inferred that the upper side of hyolithids was anatomically ventral, implying that those forms were not sediment feeders. For more conclusive evidence of the biology and affinities of the Hyolitha, we may have to await the discovery of well-preserved soft parts.

Kimberella. Neoproterozoic body fossils are quite difficult to interpret, and no undisputed bilaterians have been described. Perhaps the most promising of the

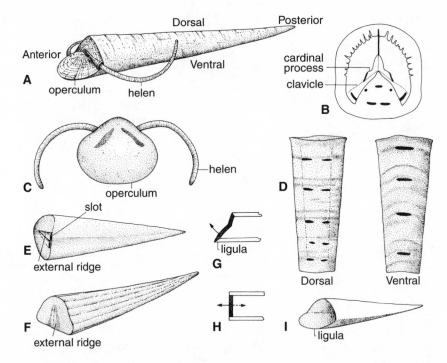

FIG. 8.14 Hyolitha. (*A*) A hyolithid; life orientation of helens is uncertain; here they point dorsally. (*B*) Opercular interior of a hyolithid with raised structures; muscle scars in black. (*C*) Exterior of the operculum of a hyolithid with helens reconstructed as pointing ventrally. (*D*) Internal mold of a hyolithid, dorsal and ventral views, showing muscle-scar impressions in black. (*E, F*) Orthothecids, showing planar operculum with ridge-and-slot structures. (*G*) Angled hyolithid aperture in cross section; the ligula is the extended lower portion of conch. (*H*) Planar orthothecid aperture in cross section. (*I*) An oblique view of a hyolithid conch showing the ligula. From Pojeta 1987.

possible Neoproterozoic bilaterians is *Kimberella* (fig. 8.15), a form originally believed to be a cnidarian from poorly preserved material. Fossils from Russian White Sea localities have formed the basis for a new, plausible interpretation of *Kimberella* as a mollusklike, creeping bilaterian (Fedonkin and Waggoner 1997). Although *Kimberella* is interpreted as having a cap-shaped shell, there is evidence only of a relatively rigid dorsal region, and assignment to Mollusca is based on general similarity of form.

Molluscan Relationships

Questions as to the ancestral bodyplan that gave rise to the mollusks, and of the relation of the mollusks to other protostome phyla, have been pursued largely without regard to the fossil record. The two main candidates for ancestral forms among crown taxa are turbellarian flatworms and annelids, although other phyla have also been suggested (table 8.3; see Haszprunar 1996a). The view that mollusks are descended from annelids or protoannelids seemed to be supported when the living

A

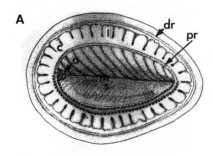

B

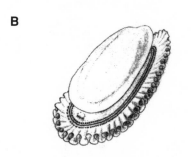

FIG. 8.15 The Vendian body fossil *Kimberella*, a possible eutrochozoan. (*A*) Sketch of fossil. (*B*) Reconstruction as a mollusklike form. *a*, anterior knoll; *c*, crenellated zone; *dr*, distal ridge; *l*, lobe; *m*, medial depression; *pr*, proximal ridge. From Fedonkin and Waggoner 1997.

tergomyans, with their highly seriated architectures, were first described (Lemche and Wingstrand 1959; and see table 8.3). However, the serial repetition of organs found in tergomyans and other mollusks would seem to have little in common with the segmented coelomic architecture of annelids (see below). An alternative view, that mollusks are descended from organisms of flatworm grade, has been more widely held in recent years (table 8.3). This notion was elaborated by Stasek (1972) to hypothesize that mollusks might have descended in three independent lines from organisms with flatwormlike bodyplans: an aplacophoran line (caudofoveates and solenogasters); a polyplacophoran line; and a conchiferan line (the remaining living classes and their extinct allies). The evolution of characteristic molluscan features, such as the hemocoel, the movement of viscera into a dorsal mass, and the presence of coelomic cavities to serve some organs, can be interpreted as independent adaptations to increasing body size. A strong case has been made that the earliest mollusks were quite small (Haszprunar 1992), which accords with the small body size of the earliest fossils.

On the other hand it is quite easy to imagine a monophyletic Mollusca and to postulate a hypothetical molluscan ancestor. A model of a body type that may represent an early molluscan bodyplan was proposed by Salvini-Plawen (1972) and a modified version (fig. 8.16) presented by Salvini-Plawen and Steiner (1996). The earlier model was somewhat reminiscent of some flatworms in that there is a regularly diverticulate (though complete) gut; however, the viscera are moved dorsally compared with flatworms, freeing the ciliated ventral sole to specialize on locomotion. In the later model the gut is not so clearly digitate but forms a complete digestive tract with a radula. The mantle cavity is chiefly posterior with a pair of ctenidia flanking the anus, and there is an open blood vascular system and a pericardial coelom with coelomoducts. This last model is chiefly based on, and closely resembles, the bodyplan of solenogasters (figs. 8.4A, C; 8.16), in the belief that they retain many of the more primitive molluscan features. It can also be argued that the cylindrical bodies of the aplacophoran mollusks are highly derived, and that the primitive mollusks were likely to be flatter-bodied, seriated creepers, resembling Polyplacophora and some of the early tergomyans.

TABLE 8.3 Views on the origin of the Mollusca since 1980. From Haszprunar 1996.

Ancestral Organization	Methods, Data	References
Acoelomate, turbellarian-like	Morphology of Aplacophora	Salvini-Plawen 1985, 1991
Segmented coelomate, annelidlike	Morphology of Conchifera	Götting 1980
Segmented coelomate, annelidlike	Theoretical biomechanics	Gutmann 1981
Not segmented, coelomate	Morphology of Neopilinidae and Polyplacophora	Wingstrand 1985
Acoelomate, nemertean-like	Morphology	Serafinksi and Strzelec 1988
Segmented coelomate, annelidlike	rRNA sequences	Ghiselin 1988
Not segmented, coelomate	Morphology of aculiferans	Scheltema 1993
Segmented coelomate, annelidlike	Theoretical biomechanics	Edlinger 1991
Not segmented, turbellarian-like	Morphology	Willmer 1990
Not segmented, gononephro-coelomate	Morphology of Aculifera and Neopilinidae	Haszprunar 1992
Not segmented, not eucoelomate	Ultrastructure of body cavity of spiralian phyla	Bartolomaeus 1993

Another possible ancestral molluscan body type, then, might display anterolateral extensions to the mantle cavity and possess more organ seriation than in the aplacophoran-based model. The positioning of the living molluscan classes on the Salvini-Plawen and Steiner tree (fig. 8.16), based on a cladistic analysis of morphological characters, agreed with the bulk of such evidence (see also Ponder and Lindberg 1997). However, there is new evidence to suggest that Scaphopoda may be more closely allied to Gastropoda and Cephalopoda than to Bivalvia (see Haszprunar 2000; Steiner and Dreyer 2002).

The early fossil record of Mollusca has a pattern that is quite similar to the record of the metazoan phyla as a whole. Most of the durably skeletonized living molluscan classes appear in the Cambrian, but clear intermediate or ancestral lineages that would permit tracking their phylogenies are lacking. There are also many molusklike fossils, such as the paragastropod groups, that may or may not belong to the living classes. Some extinct taxa, such as the rostroconchs and hyoliths, had important roles in the Paleozoic faunas, but their place in early molluscan evolution remains uncertain. As the living classes cannot be connected with each other via intermediate forms, there is no reason to expect that the extinct classes should offer any easier connections.

Annelida

The annelids, in contrast to the mollusks, are segmented and have a eucoelomic hydrostatic system. The main taxonomic divisions of the annelids as traditionally constituted are in table 8.4. Over 11,000 species have been described. The status of

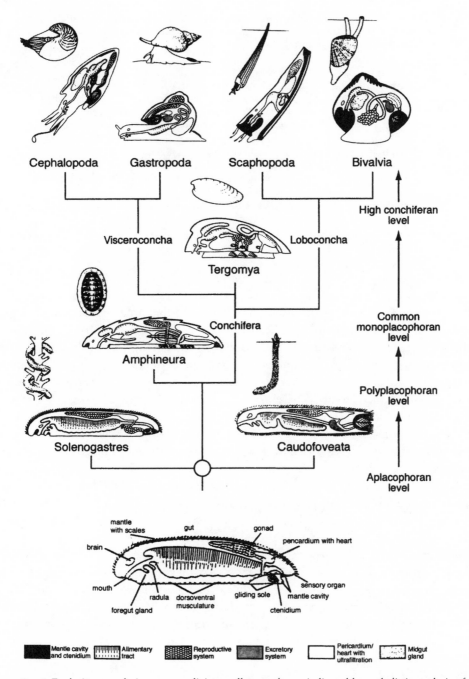

FIG. 8.16 Evolutionary relations among living molluscan classes indicated by a cladistic analysis of their morphological features (*A–H*), with a reconstruction of the hypothetical anatomy of the common molluscan ancestor (*I*). From Salvini-Plawen and Steiner 1996.

TABLE 8.4 Classes of Annelida. The myxostomiids are sometimes considered aberrant polychaetes. Oligochaetes and hirudines are sometimes united as subclasses within class Clitellata. After Brusca and Brusca 1990.

Taxon	Remarks
Class Polychaeta	Chiefly marine, with parapodia and commonly with antennae or other cephalic appendages, and bundled setae. Early Cambrian–Recent. Mostly benthic but some forms pelagic. Cambrian polychaetes do not belong among the thirteen or more living orders.
Class Myzostomiida	Flattened, with suckers, symbionts on crinoids.
Class Oligochaeta	Earthworms; chiefly terrestrial, lack parapodia and cephalic appendages, setae sparse. Carboniferous-Recent.
Class Hirudinea	Leeches; marine or terrestrial, ectoparasitic or predatory; setae sparse, with suckers. Jurassic-Recent.

crown Annelida—whether it is a monophyletic phylum, and what taxa are to be included in it—is uncertain. From a cladistic analysis of annelid morphology and ultrastructure, Rouse and Fauchald (1995) concluded that annelids are polyphyletic, formed of separate clitellate (Oligochaeta + Hirudinea) and polychaete clades that did not derive from an annelid ancestor. But another treatment of those data suggested that annelids were monophyletic, in a clade with pogonophorans (including vestimentiferans; Eibye-Jacobsen and Nielsen 1996; see also Meyer and Bartolomaeus 1996 and references therein). Sequence comparisons of elongation factor 1α have suggested that Pogonophora and Echiura each arose within Annelida, though from different ancestors (McHugh 1997; Kojima 1998). Comparisons of the arrangements and sequences of mitochondrial genomes also place Pogonophora within Annelida (Boore and Brown 2000).

Pogonophora and Echiura are treated here under Annelida, but are discussed more thoroughly than most subtaxa because their bodyplans are so derived that, based only on morphological evidence, they have been treated as phyla by most workers, and the status of Echiura in particular is by no means settled. If the positions of those groups within the annelids were confirmed, Polychaeta would become paraphyletic. Molecular evidence also suggests the derivation of the terrestrial annelids (Oligochaeta + Hirudinea) as a monophyletic group (Clitellata) from within the polychaetes (Moon et al. 1996; Kojima 1998). Therefore, Polychaeta may be paraphyletic regardless of the ancestry of pogonophorans and echiurans.

Bodyplan of Annelida

Polychaeta seems to represent the more primitive living annelid class and forms the basis for the following description (figs. 8.17, 8.18, and see Gardiner 1992). The body wall is flexible, with a cuticle that is partly collagenous, and lined internally by circular muscles. Beneath these circular muscles are longitudinal muscle blocks.

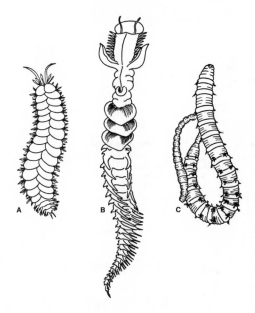

FIG. 8.17 Annelida, external views of three polychaetes. (*A*) A scale worm (Phyllodocida), a predator. (*B*) *Chaetopterus* (Chaetopterida), feeds by mucous straining. (*C*) *Arenicola* (Capitellida), feeds on sediment with an eversible proboscis. After Dales 1967.

The gut runs from an anterior mouth, usually positioned ventrally, to a terminal anus, usually ventral also, and is supported by a vertical mesentery. A coelomic body cavity lies between the gut and the body-wall muscles; it is normally lined by a peritoneum, but in very small forms the lining may be a myoepithelium (Fransen 1988). In active burrowers the trunk is usually divided into segments by transverse septa, at least partially. Within the trunk segments the body wall is usually produced laterally into paired lobate appendages, the parapodia, which contain extensions of the body coelom. The parapodia may be lined by muscles, although in many forms they are thin-walled and rely upon a system of muscles that cross the coelom diagonally to insert in the opposite body wall. The parapodia contain chitinous setae, a general characteristic of polychaetes.

An anterior, preoral prostomium may bear antennae and eyes, and is followed by the mouth-bearing peristomium, which may be composed of one or more segments that usually lack parapodia and that may bear gills in burrow- or tube-dwelling forms. Sclerotized jaws are developed within the pharynx in some forms. The prostomium and peristomium together form the head, and are followed by parapodia-bearing segments of the trunk, which in addition to the parapodial muscle systems may contain paired excretory nephridia (the ducts of which may cross segmental boundaries), paired nerve ganglia and branches, and paired gonads. Thus, although the segmental organization is dominated by a parapodial locomotory apparatus, it is echoed in a serial arrangement of internal organs. The posterior body unit, the pygidium, lacks the paired organ systems, bearing the anus. Most large forms have a closed vascular system, sometimes with a sinus lying along the gut (Fransen 1988), although many small polychaetes lack a blood vascular system altogether. Annelid respiration is commonly through the general body surface. The nervous system consists of an anterior cerebral ganglion, usually tripartite, with paired ventral nerve cords, along which lie segmentally arranged, paired ganglia that each serve parts of two segments; paired ganglia are either connected by commissures or are fused. The body wall is innervated by strands from segmental ganglia, while anterior sensory organs, jaws, and foregut are innervated from various sections of the cerebral ganglion.

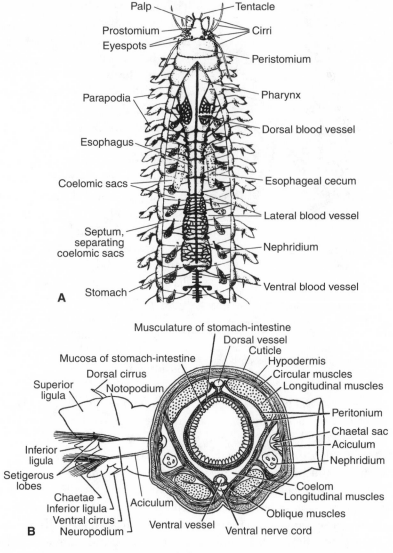

FIG. 8.18 Anatomical features of *Nereis* (Phyllodocida). (*A*) Gross anatomy of the anterior, dorsal view; an individual may have up to 200 segments. (*B*) Simplified cross section. A from Brown 1950; B from Brown 1950 after Turnbull.

There are three principle forms of benthic annelid locomotion: peristaltic, parapodial, and mucociliary. For peristalsis, locomotory mechanics are at their best when there is a relatively complete muscle coating of the body wall (see Clark 1964). In forms with parapodia, however, the continuity of body-wall muscles is broken by muscle insertions, and in forms that rely chiefly on parapodial locomotion the body-wall muscles are significantly interrupted and may be quite reduced. In forms

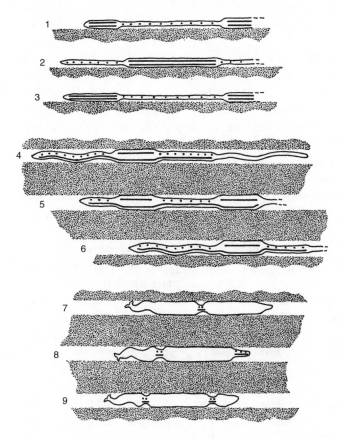

FIG. 8.19 Peristaltic locomotion in some annelids; contraction of longitudinal muscles indicated by horizontal lines, of circular muscles by dots. (1–3) Retrograde peristalsis in *Lumbricus,* an oligochaete; points d'appui (points of contact with the substrate) move in the direction of locomotion. (4–6) Retrograde peristalsis in *Arenicola,* a polychaete, which lacks complete septa; points d'appui are stationary. (7–9) Direct peristalsis in *Polyphysia* (Capitellida), a polychaete, which lacks complete septa; points d'appui, in relaxed sections, move in the direction of locomotion. From Elder 1973.

with peristaltic body movements, the septa that separate segments, while deforming to a certain extent, serve to localize the effects of contractions of body-wall muscles (Clark 1964). The muscles are antagonized by the coelomic fluid, which of course must remain at a constant volume within each segment when the septa are entire, thus producing segment lengthening or shortening to form alternating annular waves of contraction and expansion of the body diameter (fig. 8.19). By employing peristaltic waves, an annelid may creep, or may create a burrow (usually aided by some sort of probing by the head), or move within a burrow, or irrigate a burrow. Worms with complete septa, which prevent transfer of coelomic fluid between segments, can move only by retrograde peristalsis, in which waves move in the direction opposite to the progress of the body (fig. 8.19A–B). Creeping by this sort of body-wall peristalsis is certainly possible and does occur in some non-segmented taxa. On reasonably firm substrates, however, only a narrow arc of a worm's circumference actually contacts the substrate, so peristalsis should be a relatively tedious and inefficient locomotory technique. Within a burrow, however, the entire body circumference may contact the burrow walls, and retrograde peristalsis can be more effective. In actively burrowing forms the parapodia tend to be small, acting as auxiliary gripping devices.

Worms in which coelomic fluid moves easily through a significant portion of the body length may also move and even burrow by direct peristalsis—the waves moving in the same direction as the body (fig. 8.19C). When burrowing, the worm usually forms a cavity in the substrate by some probing action, and then the peristaltic wave shuffles the body wall forward into the opening (Elder and Hunter 1980). Some annelids have reduced septa over much of their body lengths, and these have been seen to use direct peristalsis in burrowing (Elder 1973). In some worm groups that entirely lack internal trunk compartments, direct peristalsis appears to be common (Elder 1980), being especially useful in organisms such as priapulans that must pull themselves forward on very soft substrates in which they are practically submerged.

In annelids with ambulatory locomotion, the septa serve to localize fluid pressure and influence segment shape (particularly the shape of the parapodia) just as they do in retrograde peristalsis, although the sequence of parapodial activity in ambulation is direct rather than retrograde (Clark 1964). The parapodia, which are usually biramous, with a dorsal notopodium and a ventral neuropodium (fig. 8.18), are provided with a capacious coelomic cavity. Chitinous setae project from the parapodia, sometimes in fascicles, to form accessory gripping mechanisms that prevent or retard parapodial slippage. In most living forms, especially active ones, each of the lobes has a chitinous rod, the aciculum, housed in a pouch and provided with special muscles, which is directly involved in the power stroke of the parapodium. In annelids with rather sedentary habits many of the septa are reduced or absent, and burrowing or ambulation is relatively slow. It appears that the more active the annelid, whether ambulating or burrowing, the more complete is the mechanical isolation of the segments.

A number of polychaetes have become minute, adapted to interstitial life. Many of these worms have been grouped as a taxonomic unit, the Archiannelida, and have sometimes been considered to be primitive forms. It is now generally conceded that Archiannelida is a highly polyphyletic assemblage descended from a variety of stocks but adapted to a common meiofaunal environment (see Swedmark 1964; Fauchald 1975). The body forms of some miniaturized annelids have been modified in a number of interesting ways (e.g., Bartolomaeus 1994). In some species the coelom is greatly reduced and partly or completely lacks a peritoneal lining. In some of these forms the reduced coelom may function as a hemocoel (Beauchamp 1910), while in others the body approximates an acoelomate condition (Fransen 1980). Despite this array of modifications, some meiofaunal annelids do retain relatively voluminous coelomic compartments that are isolated by mesodermal septa. Some very small polychaetes, including some with ample coelomic cavities and parapodia, employ mucociliary locomotion, gliding on mucous secretions via cilia (Fransen 1980).

Development in Annelida

Early polychaete development has been used as a standard for spiralians (chap. 2, fig. 2.10). Blastomere fates have been traced in several polychaete species; Anderson (1973) has generalized on such comparative data. The annelid cell fate regions mapped in fig. 2.12 are developmental territories. Anderson noted that, while fate maps of those territories that are defined at about the sixty-four-cell stage are essentially identical in all polychaetes, the cell lineages contributing to the various territories are not. Therefore, although polychaete territories are demarcated by similar cleavage planes at the sixty-four-cell stage, the earlier fates of cells and thus the relations of their cleavage planes to the developing blastula vary among species. Part of the variation is ascribed to variation in the developmental pattern of larval organs. Endomesoderm, however, is always recruited from the 4d lineage; also within this cell lineage a couple of cell lines diverge to another fate in some species (forming part of the midgut). Thus we see again that blastomere fates are subject to evolutionary change, and cell lines in different species may be assigned to different tissues. Cell fates may be specified either autonomously or conditionally after the onset of cleavage. As with mollusks, autonomous specification is associated with what are believed to be the more derived clades (Freeman and Lundelius 1992). Clitellate annelids have variously modified spiral cleavages and territory fate maps that are simplified with respect to polychaetes. These differences are associated with yolky eggs and other developmental adaptations for terrestrial life (Anderson 1973).

In polychaetes with nonyolky eggs, cleavage produces a coeloblastula; gastrulation involves invagination of presumptive endodermal and mesodermal cells; mesoderm originates from the 4d cell and the coelom arises via schizocoely; and the larval form is a trochophore. These are classic spiralian traits. The trochophore contains a growth zone just in front of the pygidium, from which a series of incipient trunk segments proliferates, usually three or so before metamorphosis. Details of segment formation have been reported for a species of the polychaete *Spirorbis* (Potswald 1981). The larval segments first develop simultaneously in ectoderm, followed by mesodermal segmentation within each segment, in an anteroposterior sequence. Polychaete larval segments are sometimes specialized for larval life and sometimes not, and in many cases few larval organs are retained into the adult stage, being resorbed during metamorphosis. In *Spirorbis,* however, larval segments give rise to adult thoracic segments. Following metamorphosis, segment formation continues, and so far as is now known the mesoderm in these segments arises from descendants of the same cells that produce larval mesoderm (Anderson 1973). After metamorphosis in *Spirorbis,* segments bud off the growth zone, with segmentation appearing first in the mesoderm. The adult thorax thus arises before, and the adult abdomen after, metamorphosis, to produce a heteronomous organism. As growth proceeds, there is concomitant elongation of the gut. Coelomic spaces arise by schizocoely within the mesodermal segments or somites and expand to form the body cavity. Muscles lining the body wall and the gut arise from the mesoderm of the somites, and the intersegmental septa develop. As the paired coelomic spaces expand, they come together dorsally and ventrally, separated only by double peritoneal linings (with their basal laminae) that form a vertical mesentery, suspending the gut with its muscle sheath. In most annelids the blood vascular system develops with the onset of segmentation at the site of the blastocoel, with longitudinal vessels (or sinuses) lying between the basal extracellular matrices of the opposing peritoneal layers inside the mesenteries. Thus the vascular system bypasses the septa without compromising their integrity (Ruppert and Carle 1983; Smith 1986).

Pogonophora

This group is largely restricted to the deep sea today. Based on early studies of incomplete material, the pogonophorans were placed with Deuterostomia (Johansson 1937), chiefly because they seemed to possess three coelomic compartments, an architecture displayed by some deuterostomes. Accordingly, the longitudinal nerve cord, now believed to be ventral, was interpreted as dorsal, inverting the dorsoventral orientation now accepted for the animal. What was first learned of their early development was also interpreted as deuterostome-like, and Ivanov (1960, 1963), who first monographed pogonophorans, was so certain of this assignment that he ridiculed a worker (Hartman 1954) who had suggested that they were polychaete derivatives. The last laugh is Hartman's. Major pogonophoran taxa are listed in table 8.5. About eighty living species are known.

Bodyplan of Pogonophora. Pogonophorans have extremely slender bodies, ranging in length from about 5 cm to over 150 cm, but in width from only about 0.1 mm to 40 mm or so. They have only rudimentary digestive tracts; trophic resources are gleaned from bacteria harbored in a specialized area of vascularized tissues, the trophosome, and perhaps also from dissolved nutrients in seawater. There are two main taxonomic subdivisions of pogonophores that have sometimes been treated as separate phyla, for their organization is rather different. The bulk of species belong to the Perviata, to which the following description applies (see especially Southward 1980). There are three external body divisions (fig. 8.20). The most

TABLE 8.5 Higher taxa of Pogonophora.

Taxon	Remarks
"Class" Perviata	No vestimentum, gut; three opisthosomal nerve cords
Order Athecanephria	Prosomal cavity sacciform
Order Thecanephria	Prosomal cavity U-shaped
"Class" Obturata (Vestimentifera)	With vestimentum, gut; one opisthosomal nerve cord
Order Axonobranchia	Branchial lamellae arise along length of plume
Order Basibranchia	Branchial lamellae arise at base of plume

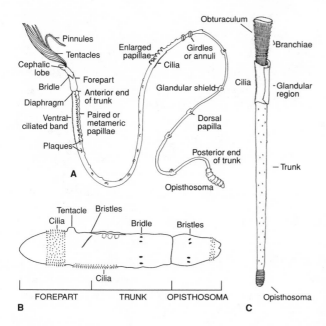

FIG. 8.20 Pogonophorans. (A) Major features of a generalized adult perviate. (B) Larva of a perviate. (C) Major features of a generalized adult vestimentiferan. A after Southward 1984; B after Southward 1980; C after Southward 1982.

anterior, the forepart, contains a cephalic lobe from which between 1 and 200 tentacles emerge dorsally, followed by a bridle or frenulum of thickened cuticular ridges. Most of the body is taken up by the trunk, which is usually separated from the forepart by a groove. The trunk bears various patterns of papillae, many of which are glandular, and ventral ciliated bands. There are rings of hooked setae (uncini) on ridges in the middle trunk section; the large trophosome is developed in the trunk. At the rear of the body is a short, slightly bulbous opisthosoma, which contains septate segments furnished with setae. This section is easily lost during collection and was not reported until after Ivanov's monograph appeared (Webb 1964). With the possible exception of the opisthosoma, the animal is contained within a flexible tube of chitin and scleroprotein, evidently secreted by the trunk glands.

The body spaces are lined with mesoderm. The tentacles contain coelomic lumina that are confluent with a small fluid reservoir at their base. Paired coelomic cavities, separated by a vertical mesentery, lie in the posterior of the forepart. These are separated by a septum from similar paired cavities that run the length of the trunk. In the opisthosoma, each segment contains an unpaired cavity. These segments are lined by mesoderm and separated from each other by muscular septa, within which the basal membranes of the mesodermal linings are apposed. The closed blood vascular system includes a heart, located dorsally behind the tentacle bases, and two longitudinal vessels running within the vertical mesentery, with sinuses in some regions, such as in the cephalic lobe and in opisthosomal septa. The nervous system consists of a nerve ring in the cephalic lobe and a longitudinal nerve cord that is assumed to be ventral, and that is elaborated into three cords and furnished with ganglia in the opisthosoma. There are paired protonephridia in the forepart, opening near the tentacle bases. Gonads are located in the trunk.

A few species are rather different from these perviates and are assigned to another taxon, the Obturata (see Land and Nørrevang 1977; Jones 1981, 1985); these are the vestimentiferans, and include large species up to 1.5 m long with up to 200,000 tentacles. Vestimentiferans live along deep-sea rifts, their trophosomes packed with symbiotic bacteria that use hydrogen sulfide from the associated vents. Folds of the body wall, termed vestimenta, lie behind the tentacles. If correctly interpreted, obturates have three segments anterior to the trunk (perviates have only one; Jones 1985). The opisthosomal segments of obturates have vertical mesenteries, unlike perviates but like annelids.

Development in Pogonophora. Cleavage is reported in vestimentiferans and is spiral and similar to polychaete cleavage; the larva is trochophorelike though lacking a metatroch or telotroch (Young et al. 1996). In species with yolky eggs, gastrulation is by epiboly (Bakke 1980). The origin of mesoderm is uncertain, but it arises posteriorly, and there is no evidence of enterocoely despite early claims.

Echiura

Echiurans have usually been regarded as a phylum of their own, although some workers (e.g., Beklemishev 1969; Nielsen 1995) have believed that echiurans are closely related to or descended from annelids, because of the developmental similarities and the common possession of setae. This relationship is now also suggested by molecular evidence (see above). About 140 living species are described.

Bodyplan of Echiura. The vermiform, rather cylindrical echiurans are often described as sausage-shaped (fig. 8.21A–C). There is a soft cuticle that is sometimes papillate. The body wall is usually lined by a circular muscle layer and by longitudinal muscle bundles, with an inner layer of oblique muscles that lie in fasciculi between longitudinal muscle bands (Stephen and Edmonds 1972). The anterior is provided with a highly extensible, richly ciliated and well-muscled proboscis (which is not an introvert), serving in feeding and respiration. There are setae, usually posteroventral to the mouth and sometimes ringing the anus. The coelom is spacious and undivided. The digestive tract begins with a mouth at the base of the proboscis and extends in a highly coiled path to a terminal anus. The vascular system is closed with one exception, and consists of dorsal and ventral vessels, with a sinus along the intestine. A longitudinal nerve cord is ventral, connecting with an anterior nerve ring in the proboscis and sending lateral nerves to the body wall. Respiration is through the body wall. Excretion is provided by anal vesicles, which are posterior sacs with funnels that open to the coelom. Metanephridia also occur, but their relative contribution to excretion is uncertain.

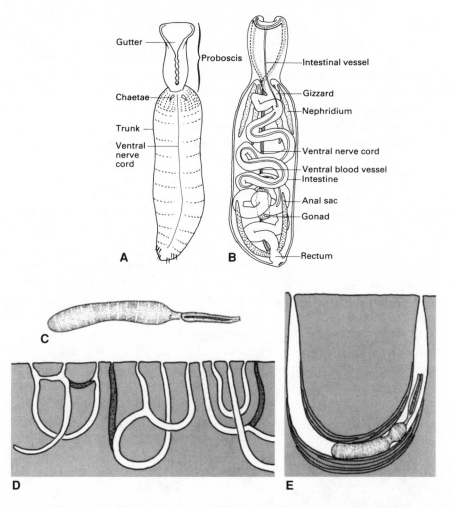

FIG. 8.21 Echiurans. (*A, B*) *Urechis caupo,* which lives in U-shaped burrows and is a suspension feeder. (*C*) *Echiurus echiurus.* (*D*) Burrows of several juvenile *E. echiurus.* (*E*) Adult *E. echiurus* at rest in burrow. A, B after Barnes et al. 1993; C–E after Bromley 1990.

Echiurans burrow by peristaltic and probing movements aided by the setae and papillae (Gislén 1940; Reineck et al. 1967; Schäfer 1972). They form U-shaped burrows, sometimes with an extensive horizontal portion, and in enlarging the burrows some species produce new burrow legs that are connected to the older shafts to create intersecting series of branching U shapes (fig. 8.21D–E). Most echiurans are selective detritus feeders, but one genus, *Urechis,* has a reduced proboscis and is a celebrated suspension feeder (fig. 8.21A–B).

Development in Echiura. It is commonly said that this group displays no sign whatever of segmentation, either during development or in the adult. Nevertheless, echiurans have an early developmental pattern that is quite similar to that of poly-chaetes, proceeding from a spiral cleavage to a trochophore larval stage. Mesoderm arises from cell 4d, and the coelom is a schizocoel (see Newby 1940). Furthermore, studies of neurogenesis in echiurans have revealed a serial repetition of ganglionic structures that closely resembles the pattern in annelids and that appears to develop in an anterior to posterior sequence as in that phylum (Hessling and Westheide 2002). Echiurans display both conditional and autonomous cell-fate determina-tion, with autonomous specification associated with more derived forms (Freeman and Lundelius 1992).

Fossil Record of Annelida

The earliest body-fossil record of a possible annelid is based on a single specimen from the Lower Cambrian Chengjiang fauna (Chen and Zhou 1997); the putative parapodia are pointed and bear sclerites at the base. Other possible annelid remains, which are poorly preserved, are reported from the late Lower Cambrian (Glaessner 1979b). The earliest body fossils of undoubted polychaetes are of middle Middle Cambrian age, the best-preserved being from the Burgess Shale (Conway Morris 1979b). These fossils are of several types (fig. 8.22); well-defined fascicles of se-tae are preserved. Although they cannot be assigned to living polychaete orders, the Cambrian annelids bear comparison with living groups in general structure and may represent both burrowing and swimming forms (Conway Morris 1979b).

FIG. 8.22 Fossil annelids from the Middle Cambrian Burgess Shale. (A) *Burgessochaeta*. (B) *Stephenoscolex*. (C) *Insolicory-pha*. (D) *Canadia*. The well-developed parapodial setae sug-gest that most of these forms were creepers or swimmers rather than active burrowers. Reconstructions A–C by Larry Isham; D by Marianne Collins. From Briggs et al. 1994.

Polychaete jaws, termed scolecodonts, appear in the Ordovician and are found throughout the subsequent Phanerozoic. The earliest undoubted oligochaetes are Carboniferous (see Conway Morris, Pickerell, and Harland 1982), and the earliest fossil hirudines known are from the Jurassic (Kozur 1970).

Some early trace fossils are suspected of being annelidan. These traces are penetrating, vertical tubular burrows, termed *Skolithos*, that first appear either in latest Neoproterozoic or earliest Cambrian rocks. Where they have been carefully studied (e.g., in the western United States; Droser and Bottjer 1988b), the earlier Cambrian (probably Tommotian) *Skolithos* are only a few centimeters deep, while later (probably early Atdabanian) burrows range to 15 cm; these are found in shallow-water terrigenous clastics. This pattern is consistent with reports of *Skolithos* from elsewhere. U-shaped burrows (*Arenicolites*) are also reported from the Early Cambrian (see Crimes 1989) and might well have been formed by annelids. Some of the Neoproterozoic creeping traces and semi-infaunal burrows may be of annelid origin, although they could have been formed by small animals with almost any vermiform bodyplan.

Pogonophora. No body fossils can be assigned to the Pogonophora, except for tubes likely to be those of vestimentiferans that have been recorded from ancient hydrothermal vent deposits. The oldest such tubes are from the Silurian of the Urals (Zaykov and Maslennikov 1987; Kuznetsov et al. 1990; Little et al. 1997), and tubes are also reported from the Cretaceous of Oman (Haymon et al. 1984). Because they are associated with fossil vents, these records are particularly convincing. Other tubelike fossils have been suggested to be pogonophorans, but the attributions are more equivocal. These include the sabellitids, described from rocks at least as old as Early Cambrian and possibly late Neoproterozoic (Sokolov 1972), mostly from eastern Europe. These are very long tubes from shallow-water deposits. Their fine structure has been investigated (Urbanek and Mierzejewska 1977; Ivantsov 1990); the fossils are more complex in internal structure than living pogonophoran tubes and exhibit other differences as well.

Echiura. The earliest body fossil attributed to the Echiura is from the late Neoproterozoic of Namibia, southern Africa (Glaessner 1979a). The specimen is a cigar-shaped cast of quartzite grossly resembling an echiuran and bearing eight longitudinal ridges that were interpreted as reflecting longitudinal muscle bands. As this specimen would be the earliest coelomate body fossil known, it deserves careful restudy; in the meantime, however, the identification is unconvincing. More convincing body fossils of echiurans do occur, in the Carboniferous Essex fauna (Westphalian C) of Illinois (Jones and Thompson 1977). No traces that can be definitively attributed to echiurans are known. As Runnegar (1982a) has noted, however, echiuranlike forms could be responsible for some Early Cambrian horizontal traces

(*Plagiogmus*), and echiurans have been suggested as producing some later Paleozoic and Mesozoic traces as well.

Annelid Ancestry

Particularly influential discussions of annelid origins are by Clark (1962, 1964), who assumed that the origin of the phylum was associated with the origin of its segmented, eucoelomic system. He noted that coelomic compartments are most completely developed in actively burrowing polychaetes, and are commonly reduced, even to a point where the trunk operates as a single compartment, in sedentary forms. Therefore, Clark concluded that the compartmentalized design was originally an adaptation for active burrowing, and that the annelid eucoelom evolved as a hydrostatic skeleton in aid of that function. Parapodia then became elaborated, first for locomotion within burrows, and of course eventually for epifaunal crawling and for swimming. Setae may have evolved early, as gripping devices, becoming associated with the parapodia and then radiating into a variety of supporting roles in locomotory systems.

Other scenarios can be proposed using Clark's approach. The same sort of contrast between active, highly compartmentalized forms and sluggish forms with incompletely divided trunk coeloms is found among epifaunal crawlers and amblers with well-developed parapodia. This situation raises the possibility that coelomic segmentation arose first in epifaunal forms, to promote rather continuous and perhaps rapid ambulation, as might be useful to predators, to prey, or to foraging feeders of many sorts. Swimming has also been suggested as the original adaptation for the annelid bodyplan, but that seems unlikely in light of Clark's discussion.

As Clark has noted, evolution of the segmental muscular system found in annelids makes sense only if the fluid skeleton were also segmented. The bodyplan within which the annelid-style segmentation arose must have already employed a coelom as a fluid skeleton, which was then segmented for active locomotion. The ancestor would presumably have employed a coelom to form a burrow, but would have been rather sedentary or at least sluggish otherwise. The reduction in compartmentalization found in many sedentary polychaetes, and presumably that in Echiura, is secondary in this scenario, a return to an earlier grade.

Sipuncula

Bodyplan of Sipuncula

Another rather cylindrical, unsegmented coelomate group, the sipunculans (fig. 8.23), were long lumped with the echiurans in an abandoned phylum, the Gephyra. Sipunculans have muscular trunks with soft flexible cuticles, the body wall consisting of a circular muscle layer lined by longitudinal muscles; either layer may be continuous or thickened into bands (see the accounts of Stephen and

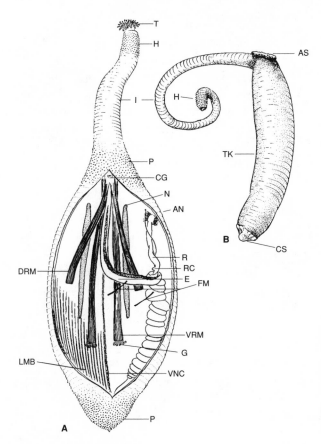

FIG. 8.23 Sipunculans. (A) Gross morphology. (B) *Aspidosiphon*. *AN*, anus; *AS*, anal shield; *CG*, cerebral ganglion; *CS*, caudal shield; *DRM*, dorsal retractor muscle; *E*, esophagus; *FM*, fixing muscle; *G*, gonad; *H*, hooks; *I*, introvert; *LMB*, longitudinal muscle bands; *N*, nephridium; *P*, papillae; *R*, rectum; *RC*, rectal cecum; *T*, tentacle; *TK*, trunk; *VNC*, ventral nerve cord; *VRM*, ventral retractor muscle. From Cutler 1994.

Edmonds 1972; Rice 1982; Cutler 1994). A slender anterior proboscis that lacks wall muscles is capable of complete retraction within the body; this introvert may bear papillae, rings of hooks or spines, and terminal tentacles. There are, however, no setae. The mouth, usually surrounded by tentacles, is central on the proboscis tip, and the alimentary tract is much coiled, describing a coiled U and returning to an anus opening dorsally on the anterior trunk or on the introvert. Like most echiurans, the sipunculans are detritus feeders. Schäfer (1972) has suggested that the lengthy, loosely suspended gut is adapted to passing heavy coarse sedimentary contents with a minimum of damage. There is no vascular system. The body coelom is spacious and lined by a peritoneum. There are coelomic spaces within the tentacles, connected to a ring canal at their bases and to a pair of compensation sacs that receive fluid from the proboscis when it is retracted (Hyman 1959). Retraction is accomplished by muscles, usually four, which begin on the esophagus and cross the coelom to insert on the trunk wall (fig. 8.23A). Extension of the proboscis is by contraction of body-wall muscles to produce fluid pressure in the coelom. There is a simple ventral nerve cord with connectives running to a dorsal bilobed ganglion overlying the esophagus and a series of lateral nerves innervating the body wall. The body cavity contains nephridia, usually a pair, and gonads. Sipunculan burrowing has been described by Schäfer (1972). The proboscis is extended and anchored by the spines and/or hooks, by a "collar" (an inflated ring), and/or by oral tentacles that can extend somewhat laterally. The proboscis is then retracted, but as it is anchored, the body moves forward around it. Burrowing sipunculans sometimes make an unlined vertical shaft like those termed *Skolithos* when fossilized (see Bromley 1990). Sipunculans also nestle in shells or in crannies and may bore into shells or carbonate rocks (Rice 1969). About 320 species are known.

Development in Sipuncula

Sipunculan development is quite similar to that of mollusks and annelids (Åkesson 1958; Rice 1985; Nielsen 2001), with spiral cleavage, a trochophore larva (usually lecithotrophic), a 4d origin of mesoderm, and a schizocoel. Sipunculan blastulae resemble mollusks more than annelids (Rice 1985). Some sipunculan larvae pass through a trochophore stage to pelagosphaera larvae, which swim by means of a postoral ring of compound cilia; some pelagosphaera larvae are planktotrophic.

Fossil Record of Sipuncula

There are no confirmed reports of sipunculan body fossils. Traces suggested as sipunculan are reported from early Paleozoic and later rocks; some Devonian records are particularly plausible (see Conway Morris 1985b for a review).

Sipunculan Relationships

A close affinity with mollusks or annelids is indicated by the developmental similarities. The utter lack of segmentation at any stage has inclined some workers to doubt an annelid affinity, although the finding that the unsegmented echiurans may be polychaetes (see above) weakens this argument. Scheltema (1993) has argued that sipunculans are sisters to Mollusca and infers an ancestral eucoelomate state for both. Trees based on SSU rRNA sequences place the sipunculans either as most closely allied to the mollusks (Lake 1990 based on data of Field et al. 1988; Winnepenninckx et al. 1996) or to the annelids (Winnepenninckx, Backeljau, and DeWachter 1995; Littlewood et al. 1998). If sipunculans are a sister clade to annelids, the perivisceral coelomic cavities of these forms may be homologous. Clark (1969) and Rice (1985) suggest a divergence of sipunculans from coelomic annelid ancestors before the annelid bodyplan was established.

Nemertea

About 900 living species of nemerteans are divided between two classes; each class has two orders (table 8.6).

TABLE 8.6 Higher taxa of Nemertea. After Gibson 1982. For molecular assessment see Sundberg et al. 2001.

Taxon	Remarks
Class Anopla	Mouth and proboscis pore separate
Order Palaeonemertea	Dermis gelatinous or absent; benthic marine
Order Heteronemertea	Dermis well developed; benthic marine, a few freshwater
Class Enopla	Mouth and proboscis pore usually confluent
Order Hoplonemertea	Proboscis armed with stylet; chiefly marine but freshwater and terrestrial as well; a few parasitic
Order Bdellonemertea	Proboscis unarmed; chiefly marine; commensal or parasitic

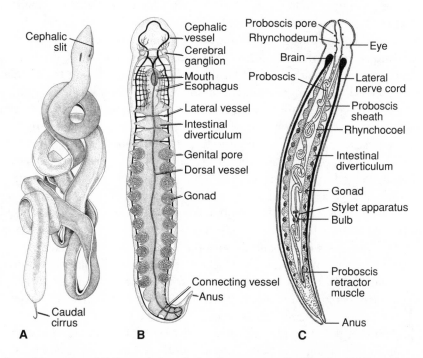

FIG. 8.24 Nemerteans. (A) External view of a heteronemertean, *Cerebratulus*. (B) General nemertean anatomy emphasizing the circulatory, nervous, and reproductive systems. (C) General anatomy of a freshwater hoplonemertean, *Prostoma*, emphasizing the stylet system and the proboscis, which can be everted to capture prey. A, B from Bayer and Owre 1968 after Joubin; C after Hyman 1951a.

Bodyplan of Nemertea

Nemerteans have an elongated vermiform body (fig. 8.24), flattened to roundish, usually several centimeters long but ranging up to several meters. The gut extends from a mouth near or at the anterior end to a posterior anus. In many forms there is a paired series of lateral digestive outpockets between which gonads may be situated, producing an architecture reminiscent of some large triclads. As the gonads are located within mesoderm, any spaces around them, together with their ducts, may be considered coelomic. More important is the eversible proboscis, a major organ system that may run much of the body length when retracted and that lies within a fluid-filled cavity, the rhynchocoel, which is sheathed by mesodermal tissue (Turbeville and Ruppert 1985; Turbeville 1991). The proboscis cavity, then, is coelomic. The proboscis in some orders is equipped with a stylet to puncture prey; toxic secretions are produced by proboscis glands. In forms with an armed proboscis, the mouth and the proboscis share a common exterior opening. The two classes are distinguished by whether the mouth and proboscis have separate or common openings.

Nemerteans possess a circulatory system of lateral vessels, and sometimes a dorsal vessel, commonly elaborated into branch vessels and sinuses. There is no heart, but the main vessels have muscular linings; their epithelia and mode of development suggest that they are coelomic rather than hemocoelic (Turbeville 1986; Jespersen and Lützen 1988). On the other hand, excretion involves abundant protonephridia associated with the circulatory vessels, suggesting that the circulatory system is a blood vascular system. The brain consists of two bilobed parts connected by a large commissure. A longitudinal nerve cord arises from each ventral lobe; these cords are lateral or ventrolateral in position and connected by a series of commissures. In many forms paired longitudinal dorsolateral nerves arise from dorsal lobes, and a medial dorsal nerve from the brain commissure; these nerves do not generally extend to the posterior, though there is much variation among taxa.

A serial arrangement of organs is carried to an extreme in an unusual marine nemertean from Norway that is essentially segmented (Berg 1985). The body is markedly annulated, with reduced body-wall muscles and transverse tissue membranes at the junctions of the annuli, and the intestine is constricted at a regular series of narrow muscular necks that correlate with the annulations. Individual nemerteans have on the order of thirty-five to forty cell types (Turbeville 1991).

Nemertean locomotion is chiefly by mucociliary creeping in small forms. Peristaltic locomotion and burrowing is recorded in some large species (Eggers 1924; Coe 1943) and has been thought to chiefly involve the gelatinous parenchyma (Clark 1964), but Turbeville and Ruppert (1983) have shown that very active burrowing involves a specialized mesodermal musculature surrounding the rhynchocoel. In some large terrestrial nemerteans, the proboscis is involved in locomotion (Pantin 1950). There are also a few pelagic nemerteans.

Development in Nemertea

Nemertean development has been reviewed by Henry and Martindale (1997, 1999). Cleavage is of the classic spiral pattern and typically equal and lacking the cross furrow seen in rhabditophoran flatworms and mollusks. Usually a coeloblastula is formed. Most nemerteans show direct development to a nonfeeding larva that resembles the adult. Endomesoderm arises from the 4d blastomere, ectomesoderm from other quadrants. A number of larval types have been named (see Nielsen 2001). Most of the larvae of indirect developers are feeding pilidium larvae (fig. 2.18), which have ciliated lateral lobes for swimming; most other nemertean larval types resemble the pilidium to some extent. At metamorphosis much of the adult body develops from cells in imaginal disks within the pilidium, and the source of cells responsible for adult mesoderm is uncertain. Feeding larvae appear to be restricted to Heteronemertea, believed to be a derived clade, in which case direct development may be the primitive nemertean condition.

Fossil Record of Nemertea

There are two occurrences of fossil worms that have been plausibly assigned to the Nemertea, both Carboniferous, from the Namurian Bear Gulch Limestone of Montana and from the Westphalian C Essex fauna of Illinois (Schram 1973, 1979). The features that permit assignment of nemerteans to class or order, which include details of the nervous system and of body-wall structure, are not preserved on these fossils. Nevertheless, the gross morphology of the fossils is entirely in keeping with that of living nemerteans, giving them a generally modern appearance. Little is known of the traces produced by nemerteans; some living species produce copious mucus and leave deep grooves when creeping, and it is possible that nemerteans are represented among the trace fossils of the Neoproterozoic or the Cambrian. There is a possibility that *Amiskwia*, an enigmatic fossil from the Burgess Shale, is a stem nemertean (Conway Morris 1977c).

Nemertean Relationships

Nemertean development allies them with the classic spiralians, and SSU rRNA data support this position (Turbeville et al. 1992). It has been common for nemerteans to be regarded as somewhat advanced derivatives of the flatworms, perhaps owing their novel features to their greater size (Hyman 1951a; Gibson 1972; Willmer 1990). However, Turbeville (1991), who has shown that nemerteans have a true hydrostatic coelom, has aptly argued that nemerteans and flatworms do not have synapomorphies.

Mesozoans: Rhombozoa and Orthonectida

Among the more perplexing of the living metazoan phyla are the parasitic Orthonectida and Rhombozoa (fig. 8.25). Orthonectids have not been subdivided into higher taxa, but Stunkard (1982) has separated two rhombozoan species that are bizarre even for rhombozoans, to form the order Heterocyemida, and considers the remaining rhombozoans to form another order, the Dicyemida. Orthonectids and rhombozoans are commonly united as the phylum Mesozoa, but as Hochberg (1982) has emphasized, their anatomies are distinctive and their life cycles lack homologies, so perhaps they should be treated as separate phyla but at a "mesozoan structural grade." This suggestion has been supported by 18S rRNA sequence comparisons (Pawlowski et al. 1996; Hanelt et al. 1996), and though the evidence is not

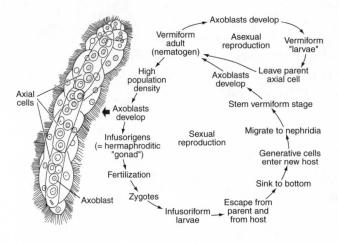

FIG. 8.25 An adult dicyemid during one phase of the complex life cycle; these organisms are endoparasitic in cephalopods. From Brusca and Brusca 1990.

absolutely conclusive, they are likely to be separate phyla. There are about seventy rhombozoan and ten orthonectid species described, probably a gross undersampling of the living fauna.

Bodyplans of Mesozoans

Orthonectids are parasitic on a variety of marine invertebrate phyla, and rhombozoans are parasitic in the renal organs of benthic or epibenthic cephalopods. Orthonectid adults are free-swimming, with an outer layer of ciliated jacket cells surrounding inner contractile cells; gonads are interior to these. They produce ciliated larvae that enter the tissue spaces of hosts, where they form amoeboid syncytia that multiply. Adults arise from within the syncytia and return to the sea. Rhombozoans are also simple, with ciliated jacket cells encasing a long axial cell, and modified cells that form an attachment structure, the calotte, at one end. The rhombozoan life cycle is complicated and perhaps incompletely known; the axial cell acts as a nursery cell, larvae growing and reproducing sexually (but by selfing) still within the cell (see McConnaughey 1963). There is no indication of specialized nerve cells in rhombozoans. Dicyemid rhombozoans have two adult and two larval forms; one of the larval forms, the infusoriform larva, is released in its host's urine and presumably can then infect other cephalopods.

Development in Mesozoans

Rhombozoan development has been reviewed by Furuya et al. (1996). Cleavage is spiral in quartets in both larval forms and is reminiscent of flatworm quartet cleavage through the fourth division, though with its own peculiarities. Orthonectid cleavage was described and sketched by Caullery and Lavellée (1908); there does not seem to be synchronous cleavage of quartets, but the pattern is apparently spirallike, early micromeres lying on cleavage sutures of macromeres. The blastula is morulalike.

Mesozoan Relationships

The sorts of life cycles exhibited by mesozoans are characteristic of forms adapted to parasitism. Endoparasites are commonly much simplified morphologically, to the point of absence of some ancestral cell types, tissue types, or organs, with which they can dispense because of their small size or because the functions are provided by their hosts. There is every chance that mesozoans are descended from more complex forms. Indeed, SSU 18S rRNA sequences suggest that both mesozoan groups are bilaterians (Katayama et al. 1995; Pawlowski et al. 1996; Hanelt et al. 1996) and probably eutrochozoans (Winnepenninckx, Van de Peer, and Backeljau 1998). Further, comparisons of Hox gene sequence are consistent with an assignment of dicyemids to lophotrochozoans (Kobayashi et al. 2000; but see Telford 2000). Stunkard (1954) argued for mesozoan descent from parasitic flatworms, partly on the basis that there are complex life cycles in both groups. It is also possible that clades to which the free-living ancestors of mesozoans belonged are extinct. The times of origin of mesozoans are quite uncertain; there are no known fossils, and as mesozoans do not produce traces, it would take exceptional circumstances to preserve them. Some groups of organisms now parasitized by orthonectids surely date from the late Neoproterozoic, but cephalopods probably arose only in Late Cambrian time (although rhombozoans could have infested some other form previously).

Fossil Groups That May Be Eutrochozoans

Coeloscleritophora

Among the small shelly fossils of the Early Cambrian are a variety of minute spines, cones, and plates that are evidently sclerites, forming elements in composite skeletons of larger organisms. Some of the hollow, calcareous sclerites that lack indications of incremental growth have been assembled into a higher taxon of uncertain rank and affinities, Coeloscleritophora (Bengtson and Missarzhevsky 1981). The chancelloriids were originally included in this assemblage, but are considered here with the sponges (chap. 6). In a few instances, coeloscleritophoran sclerites have been found in life associations, permitting reconstruction of their original body form. The most spectacular of these finds is *Halkieria* from the Lower Cambrian of Greenland (see Conway Morris and Peel 1995). *Halkieria* has a sluglike body with a covering of several types of calcareous spines and plates, and two cap-shaped shells on the dorsal surface that appear to be formed from the amalgamation of numerous small spines or spicules (fig. 8.26). A similar form, though with unmineralized sclerites and lacking the shells, is *Wiwaxia* (fig. 8.27), known chiefly from the Burgess Shale (Bengtson and Conway Morris 1984; Conway Morris 1985a). Such finds have encouraged paleontologists to interpret morphologically disparate sclerites that co-occur as parts of the same animal. Thus Bengtson (1992a) suggested that a species of an earlier, cap-shaped sclerite, *Maikhanella* (fig. 8.8H), is part of a scleritome

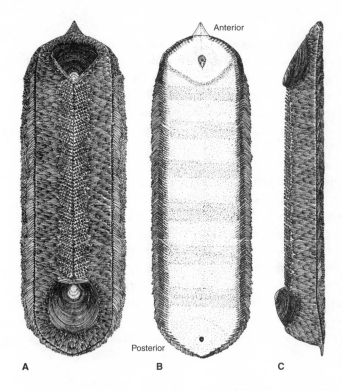

FIG. 8.26 The coelosclerito-phoran *Halkieria evangelista* from the Lower Cambrian of north Greenland. (*A*) Dorsal view. (*B*) Ventral view. (*C*) Lateral view. The bands crossing the sole are meant to suggest pedal locomotory waves. From Conway Morris and Peel 1995.

that includes narrow curved spines assigned to a group termed siphonoguchitids. He has pointed the way to other associations of morphologically diverse sclerite assemblages that may have adorned ancient sluglike coeloscleritophorans (review in Conway Morris and Peel 1995).

The possible affinity of Coeloscleritophora has engendered lively debate. *Wiwaxia* has a structure similar enough in form and position to the gastropod radula to be regarded as at least an analogous feeding device. Partly on this basis, Conway Morris (1985a) thought wiwaxids had emerged from a flatworm ancestor that was closely related to ancestral mollusks (and hyolithids, if those were not mollusks themselves). The presence in *Halkieria* of the cap-shaped shells formed from spicules suggested to Bengtson (1992a) that other molluscan shells secreted by tissues may have evolved from the consolidation of spicular skeletons. For example, the shells of polyplacophorans, which produce spicules in the girdle, may have arisen in this manner. Thus polyplacophorans might be descended from Coeloscleritophora, which perhaps had some place within stem-group mollusks.

On another tack, Butterfield (1990) studied both wiwaxid and annelid preservation in the Burgess Shale, extracting sclerites and studying them with scanning electron microscopy. He concluded that wiwaxid sclerites were identical in structure to fossil annelid sclerites, and were similar to flattened setae (paleae) found

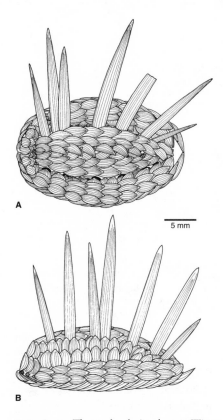

FIG. 8.27 The coeloscleritophoran *Wiwaxia* from the Burgess Shale. A number of discrete, spinose sclerite types are present. (*A*) Dorsal view; large spines removed from left side to display spine pattern along body. (*B*) Left lateral view. From Conway Morris 1985a.

in some living annelids. Therefore he considered wiwaxids to be jawed annelids. With the complete description of the scleritome of *Halkieria,* sclerites of which resemble mineralized wiwaxid sclerites, Conway Morris and Peel (1995) suggested that the halkierids and wiwaxids, while not annelids themselves, did represent the group from which both mollusks and annelids arose. Indeed, for those authors, the coeloscleritophorans may have also given rise to brachiopods and perhaps to other lophophorates (coeloscleritophoran sclerites being homologized with brachiopod setae, and the cap-shaped shells of halkierids with brachiopod shells).

The purported homologies that permit coeloscleritophorans such a central position in protostome evolution are quite speculative. There is no special reason to believe that chiton valves or conchiferan shells evolved from fused spicules, and even if chiton plates were originally spicular, homology with halkierid shells is to say the least uncertain. Based on their general bodyplan, it does seem likely that coeloscleritophorans were eutrochozoans.

Turrilepadida

There is another group of Paleozoic organisms, ranging from the Lower Ordovician to the Upper Carboniferous, that are probably eutrochozoans and that may even be allied with coeloscleritophorans. This is Turrilepadida (see Dzik 1986), consisting of elongate bilaterians covered by sclerites arranged in longitudinal rows (fig. 8.28). In most forms the sclerites are secreted incrementally from their bases. Forms with flattened bodies such as *Plumulites* (fig. 8.28A) are assumed to have been epifaunal creepers, but most turrilepadids have been interpreted as burrowers in loose sediments by virtue of their cross-sectional shapes and rugose skeletons (Dzik 1986). The sorts of uncertainties that surround the taxonomic assignment of the coeloscleritophorans plague the turrilepadids as well. Turrilepadids, along with somewhat similar Paleozoic fossil groups, are usually assigned to the extinct class Machaeridia (Withers 1926).

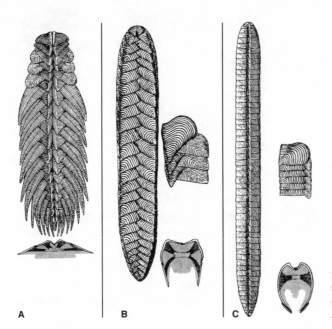

A   B   C

FIG. 8.28 Representatives of three families of Turrilepadida. (A) *Plumulites*. (B) *Turrilepas*. (C) *Aulakolepos*. From Dzik 1986.

Fascivermis

Fascivermis, from the Chengjiang fauna, is a finely annulated worm that may be allied to annelids but which evidently lacks parapodia, and there is no evidence of setae (Hou and Chen 1989b). This form has five annulated tentacles, borne dorsally on the five anterior annulations. A terminal anterior structure is interpreted as a proboscis. The putative proboscis is papillate, and the trunk also bears papillae, some arranged in annulations, some scattered. The gut is straight. *Fascivermis* is about 5.5 cm long, so by the time this form appeared there is evidence of burrows large enough to contain it, although the body fossils cannot be associated with particular traces—the usual case with trace fossils. The tentaculate structure is quite puzzling, and the affinity of this form is obscure.

Other Groups of Problematica

Two vermiform fossils from the Lower Cambrian Kinzers Formation of Pennsylvania (Garcia-Bellido Capdevila and Conway Morris 1999) may represent lophotrochozoans and possibly eutrochozoans. One, *Kinzeria crinita*, is divided into a tail, trunk, and tentacular crown, and a saclike internal organ that could be a compensation sac. The other, *Atalotaenia adela*, is annulate and may be segmented; the specimen is quite incomplete, lacking at least a head region; no setae were observed.

There are other fossils that may well represent eutrochozoan taxa, especially among the small shelly fossils of the Lower Cambrian, and in the later Paleozoic faunas. The relations of these problematic groups are even more obscure than those reviewed here, however, and their significance for the origins of bodyplans

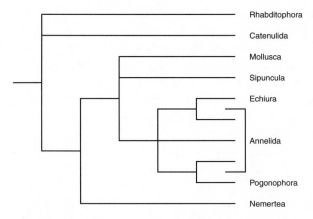

FIG. 8.29 Tentative hypothesis of eutrochozoan relationships, based partly on molecular data, but supplemented by developmental and morphological data. Sorting out the eutrochozoan branching pattern has proven to be especially difficult from molecular evidence (see text).

is unknown. In general, they are most common early in the Paleozoic, when the unusual extinct groups that are better known are also most common. If they carry some sort of message, it would seem to be that the patterns that we can document would probably simply be amplified if we understood the fossil record thoroughly.

Possible Branch Points within Eutrochozoa

Informative differences among the SSU rRNA sequences of the eutrochozoans are few, and although structure within the eutrochozoan tree can be found in some data sets under some algorithms, the relations of the phyla are in general not very stable. The classic spiralian phyla do tend to be closely allied in most SSU rRNA trees, however, reflecting some weak molecular signal. A number of paracoelomate phyla (discussed in chap. 10) also appear either to fall within Eutrochozoa or to be quite closely allied with it. The topology in fig. 8.29 (and see fig. 4.17) is an effort to represent eutrochozoan branch points suggested by, first, the best of the weak molecular data and, second, morphological and developmental patterns, but considerable uncertainty remains as to the history of these phyla.

The features that made platyhelminths popular as the earliest bilaterians also make rhabditophorans seem likely to be the earliest classic spiralians: they are relatively simple, they display classic spiralian features, and there are seemingly plausible scenarios to derive more complex spiralians from them. Perhaps rhabditophorans descended from a paracoelomate relative of the ecdysozoan/lophotrochozoan common ancestor. The possibility that rhabditophorans are reduced from eucoelomate or at least more complex ancestors cannot be ruled out, however. Mollusks are likely to be descended either from early flatworms or from a paracoelomate form, radiating to exploit a pseudocoel as a hemocoel or blood vascular system. Organ coeloms, possibly homologous among living mollusks, presumably originated as required during body-size increases within the phylum. The visceral coelomic cavity in cephalopods is clearly a derived feature with respect to basal mollusks. There does not seem to be any hint in morphology or development that mollusks descended from a eucoelomate.

Several spiralian phyla—Annelida with its likely derivative pogonophorans and perhaps echiuroids, the Sipuncula, and the Nemertea—have hydrostatic coeloms. These eucoeloms are all schizocoels, and it may be that they are homologous. A parsimonious phylogeny with regard to the eutrochozoan coelom is for the ancestral form to be unsegmented, giving rise first to two branches, one leading to nemerteans, the other to sipunculans and annelids.

The affinities of many of the numerous early fossil groups interpreted as eutrochozoans are not very securely established. Certainly the helcionelloids, and probably the rostroconchs and the hyoliths, were indeed mollusks or mollusk relatives, perhaps stem groups. However, the coeloscleritophorans and the turrilepadids are more difficult to place with crown groups, and either or both may be stem eutrochozoans. Some of these extinct groups were not uncommon members of the benthos well into the Paleozoic, joining ecdysozoan taxa such as trilobites and paleoscolecids in lending an exotically archaic aspect to those early faunas. Others of the extinct eutrochozoan groups, such as pelagiellids and stenothecoids, disappear from the record quite soon, geologically speaking, but along with such ecdysozoan groups as the segmented forms reminiscent of arthropods (e.g., anomalocarids, *Kerygmachela*), they serve to emphasize the great disparity displayed by Protostomia during the Cambrian explosion.

CHAPTER 9

Protostomes: Lophotrochozoa 2: Lophophorates

The three living lophophorate phyla, Bryozoa, Phoronida, and Brachiopoda, were first hypothesized to be close relatives by Caldwell (1882) and are sometimes considered to be so closely similar morphologically as to represent a single phylum (Hatschek 1888–1891; Emig 1977). As a group they have been shunted back and forth by different workers from the protostome to the deuterostome branch of the metazoan tree, or sometimes placed as a third branch, intermediate between those two. Among the difficulties in placing these phyla is that their developmental stages share both protostomian and deuterostomian features but not classically so, while their adult bodyplans share features with the hemichordate deuterostomes but not definitively so (see Zimmer 1997). Molecular evidence, based first on brachiopods (Field et al. 1988) and then expanded to include bryozoans and phoronids (Halanych et al. 1995; Mackey et al. 1996; Cohen and Gawthrop 1997; Cohen, Gawthrop, and Cavalier-Smith 1998; Cohen 2000) indicate that all three lophophorates are protostomes, somehow related to Eutrochozoa; this is one of the surprises of the molecular tree. Accordingly, it has been proposed that there is a clade that includes the lophophorates and the eutrochozoans (Lophotrochozoa; Halanych et al. 1995). However, it is not clear that the lophophorate phyla form a clade themselves; bryozoans may not be closely related to the others. Nevertheless, as all three of these phyla have developmental features (especially their radial or biradial cleavages) that set them apart from classic spiralians, and as their bodyplans have much in common, they are treated together here.

Bryozoa

The main bryozoan taxa are listed in table 9.1; most have calcareous exoskeletons, but one marine order (Ctenostomata) and the one nonmarine class (Phylactolaemata) are soft-bodied. Approximately 4,000 living species have been described.

Bodyplan of Bryozoa
The minute (1 mm or less), colonial bryozoans (figs. 9.1, 9.2) possess a tentacular crown or lophophore that has been considered homologous with phoronid and brachiopod lophophores by many workers (Hyman 1959). The individual colonists are termed zooids, and the skeletal chamber associated with each zooid is a zooecium.

TABLE 9.1 Higher taxa of Bryozoa. After Boardman et al. 1983; and see McKinney and Jackson 1989.

Taxon	Fossil Record	Remarks
Class Stenolaemata	Lower Ordovician–Recent	Marine; mineralized, frontal wall rigid
Order Trepostomata	Lower Ordovician–Upper Triassic; later?	Living?
Order Cystoporata	Lower Ordovician–Upper Permian; later?	Extinct
Order Cryptostomata	Lower Ordovician–Upper Permian; later?	Extinct
Order Fenestrata	Lower Ordovician–Upper Permian; Triassic?	Extinct
Order Cyclostomata	Lower Ordovician–Recent	
Class Gymnolaemata	Upper Ordovician–Recent	Marine; unmineralized or mineralized, frontal wall flexible
Order Ctenostomata	Upper Ordovician–Recent	Unmineralized
Order Cheilostomata	Upper Jurassic–Recent	Mineralized
Class Phylactolaemata	Neogene-Recent	Freshwater;
Order Plumatellida		unmineralized

FIG. 9.1 Modern bryozoan colonies. (*A*) *Microporina*. (*B*) *Idmidronea*. (*C*) *Cupuladria*. (*D*) *Cystisella*. (*E*) *Frondipora*. (*F*) *Diaperoecia*. (*G*) *Hornera*. (*H*) *Lichenopora* (encrusting a twig). From Boardman et al. 1987.

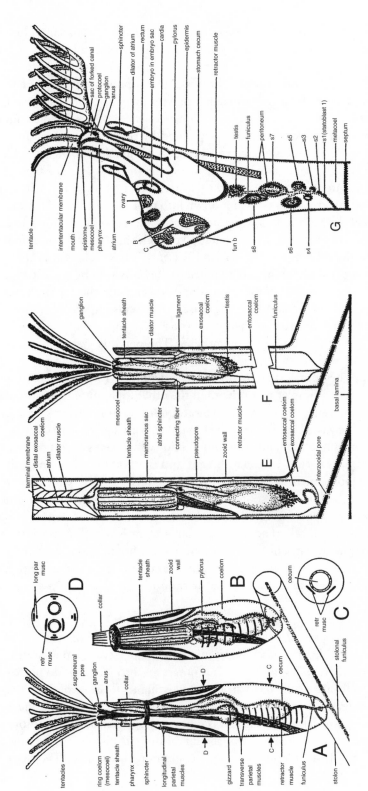

FIG. 9.2 Gross anatomy of bryozoan classes. (A–D) Gymnolaemata; *Bowerbankia*, a ctenostome showing two zooids rising from a stolon, tentacles everted and withdrawn, respectively. Boxy or sac-shaped zooids typify this class. (E, F) Stenolaemata; schematic zooids with tentacles everted and withdrawn. Elongate zooids typify this class. (G) Phylactolaemata; note horseshoe lophophore and epistome, possibly homologous with phoronid epistome. From Ryland 1970.

Three divisions of the body cavity have been identified in bryozoans; the trunk and lophophoral cavities plus, in one group (the Phylactolaemata), a body space in an anterior "lip" or epistome. These body divisions are usually homologized with coelomic regions of other lophophorates. The lophophore usually encircles the mouth (it is U-shaped in phylactolaemates; Mundy et al. [1981] suggest this is because those zooids are relatively large), and the tentacles possess coelomic lumina; the lophophoral and trunk coeloms are separated by only a partial septum at the lophophoral base. Three tracts of cilia run along each tentacle; one on each side, the lateral cilia, and one down the inside of the cone, the frontal cilia. The frontal cilia arise from multiciliated cells and beat toward the tentacle tips, away from the mouth. The lateral cilia beat so as to draw water down into the cone. As the area of the cone decreases toward the mouth, this water accelerates and passes continuously out of the bowl between the tentacles; suspended food particles are collected on lateral cilia and passed into the mouth (see Nielsen and Riisgård 1998). The gut is U-shaped, hanging within the trunk coelom and attached to the body wall by mesenteric cords (fig. 9.2); the anus is dorsal to the mouth and lies outside the lophophore. The lophophore is everted hydrostatically; the muscular systems responsible for eversion differ significantly among the classes (Taylor 1981). Withdrawal of the lophophore into the trunk is effected by retractor muscles. The nervous system consists of a plexus in the body wall, with a ganglion that lies between the mouth and the anus and is assumed to be dorsal, thereby providing orientation for zooidal anatomy.

Bryozoans have a simpler anatomy than the other lophophorates, presumably because of their small size. There are no respiratory or excretory systems, and there is no heart. A possible vascular structure, the funiculus, consists of a cord of mesoderm surrounded by an epithelium and containing fluid (which lacks blood cells) within a lumen. The funiculus runs between the gut and the body wall, often at the site of gonadal development. In most groups the funiculus connects with an external pore of some sort that puts it in contact with other zooids; in at least some stolonal forms the zooidal funiculi connect in a stolonal funicular system (fig. 9.2). Carle and Ruppert (1983) suggest that the funicular system is a rudimentary vascular system concerned with nutrient transport and may be homologous with the blood vascular system of the other lophophorate phyla, and thus the heart and blood cells may have been lost, perhaps during miniaturization.

Development in Bryozoa
Ontogeny. Cleavage is radial, with the form of the blastula and the process of gastrulation varying among the classes. Some gymnolaemate larvae (cyphonautes) have functional guts and are planktotrophic, but others are brooded and do not feed. In stenolaemates, cells are pinched off from the blastula to form secondary embryos that develop into larvae that do not feed. Larval mesoderm arises from the ingression of cells into the blastocoel. All of the larval types undergo a striking metamorphosis that begins after settling when most of the larval body is histolyzed, leaving a cuticularized, essentially ectodermal envelope containing a few cells. The site of the

blastopore is thus obscured, and it is not possible, or perhaps even meaningful, to know whether the adult mouth originates there or elsewhere. The ectodermal envelope invaginates to produce a space that divides to form the lophophore, oral structures, and gut. The outer wall of the invagination is coated by mesoderm that lines coelomic spaces as the zooid develops. In the strictest sense, the bryozoan coelom is neither schizocoel nor enterocoel.

Astogeny (Colony Development). A zooid (the ancestrula) may found a colony, which enlarges by budding; colonies run from a couple of zooids to (more commonly) tens of thousands, to tens of millions. In phylactolaemates the colonies are usually founded from resting bodies, termed statoblasts, produced asexually from previous colonies to survive inclement conditions (Brien 1953). Colony form depends upon the budding pattern (well discussed by McKinney and Jackson 1989). Specialized zooids (heterozooids) that function in reproduction, defense, and many other activities occur in most groups; up to a dozen or so different sorts may be present in a given colony. The specialized zooids receive nutrients from unspecialized feeding zooids (autozooids) with which they are in physical communication. There is a tendency for colonies of the more recently evolved clades (such as the cheilostomes) to exhibit the largest number of specialized zooid types and to display the greatest amount of communication between zooids (although even the earliest fossil bryozoans display polymorphic zooecia, presumably indicating polymorphic zooids). Specialization among individuals thus correlates with increased integration of the colony.

Fossil Record of Bryozoa

Bryozoa are not known in the fossil record before the Lower Ordovician, but by the latter part of that epoch all five orders of the stenolaemates have appeared (Taylor and Curry 1985), certainly an explosive appearance of higher taxa. Some common Ordovician types are shown in fig. 9.3. As bryozoans do not require mineralized skeletons as part of their bodyplan, it is possible that the phylum was extant for some time before evolving durable skeletons. The Lower Ordovician events may reflect a near-contemporaneous radiation, the parallel development of skeletons in several lineages, or some mixture of these processes. As Taylor and Curry note, the Lower Ordovician occurrences were geographically widespread, both within and among orders, which suggests that those clades had something of a previous history. The earliest records assigned to the Gymnolaemata are of borings that appear in the Upper Ordovician and that are similar to those caused by ctenostomes; it is not until the Jurassic that carbonate skeletons of cheilostomes are first found. The freshwater, soft-bodied phylactolaemates are known as fossils only from statoblasts of late Cenozoic age.

Bryozoan Relationships

Although early developmental patterns of bryozoans are not yet known in satisfactory detail, they certainly bear little similarity to the classic spiralian pattern,

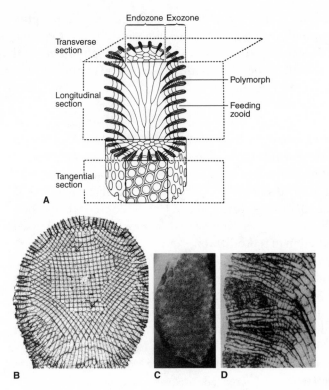

FIG. 9.3 Some Ordovician Stenolaemata. (*A*) Sectioning convention for bryozoans; the form illustrated has a dendroid geometry. (*B*) A dendroid trepostome, *Rhombotrypa,* in transverse section, about 1 cm wide; arrows indicate new buds. (*C, D*) A cystoporate, *Constellaria,* exterior (× 1) and longitudinal section (× 13); note the star-shaped structures in C, termed maculae, that probably mark the location of excurrent chimneys, and the basal diaphragms crossing the zooecia in D. From Boardman et al. 1987.

cleavage being in fact radial. Their close alliance with the other lophophorate phyla has been based more on their overall similarity of body architecture, rather than on similarities of developmental or histological details. The SSU rRNA data suggest that the bryozoans and other lophophorates are protostomes allied with Eutrochozoa, but not that Bryozoa are most closely allied to the other lophophorates (Mackey et al. 1996; Cohen 2000). However, it is not difficult to generate a scenario that derives Bryozoa from phoronids or vice versa, with many of the peculiar bryozoan traits explained as coadaptations to their small size and coloniality (see below). Which of the bryozoan classes displays the more ancestral features is not certain; each has its champions. The stenolaemates, with their early diversification, would seem to be a leading candidate, though there is a stenolaemate feature, a body space that develops between mesoderm and body wall (a sort of pseudocoel involved in eversion), that has been interpreted as being highly derived (Nielsen 1985; see also Jägersten 1972). Gymnolaemates are favored as ancestral forms by some workers partly because they possess the cyphonautes, considered to resemble the basal bryozoan larva, but that could indicate larval retention but not necessarily origin. Phylactolaemates, with an epistome and a crescentic lophophore, have also been suggested as basal.

Phoronida

Phoronida contains only about twelve or so living species, usually classed in two genera, *Phoronis* and *Phoronopsis,* within a single family.

Bodyplan of Phoronida

Phoronids (fig. 9.4) are slender worms with two clear coelomic regions, a trunk that contains the viscera, and the lophophore, which functions in feeding and probably respiration. The lophophoral coelom extends into a preoral dorsal lip or epistome and thence laterally as a tube over the lophophoral arms, most extensively in large species (Emig and Siewing 1975). The mouth is located anteriorly within and at the base of the circlet of tentacles, the digestive tract then running posteriorly but recurving within the trunk to form a U and ending in an anus dorsal to the mouth. The tentacles arise in a single row from a basal ridge that has the form of a horseshoe, with the open side dorsal, facing the anus; the mouth is elongate and lies just within the ventral curve of the ridge (figs. 9.4, 9.5). In small species the basal ridge is single. In larger species the ridge is doubled, so that there are two rows of tentacles around the horseshoe; and in still larger species the arms of the horseshoe are coiled into spirals (fig. 9.5). Thus the trophic and respiratory requirements of larger body sizes are accommodated by increased numbers of tentacles. When feeding, the incurrent flows down the inside of the horseshoe—now a deformed lophophoral cone—exiting between the tentacles.

The body wall has a soft flexible epidermis underlaid by circular and then longitudinal muscle layers; the latter are somewhat bundled, with considerable variation among species. The coelomic cavity is lined by a peritoneum. The trunk coelom is separated from the lophophoral coelom by a septum, but the two anterior coelomic regions are interconnected. The trunk coelom is furnished with a ventral mesentery that arises during development when an unpaired embryonic dorsal coelomic cavity expands around the gut, meeting ventrally to produce the mesentery; later, a dorsal and two lateral mesenteries develop secondarily in most species. The tentacles are furnished with intrinsic wall muscles that are antagonized by fluid of coelomic lumina. The vascular system is closed, with two or three longitudinal vessels, a digestive plexus, and a system of vessels that serve the lophophore. One or more longitudinal nerve fibers connect to a ganglion in the epistome; a nerve ring serves the lophophoral ridge and from this the tentacles are innervated, while a nerve plexus pervades the epithelium. A pair of nephridia within the trunk coelom communicates via coelomoducts with the exterior. All species are tubicolous, secreting a soft chitinous substance that hardens to form a living tube in which sand grains become embedded as an outer layer. Most species form vertical burrows, although a few bore into shells, while one species nestles or forms masses of tangled tubes, more or less horizontal, on solid substrates, sometimes proliferating asexually.

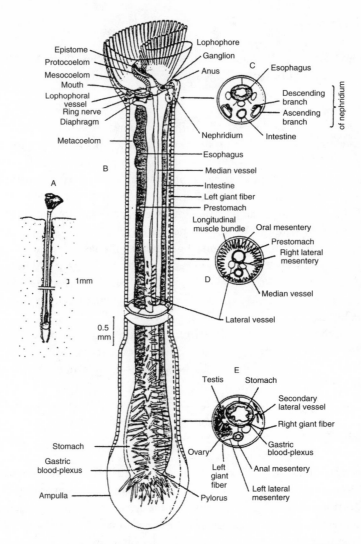

FIG. 9.4 Bodyplan of Phoronida. (A) Phoronid in burrow. (B) Diagram of anatomy. (C–E) Cross sections of body. From Emig 1979.

Development in Phoronida

Early phoronid development has been studied and reviewed by Zimmer (1980, 1997). The cleavage pattern is somewhat variable and has been called spiral by Rattenbury (1954). Zimmer has shown that the pattern is not a stereotypical spiralian pattern, however, and is in fact commonly radial, and concluded that phoronid cleavage is best termed biradial. Gastrulation is by invagination at the vegetal pole to produce a cup-shaped archenteron that remains connected to the

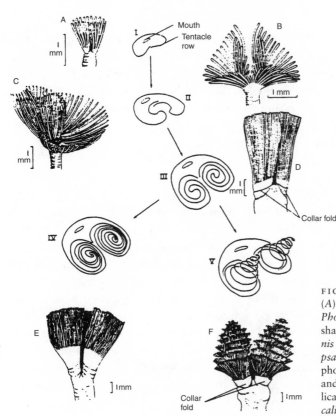

FIG. 9.5 Phoronid lophophores. (A) Oval lophophore in the minute *Phoronis ovalis*. (B, C) Horseshoe-shaped lophophores in (B) *Phoronis muelleri* and (C) *Phoronis psammophila*. (D, E) Spiral lophophores in (D) *Phoronopsis harmeri* and (E) *Phoronis australis*. (F) Helical lophophore in *Phoronopsis californica*. From Emig 1979.

anterior portion of the blastopore as the larva develops. The blastopore region becomes the mouth, so phoronids are literally protostomous. Two categories of mesoderm formation have been described: mesoderm may be budded off from cells of the archenteron, usually in paired masses just anterior to the blastopore, and from these the pair of trunk coeloms originate, perhaps as schizocoels. Mesoderm is sometimes budded off before invagination of the archenteron, in which case it is not clear that these earliest cells are of bona fide endodermal origin (see Lüter 2000). The planktotrophic larvae, termed actinotrochs, are quite different from the trochophore larvae found in many classic spiralians, particularly in the structure of their ciliary bands (Nielsen 1987). Larval phoronids have three coelomic compartments. The most anterior coelomic space is in a preoral hood that becomes the epistome of the adult. The midregion possesses larval tentacles, and the archenteron extends into a larval trunk region behind them. At metamorphosis the larval trunk collapses, and the anus comes to lie below the tentacles as the adult trunk elongates and the gut forms a U-shaped tube. Phoronid development is certainly not very similar to that displayed by classic spiralians.

Fossil Record of Phoronida

Iotuba chengjiangensis, a form known from three specimens from the Lower Cambrian Chengjiang fauna, has been interpreted as a phoronid (Chen and Zhou 1997). It appears to have a U-shaped gut and is tentaculate. No other body fossils have been assigned to the phoronids. Vertical fossil burrows (*Skolithos*) are sometimes suggested to be phoronid burrows, but there are no criteria known that can assign such burrows uniquely to phoronids. Joysey (1959) tentatively assigned a small curved boring in a Cretaceous echinoid to a phoronid, a possibility, though the trace could have been formed by another kind of organism.

Phoronid Relationships

A number of workers have considered phoronids to be basal among crown lophophorates (Hyman 1959; Clark 1964; Wright 1979). If this is the case, then phoronids must have originated in the Neoproterozoic or Manykaian, as brachiopods are known from the lowest beds of the Tommotian Stage (Zhuravleva 1970a). If Neoproterozoic phoronids existed, they were most likely epifaunal forms, judging from the paucity of burrows at that time. It has also been argued that the last common phoronid/brachiopod ancestor was vermiform (Nielsen 1991). Phoronids are certainly closely related to brachiopods, and some molecular evidence suggests that they are sisters to brachiopods as a whole or to one of the inarticulate clades (Cohen, Gawthrop, and Cavalier-Smith 1998; Cohen 2000). Brachiopoda + Phoronida may be derived from some sort of protophoronid—a nonshelled tentaculate suspension feeder.

Brachiopoda

Brachiopods, bivalved lophophorates, have traditionally been divided into two classes, Inarticulata, which lack articulating structures on their valves, and Articulata, which possess such structures. However, a clade-based revision of brachiopod groups has resulted in the erection of three new subphyla (fig. 9.6; Williams, Carlson, Brunton, Holmer, and Popov 1996). The subphylum Linguliformea lacks valve articulations. The subphylum Craniformea chiefly lacks valve articulation but includes a group (Trimerellida) that has articulating denticles. Some early members of the other subphylum, Rhynchonelliformea, have articulating structures, but each group may have evolved them independently. The later orders of Rhynchonelliformea—the Strophomenata and Rhynchonellata—include the classically articulate brachiopods that dominate brachiopod species richness after the Cambrian. Brachiopods first appear in the Early Cambrian (Tommotian; see above) and became major elements in many benthic fossil communities of the post-Cambrian Paleozoic; there are about 345 living species, of which about 300 are rhynchonelliforms.

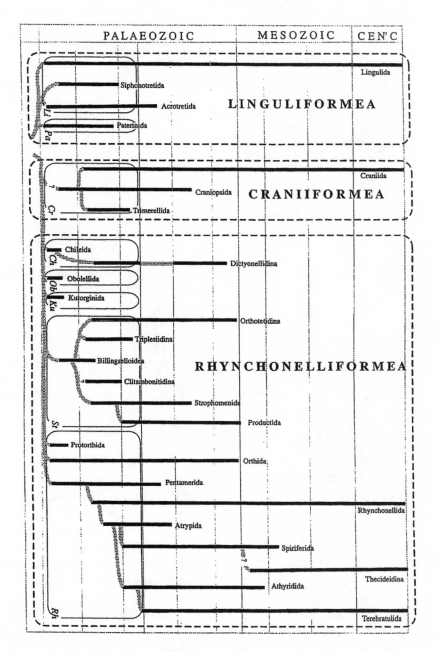

FIG. 9.6 Supraordinal classification of brachiopods proposed by Williams et al., with an inferred phylogenetic tree scaled against Phanerozoic time (vertical lines are period boundaries). Areas within dashed lines are subphyla. Classes are abbreviated: *Li*, Lingulata; *Pa*, Paterinata; *Cr*, Craniata; *Ch*, Chileata; *Ob*, Obolellata; *Ku*, Kutorginata; *St*, Strophomenata; *Rh*, Rhynchonellata. From Williams, Carlson, Brunton, Holmer, and Popov 1996.

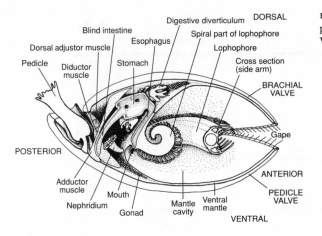

FIG. 9.7 Rhynchonellate brachiopod, generalized anatomy. From Williams and Rowell 1965a.

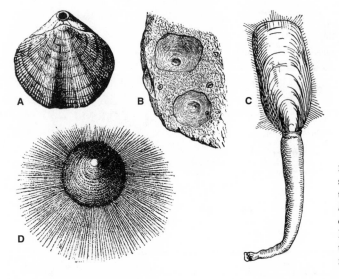

FIG. 9.8 Representatives of some living brachiopod higher taxa. (*A*) Rhynchonellate (*Terebratalia*). (*B*) Craniid (*Crania*), cemented to a substrate. (*C*) Lingulid (*Lingulida*). (*D*) Acrotretid (*Pelagodiscus*). After Hyman 1959.

Bodyplan of Brachiopoda

Brachiopods (figs. 9.7, 9.8) are enclosed within a bivalved shell, either chitino-phosphatic or calcareous. Articulate valves are usually connected by tooth-and-socket structures. The valves are opened by diductor muscles that rotate one valve at the hinge by acting on a leverlike structure. Inarticulate valves are opened by hydrostatic pressures generated by contractions of some of the adductors and other muscles peripheral to the body (see Gutmann et al. 1978). The visceral sac lies in the posterior third or so of the shell, and from it a fleshy pedicle extends posteriorly. The pedicle in many forms anchors the brachiopod to the substrate, but in others it is rather obsolete and may be entirely enclosed within the valves. A thin fold of the body wall, the mantle, extends anteriorly to enclose the lophophore within a mantle cavity. Coelomic mantle canals have a respiratory function. The adult shell is secreted by the margins of the mantle, enlarging at the commissure as body size increases. Some brachiopod taxa also have internal mineralized skeletal struts or ribbons, secreted by mesoderm, that may be rather extensive and that support the lophophore; mesodermal skeletons are otherwise unknown in protostomes (Schumann 1973).

Brachiopods are usually interpreted as having a triregionated coelomic architecture. A visceral or trunk coelom lies posteriorly; coelomic canals within the mantle sometimes harbor egg masses. The lophophoral coelom, which is commonly divided by a mesentery into two compartments, is usually confluent with the trunk coelom, although in adult *Crania* it is described as separated from it by a septum (see Hyman 1959). Anatomically anterior to the lophophoral coelom is a narrow lip, the epistome, that in some inarticulates contains small coelomic spaces, confluent with the lophophoral coelom; it is sometimes considered to represent a third coelomic region. However, in adult rhynchonellates the epistome is usually solid.

The brachiopod lophophore is borne on a brachial axis, presumably homologous with the basal ridge of phoronid lophophores. The axis has two arms that in young individuals are curved so as to form an oval (fig. 9.9), separated by a gap

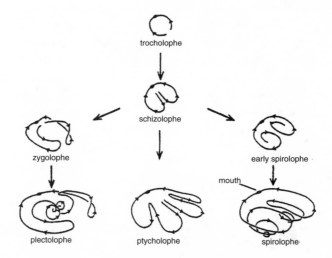

FIG. 9.9 Lophophore patterns in rhynchonellate brachiopods. The lines are traces of the brachial axes, with filaments omitted; the arrowheads indicate direction of feeding currents along the axes. The large arrows indicate the alternative ontogenetic pathways followed by lophophores during growth in different clades; larger brachiopods have more complexly folded lophophores, and several methods have been followed to achieve increased lophophore length. Very small brachiopods have trocholophe or schizolophe lophophores even as adults, and it is possible that the arrows indicate evolutionary as well as ontogenetic paths. From Rudwick 1970.

at their tips (their growing points), reminiscent of the simple horseshoe of small phoronids. As body size increases, the lophophore becomes folded in various patterns that are characteristic for different brachiopod taxa (fig. 9.9), thus becoming longer and providing more tentacles. The tentacles, termed filaments, lie parallel to the lophophore axis behind a food groove (fig. 9.7). The filaments do not always contain hydrostatically functional coelomic lumina (Hyman 1959) and are essentially cartilaginous in some forms (Reed and Cloney 1977). The canals in the brachial axis evidently do not always communicate with the tentacular filaments as a hydrostatic skeleton but may be somewhat occluded by connective tissue, as reported in *Terebratalia* (Reed and Cloney 1977). The lophophores are stiffened, and at least in some adults the filaments seem not to be capable of manipulation as hydrostatic tentacles, but do possess myoepithelial cells and may be flicked (Reed and Cloney 1977). Unlike most bryozoans and phoronids, brachiopod tentacles do not channel food to the mouth, but to a food groove that runs along the frontal side of the brachial axis between the filament bases and a muscular brachial fold or lip that is presumably homologous with the epistome of other lophophorates.

The brachiopod gut extends backward or upward from the mouth but then loops back so that the aboral section is the ventral leg of a U. In living rhynchonellates the gut ends blindly; in living lingulids and craniids, the anus is ventral or posterior (fig. 9.10). Incomplete dorsal and ventral mesenteries cross the visceral coelom from the gut to the body wall, and lateral mesenteries are sometimes present. The circulatory system is poorly known, but is open and evidently lies within the mesenteries. Contractions of a dorsal vascular channel are localized to form a heart dorsal to the gut. There is a pair of nephridia (two pair in rhynchonellids) opening via coelomoducts into the mantle cavity ventral to the mouth. The gonads, usually

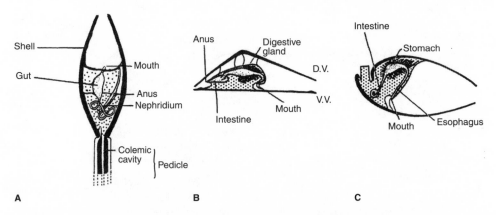

FIG. 9.10 Digestive systems of some brachiopods. (A) A lingulid; note anus. (B) A craniid; note anus. (C) A rhynchonellid; note blind intestine. Stippling indicates viscera exclusive of digestive system; lophophores omitted. D.V., dorsal valve; V.V., ventral valve. A after Barnes et al. 1993; B, C from Rudwick 1970.

two pairs, are located in the visceral cavity or mantle canals; gametes escape via coelomoducts. The nervous system is simple. Nerve thickenings occur above and below the esophagus, at the base of the epithelium or just below. The lophophore is innervated from the supraesophageal ganglion, while most muscles and the mantle lobes are innervated from the subesophageal ganglion.

Development in Brachiopoda

Brachiopod development involves many features that are found in deuterostomes (Conklin 1902); Nielsen (2001) gives an outstanding review. Cleavage is generally radial and produces a hollow blastula that undergoes gastrulation by invagination (Nislow 1994). The blastocoel is entirely filled with the developing archenteron, and mesoderm is derived from cells along the archenteron, and not from a 4d mesentoblast (Lüter 2000). In articulates, coelomic spaces may arise as pockets cut off from the gut—an enterocoelic pattern (see Long and Stricker 1991; Lüter 2000). In inarticulates, however, masses of mesoderm that proliferate along the enteron may split to form a schizocoel, or fold to produce the coelom. In *Crania* four paired coelomic spaces are produced by folding of mesodermal tissues; the second and third become lophophoral and trunk coeloms, while the fates of the first and fourth are uncertain (Nielsen 1991). As brachiopod mesoderm arises from the gut, these variations of coelom formation are sometimes considered modified forms of enterocoely (see Nielsen 1991; Zimmer 1997). However, the mouth of most brachiopods arises near the site where the blastopore has closed, a protostome character. The mouth of the inarticulate *Crania*, though, arises far from the former blastopore.

All living rhynchonelliforms and craniforms have lecithotrophic larvae, while linguliform larvae are planktotrophic. Freeman and Lundelius (1999) have studied earliest shell deposition in living and fossil brachiopods and have been able to interpret changes in early shell growth as indicating embryonic, larval (as either lecithotrophic or planktotrophic), and adult stages. They have shown that some craniforms probably had planktotrophic larvae during the Mesozoic, and that the onsets of various stages have undergone ontogenetic shifts in both craniforms and linguliforms. Embryonic linguliform shells from the Devonian are only about one-third the size of those of living forms and differ in other ways, also suggesting significant developmental evolution (Balinski 2001). Among rhynchonelliforms, some extinct Early Cambrian clades also show evidence that they may have been planktotrophic. Living brachiopod planktotrophic larvae somewhat resemble small adults, with two valves and a set of ciliated tentacles used in swimming as well as feeding; the nature of the imputed rhynchonelliform planktotrophs is unknown. Carlson (1995) and others have regarded modern planktotrophic larvae as modified trochophores. However, Jägersten (1972), who discussed the linguliform larva of *Lingula,* regarded it as a product of an essentially direct development that happened to have evolved a pelagic mode of life. An origin independent of the trochophores of classic spiralians seems likely. Although evidence of the larval mode ancestral to brachiopods has been obscured by subsequent evolutionary events, it seems likely to have been lecithotrophic (see Lüter 2001).

Fossil Record of Brachiopoda

The earliest brachiopods known seem to be paterinid linguliforms from the Early Cambrian (Lower Tommotian); a large number of brachiopod skeletal types appear nearly contemporaneously (fig. 9.6), suggesting a broad and rather abrupt radiation (see Wright 1979). Some of the earliest forms are difficult to assign to living clades or even to those extinct groups that have good fossil records, and relationships among some of the extinct groups are uncertain (e.g., see Rowell 1977; Rowell and Caruso 1985; Andreyeva 1987; Ushatinskaya 1987; Popov 1992; Jin and Wang 1992). From the Ordovician through the Permian, Strophomenata and Rhynchonellata, evidently composing a monophyletic clade of articulates, dominated many benthic marine communities, but at most times did not display a particularly great range of morphological diversity. Many morphological themes are revisited during the evolutionary history of this clade, which shows large numbers of homoplasies and has highly polythetic assemblages of characters (see Williams and Rowell 1965b; Carlson 1995).

Extinct Brachiopod-like Forms

There are some Early Cambrian bivalved fossils that, while resembling brachiopods, are so distinctive that it is unlikely that they are included in that phylum. These include tianzhushanellids, known from the Atdabanian of China and Australia (see Bengtson et al. 1990). One genus, *Apistoconcha*, has elongate toothlike structures, a single median one on one valve and a lateral pair on the other that evidently articulate, but that bear little resemblance to the teeth and sockets found in rhynchonelliforms (fig. 9.11). Another genus, *Aroonia*, has contrasting pits and bosses along the presumably posterior valve margins, but these structures do not articulate, and it has been suggested that the valves were connected by a ligament or muscle inserted on these structures. Conway Morris and Bengtson (in Bengtson et al. 1990)

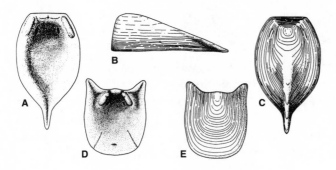

FIG. 9.11 *Apistoconcha siphonalis,* a minute bivalved form that may be related to the brachiopods, from the Early Cambrian of South Australia. (*A–C*) Ventral valve; interior, lateral, and exterior views (*D, E*). Dorsal valve; interior and exterior views (<1 mm). From Bengtson et al. 1990.

suggest that these forms may be products of an early lophophorate radiation, perhaps allied to the brachiopods as a sister clade. If the nature of valve articulations is an important systematic character, these genera are not very closely related.

Brachiopod Relationships

The brachiopods are the earliest, unequivocally eucoelomate phylum known to appear in the body-fossil record. Within a few million years of their first appearance they are represented by about seven clades so distinctive that the surviving ones are considered to be orders, classes, and subphyla. There are also several enigmatic early taxa, and the brachiopod-like tianzhushanellids.

Whether brachiopods are more likely to be monophyletic or polyphyletic has long been debated. Monophyly has been strongly argued by Rowell (1982; see Carlson 1991), who identified likely synapomorphies of all brachiopods (table 9.2), including tentacles arranged along one side only of the lophophore axis, two coelomic canals in the lophophore, and the presence of mantle canals. Carlson (1995) subsequently addressed this question in an extensive cladistic analysis of the living brachiopod superfamilies, based on morphological and developmental criteria, and using phoronids and bryozoans among the outgroups (table 9.2). Her conclusions were that lophophorate monophyly was tentatively supported, brachiopod monophyly well supported, articulate monophyly strongly supported, but inarticulate monophyly only weakly supported. See also Holmer et al. 1995 for further morphological support for brachiopod monophyly.

TABLE 9.2 Possible brachiopod soft-part synapomorphies proposed by Rowell (1982) and Carlson (1995).

Rowell
    Filaments in a single palisade about lophophore axis
    Double row of filaments on adult lophophore
    Brachial lip bounding food groove
    Mantle canals
Carlson
    Sexes always separate
    Highly modified trochophore larvae
    Short free-swimming larval stage
    Two pairs of larval setae
    Ventral and dorsal mantles, with valves, mantle canals
    Lophophore tentacles on only one side of arm axis
    Paired lophophore tentacles
    Two coelomic spaces per lophophore arm
    Anus ventral to mouth, gut curves down
    One pair of digestive diverticula
    Single contractile heart
    Respiratory pigments as hemerythrin
    Primary nervous ganglion subenteric
    Both smooth and striated muscle fibers

Brachiopod polyphyly, based on the differences between articulates and the inarticulates as a group, has also been argued (Percival 1944; Valentine 1975), while Wright (1979) suggested that each of the major inarticulate groups may have arisen independently. Both Valentine and Wright suggested that the ancestral bodyplan was phoronidlike. As a bivalved shell is a feature of brachiopods, it seems that some similarities among brachiopod clades might be homoplasies arising from common requirements of encasing a lophophorate within a bivalved shell—"brachiopodization."

Lophophorate Relationships

Minute or large, colonial or solitary, burrowing or epifaunal, the lophophorate phyla display a strong similarity of bodyplan. To some workers it has seemed that they are no more distinctive from each other than, say, the molluscan classes (Emig 1977). All of the lophophorates can be interpreted as originally sharing a triregionated bodyplan, commonly reduced to a biregionated one. Many of the differences among these phyla can be explained as adaptations to distinctive modes of life; each bodyplan seems to involve a harmonious coadaptation of organ systems designed for a particular adaptive zone.

On the other hand, some workers have concluded that the lophophorates do not represent a coherent, related group, but display morphological convergence. Nielsen (1971, 1995) has regarded Bryozoa, for which he uses the term Ectoprocta, as allied to the Entoprocta, and joins them into a phylum, this termed Bryozoa, which is placed in Spiralia, while the other lophophorates are placed in Deuterostomia. Nielsen does not regard the tentacular structure of his Ectoprocta to be a lophophore, nor their body cavity to be a triregionated coelom. Other features that are used to separate ectoprocts from other lophophorates include the larval types (Nielsen 1971). The features used by Nielsen to relate ectoprocts (Bryozoa here), and entoprocts are chiefly ontogenetic and associated with metamorphosis.

Nielsen (1991) has pointed out that the U-shaped digestive tracts of phoronids loop dorsally to the anus, while brachiopod guts loop ventrally (fig. 9.12); in other words, phoronids are endogastric while brachiopods are exogastric. Thus the common phoronid/brachiopod ancestor might have had a bodyplan unlike either of them. Nielsen favors a creeping worm with a straight gut as a progenitor of both phyla. Zimmer (1997) postulates a small tentaculate worm with a U-shaped gut, closely resembling a small phoronid, as a hypothetical ancestor.

Even though the developmental features of lophophorates arguably have more in common with deuterostomes than with protostomes, all of the molecular evidence that has been brought to bear on the lophophorates confirms that lophophorates are protostomes, presumably ending that dispute. It is not yet clear, however, just how the lophophorates are interrelated, and just how they are related to the classic

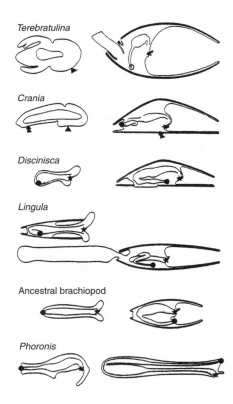

Terebratulina

Crania

Discinisca

Lingula

Ancestral brachiopod

Phoronis

FIG. 9.12 The position of some landmark features in larvae (left column) and adults (right column) of four brachiopod orders plus a hypothetical ancestral brachiopod, contrasted with a phoronid. The brachiopod and phoronid guts curve in different directions (ventral versus dorsal, respectively). *Single arrowhead,* position of future mouth; *double arrowhead,* position of blastopore; *star,* mouth; *filled circle,* anus; *open circle,* end of blind gut. From Nielsen 1991.

spiralians. SSU rRNA evidence suggests that bryozoans are not closely related to the other lophophorates (Halanych et al. 1995; Mackey et al. 1996; Littlewood et al. 1998; Cohen 2000). On the other hand, SSU rRNA evidence suggests that phoronids are indeed closely related to brachiopods (see Cohen 2000). Somewhat different phylogenies have been indicated by different analyses. Halanych et al. (1995) produced a tree in which phoronids and "articulate" brachiopods are sisters. Mackey et al. (1996), with additional data, presented two trees. One tree, from a neighbor-joining algorithm, indicated that brachiopods are paraphyletic, the articulate branch being deepest and giving rise to inarticulate branches, from which phoronids descend. A second tree, a maximum parsimony analysis, indicated that the phoronids are paraphyletic and branch deepest, while the brachiopods are polyphyletic, arising from different phoronid ancestors (see also Littlewood et al. 1998). Finally, additional sequencing (Cohen 2000) has provided support for the monophyly of a brachiopod + phoronid clade, and moderate support for a clade of phoronids + inarticulates, to which other brachiopods are sisters (fig. 9.13). Cohen placed phoronids as a subphylum (Phoroniformea) under the phylum Brachiopoda. The uncertainties in molecular phylogenies of the lophophorates mirror the difficulties with molecular phylogenies within eutrochozoans (chap. 8).

Thus, although progress has been made on lophophorate phylogenetics, their interrelationships have not been entirely clarified. Morphologically based phylogenies such as that of Williams, Carlson, Brunton, Holmer, and Popov (1996) place both phoronids and Bryozoa outside of the brachiopod clade—the morphological characters common to the bivalved brachiopods exclude those taxa. Many of those characters involve the lophophore as well as the shell; perhaps those features, at least the tentacular characters, are coadaptive with a feeding system that operates inside

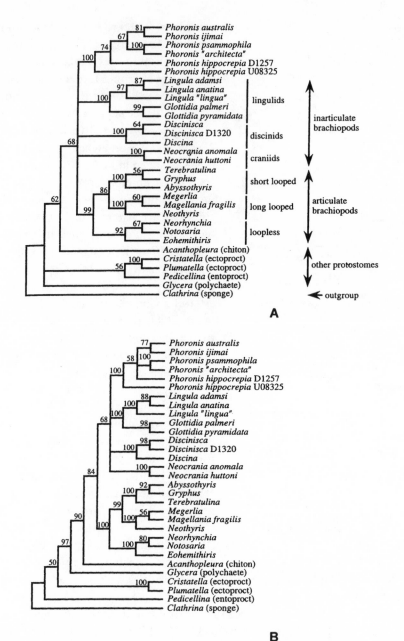

FIG. 9.13 SSU rRNA phylogenies of lophophorates and some other protostomes. (A) Consensus tree from a maximum parsimony analysis, with nodes with less than 50% bootstrap support collapsed. Relations between phoronids and brachiopod clades are unresolved. (B) Consensus tree from a neighbor-joining minimum evolution algorithm (see Cohen 2000 for further specifications). Moderate support for a phoronid-inarticulate clade is indicated. Both trees suggest that phoronids + brachiopods are monophyletic with respect to the other protostomes, but that bryozoans (ectoprocts) are only distantly related to that clade, being closer to the entoprocts. From Cohen 2000.

a shell. Judging by experience with phylogenies wherein morphological and molecular data are combined (see chap. 4), the topology of the lophophorate tree would tend to be most influenced by morphological features in a combined phylogeny. The assumptions made about morphological homologies thus become critical, for they can have a controlling influence on the topology. Lophophorate interrelations remain open questions.

CHAPTER 10

Protostomes: Paracoelomates

Small-bodied triploblastic phyla that lack secondary body cavities (paracoelomates) have been extremely difficult to place on the phylogenetic tree with morphological data, and many have been lumped together into the sort of trash-can assemblage of aschelminths, for want of better information. As fossil and molecular evidence indicates that the late Proterozoic benthos included small-bodied creeping Bilateria that were probably of paracoelomate grade, these understudied phyla have become extraordinarily interesting. Some of the living phyla may be either crown members of late Proterozoic phyla, or collateral relatives of those early Bilateria, or at least are likely representatives of a structural grade that characterized early bilaterian faunas, and therefore we stand to learn much about early metazoans from these groups. Unfortunately, their fossil records are poor.

SSU rRNA sequences have strongly suggested that Kinorhyncha, Priapulida, Nematoda, and Nematomorpha belong to the ecdysozoan protostome alliance, and accordingly they were treated in chapter 7. Six more paracoelomate groups, Gastrotricha, Rotifera, Acanthocephala, Entoprocta, Gnathostomulida, and Cycliophora, appear to be protostomes as well, but none lies within Ecdysozoa. Entoprocta appears to lie within Eutrochozoa, but while SSU rRNA data sometimes ally the others to Eutrochozoa as well, they also appear in a variety of topologies that differ from study to study. Giribet et al. (2000), using a large combined morphological and molecular database, find a clade of paracoelomates (Platyzoa) that includes Gastrotricha, Rotifera, Acanthocephala, Cycliophora, Gnathostomulida, and flatworm groups, and that is sister to Lophotrochozoa. This clade is proposed as a major protostome alliance, and except for flatworms (chaps. 6, 8), these phyla are treated here, together with Entoprocta and the enigmatic Chaetognatha.

Although all these phyla have sometimes been lumped as pseudocoelomates, four of them (Gastrotricha, Entoprocta, Cycliophora, and Gnathostomulida) do not display any internal spaces at all except for reproductive or excretory housings or ducts. These phyla are thus at about the same structural grade as the flatworms, and could just as easily be termed acoelomate. Rotifera, however, has a capacious fluid-filled body cavity and is at the classic pseudocoelomate grade. Acanthocephala is exclusively parasitic and appears to be derived from within the Rotifera. These five taxa share many features, but each has distinctive apomorphies. Among flatworms, the rhabditophorans and catenulids appear to be allied in some way to the platyzoan phyla on the basis of SSU rRNA data (Carranza et al. 1997; Littlewood et al. 1998; Giribet et al. 2000).

Gastrotricha

Bodyplan of Gastrotricha

Gastrotrichs (fig. 10.1; Hummon 1982; Ruppert 1991a) are minute (0.1–3 mm, chiefly <1 mm) unsegmented worms that lack body cavities except for ducts and interstices within the mesoderm, and thus could be termed acoelomate. The ventral surface tends to be flattened and ciliate; the dorsal surface is highly convex. There are two classes, an exclusively marine Macrodasyida and an essentially nonmarine Chaetonotida. The body is divided into head and trunk regions, sometimes separated by a lateral constriction, and many forms (particularly freshwater groups) have characteristically forked "tails." The cuticle, while featureless in some forms, especially pelagic ones, is often elaborated into spines or scales that are presumably adaptive to interstitial life in sands (Boaden 1985). The epidermis is partially or entirely syncytial. Apomorphic features include minute adhesive tubes that project from the cuticle and emit a sticky material, secreted by gland cells, by which gastrotrichs may adhere to their substrates. There may be up to 250 of these tubes arranged in rows along or across the trunk or clustered on the head or tail in marine forms. Nonmarine gastrotrichs, however, have only a few adhesive tubes, or, particularly in pelagic forms, the tubes are lacking altogether.

The anterior mouth leads into a shallow buccal capsule and thence to a pharynx that is triangular in cross section with a Y-shaped lumen and may be as long as one-third of the body length, leading into a straight gut extending to a subterminal anus. Food is moved into the midgut via pharyngeal pumping. Musculature consists of circular fibers, usually beneath the epidermis, and longitudinal muscle bundles that are most pronounced ventrolaterally; both sorts of muscles are somewhat oblique. In benthic marine forms many of the muscles are involved in manipulating the adhesive tubes. Ventrolateral nerve tracts terminate anteriorly in ganglia that are joined by a commissure. Macrodasyids lack excretory organs; chaetonotids have protonephridia. Most marine gastrotrichs are benthic, most living meiofaunally within loosely compacted sediments, while a few are epibenthic on compacted muds or larger organisms. There are about 450 described species.

Development in Gastrotricha

Gastrotrich development has been described by Beauchamp (1929), Sachs (1955), and Teuchert (1968). The cleavage was interpreted as spirallike by Costello and Henley (1976), distorted but "vaguely reminiscent of spiral" by Brusca (1975), and "essentially radial" by Brusca and Brusca (1990). A clear evaluation of gastrotrich cleavage is given by Nielsen (1995). Cleavage is holoblastic and unequal,

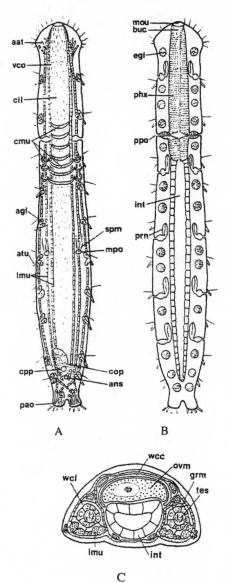

FIG. 10.1 Gastrotricha. (*A*) Gross anatomy from ventral view. (*B*) Dorsal view emphasizing the digestive system. (*C*) Transverse section across region of egg. *aat*, anterior adhesive tubule; *agl*, adhesive gland; *ans*, anus; *atu*, adhesive tubule; *buc*, buccal cavity; *cil*, cilia; *cmu*, circular musculature; *cop*, copulatory organ; *cpp*, copulatory organ pore; *egl*, epidermal gland; *grm*, germinal epithelium; *int*, midgut; *lmu*, longitudinal musculature; *mou*, mouth; *mpo*, male gonopore; *ovm*, egg; *pao*, posterior adhesive organ; *phx*, pharynx; *ppo*, pharyngeal cleft; *prn*, protonephridium; *spm*, sperm; *tes*, testis; *vco*, ventral brain commissure; *wcc*, wall cell of central body; *wcl*, wall cell of lateral body region. From Ruppert 1991a.

and although there is some displacement among the blastomeres, it is clearly not spiral, being unique and involving the movement of blastomeres; in this feature it resembles nematodes although the pattern of movement is different. Whether this cleavage is reminiscent of a radial ancestor is uncertain. A coeloblastula is formed; gastrulation involves the ingression of two cells into the blastocoel, which then produce the archenteron. Mouth and anus are both secondary. Development is direct. Many gastrotrichs are parthenogenetic.

Fossil Record of Gastrotricha

No gastrotrichs have been reported as fossils.

Gastrotrich Relationships

Based on morphological evaluations, many workers have concluded that the gastrotrichs belong to a clade that includes nematodes and other phyla that are ecdysozoans on molecular evidence (e.g., Conway Morris and Crompton 1982; Boaden 1985; Lorenzen 1985; Nebelsick 1993; Neuhaus 1994; Malakhov 1994; Nielsen 1995). However, SSU rRNA sequences ally gastrotrichs with the lophotrochozoans (Winnepenninckx, Backeljau, Mackey, Brooks, DeWachter, Kumar, and Garey 1995; Littlewood et al. 1998; Giribet et al. 2000). A lophotrochozoan assignment is supported by the presence of ectodermal cilia and lack of a molting habit.

Rotifera

Although this phylum comprises perhaps 2,000 described species, few are marine, and these few are chiefly pelagic, though several, forming the monogeneric class Seisonidea, are epizoonts on crustacean gills. Accounts of this group include those of Hyman (1951b) and Clément and Wurdak (1991). There are three classes: Seisonidea (with a single genus), Bdelloidea, and Monogononta.

Bodyplan of Rotifera

Rotifers are minute, ranging between 40 μm and 3 mm (usually below 0.5 mm) in length. Their bodies (fig. 10.2) can usually be divided into head, trunk, and foot (with extensions known as toes). The cuticle is gelatinous and lacks collagen or chitin, but is sometimes thickened to form a lorica that serves for support and muscle attachment. The body wall is not coated by muscles, but the adult body-wall tissue is syncytial (as are most adult tissues), and an internal skeletal lamina lies just within the outer membrane of that syncytium. Anteriorly a ciliated, lobed corona produces feeding currents in most forms, and also serves for locomotion. The mouth leads into a muscular pharynx, the mastax, which bears internal jaws or trophi, worked by special muscles. The gut is fairly straight and usually ends subterminally. The foot lies posterior to the anus, serving for attachment in sessile forms. Muscles are reported to be ectomesodermal and on the body wall are arrayed in narrow bands, circular and longitudinal, often irregular and incomplete; there is very little connective tissue. The fluid-filled body cavity is capacious relative to body size and is not lined by a cellular membrane and only sparsely by basal laminae; body-cavity fluid is eliminated through protonephridia and replenished from the digestive tract (Clément and Wurdak 1991). The nervous system consists of a brain mass lying dorsal to the mastax, from which nerves proceed to anterior sense organs, to the mastax, and to the posterior chiefly via paired ventral nerve cords, usually placed

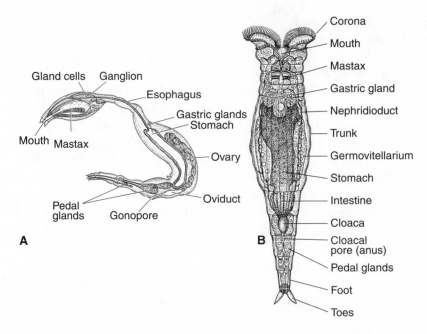

FIG. 10.2 Rotifera, gross anatomy. (*A*) A seisonid, epizoic on marine crustacean gills. (*B*) *Philodina*, a bdelloid. From Nogrady 1982.

ventrolaterally; the number and placement of longitudinal nerves vary among taxa. There is usually at least one eye (ocellus), and a number of other sensory cells—tactile sensors and chemoreceptors—are placed anteriorly. No collagenous matrix is known to exist within any rotifer tissues.

Development in Rotifera

Development is unusual, beginning with what has variously been called modified spiral or modified radial cleavage. Costello and Henley (1976) consider cleavage to be spiral through the four-cell stage but not after (based on work by Jennings 1896). Nielsen (2001) reviewed recent evidence (see Lechner 1966) and concluded that, though the cleavage pattern is not actually spiral, the fates of early blastomeres are similar to those of classic spiralian clades. A stereogastrula is produced. The midgut arises from dorsal cells in the blastula wall. Muscles are evidently derived from the epidermis. After the tissues form, most become syncytial; there is a fixed number of nuclei from the late embryonic stage on, which in turn fixes the number of potential eggs. Many rotifers, perhaps most, are parthenogenetic; no males are known for many forms, and none at all among bdelloids. Meiosis is unknown in bdelloids, and their genomes lack pairs of closely similar haplotypes, consistent with the maintenance of asexual reproduction over millions of years (Welch and Meselson 2000).

Fossil Record of Rotifera

Rotifers are reported from nonmarine Eocene deposits in South Australia (Southcott and Lange 1971), in amber of Late Eocene or Early Oligocene age from the Dominican Republic (Poinar and Ricci 1992; Waggoner and Poinar 1993), and from Holocene peat deposits in northern Ontario, Canada (Warner and Chengalath 1988). The fossils in amber all seem to be bdelloids, which is interesting as bdelloids today are entirely parthenogenetic, and as the investigators note, the early Cenozoic records may indicate retention of this asexual reproductive mode for between 30 to 45 million years.

Rotiferan Relationships

Clément (1985) has reviewed the evidence for the relationships of the rotifers, and concludes that they are quite distinctive and separate from other paracoelomates with the possible exception of the acanthocephalans (see below). Again with this exception, he considers rotifers to share more features with the flatworms than with other paracoelomates, suggesting that rotifers have evolved from the flatworm planula larva (see also Clément and Wurdak 1991). Nielsen (1995) also believes that rotifers have a larval morphology and suggests that they are neotenic descendants of the ancestral protostomian, a form that he visualizes as having both larval and adult stages, with a planktonic larval form being ancestral to an adult benthic stage. SSU rRNA evidence suggests that rotifers (and acanthocephalans) are lophotrochozoans and possibly belong to a clade that is sister to Eutrochozoa (see Winnepenninckx, Backeljau, Mackey, Brooks, DeWachter, Kumar, and Garey 1995; Garey, Near, Nonnemacher, and Nadler 1996; Aguinaldo et al. 1997; Giribet et al. 2000), but the relationships within this putative alliance remain uncertain.

Acanthocephala

The acanthocephalans are entirely parasitic; adult acanthocephalans (fig. 10.3) live in the small intestines of vertebrates, chiefly freshwater teleosts, and vary in length from a few millimeters to 60 cm or so. About 1,250 species have been described. Van Cleve (1948) made the case that acanthocephalans represent a separate phylum, which is often divided into three classes. General accounts include those by Hyman (1951b) and Crompton and Nickol (1985) for morphology and biology, Dunagan and Miller (1991) for fine structure, and Bullock (1969) for systematics.

Bodyplan of Acanthocephala

Anteriorly there is a retractable proboscis armed with hooks, serving to fasten the worm to the host intestinal wall. The proboscis is followed by a smooth neck, which is followed in turn by a trunk containing the viscera. Within the anterior end of the trunk lie two spoon-shaped organs, the lemnisci, of ectodermal origin. The tissues

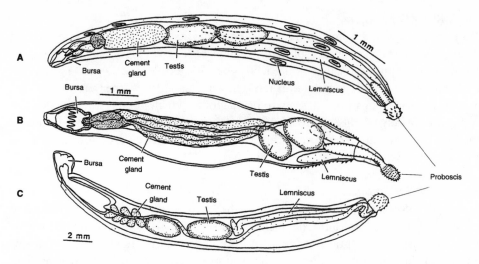

FIG. 10.3 Acanthocephalans, gross anatomy. (A) *Neoechinorhynchus,* an eoacanthocephalan. (B) *Polymorphus,* a palaeacanthocephalan. (C) *Prosthenorchis,* an archiacanthocephalan. From Dunagan and Miller 1991.

of the lemnisci contain a system of anastomosing lacunae that are confluent with a similar system in the proboscis. It has been speculated that the fluid in these lacunae forms part of a hydraulic system for manipulation of the proboscis, or alternatively that the fluid has a chiefly physiological function.

Acanthocephalans entirely lack guts, absorbing nutriment from their hosts through the body wall, which is a syncytium in which numerous minute invaginations, lacunae, and canals are elaborated so as to produce a vast absorptive surface, between one and two orders of magnitude greater than the areal dimensions of the body. A tegmen lies inside the syncytial membrane, quite analogous and perhaps homologous with the lorica of rotifers. Beneath the epidermal syncytium are two muscle coats, the outer circular, the inner longitudinal, both pervaded by extensions of the canals, which coalesce into a system of circular and longitudinal canals on the inner body wall. On the usual site of the digestive tract is a ligament that supports the reproductive organs and, when present, protonephridia. The body cavity is lined along the body wall by a muscular epithelium, and as there is no gut the nature of an inner lining cannot be specified; nevertheless, the cavity is regarded as a pseudocoel. There are up to three nerve ganglia with two main longitudinal nerve cords. In common with many parasites, these worms have a complex life history (see the references cited above).

Development in Acanthocephala
Development has been reviewed by Meyer (1933, 1938) and Schmidt (1985). Embryonic cell membranes are lost at various times, and blastomere relations are difficult to specify. Though cleavage has been considered a deformed spiral pattern

by some workers, this interpretation is rejected by Nielsen (1995). Development is determinate, and the nuclear number appears to be fixed in adults. Vertebrates are the definitive hosts for these endoparasites. Early acanthocephalan development proceeds within the body cavity of the mother, from which juveniles are released as shelled "acanthors," to be shed in host feces. The acanthors are ingested by arthropods as intermediate hosts, wherein development continues by stages. Vertebrates are infected by eating the arthropods. Life cycles are still more complex in some forms.

Fossil Record of Acanthocephala
There is no record of fossil acanthocephalans.

Acanthocephalan Relationships
Acanthocephala have been allied by morphological criteria to a number of phyla, earlier to Platyhelminthes, and more recently to Rotifera, with which they share some detailed similarities in features of the epidermis (see Storch 1979) and musculature. There are also similarities to Priapulida, with which they share proboscises with hooks, and similar placements of reproductive and renal structures (Conway Morris and Crompton 1982 and references therein; Lorenzen 1985). Golvan (1958) reviewed the similarities between acanthocephalans and other paracoelomates that he believed to form an early metazoan clade; he concluded that acanthocephalans lie closest to priapulids and nematomorphs. Assuming that the ancestor of acanthocephalans was free-living, Golvan reconstructed the gross anatomy of such an ancestor by restoring the digestive tract and other features that may have been lost during the evolution of a parasitic life mode (fig. 10.4). Conway Morris (1981) and Conway Morris and Crompton (1982) have noted that a fossil priapulid from the Middle Cambrian Burgess Shale, *Ancalagon,* is similar to this hypothetical free-living acanthocephalan ancestor. Therefore they favored either a priapulid ancestor for acanthocephalans, or, alternatively, a common ancestor that gave rise to both of those phyla.

Early SSU rRNA sequence comparisons suggested that the acanthocephalans are closely related to flatworms (Telford and Holland 1993) or rotifers (Telford and Holland 1993; Winnepenninckx, Backeljau, Mackey, Brooks, DeWachter, Kumar, and Garey 1995). Study of additional sequences from all acanthocephalan and rotiferan classes (Garey, Near, Nonnemacher, and Nadler 1996) have not only confirmed an alliance with rotifers but place acanthocephalans as sister to the bdelloids; a study using regions of the *hsp82* gene has placed acanthocephalans within Monogononta (Welch 2000). Thus the acanthocephalans are likely to be parasitic derivatives of the rotifers. It has been suggested that Rotifera and Acanthocephala be united in a clade Syndermata (see Haszprunar 1996b). The classification of acanthocephalans commonly recognized on morphological grounds (as for example by Golvan) is generally supported by SSU rRNA sequence comparisons (Near et al. 1998).

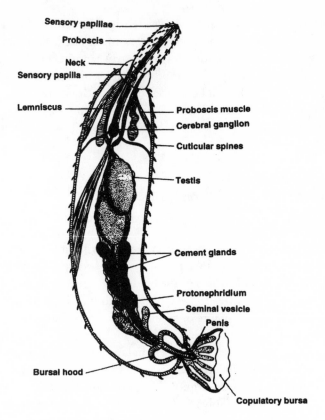

Sensory papillae

Proboscis

Neck

Sensory papilla

Lemniscus

Proboscis muscle

Cerebral ganglion

Cuticular spines

Testis

Cement glands

Protonephridium

Seminal vesicle

Penis

Bursal hood

Copulatory bursa

FIG. 10.4 Major features of a hypothetical acanthocephalan ancestor proposed by Golvan (1958).

Entoprocta

Bodyplan of Entoprocta

Entoprocts (see Hyman 1951b; Nielsen 1971; Emschermann 1982) are either solitary or colonial, and resemble bryozoans (ectoprocts) in their external characters, and partly in anatomy (fig. 10.5). Entoprocta contains about 150 species in two orders. The body is small, consisting of a calyx that is slightly flattened laterally, with an elliptical crown of tentacles around the body on the ventral side, which faces upward, and an attachment stalk, which connects to the dorsal side. Tentacles and stalk are formed by simple extensions of the body wall; the stalk may be a simple adhesion disk or may be complex and jointed, and in colonies the stalks are joined by stolons. The space between the adult body wall and gut is packed with mesenchyme, as are the interiors of the tentacles. The mouth and anus lie inside the tentacular ellipse at opposite ends of the long axis, the digestive tract forming a U within the calyx. The ventral surface of the calyx inside the tentacles forms the

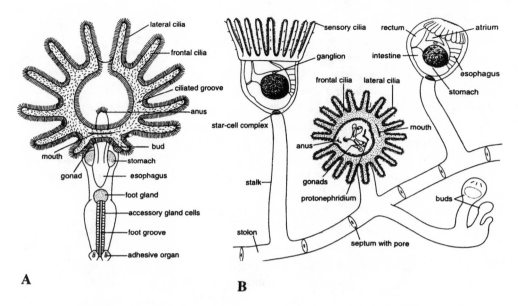

FIG. 10.5 Entoprocta, a solitary and a colonial form. (*A*) *Loxosomella*. (*B*) *Pedicellina*. Note that both mouth and anus lie within the tentacular crown. From Nielsen and Jespersen 1997.

vestibule. The inner surfaces of the tentacles bear three rows of cilia, two lateral and one frontal, and paired ciliated feeding tracts along the margins of the vestibule (the vestibular groove) lead from the tentacular tracts to the mouth. Feeding currents are directed inward between the tentacles by lateral cilia, and escape upward. The current pattern is thus different from that in the bryozoans (chap. 9), but as the anus lies within the tentacular crown in entoprocts, this pattern would seem best for removal of metabolic waste from the vestibular surface. Food items are captured by lateral cilia and transferred to frontal cilia that pass them to the vestibular groove and thence to the mouth.

The body wall is a cellular epidermis overlaid by a cuticle except on tentacles and vestibule. Neither the epidermis nor the digestive tract is lined by a muscle coat, although longitudinal muscle fibers are present locally on the inner calyx wall, and there are sphincters and other fibers localized on the gut. A transverse muscle bundle crosses the narrow axis of the calyx beneath the gut. The tentacles contain muscle fibers, and muscles may sheath the stalk or may be restricted to localized bands there. A ganglion lies ventral to the stomach (i.e., above it in life) and gives off pairs of nerves that innervate tentacles, body wall, stalk, and gonads. Excretion is via a pair of protonephridia.

Development in Entoprocta
Cleavage is spiral and determinate in most species of entoprocts, which are protostomous. The embryo may receive nutrients directly from the parent while developing

(Mariscal 1975). The larvae are ciliated and at least some feed; Nielsen (1971, 1995) considers them to be a trochophore type, although they have idiosyncratic features. Upon settling, some larvae attach to the substrate by a frontal organ, and therefore the mouth and anus, which have been ventral in the larva, are brought into a lateral position. During subsequent metamorphosis a rotation of 90° causes the developing tentacles, and the enclosed vestibular region containing the mouth and anus, to face upward, so that the stalk is regarded as dorsal. In some forms larvae bud before or during metamorphosis, and it is the buds that develop into the adult, but these patterns are considered to be derived (Nielsen 1971). Colonies develop via budding, details of which are reviewed in Hyman 1951b and Nielsen 1971.

Fossil Record of Entoprocta

The single fossil record of entoprocts is of a bioimmured colony, overgrown by the shell of an oyster, from the Upper Jurassic of England (Todd and Taylor 1992). These fossils could be assigned to the living genus *Barentsia*.

Entoproctan Relationships

The similar bodyplans of bryozoans and entoprocts has led to suggestions that these phyla are closely allied, a view particularly championed by Nielsen (1971). An objection to this association is that the bryozoans are interpreted as coelomate (although this is disputed by Nielsen) while the entoprocts do not have any clear adult body cavity, for the space where a pseudocoel would be developed is filled with a gelatinous mesenchyme. It might be argued that filling of the primary body cavity of entoprocts is a consequence of their particular biomechanics or of their small size, although entoproct bodies are significantly larger than bryozoans. Entoproct development is typically spiralian, and does not suggest any close alliance with the radially cleaving bryozoans.

SSU rRNA data are equivocal about the position of Entoprocta. The study of Mackey et al. (1996) suggests that entoprocts are members of the Eutrochozoa, perhaps allied to the annelids, but are not sisters to the bryozoans. The phylogenetic position of the entoprocts within Lophotrochozoa is unstable when explored by different algorithms, however, and in some molecular trees they do show a relation to bryozoans (see Cohen 2000). The exact relationship of Entoprocta remains one of the many enigmas of invertebrate phylogeny.

Cycliophora

The phylum Cycliophora was first described by Funch and Kristensen as recently as 1995. The single species known from this phylum was discovered living on the mouthparts of the Norway lobster, giving rise to dreams that other yet-undiscovered bodyplans might still be lurking in unusual places.

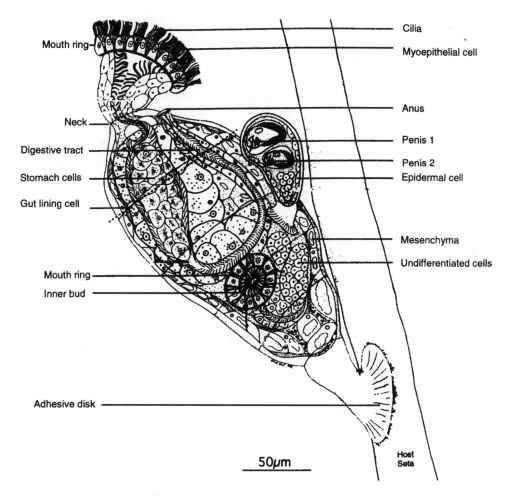

Cilia

Mouth ring

Myoepithelial cell

Anus

Neck

Penis 1

Digestive tract

Penis 2

Stomach cells

Epidermal cell

Gut lining cell

Mesenchyma

Undifferentiated cells

Mouth ring

Inner bud

Adhesive disk

50μm

Host
Seta

FIG. 10.6 Cycliophora, gross anatomy. *Symbion pandora,* the only species now known from this phylum. From Funch and Kristensen 1995.

Bodyplan of Cycliophora

Adult cycliophorans are distinctively dimorphic, with sessile females to which dwarf males may be attached (fig. 10.6). The females are only about 350 μm in length, the males about half that. The female body is ovoid with a prominent anterior feeding structure that may be replaced several times by internal budding. The feeding structure is separated from the main body by a constriction. Atop the feeding organ is a mouth ring of alternating myoepithelial and multiciliated cells leading into a funnel-shaped, ciliated buccal cavity and thence to an alimentary tract. Muscle strands connect the base of the buccal funnel to the body wall. The gut is U-shaped, the anus opening to one side of the feeding structure, with the stomach greatly expanded to fill most of the body anteriorly; in fig. 10.6 an inner bud lies posteriorly. A nerve ganglion lies between the base of the funnel and the anus. Mesenchyme fills the compartment between endoderm and ectoderm; there is no fluid-filled body cavity. Attachment to the substrate is by a pedal disk. The males attach to the females via a stalked attachment disk. The male body is divided into two compartments that are essentially gonads, containing developing sperm and two penises; males lack alimentary structures.

Development in Cycliophora
Cleavage is unknown. Although free-living, cycliophorans are inferred to have alternate sexual and asexual phases, each involving unusual stages and rivaling the complex life cycles of many endoparasites (Funch and Kristensen 1995). At one feeding stage, for example, a larva (Pandora larva) develops asexually within the female brood chamber and, while still brooded, buds off a new feeding stage internally. A sexually produced larva has a rod of mesodermal cells ventrally, and is termed the chordoid larva. Funch (1996) made a detailed study of this larva by electron microscopy and concluded that it is a modified trochophore.

Fossil Record of Cycliophora
Cycliophorans are unknown from the fossil record.

Cycliophoran Relationships
Epidermal cells of cycliophorans are multiciliated, recalling protostome larvae, and there is a general resemblance to rotifers. Otherwise, the ciliary collecting system and the importance of budding in this minute acoelomate form suggest alliance with the entoprocts. An SSU rRNA sequence (Winnepenninckx, Backeljau, and Kristensen 1998) is consistent with an alliance between rotifers and cycliophorans.

Gnathostomulida

Bodyplan of Gnathostomulida
Gnathostomulida is a morphologically compact acoelomate phylum (Ax 1965, 1985) with a simple bodyplan (Riedl 1969; Sterrer 1972; Sterrer et al. 1985). Gnathostomulids are minute (chiefly 0.3 to 3 mm long) and are characterized by a specialized pharynx that includes hardened, toothed jaws and a basal plate said to be formed of cuticlelike material (fig. 10.7). There are fewer than 100 described species, in two orders, but the phylum must be very incompletely known. The anterior is expanded and bears sensory cilia; the jaws are just posterior to this expansion, and the gut is unbranched and usually described as blind, although it seems that an anal pore is sometimes present (Knauss 1979). Ovaries, a bursa, and a testis occur in the central and posterior parts of the body. Parenchyma is reduced or absent. The epidermis consists largely of monociliated cells, unlike the multiciliate flatworms. There are a ganglionic brain and one to three pairs of longitudinal

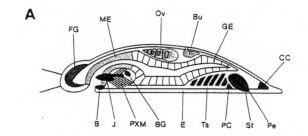

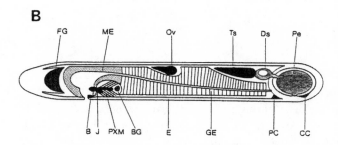

FIG. 10.7 The bodyplan of Gnathostomulida. (A) *Gnathostomula paradoxa*. (B) *Pterognatha meixneri*. B, basal plate; BG, buccal ganglion; BU, bursa; CC, caudal neuronal ganglion; Ds, sperm duct; E, epidermis; GE, gut epithelium; FG, frontal ganglion; J, jaw; ME, mouth epithelium; Ov, ovary; PC, penal neuronal connection; Pe, penis, PXM, pharyngeal musculature; ST, stylet; Ts, testes. From Lammert 1991.

nerves (Kristensen and Nørrevang 1977). There appear to be about fifteen major cell types in a given individual, with perhaps fifteen or more additional sensory cell types scattered among the taxa (Sterrer et al. 1985; Lammert 1991). A similar proliferation of sensory cell types is found among the flatworms. Gnathostomulids are meiofaunal, commonly inhabiting anoxic marine sands.

Development in Gnathostomulida
Cleavage is spiral, and development is direct; otherwise little is known of gnathostomulid development.

Fossil Record of Gnathostomulida
When gnathostomulid jaws became well-illustrated (see Riedl 1969), it was noticed that they resembled both euconodonts and minute jawlike objects ("microconodonts") from the Cretaceous. This observation led to speculation that they were allied or conphyletic with one or the other of these fossil groups (Durden 1969; Rodgers 1969), although they are compositionally distinct from those forms. Euconodonts are now best interpreted as chordates and microconodonts as annelids, and no jaws or other fossil remains or traces have been confirmed as belonging to gnathostomulids.

Gnathostomulid Relationships
Gnathostomulida is sometimes considered to be a sister group to the flatworms (Ax 1985), which gnathostomulids resemble in general body form, gut, reproductive systems, and some other respects, but these characters are not known to be

synapomorphies. Gnathostomulids resemble some paracoelomates in other characters: for example, they are monociliated and have body walls that resemble gastrotrichs, kinorhynchs, and rotifers, and not flatworms (see Rieger and Mainitz 1977), and gnathostomulid jaws resemble the trophi of rotifers (Rieger and Tyler 1995). SSU rRNA sequences from gnathostomulid species were studied by Littlewood et al. (1998); they formed a clade, chaetognaths + gnathostomulids, that is sister to the nematodes and that lies within the Ecdysozoa. These associations may be due to long-branch attraction. Additional taxon sampling by Giribet et al. (2000) placed gnathostomulids as allies of the nonecdysozoan paracoelomates, including flatworms.

Chaetognatha

The chaetognaths, or arrowworms, are among the more common planktonic metazoans, especially in the open oceans, although a few estuarine forms and even some benthic species are known. They are small-bodied, ranging from 2 mm to about 12 cm in length. About 100 living species are known.

Bodyplan of Chaetognatha

The chaetognath body (fig. 10.8) is divided into head and trunk regions and can be considered biregionated (see Bone et al. 1991; Kapp 1991). Many workers identify the tail as a third region, however, and consider the body "tripartite" or "trimeric" (though it is uncertain whether there are adult coelomic spaces). Lateral and caudal fins are supported by "rays." The head can be encased within a hood, an extension of the head epidermis, and has an array of chitinous grasping spines with supporting musculature for prey capture; the anteroventral mouth leads into a pharynx. The epidermis is locally covered by a thin cuticle and is underlaid by a firm, flexible basement membrane, which in turn is lined by four longitudinal muscle bands. The gut runs to a ventral anus at the posterior end of the trunk. In some chaetognaths, myocytes have been identified overlying the gut, and the gut epithelium contains myofilaments (see Kapp 1991).

There are three body cavities, head, trunk, and tail, separated by transverse septa. The trunk and tail cavities are further divided by pairs of longitudinal septa. The cavities are lined by a thin epithelium of uncertain derivation. These fluid-filled cavities, surrounded by the tough, muscle-lined basal membrane of the epidermis, form a hydrostatic system. There is no circulatory or excretory system, and the extent to which the body fluids are involved in such functions is not known. A cerebral ganglion lies dorsal to the pharynx; a generalized nervous system is shown in fig. 10.8B. Chaetognaths are hermaphroditic, with ovaries in the trunk and testes in the tail. Locomotion is chiefly by swimming via dorsoventral oscillations of the body. Prey are commonly copepods or other small zooplankters. The prey are paralyzed

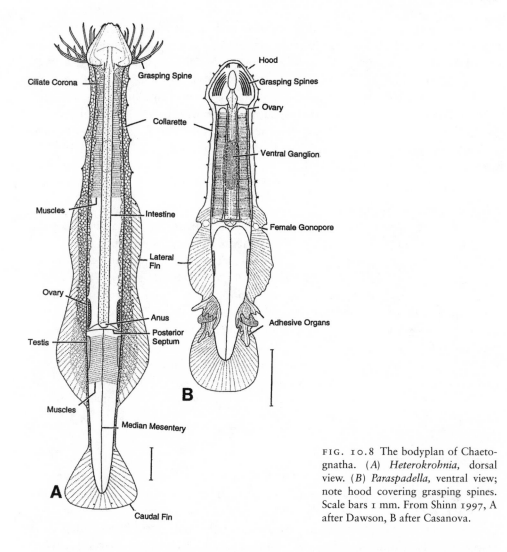

FIG. 10.8 The bodyplan of Chaeto-gnatha. (*A*) *Heterokrohnia*, dorsal view. (*B*) *Paraspadella*, ventral view; note hood covering grasping spines. Scale bars 1 mm. From Shinn 1997, A after Dawson, B after Casanova.

by a type of toxin (tetrodotoxin) that is known only in one other invertebrate, a species of octopus (Theusen 1991); it is produced by bacterial symbionts.

Development in Chaetognatha

Chaetognath cleavage is radial, producing a coeloblastula that invaginates during gastrulation. Development is direct. Although the blastopore closes, it is situated posteriorly in adult orientation; the mouth naturally opens anteriorly, with the anus eventually appearing near the former blastopore site; thus chaetognaths are developmentally deuterostomous. However, chaetognaths are not enterocoelic (Kapp 1991), the origin of their body cavities being unique according to Kapp (2000). She reports that space within the endodermal lining of the gastrula develops into a larval body cavity rather than a gut (which forms from a stomadaeum opposite the blastopore). Tissues proliferate from folds of endoderm near the anterior end of the larva; the outside of the folds come to line portions of the anterior body cavity, and could be considered endomesoderm, while inside portions of the fold develop into gut tissues. The body cavity, then, ends up being lined by essentially an endodermal epithelium, except perhaps for the outside of the anterior folds. The source of tissues for the head coelom is uncertain. Kapp has called this type of body-cavity formation heterocoely. It is not clear that such a cavity should be considered a coelom. Further work is badly needed.

Fossil Record of Chaetognatha

A possible chaetognath has been described from a single specimen, about 2.5 cm long, from the Lower Cambrian (late Atdabanian?) of south China (Chen and Huang 2002). Specimens assigned to Chaetognatha have been described from the Late Carboniferous Francis Creek Shale of Illinois (Schram 1973). These fossils range from 1.5 to 3 cm in length and have poorly developed lateral fins. Certain protoconodonts, which are hook-shaped spines found from the Lower Cambrian to the Lower Ordovician, may represent chaetognaths. These spines are sometimes preserved as a functional apparatus and have been closely compared with chaetognaths by Szaniawski (1982). The similarities of those apparatuses is compelling, and if the assignment of protoconodonts to Chaetognatha is correct, then protoconodonts are unrelated to paraconodonts and conodonts proper, which are almost surely chordates (see chap. 11).

Chaetognath Relationships

Darwin (1844) remarked that chaetognaths were "remarkable for the obscurity of [their] affinities." The developmental features of radial cleavage and deuterostomy, and the perception of enterocoely, have caused many workers to consider chaetognaths an early branch of the Deuterostomia (e.g., Hyman 1959; Brusca and Brusca 1990). The developmental and structural oddities of chaetognath body cavities have engendered skepticism with this interpretation, and SSU rRNA sequences do not support an affinity with deuterostomes (Telford and Holland 1993; Wada and Satoh 1994). Molecular data have been used to place chaetognaths in a variety of positions: deeply within Metazoa (Telford and Holland 1993), within Ecdysozoa as sisters to Nematoda (Halanych 1996), and with Gnathostomulida in a sister clade to Nematoda (Littlewood et al. 1998). The chaetognaths studied are quite long-branched (Halanych 1996; Aguinaldo et al. 1997), as are gnathostomulids and nematodes, which may well account for these last two results, although Halanych found a chaetognath-nematode association to be possible. The bi- or triregionated bodyplan is reminiscent of the deuterostomes but also of the lophophorates, and they share some early developmental features with both groups. Perhaps a position near the protostome-deuterostome branch point is most likely on present evidence. Chaetognath relationships remain remarkably obscure.

Phylogenetic Schemes for Paracoelomates

Major Schemes for Vermiform Paracoelomates

Assuming that the SSU rRNA evidence is successful at least in assigning phyla to their major clades, then the paracoelomates are all protostomes with the exception of the Acoelomorpha, presumed stem bilaterians, and possibly of chaetognaths. Recent schemes proposed for paracoelomate interrelationships illuminate the difficulties that must still be confronted in accounting for the origins of these phyla.

Two contrasting, morphologically based phylogenetic schemes have been proposed for different groupings of wormlike paracoelomates, one using overall similarity, the others using cladistic methodology; the schemes are associated with strikingly different evolutionary scenarios. Some workers have proposed that these phyla arose ultimately from free-living turbellarian flatworms. Conway Morris and Crompton (1982) suggested the relationships shown in fig. 10.9A, using gnathostomulids and platyhelminths as outgroups. There are two major clades recognized, the nematode + nematomorph + gastrotrich alliance being considered the more homogeneous. For the other clade, an overall similarity in appearance of priapulids and acanthocephalans was a major factor in proposing their sisterhood.

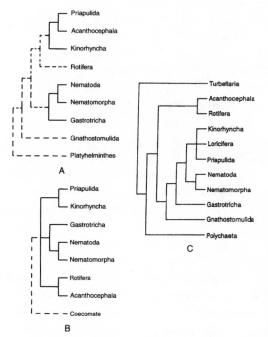

FIG. 10.9 Hypotheses of the phylogenetic arrangement of some paracoelomate phyla based on morphology. (A) As suggested on the basis of general resemblance (Conway Morris and Crompton 1982); dashed lines indicate less confidence in the relationships than solid lines. (B) As suggested on the basis of a cladistic analysis of morphological features (Lorenzen 1985). (C) As suggested by another cladistic analysis of morphological features (Wallace et al. 1996).

The scheme of Lorenzen (1985), a morphologically based cladistic analysis, is illustrated in fig. 10.9B. These phyla are assumed by Lorenzen to have descended not from flatworms but from coelomate ancestors, whether from one or more being uncertain, and to owe their similar grades of construction to their small body size. There are three major clades in this scheme. The rotifers and acanthocephalans are united by a presumed synapomorphy of epidermal structures, both possessing a syncytial epidermis with intracellular filamentary skeletons invaginated by crypts (see Ruppert 1991a). Gastrotrichs and nematodes share bilayered cuticles, and the nematomorph cuticle is said to resemble the cuticle of some nematodes, so these groups can be allied, although on the basis of a somewhat weak criterion, but here in agreement with Conway Morris and Crompton (1982). Other features, such as the dominance or exclusivity of longitudinal body-wall muscles, are also used to consider nematomorphs and nematodes as sisters. Similarities in the arrangement of scalids on the introverts of kinorhynchs and priapulids is invoked as a possible synapomorphy associating those phyla. This clade is placed as a sister to the nematode/gastrotrich/nematomorph clade because introverts occur in most of these groups (though not in gastrotrichs, for which it is necessary to assume loss of this feature). Loriciferans, not treated by Lorenzen, are considered allied to kinorhynchs and priapulids by Nielsen (1985, 1995).

Evidently, morphological evidence is insufficient to polarize this group of paracoelomates with respect to the morphological grade of their ancestor(s); that is, they could be more primitive than coelomates or they could be reduced coelomates, a choice that at present depends on one's overview of metazoan evolution.

A third scheme, based on a comprehensive cladistic analysis of morphology, was presented by Wallace et al. (1996; fig. 10.9C). SSU rRNA data have also been brought to bear on paracoelomate relationships (Telford and Holland 1993; Raff et al. 1994; Winnepenninckx, Backeljau, Mackey, Brooks, DeWachter, Kumar, and Garey 1995; Garey, Near, Nonnemacher, and Nadler 1996; Aguinaldo et al. 1997; Littlewood et al. 1998; Giribet et al. 2000). The molecular trees have not been stable, varying with exemplars and with the included groups, but the relations suggested by the best data (more taxa and/or more complete sequences) agree rather well with Wallace et al.'s morphological tree. The chief difference is in the placement of gastrotrichs, which are allied with the ecdysozoans by Wallace et al. Neuhaus (1994) and Nielsen (1995) also ally the gastrotrichs with ecdysozoans on morphological criteria. However, gastrotrichs are usually allied with eutrochozoans by SSU rRNA data and are treated with that clade here.

The Fossil Record and Paracoelomate Histories

Among the major questions raised by paracoelomate phyla is whether they originated from simple ancestors or whether they have been reduced from more complex and perhaps coelomate ancestors. Experienced workers are divided on this question. Some authorities on the morphology and fine structure of these groups contend that they have arisen by progenesis from ancestors that were coelomate as adults, hypothesized to be basal to Bilateria (e.g., Rieger 1986; Tyler 2001). Dewel (2000) has listed twenty (of the twenty-nine phyla she recognizes) as having been simplified from coelomate ancestors. Other workers have preferred evolution from simpler, planuloid ancestors that are hypothesized to be basal to Bilateria (e.g., Salvini-Plawen 1978). In either of these two hypotheses it could be possible that at least the wormlike paracoelomates, together with their last common ancestor, form a clade or paraclade.

Many workers, using either morphological or molecular criteria, have concluded that those paracoelomate groups are likely to be polyphyletic (e.g., Winnepenninckx, Backeljau, Mackey, Brooks, DeWachter, Kumar, and Garey 1995; Nielsen 1995; Wallace et al. 1996; Aguinaldo et al. 1997; Littlewood et al. 1998). The paracoelomate phyla form two distinctive groups on the basis of general resemblance with respect to morphological features. It is possible to interpret the SSU rRNA data as suggesting that a paraclade descended from the last common protostome/deuterostome ancestor branched into two major clades, one giving rise to ecdysozoans. The history of the other clade is ambiguous; perhaps it branched to give rise to a clade of crown paracoelomates (Platyzoa) on one hand and to Lophotrochozoa on the other. In this case most of the wormlike paracoelomates, including ecdysozoans, could be monophyletic although not holophyletic, and their last common ancestor would have also been a paracoelomate—the last common ancestor of the major protostome clades. The most likely exception to this notion would be the nematodes, which may well be simplified from a more complex ecdysozoan ancestor. The position of chaetognaths and to some extent of gnathostomulids remains problematic. Problems associated with these small-bodied phyla form a microcosm, so to speak, of difficulties with the phylogeny of metazoans as a whole.

At present there is no way to reconcile the protostome relationships suggested by the available molecular data with those suggested by the pattern of their developmental similarities. In large measure this problem arises from the presence of radial or radiallike cleavage and mesoderm origin in the lophophorate phyla, and the uncertain position of rhabdocoels relative to Lophotrochozoa. Everything about the early development of rhabdocoels suggests that they are basal to the classic spiralians, yet they usually fall outside the lophotrochozoans on molecular evidence, while the radially cleaving lophophorates fall within that clade. Perhaps denser taxon sampling and more molecular trees will clarify their histories.

Although none of the crown paracoelomate phyla is known from Neoproterozoic body fossils, the trace-fossil record of small-bodied creeping and burrowing metazoans begins about 40 million years or so before the confirmed appearance of complex metazoan bodyplans in the fossil record (chap. 5). Small-bodied derived members of complex groups commonly have simplified bodyplans. For example, in the case of annelids, some of the reduced body architectures are at a paracoelomate grade (chap. 8). Although such reductions are sometimes taken to suggest that crown paracoelomates are reduced (e.g., Rieger 1986), they can also be taken to indicate that the bodyplans of paracoelomates are well-adapted to life at small sizes. The presence of a fauna of small-bodied epifaunal creepers and shallow infaunal burrowers during some tens of millions of years during the late Neoproterozoic suggests that those forms were most likely paracoelomates. Perhaps this postulated Neoproterozoic paracoelomate fauna consisted mostly of stem groups, for crown paracoelomates usually do not show deeper branching than do complex phyla, although more basal positions for some paracoelomates were indicated in some early studies. Some SSU rRNA trees suggest sister relations between crown paracoelomates and their complex allies (as in fig. 4.15, from Giribet et al. 2000, which is, however, a combined morphological-molecular tree and may be influenced by morphological grade). Lineages linking the crown paracoelomate phyla are presumably extinct, including the stem and crown ancestors of major protostome clades. In this view the eucoelomic and hydrostatic hemocoelic cavities found in complex metazoans evolved as adaptations to their increased complexities and body sizes. That some of the crown paracoelomates are reduced forms, however, certainly remains possible.

CHAPTER 11

Deuterostomes

At one time, Deuterostomia was regarded as a large branch of the animal king-dom, characterized by a series of developmental and morphological features that embraced up to ten phyla in some schemes. With the advent of molecular phylo-genetics, however, all but three of those phyla have been assigned to other clades. The remaining crown deuterostome phyla are Hemichordata, Echinodermata, and Chordata, which usually includes Urochordata as a subphylum.

All but one of the crown deuterostome groups are coelomate, the exception being the urochordates. For the deuterostomes that have coelomic spaces, at least one region of the coelom serves a hydrostatic function. Other features that deutero-stomes ordinarily share include radial cleavage, formation of the mouth at some distance from the blastopore, derivation of mesoderm from endoderm lining the enteron, and origination of the coelomic spaces, when present, as outpockets from the archenteron (enterocoely). These features had been considered virtual hallmarks of the deuterostomes, but none is exclusive to the phyla remaining in the deutero-stomes. On the other hand, many of these features are not found even in all members of the deuterostomes; for example, some or all coelomic spaces arise as a split within mesodermal tissue—via schizocoely—in a number of deuterostome groups. Nevertheless, as the mesoderm in these forms arises from the walls of the enteron (as in all, but not only, deuterostomes), their coelom formation is considered to represent a "modified enterocoely"; in other words, it is considered to have evolved from a primitive enterocoelic condition (chap. 2).

Another feature that characterizes some deuterostomes is that they possess three coelomic regions in an anteroposterior series, at least in early development, and one phylum (Hemichordata) displays a tripartite bodyplan in the adult. From anterior to posterior the coelomic regions are termed protocoel, mesocoel, and metacoel, and the corresponding body regions are prosoma, mesosoma, and meta-soma. The systematic importance that has been given to triregionated bodyplans is illustrated by the early assignment of the pogonophorans to Deuterostomia chiefly because they appeared to be tripartite (a segmented posterior region was miss-ing from early specimens; chap. 8). For example Hyman (1959), usually cautious about phylogenetic relationships, declared that "it is not open to doubt that the Pogonophora belong to the Deuterostomia." Somewhat similarly, the pattern of coelomic regionation of the lophophorates has been used (along with other features)

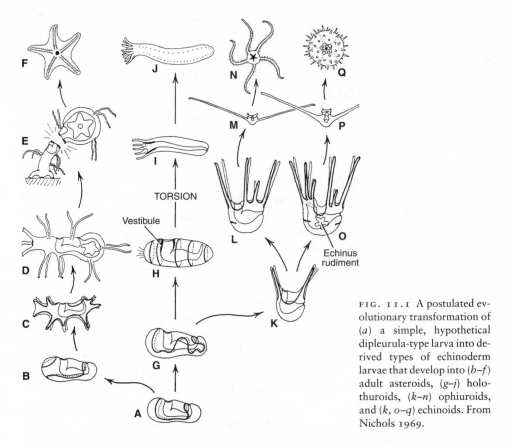

FIG. 11.1 A postulated evolutionary transformation of (*a*) a simple, hypothetical dipleurula-type larva into derived types of echinoderm larvae that develop into (*b–f*) adult asteroids, (*g–j*) holothuroids, (*k–n*) ophiuroids, and (*k, o–q*) echinoids. From Nichols 1969.

to argue that they are deuterostomes, and regions of the lophophorate coelom and body are commonly described by the proto-, meso-, and meta-terminology of triregionated protostomes. As the homology of coelomic regions in deuterostomes and lophophorates is unlikely on present evidence, these terms are restricted to deuterostomes here.

Finally, another adult feature found in most deuterostomes is pharyngotremy—the presence of spiracles or gill slits into the pharynx. Derived chordates such as mammals lack gill slits in adults, although they appear fleetingly in embryos, and crown echinoderms lack pharyngeal openings altogether, though there are indications that early lineages may have possessed them.

A larval type considered to be characteristic of invertebrate deuterostomes is the dipleurula (fig. 11.1), a hypothetical type from which the actual deuterostome larvae can be imagined to have evolved. Dipleurula-style larvae have a single ciliated band encircling the mouth. These cilia capture food particles, presumably retaining them on the upstream side of the feeding current (see Strathmann 1978); the particles are then conveyed to the mouth. As dipleurula-style larvae grow or become large evolutionarily, the prototroch often becomes very convoluted, evidently to accommodate more cilia than is possible in a simple band (fig. 11.1). This feeding arrangement contrasts with that of the trochophore-style larvae of most classic spiralians, which have two ciliary bands near the mouth, the prototroch and metatroch, commonly with feeding cilia between them.

A number of molecular phylogenies of deuterostomes based on comparisons of SSU rRNA sequences are now available (e.g., Field et al. 1988; Holland et al. 1991; Wada and Satoh 1994; Turbeville et al. 1994; Halanych 1995; Littlewood et al. 1998; Cameron, Garey, and Swalla 2000; Winchell et al. 2002). While the branching patterns have varied somewhat among methods, the more recent studies using more complete data sets identify two major deuterostome clades: Hemichordata + Echinodermata and Chordata (including urochordates). This arrangement had not been supported in all morphologically based phylogenies, but it is not inconsistent with developmental and morphological evidence.

Hemichordata

Hemichordates are clearly deuterostomes and appear to be sisters to the echinoderms. Adult hemichordates display a triregionated plan with particular clarity. There are about ninety living species divided into two classes, Pterobranchia and Enteropneusta (Hyman 1959; Barrington 1965; Benito 1982; Benito and Pardos 1997). There have been several SSU rRNA sequences produced for enteropneusts (Holland et al. 1991; Wada and Satoh 1994; Turbeville et al. 1994; Cameron, Garey, and Swalla 2000) and two for pterobranchs (Halanych 1995; Cameron, Garey, and Swalla 2000). The molecular evidence suggests that enteropneusts are basal and that pterobranchs have arisen from within the enteropneust clade (see Winchell et al. 2002).

Bodyplan of Enteropneust Hemichordata

The enteropneusts are vermiform and much larger than pterobranchs, growing to 1.5 m or so (fig. 11.2). Some are infaunal, forming U-shaped burrows (fig. 11.2D), while others are found epifaunally among algae, rocks, or shells. The prosoma is a proboscis, heavily ciliated, that is used for food gathering and locomotion. The interior of the proboscis is surrounded by a thin layer of circular muscles and is nearly filled with longitudinal muscle fibers interspersed within connective tissue. The protocoel is thus chiefly reduced to a narrow lumen, expanding into pouches posteriorly where the proboscis is supported by internal collagenous plates, the proboscis skeleton. Although it is not heavily muscled, the proboscis can be thrown into peristaltic waves, which, with the aid of ciliary activity, can provide for burrowing or creeping.

The mesosomal collar is heavily ciliated and glandular; it is cut off from the trunk by a transverse septum and contains longitudinal muscle bundles that insert on the septum or on the body wall, and that run forward to the proboscis skeleton

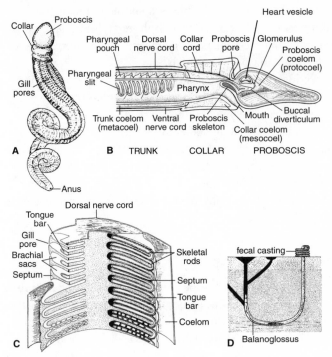

FIG. 11.2 Enteropneust hemichordates. (*A*) External view of *Harrimania*. (*B*) Sagittal section of anterior portion of a typical enteropneust to show gross anatomy. (*C*) Detail of gills of *Saccoglossus*, showing coelomic spaces within the tongue bars; partially cut away on left, different details emphasized on different sides. (*D*) A burrowing enteropneust, *Balanoglossus*, and its burrow system. A, C, D from Hyman 1959 after Spengel, Delage and Herouard, and Stiasny, respectively; B from Pearse et al. 1987.

or other insertion sites in the proboscis. The mesocoel is extensive. The mouth lies at the anteroventral border of the collar and leads into a buccal tube. From the position of the collar septum a pharynx extends posteriorly into the anterior trunk, followed by the esophagus and then the intestine proper, extending to a terminal anus (fig. 11.2A–B). The alimentary tract is supported by dorsal and ventral mesenteries that cross the metacoel. From the pharynx a paired series of pouches extend dorsolaterally to the body wall, bearing U-shaped slits that communicate with the exterior through ciliated pores. The basement membrane of the pharyngeal epithelium is elaborated into a paired series of tongue bars to support the pouch-and-slit structure; these contain coelomic spaces (fig. 11.2C).

The blood vascular system is open, with dorsal and ventral longitudinal vessels running within the mesenteries and widening into sinuses in the collar, in the proboscis, and along the gut and body walls—in the positions of a pseudocoel. A contractile heart and a glomerulus (a unique hemichordate excretory organ) are located in the proboscis. The walls of the pharyngeal pouches are richly vascularized and presumably serve as gills. The nervous system is quite simple, consisting chiefly of a matted plexus of nerve fibers lying above the basement membrane of the epithelium. There are also both dorsal and ventral longitudinal cords in the trunk, sunk below the epidermis; the dorsal cord also extends into the proboscis. These cords consist of bundled axons but usually lack neuronal cell bodies, which

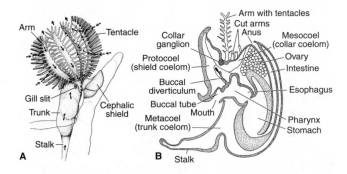

FIG. 11.3 The pterobranch hemichordate *Cephalodiscus*. (A) External view with tentacles extended to form a feeding basket. Around tentacles, dashed arrows indicate incurrents, downward arrows movement of food in ciliated grooves toward mouth, upward arrows rejection currents. On stalk and trunk, arrows indicate food movement via cilia. (B) Cross section, tentacles cut. A from Lester 1985; B from Brusca and Brusca 1990 after Schepotieff.

are instead associated with the basiepithelial axon mat. The nerve cords are thus conductors but not integrators or coordinators, and so do not constitute a central nervous system (Bullock 1965). Food items are chiefly sedimentary particles collected by the proboscis; they are trapped on mucus and transported to the mouth along ciliary tracts. Unwanted items are moved to the collar for rejection. Ingestion may be aided by the stream of water, partly respiratory in function, which enters the mouth and exits through the gill pores.

Bodyplan of Pterobranch Hemichordata

Pterobranchs (fig. 11.3) are small, 1 to 12 mm long, and are known from only three living genera. Individual zooids of *Cephalodiscus*, the largest pterobranch, inhabit a collagenous structure of chiefly horizontal branching tubes, the coenecium, secreted in a series of rings or fuselli; vertical tubes rise at intervals from the horizontal stolons. The living associations are aggregated and sometimes are colonial. Zooids sometimes bud to produce a group of interconnected zooids in varying stages of maturity; they may have a common muscle stalk that adheres to the coenecium. The stalk is extremely extensible, varying from essentially complete contraction to an extension over seven times the body length of a zooid. Thus zooids can and do leave the coenecium and wander over its surface while the tip of the stalk is still adherent deep within the coenecium (Lester 1985).

The prosoma is a cephalic shield, ciliated and bearing a glandular tract from which the coenecium is secreted or repaired (Dilly 1986). Locomotion is accomplished by the cephalic shield, which acts as a creeping foot; evidently creeping is chiefly via ciliary activity, but shape changes have been observed, and the hydrostatic skeleton of the protocoel may sometimes be involved. The mesosoma, or collar, bears tentaculate arms, from five to nine pair, containing extensions of the mesocoel; the tentacles collect food. The bases of the arms are surrounded by a groove that runs toward the mouth on each side, disappearing beneath one of the paired oral lamellae. The anterior portion of the digestive tract is a richly ciliated ectodermal invagination consisting of a short buccal tube succeeded by the pharynx, which is perforated by a pair of gill slits that pierce the wall of the mesosoma laterally (fig. 11.3). The metasoma, or trunk, contains a U-shaped gut suspended in the metacoel and ending in an anus anterodorsally. Posteriorly the zooid body proper terminates in a stalk containing an extension of the metacoel. The metasoma is covered with cilia that may pass particles to the tentacles for possible ingestion. The open vascular system consists of a series of pseudocoelomic sinuses without discrete vessels, although there is a localized heart. Excretion is via the glomerulus. The nervous system consists chiefly of a plexus within the epithelium of the body wall. Reproduction by budding is common. Gonads are borne in the mesodermal tissues of the metasoma, and gametes are released through pores.

In feeding, adult *Cephalodiscus* zooids extend from the apertures of the vertical tubes, which are furnished with collagenous spines that form perches (fig. 11.3A). The arms are extended so as to form a spheroid basket, with the tentacles crossing at right angles between the arms and overlapping in the interspaces to create a remarkable feeding structure (Lester 1985; Dilly 1985). The tentacles and arms are densely ciliated. Ciliary activity draws water into the feeding sphere and out through an opening at the distal end of the tentacles. Ciliary activity on the inside of the arms creates rejection currents that flow inside the sphere and away from the pterobranch body. By contrast, the broad outer surfaces of the arms are washed by incoming currents. Particles are collected on the outer surface of the tentacles and arms; particles on the tentacles are passed to the arms via cilia or by tentacle flicking. The arms act as food grooves and pass accepted particles in mucous strands to grooves at their base, whence they are conveyed beneath the lamellae and into the mouth by cilia that are placed on either side of the groove, on the lamellae, and on the base of the cephalic shield. Water that enters the pharynx with the food strand flows out the gill slits.

A second genus, *Atubaria,* is known only from the deep sea; it is quite similar to *Cephalodiscus* in most anatomical details, but has a longer stalk that may be prehensile, and is not known to inhabit a coenecium. The third genus, *Rhabdopleura,* not only has a coenecium but is truly colonial somatically, all individuals being joined by a so-called black stolon, a communal stalk. Each colony is founded by an individual that is produced sexually, but then multiplies through budding. Individuals are on the order of 1 mm, and the colonies reach less than 10 cm in height. These small pterobranchs have only one pair of arms and entirely lack gill slits, though there are pharyngeal grooves internally at the position of gill slits in their larger relatives. In feeding, the arms are extended in straight lines about 30° apart (Stebbing and Dilly 1972). Feeding currents (fig. 11.4) are on the oral face of

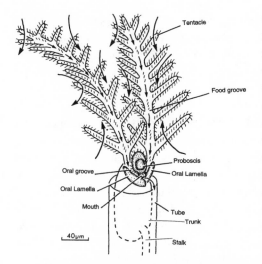

the tentacles and arms, and particles in mucous strands are conveyed to the mouth in grooves that run under the oral lamellae. Rejection tracts are on the adoral (back) sides of the tentacular apparatus, while cilia on the cephalic disk above the basal grooves, and on the exterior of the oral lamellae, also sort and reject particles. It has been suggested that with only two arms and relatively few tentacles, the amount of water delivered to the mouth is sufficiently small that gill slits are not required (Gilmour 1982).

FIG. 11.4 The pterobranch *Rhabdopleura*; feeding currents and ciliary tracts indicated. From Gilmour 1979.

Development in Hemichordata

Enteropneusts have the classic deuterostome features of early development (Burdon-Jones 1952; Hyman 1959). Cleavage is radial and holoblastic, producing a coeloblastula. Following gastrulation the archenteron appears, from which an outpocketing gives rise to an enterocoel. In many cases this enterocoel becomes a protocoel and gives rise to more posterior mesocoels and metacoels. In indirect developers the larva, termed a tornaria, is dipleurula-like with a convoluted ciliary feeding band encircling the mouth, and an apical tuft used in swimming.

Molecular investigation of neurogenesis in the direct-developing enteropneust *Saccoglossus kowalevskii* (Lowe et al. 2003) has served to establish the basal state of hemichordate nervous systems within crown deuterostomes, which bears importantly on hypotheses of chordate origins (see below). The early expression patterns of twenty-one genes, chosen as being orthologs of chordate genes involved in patterning the central nervous system, were studied. Expression maps of these hemichordate genes were nearly identical in anteroposterior patterns to those in chordates; their expressions were not restricted to nerve tracts, however, but encircled the entire body within the epidermis. Thus neurogenesis in enteropneusts is diffuse throughout the nerve net and not concentrated in the nerve cords, consistent with Bullock's conclusion that coordinating functions arise from within a generalized neurectoderm and that the nerve cords do not form a central nervous system.

In pterobranchs with indirect development, mesoderm is derived from the archenteron, and the coelom forms from a number of archenteric outpocketings, with perhaps as many as five pairs of compartments being present at one stage (Hyman 1959; Brusca 1975). Early pterobranch larvae are often described as planulalike; they are originally spherical but become flattened and are

ciliated over their entire surface. As they grow, they develop a tornaria-like apical tuft, and somewhat resemble young enteropneusts until settling and undergoing metamorphosis. Development in some forms is direct or proceeds from
buds.

Fossil Record of Hemichordata

Enteropneusta. An unusual fossil from the Lower Cambrian Chengjiang fauna, *Yunnanozoon*, has been reconstructed as having affinities with enteropneusts (Shu,
Zhang, and Chen 1996), but may be allied to the chordates (Chen and Li 1997;
see below). The earliest undoubted body fossils of enteropneusts are from the Essex
fauna of Pennsylvanian age, from Illinois and Missouri (Bardack 1997), and younger
body fossils are known from Jurassic beds at Osteno in northern Italy (Arduini et al.
1981). A number of traces have been assigned to enteropneusts, but the only trace
that possesses possibly unique enteropneust characters is from Lower Triassic Wer-
fen beds in the Dolomites of northern Italy (Twitchett 1996). U-shaped burrows,
sometimes probing upward to form multiple burrow extensions to the surface, are
found from the Early Cambrian onward, and some may have been formed by enteropneusts.

Pterobranchia. There is a surprisingly good early record of fossil pterobranchs,
thanks to the relative durability of their coenecia. The earliest known pterobranch is from the Middle Cambrian of
Sweden (Bengtson and Urbanek 1986), an
encrusting form assigned to the Rhabdo-
pleuridae on the basis of the colonylike
pattern of the tubes, which clearly display
a fusellar structure. Other scattered fossil occurrences of Paleozoic and later pterobranchs are known (see Bulman 1970;
Rickards et al. 1984).

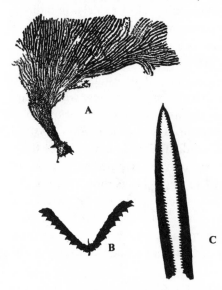

FIG. 11.5 Specimens of graptolites. (*A*) *Dic-
tyonema*, order Dendroidea, a benthic group.
(*B, C*) *Didymograptus* and *Maeandrograptus*,
order Graptoloidea, a planktonic group. From
Bulman 1970.

Graptolithina. The most important fossil
hemichordates are the graptolites (fig.
11.5), which range from the Middle Cambrian to the Late Carboniferous, being
especially widespread and abundant from
the late Early Ordovician to the Early Devonian (Bulman 1970; Berry 1987; Palmer
and Rickards 1991). With few exceptions
graptolites are known only by the remains of their collagenous skeletons, which
are fusellar, closely similar in structure to

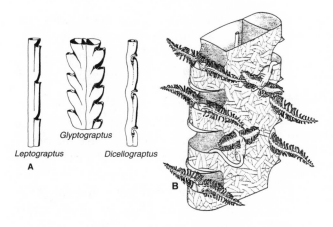

Glyptograptus

Leptograptus   *Dicellograptus*

A

B

FIG. 11.6 Graptolites. (*A*) Three graptoloid genera illustrating a variety of thecal shapes and arrangements. (*B*) A rhabdopleurid-inspired reconstruction of several zooids of *Climacograptus*. Note the bandages covering the rhabdosomal surface. A from Bulman 1970; B from Crowther and Rickards 1977.

pterobranch coenecia (Kozlowski 1948). Fortunately, a Silurian graptolite has yielded remains of zooids. While lacking in detail, these fossils appear to be similar to rhabdopleurid pterobranch zooids and display structures interpretable as black stolons, which are known in a few other specimens also (Rickards and Stait 1984). It would be of interest to know if graptolites, which are generally larger than the black stolon–bearing rhabdopleurids, had gill slits, but as yet there is no information on this point. Graptolites are usually treated as a class of hemichordates, Graptolithina, with eight orders.

Two orders are common and relatively well known, the Dendroidea and the Graptoloidea (fig. 11.5). Colonies begin as a conical structure termed a sicula, which presumably housed a sexually produced individual. A thin tubular structure, the nema, is produced from the apex of the sicula, while a series of thecae, each of which probably lodged a zooid, buds off from the sicula in an adapical direction (fig. 11.6A). The skeleton or rhabdosome consists of two layers, the fusellar rings that line the interior, and the cortex, an outer laminated layer that is sometimes built up of crisscrossing units termed bandages (fig. 11.6B). Dendroid thecae are sometimes dimorphic, possibly indicating a sex difference. Dendroid rhabdosomes are multiply branched, sometimes with cross supports between branches, and were evidently chiefly benthic; some have cortical holdfast systems. Graptoloids had fewer branches and were chiefly pelagic and often abundant and widespread; they evolved rapidly enough to make splendid zone fossils. The earliest records of each of these groups are from the Middle Cambrian. Several of the other graptolite orders seem to have had an encrusting colonial habit.

There has been a long controversy over the method of secretion of the cortex, but if the graptolite zooids were able to detach their stolons from the coenecia, or at least extend their stolons in the way that some pterobranch zooids can, there is no reason to believe that the zooids could not simply leave the thecae and deposit

FIG. 11.7 Interpretation of a hemichordate zooid leaving its coenecial tube and constructing a spine. From Dilly 1993.

cortical material on the rhabdosomal surface to form a nema or, for benthic forms, a holdfast. Dilly (1993) has described a species of *Cephalodiscus* from deep water off New Caledonia that bears coenecial spines up to thirty times the length of the zooids. The zooids evidently leave the thecae and feed while resting on the spines (fig. 11.7). One can imagine a graptolite colony swarming with zooids, feeding or secreting holdfasts or some other cortical feature.

Hemichordate Relationships and Ancestry

Hemichordates seem to have undergone an extensive early radiation, as did so many crown phyla, probably producing by at least Middle Cambrian time the enteropneusts (suggested by their branching before pterobranchs according to SSU rRNA data), the pterobranchs (both rhabdopleurids and cephalodiscids, so far as we can tell), and a number of distinctive graptolite groups.

In hemichordates the feeding system is of special interest, bearing on questions of their ancestry, their radiation, their relations to other deuterostomes, and their relations to lophophorates. The pterobranch tentacular system is sometimes said to differ from a lophophore only in not surrounding the mouth, implying that this difference is relatively unimportant. However, it can be argued that the tentacular apparatuses of hemichordates and lophophorates display the sort of differences one expects to find when similar adaptations have evolved independently—that they can represent convergences at least as easily as homologues. Certainly the known hemichordate arms with their double rows of tentacles are not like the known lophophores. The more complex hemichordate apparatuses have feeding tracts on the outside of the tentacles, while most lophophorates have feeding tracts on the inside. More extensive use of mucus is found in pterobranch feeding, perhaps because the feeding streams are more exposed to ambient currents. It is possible that final particle-sorting procedures were internalized in hemichordates, as the mucous feeding strands were lubricated and driven into the mouth partially by water streams. Because of the conical incurrent chambers created within the lophophoral curtains, the water volume of the feeding stream is progressively reduced in lophophores, which need not deliver an excess of water to the mouth.

Although there are no living paracoelomates known to be deuterostomes, the ancestral hemichordate evolved (probably not directly) from the protostome/deuterostome ancestor, which is hypothesized to have been a benthic vermiform paracoelomate (see chap. 13). Tyler (2001) has explored the similarities between juveniles of enteropneusts and the structurally paracoelomate flatworms.

The juvenile enteropneusts have an acoelomate trunk and collar, and like flat-worms possess a ciliated epidermis lacking a cuticle, underlaid by a muscle grid similar to that of some flatworms. The juvenile enteropneust pharynx is similar to those in the more basal flatworms, and there is a basiepithelial nerve plexus similar to that in some flatworms. Tyler believes that flatworms, and by extension some paracoelomates, are likely to be descendants of coelomates (see also Rieger 1986), and that such similarities may indicate coelomate features that have been retained in those clades. Those characters are not polarized by his analysis, however, and they are more likely to be plesiomorphies of both deuterostomes and protostomes, which were retained in juveniles, and in some cases in adults, of more complex phyla.

Echinodermata

Echinoderms display the classic developmental features that characterize deutero-stomes. There are five, possibly six living classes (fig. 11.8), and numerous fossil echinoderm groups are organized so uniquely that they are usually considered to deserve high taxonomic status (table 11.1). Morphological and molecular evidence suggest that Crinoidea is the most basal of the five major living classes, and that the classes Echinoidea and Holothuroidea are sisters and form the most derived clade (Littlewood et al. 1997).

Bodyplan of Echinodermata

In early development, echinoderms are bilateral and triregionated much like hemi-chordates, and while living echinoderm groups undergo a metamorphosis that usu-ally results in a pentaradial symmetry, stem echinoderms do not show radial sym-metry. Basic to the echinoderm anatomy is a unique system of coelomic canals, the hydrostatic water vascular system (fig. 11.9), which connects to the exterior by tubes or canals, sometimes very indirectly. This system consists of a ring canal around the anterior portion of the digestive tract and, usually, five radial canals. The ring canal may be connected to the exterior though one or more stone canals, so-called because of their calcified walls, which terminate in a specialized swelling or plate termed the madreporite. In some groups the madreporite lies, not at the surface, but within a coelomic space. In at least most classes there are numerous short offshoots of the radial canals, termed lateral canals, which bear tentacular podia or tube feet, and which function variously in locomotion, food gathering, and respiration. Compensation sacs of some sort, termed ampullae (usually when associated with tube feet) or polian vessels (otherwise), function as reservoirs for the fluid in the water vascular system. The radial canals and their associated tube feet define ambulacra, while the regions between ambulacra, in interradial positions, are interambulacra. Echinoderm skeletons, composed of calcite, are mesodermal and have a texture (stereom) that is unique among invertebrates and is easily rec-ognized even in small fragments. Mere scraps of fossil echinoderm skeletons can

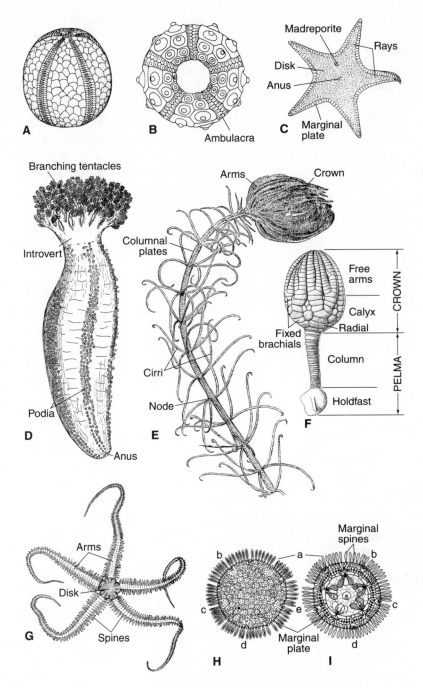

FIG. 11.8 The six living classes of Echinodermata, external views. (*A*) Echinoidea, a lepidocentrid, a stem form. (*B*) Echinoidea, a cidaroid, a crown form. (*C*) Asteroidea, the starfish *Ctenodiscus*. (*D*) Holothuroidea, *Cucumaria*. (*E*) Crinoidea, a crown form, *Cenocrinus*. (*F*) Crinoidea, a stem form, *Calpiocrinus*. (*G*) Ophiuroidea, *Ophiocoma*. (*H*, *I*) Concentricycloidea, dorsal and ventral views of *Xyloplax*; *a–e*, radii. A, B from Smith 1984; C–E, G from Hyman 1955, C after Gregory; F from Ubaghs 1978; H, I from Baker et al. 1986.

TABLE 11.1 Subphyla and classes of Echinodermata. Asterisk indicates extinct groups. The number of genera is approximate. Concentricycloids are considered to belong to Asteroidea by Smith (1988b). Chiefly after Sprinkle and Kier 1987.

Taxon	Remarks
Subphylum Crinozoa	
Class Crinoidea	Middle Cambrian–Recent; 1,000+ genera
*Class Paracrinoidea	Lower Ordovician–Lower Silurian; 15 genera
*Subphylum Blastozoa	
*Class Eocrinoidea	Lower Cambrian–Lower Silurian; 32 genera
*Class Rhombifera	Lower Ordovician–Lower Devonian; 60 genera
*Class Diploporita	Lower Ordovician–Lower Devonian; 42 genera
*Class Parablastoidea	Lower-Middle Ordovician; 3 genera
*Class Blastoidea	Middle Ordovician? Middle Silurian–Lower Permian; 95 genera
*Subphylum Asterozoa	
Class Asteroidea	Lower Ordovician–Recent; 430 genera
Class Ophiuroidea	Lower Ordovician–Recent; 325 genera
Class Concentricycloidea	Recent; 1 genus
*Subphylum Homalozoa	
*Class Stylophora	Middle Cambrian–Middle Devonian; 32 genera
*Class Homoiostelea	Middle Cambrian–Lower Devonian; 13 genera
*Class Homostelea	Middle Cambrian; 3 genera
*Class Ctenocystoidea	Middle Cambrian; 2 genera
Subphylum Echinozoa	
Class Echinoidea	Lower Ordovician–Recent; 765 genera
Class Holothuroidea	Middle Cambrian? Middle Ordovician–Recent; 200 genera
*Class Edrioasteroidea	Lower Cambrian–Upper Carboniferous; 37 genera
*Class Helicoplacoidea	Lower Cambrian; 3 genera
*Class Ophiocistoidea	Lower Ordovician–Lower Carboniferous; 6 genera
*Class Cyclocystoidea	Middle Ordovician–Middle Devonian; 8 genera
*Class Edrioblastoidea	Middle Ordovician; 1 genus

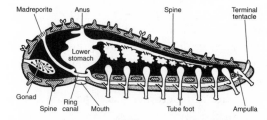

FIG. 11.9 Generalized cross section of an asteroid arm to show the water vascular system. From Pearse et al. 1987.

be identified with the phylum, and extinct groups that might otherwise be assigned elsewhere can be placed as echinoderms with reasonable confidence, if they possess the stereom texture. As Budd and Jensen (2000) point out, it is possible that stereom could be plesiomorphic for echinoderms and could have been found in sister or ancestral taxa, but all bodyplans in which it is known are plausibly assigned to stem or crown Echinodermata. Despite this easy identification of echinoderm skeletons, most of the nominal classes are rather distinctive, appear abruptly, and are separated from one another by morphological gaps at their first appearances. There are, however, a number of important attempts to produce phylogenies with plausible scenarios of evolutionary pathways between the classes (e.g., Smith 1984; Paul and Smith 1984).

The most unusual of living echinoderm bodyplans is that of the concentricycloids (fig. 11.8H–I; Baker et al. 1986). The water vascular system consists of two concentric ring canals connected by five radial canals from which polian vessels extend inward. Podia lie, not on the radial canals, but along the outer side of the outer ring canal. The developmental pattern is not yet known. It has been suggested that concentricycloids evolved from asteroids (see Rowe et al. 1988), and indeed Smith (1988b) believes they should be retained within that group.

Whether or not the echinoderms have a blood vascular system is not clear. There is a so-called hemal system but no hemal vessels; the hemal fluid is found in lacunae that encircle the mouth with the ring canal and radiate into the ambulacra. A so-called axial gland, borne on a mesentery that supports the stone canal, connects the oral hemal ring to an aboral ring from which lacunae radiate to the gonads; sometimes there is also an intermediate gastric ring. An astonishing feature of this system is that the axial gland and much of the hemal ring systems are encased in tubelike spaces, coelomic canals that originate from the oral somatocoel (see below). Whether the hemal system itself is coelomic or is reminiscent of some ancestral pseudocoelomic topology seems to be an open question. The function of the perihemal tubes is uncertain. As nerve cords also run within these coelomic canals, some of which are termed hyponeural canals, Hyman (1955) suggested that they may function to cushion nerves against injury. In some taxa the function of the hemal system is believed to be circulatory and associated with the distribution of digestive products, but there is no gastric hemal ring in some classes. The nervous system lacks ganglia, consisting of a basiepithelial plexus thickened into radial nerve tracts, and with a nerve ring around the esophagus.

Development in Echinodermata

Echinoderms display classic radial cleavage. Development is best known among the sea urchins (Echinoidea). The first two cleavages are longitudinal, but then in indirect developing forms, meridional cleavages begin to separate tiers of blastomeres, including a vegetal tier of very small cells or micromeres (fig. 11.10), which are specified autonomously. However, the lower tier of macromeres are specified conditionally, at least in part, by signaling from the micromeres (see Cameron and Davidson 1991); these cells produce the archenteron.

Fig. 11.11 depicts cell fates and specification patterns in the early embryos of two species of sea urchin, one of which is planktotrophic (fig. 11.11A) and the other a direct developer, a derived condition in this species (fig. 11.11B; see Wray and Raff 1990; Henry et al. 1990; and references therein). The derived form displays earlier specification of the dorsoventral axis (indeed the evidence suggests that it occurs before fertilization), altered cleavage patterns (the vegetal micromeres are not produced until the sixty-four-cell stage), and rearranged cell lineages, so that,

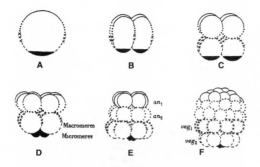

FIG. 11.10 Cleavage in an indirect-developing echinoid, *Paracentrotus*. (*A*) Uncleaved zygote. (*B*) Four-cell stage; first two cleavages are meridional. (*C*) Eight-cell stage, third cleavage is equatorial. (*D–F*) Sixteen-, thirty-two-, and sixty-four-cell stages. Note the vegetal sequestering of materials that eventually segregate in the micromeres. *an,* animal; *veg,* vegetal; *crosses,* regions of prospective or actual veg_1 cells; *dots,* regions of prospective or actual an_2 cells; *dashes,* regions of prospective or actual veg_2 cells. From Hörstadius 1939.

for example, more of the internalized cells produce mesoderm. In general there is an accelerated appearance and proliferation of adult tissues. The lack of developmental homology between cells and tissues in these forms led Wray and Raff to conclude that the direct-developing embryo is a novel type, and not simply the degenerate embryo of indirect developers.

Gastrulation is by invagination, and in indirect developers, growth to the larval stage involves cell rearrangements. The archenteron grows forward to open as a mouth anteriorly, while the blastopore becomes the anus. Coelomic cavities form from outpocketings of the archenteron. The larvae are bilaterally symmetrical forms of several types that are collectively termed dipleurulae (fig. 11.1). The larvae undergo a complicated metamorphosis involving differential growth and rotation to produce a radial symmetry. Thus echinoderm architectures are highly modified from such bilaterian bodyplans as displayed by hemichordates; their adult anatomies are perhaps easiest to appreciate in the light of their coelomic development. A clear review of coelom formation in some echinoderms is given by Jefferies (1986), from which much of this account has been abstracted. Although the form and distinctiveness of the coelomic divisions vary among taxa, there are usually three sections, termed (from the anterior) the axocoel, hydrocoel, and somatocoel (fig. 11.12), presumably homologous with the triregionated coelom of the hemichordates.

The axocoel is commonly unpaired and broadly confluent with the hydrocoel in early development; these two coelomic regions are sometimes termed the axohydrocoel, and in some taxa there is no demarcation between them. The other coelomic compartments are paired. As development proceeds, the left hydrocoel area gives rise to five lobes posteriorly (fig. 11.12B); this rudiment gives rise to the radial canals of the water vascular system, which thus arises entirely from the left hydrocoel. The right hydrocoel is lost or becomes greatly reduced. Then the larva rotates, often through as much as 90°, so that the plane in which the water vascular system is developing, and in which there is a pentameral symmetry, has an oral-aboral axis (fig. 11.12C). The left somatocoel pouch, which is enlarging, is thus rotated and brought to the oral side of the right somatocoel pouch; the former is then an oral coelom, the latter an aboral coelom, but later they coalesce, sometimes with the axocoel, to form a capacious perivisceral coelom. As for orientation

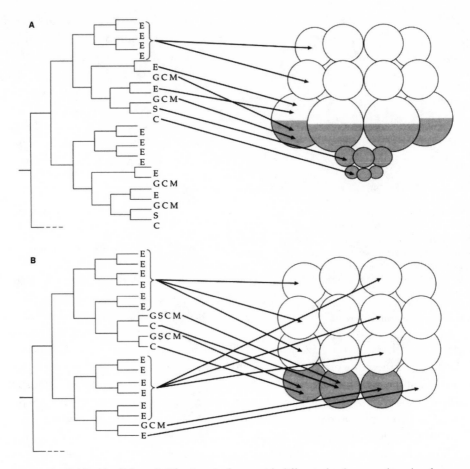

FIG. 11.11 Echinoid cell-fate modifications in forms with different developmental modes; fate maps are for two quadrants of each species. (A) *Strongylocentrotus purpuratus*, a species with planktotrophic larvae; note that the lower quadrant has nearly the same cell fates as the upper, so mapping is omitted for clarity. (B) *Heliocidaris erythrogramma*, a direct-developing species. The cell fates are different in each quadrant, and each is different from cell fates in the planktotrophic form. Fates: *E*, ectoderm; *G*, gut; *C*, coelom; *M*, muscle, pigment cells, and coelomocytes; *S*, skeletogenic cells. From Wray 1997.

in living positions, the oral side is ventral in asteroids, ophiuroids, and echinoids; in crinoids (and in some extinct classes) the mouth is dorsal; and in holothuroids, which rest on their sides, it is anterior. Usually quite early in development, the left axocoel or axohydrocoel becomes connected to the surface of the developing larva to form a hydropore. The hydrocoel communicates with the hydropore, but indirectly. For example, in crinoids a stone canal grows from the hydrocoel to communicate with the axocoel and hence the hydropore; later, the somatocoel and axocoel unite, forming a perivisceral coelom into which the stone canal then opens. Details of the developmental sequence vary among taxa, but the net effect in all of them is to produce a pentameral bodyplan from a bilateral one.

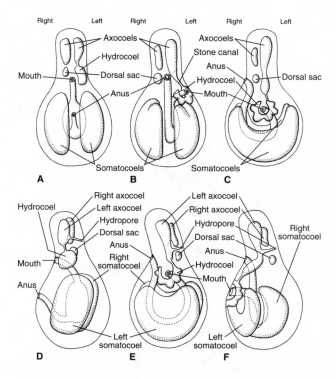

FIG. 11.12 Early metamorphic phases in a generalized echinoderm. (A–C) Ventral view of larvae, showing (A) an initial essentially symmetrical phase, (B) an early asymmetrical phase, and (C) a phase with developing secondary symmetry. (D–F) Same phases, viewed from left side of larva. From Ubaghs 1967b after Heider.

Fossil Record of Echinodermata

A large number of extinct echinoderm bodyplans are known, many of which are ranked as classes. Sprinkle (1980) recognized twenty classes that are found as fossils, including five of the six living ones; the classes are sometimes grouped into five subphyla (table 11.1). A form from Neoproterozoic beds in South Australia, *Arkarua,* has been suggested to be the earliest known echinoderm (Gehling 1987). *Arkarua* is a minute discoidal nonskeletonized form with five radiating rays on the upper surface, resembling a tiny soft-bodied edrioasteroid (see below). Without more convincing echinoderm synapomorphies, the relationship of *Arkarua* remains uncertain. The earliest undisputed echinoderms are from the middle Early Cambrian (Atdabanian Stage), and include the helicoplacoids and edrioasteroids, both of the subphylum Echinozoa, and the eocrinoids of the subphylum Blastozoa.

Crinozoa. Crinoids are believed to be basal within crown echinoderms, and the earliest known member of the Crinoidea may be *Echmatocrinus,* described from several Burgess Shale specimens (Sprinkle 1973). The characteristic stereom plate structure has not been observed, and assignment of this fossil has been questioned—for example, Conway Morris (1993b) suggested that it might be a cnidarian. In a restudy, Sprinkle and Collins (1998) supported its echinoderm affinities, but Ausich and Babcock (1998, 2000) have subsequently suggested an octocoral affinity. Evidently *Echmatocrinus* cannot be placed definitively as yet.

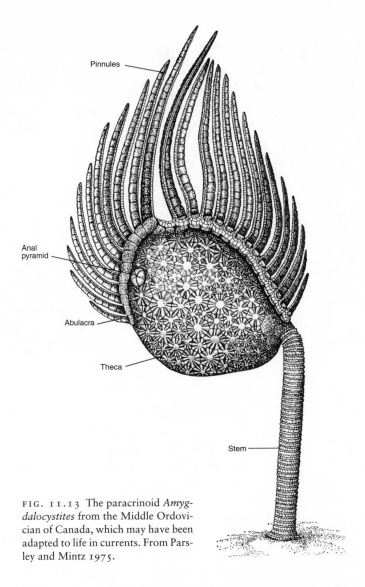

FIG. 11.13 The paracrinoid *Amyg-dalocystites* from the Middle Ordovician of Canada, which may have been adapted to life in currents. From Parsley and Mintz 1975.

An extinct class of Crinozoa, the Paracrinoidea, has been described from the Ordovician (fig. 11.13). Some authorities believe that paracrinoids are distinctive enough to be regarded as a subphylum of their own (Parsley and Mintz 1975). The test is composed of many irregularly arranged plates, and although there is an extensive and complex water vascular system, it may not have extended into the arms, which are recumbent with numerous pinnules as in the figured specimen.

Blastozoa. In contrast to crinozoans, blastozoans lack arms (which are extensions of the calyx wall), but bear food-gathering structures termed brachioles, light ap-

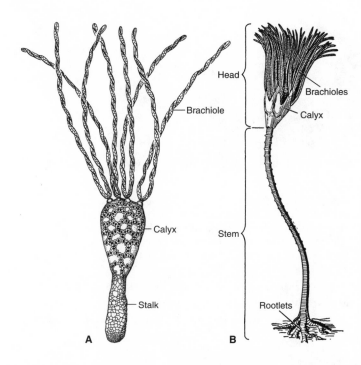

FIG. 11.14 Blastozoan echinoderms. (*A*) Reconstruction of an early eocrinoid, *Gogia,* from the Middle Cambrian of Canada. Note the numerous irregularly arranged plates. (*B*) Reconstruction of a blastoid, *Orophocrinus,* from the Lower Carboniferous of the United States. A from Ubaghs 1967a; B from Fay 1967.

pendages mounted on ambulacral plates that overlie the calyx (fig. 11.14). Brachioles have been modeled as containing water vascular canals with tube feet (Breimer and Macurda 1972) and without tube feet (Sprinkle 1973). Eocrinoids, the earliest blastozoans known, usually have globular calyxes attached to the substrate by a stemlike holdfast, with brachioles emanating from the upper surface of the calyx (fig. 11.14A). The mouth is often central on the upper surface, with from three to five radiating ambulacra, while the anus, when known, is offset toward the margin of the upper surface, indicating a U-shaped (or perhaps more convoluted) digestive tract. In early species, such as the form pictured in fig. 11.14A, the calyx is covered by numerous irregular plates, but as time goes on, the pattern evolves and plates become fewer and more regular. Similarly, early holdfasts are covered by numerous small imbricating plates, while later ones have fewer larger plates and eventually are composed of disk-shaped columnar plates stacked like coins. The earliest eocrinoids are known from fragments from the middle Lower Cambrian, and a specimen has been described from the Chengjiang fauna (*Cambrofengia;* Hou et al. 1999). Well-preserved specimens of late Lower Cambrian age have five ambulacra. Nearly universal throughout the blastozoans are folds or pores of various geometries that lie within or across calyx plates. In eocrinoids the pores are termed epispires; they usually run normal to the plate surfaces to the calyx interior, and are found at the margins of two or more plates. It is believed that epispires represent the sites of respiratory papillae. The papillae may have contained extensions of the hydrocoel,

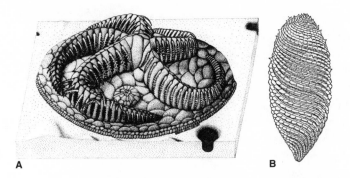

FIG. 11.15 Representatives of two extinct classes of echinozoan echinoderms. (A) Reconstruction of the edrioasteroid *Lebetodiscus* living on a firm seafloor; anal structure in center, podia extended on some ambulacra. (B) The helicoplacoid genus *Helicoplacus* as reconstructed by Durham. A from Bell 1980; B from Durham and Caster 1966.

in which case they might well be homologous with tube feet (see Sprinkle 1973). On the other hand, dermal respiratory papillae in living asteroids are invested by the somatocoel. A later blastozoan, the Carboniferous blastoid *Orophocrinus,* is shown in fig. 11.14B.

Echinozoa. The echinozoans (fig. 11.15), which include the living echinoids and holothuroids, are not stemmed. The extinct edrioasteroids lay directly on firm substrates or were sometimes attached to invertebrate skeletal debris (Regnell 1966; Bell 1980). Edrioasteroid skeletons (fig. 11.15A) tend to be discoidal to cylindrical, with upper surfaces bearing five usually arcuate ambulacra radiating from a central mouth; the anus lies in an interambulacrum on the oral side. There is a hydropore near the mouth, presumably opening into a water vascular system, and there were evidently tube feet along the ambulacra, at least in most forms, which possess ambulacral pores. The skeleton of the oral surface was composed of numerous plates, and commonly encircled by a rim of small plates that was firmly secured to the substrate. In some forms the aboral skeleton is known, often of many imbricating plates also, a pattern that must have promoted skeletal flexibility. In some taxa numerous pores are found between the interambulacral plates, suggesting the presence of papillae, perhaps respiratory or possibly sensory; these same forms also have ambulacral pores that suggest the presence of tube feet.

Helicoplacoids (Durham and Caster 1966; Durham 1967) are an exclusively Early Cambrian group of small, fusiform to pyriform echinoderms with tests of spirally arranged columns of imbricating plates (fig. 11.15B). A single ambulacral groove begins near the pole of the larger end and spirals down the test to split into two grooves near the midpoint of the test. The position of the mouth has been interpreted by Derstler (1981) as lying at the point of ambulacral splitting; in this case, there would be a triradiate ambulacral system, one ray going toward the large and two toward the small end. However, Durham (1993) restudied this class and concluded that the mouth lies at the larger pole. Flooring plates penetrated by pores lie within ambulacra, with covering plates along ambulacra. Durham (1967) suggested that the pores were for tube feet, and that radial water vessels thus lay beneath the flooring plates. Paul and Smith (1984) suggested, however, that the radial vessels lay above the flooring plates and operated the covering plates hydraulically, perhaps by short extensions homologous with tube feet; the pores are interpreted as the sites of canals that led to internal compensation sacs. The helicoplacoid mode of life is unknown. Because the imbricating plate rows can be interpreted as being folded in such a way as to permit considerable extension when straightened, it has been suggested that helicoplacoids could expand their body volume (Durham 1967). Paul and Smith (1984) concede that they could change shape but doubt they could change volume, and suggest that they lived with one end buried in sediment, while Jefferies (1990) postulates that they were infaunal, burrowing by expansion and contraction of the test.

Three of the many other distinctive echinoderm groups that appear during the Ordovician are recognized by some workers as extinct echinozoan classes (Ophiocistoidea, Edrioblastoidea, and Cyclocystoidea; see Moore 1966). They are quite rare, but they form interesting elements in a radiation that produced a large number of disparate echinoderm architectures.

Homalozoa. The homalozoans are an extinct group of unusual fossils, sometimes called carpoids, that appear in the late Lower Cambrian and that are commonly considered to represent a subphylum of echinoderms (e.g., by Sprinkle in Boardman et al. 1987). One homalozoan group, the Stylophora (fig. 11.16A–D), has been hypothesized to be a likely chordate ancestor and elevated to a subphylum, the Calcichordata, by Jefferies (1967, 1968). In later works Jefferies has withdrawn the subphylum assignment in deference to a cladistic approach to classification (see Jefferies 1986) but has continued to support a calcichordate ancestry of chordates. The calcichordate hypothesis has not been generally accepted; it involves morphological details that are capable of alternative interpretations.

Homalozoans were flattened, clearly benthic organisms with tests of calcite plates that preserve a stereom structure. There is no pentameral symmetry, however; they are bilateral or somewhat irregular. Two of the classes are of particular interest, the Stylophora with two orders, Cornuta (fig. 11.16A–B) and Mitrata (fig. 11.16C–D), and the Homoiostelea with a single order, Soluta (fig. 11.16E). The solutes have the usual flattened body covered by calcite plates, from which extend two appendagelike organs. One of these appendages is generally interpreted as a stele or tail, the other as a food-gathering arm, presumably housing an ambulacrum. The arm is covered by rows of plates; a pair of plate rows may have opened to expose tube feet. The mouth has not been identified but may have lain under cover plates at the base of the arm. A sievelike hydropore plate and a node interpreted as

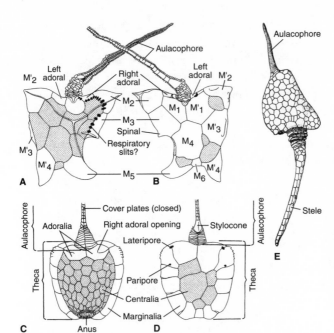

FIG. 11.16 Representatives of extinct classes of homalozoan echinoderms. (*A–D*) Stylophorans, marginal plates white, some labeled; central plates shaded. (*A*) Upper surface of the cornute *Ceratocystis*. (*B*) Lower surface of *Ceratocystis*. (*C*) Upper surface of the mitrate *Microcystites*. (*D*) Lower surface of *Microcystites*. (*E*) The solute *Dendrocystites* from the Middle Ordovician of Central Europe. After Ubaghs and Caster 1967.

the site of a gonopore are found near the arm base; an anal opening surrounded by a low pyramid of small radial plates is found near the tail (Caster 1967; Jefferies 1990). Jefferies has reconstructed the soft-part anatomy and believes there was a branchial slit at the left posterolateral corner of the body, partly by analogy with cornutes (see below).

The stylophorans have only a single appendage, the aulacophore (fig. 11.16A–D). This has been considered an anterior feeding appendage by a number of workers (e.g., Ubaghs 1961, 1967c; Parsley 1988; Lefebvre 2003). Jefferies (1990) has argued that this appendage is homologous with the stele of the solutes, and thus posterior, in which case the orientation of the animal should be reversed; what has been interpreted as an anal pyramid becomes a mouth structure, and so forth. Therefore stylophorans would not possess an ambulacrum. However, the aulacophore has been shown to be divided into three regions, the more distal two containing an ambulacral groove, with a mouth at the proximal end (see Lefebvre 2003). Cornute stylophorans clearly possess a series of openings that can be interpreted as respiratory structures, possibly homologous with gill slits (fig. 11.16A). Such slits are lacking in mitrate stylophorans, but Jefferies has reconstructed the internal plate geometries of some mitrates and cites evidence for openings, interpreted as excurrent atrial openings, posteriorly on either side of the aulacophore. Trace fossils associated with mitrates suggest that they moved with the aulacophore forward (Sutcliffe et al. 2000).

Of the two remaining homalozoan classes, one, the Ctenocystoidea, is poorly known; evidently it lacks both feeding arm and stele (Robison and Sprinkle 1969).

The other class, the Homostelea, has a single order, Cincta, which has a single appendage. This has been nearly universally interpreted as a stele, and as Paul (1990) notes, it is likely to be homologous with the solute appendage and the so-called aulacophores of stylophorans. Cinctans have openings on their tests that can be interpreted as mouth and anus; the mouth is bordered by two ambulacral grooves that run along the anterolateral test margins. Possible branchial slits border an anterior plate or operculum (Jefferies 1990).

The homalozoans were evidently bottom-dwelling organisms with somewhat flexible tests. Most of them probably moved via the stele or tail and employed water currents for feeding and probably respiration, the feeding stream entering by the mouth, possibly passing through pharyngeal openings and exiting via slits or pores of some type. Some, perhaps all, may have possessed a water vascular system. In the light of their echinoderm-like tests of stereom-textured plates, the position of homalozoans as a subphylum of echinoderms seems entirely plausible (see Philip 1979; David et al. 2000). The echinoderms do not appear to lie in the ancestry of chordates, and there is therefore no reason at present to regard any of the homalozoans as chordate ancestors.

The Echinoderm Skeleton. Despite the disparate structure of the echinoderm classes, the identification of two skeletal types that originate from separate larval sources has permitted the recognition of architectural trends within the phylum (Mooi et al. 1994; Mooi and David 1998 and references therein; David et al. 2000). One skeletal type is termed axial (in reference to its alignment with the axis defined by the ambulacra). The axial skeleton originates from the lobed rudiment that arises from the left hydrocoel (see above) and comprises the plates associated with the radial ambulacral system. Radial plates are added distally, at the tip of the ambulacrum; usually they are biserial, staggered so as to produce a zigzag suture between series. The extra-axial skeleton composes the remainder of the test, and arises from nonrudiment portions of the larval body, chiefly surrounding the somatocoel. Some parts of the echinoderm skeleton are perforated by a periproct, and by gonopores, hydropores, and so forth, depending on the taxon; the perforate skeletal elements evidently arise from the left somatocoel region and are extra-axial.

The echinoderm taxa that appear during the Cambrian have relatively small contributions from axial skeleton (e.g., edrioasteroids and helicoplacoids, fig. 11.15). In these forms the bulk of the test wall is composed of extra-axial skeleton, much of it imperforate, implying that development of the adult body wall proceeded from cell lines derived from larval body-wall tissues and therefore that metamorphosis was probably rather inconspicuous. However, later branches of the echinoderms are more or less progressively dominated by plate series derived axially. Among crown groups, only crinoids have imperforate extra-axial skeletal elements (stems); asteroids and ophiuroids have more important axial skeletal elements and only perforate extra-axial skeleton; and echinoid skeletons are almost entirely axial. Accordingly, these groups show increasingly more conspicuous metamorphosis as the rudiment contributes a greater portion to adult development (see Mooi and David 1998).

Echinoderm Ancestry

The early radiation of echinoderm bodyplans was chiefly an Early to Middle Cambrian event, although a very important diversification produced many more classes during the Ordovician (table 11.1). As early echinoderm tests are not the most easily preserved of skeletons, it is likely that the evidence of the radiation is systematically biased to appear later than it occurred. Even taking the fossil evidence at face value, the major body types of echinoderms, as of so many phyla, appeared very early in Phanerozoic history, and their subsequent record chiefly involves the elaboration of some of those types and the elimination of many others. Helicoplacoids and homalozoans appear first in the record; both of them had water vascular systems and the characteristic stereom skeletal texture, features that are plausibly considered shared derived characters that serve to unite these forms as conphyletic with each other and with the other classes in table 11.1.

As echinoderm larvae have a triregionated plan like the bodyplan of pterobranch hemichordates, and as the echinoderm mesocoel becomes elaborated as a water vascular system that is likely to be homologous with the mesocoelic tentaculate system of hemichordates, it is reasonable to hypothesize that the echinoderms and the hemichordates share a triregionated common ancestor that was a deuterostome. If one accepts that the homalozoans displayed pharyngotremy, then that common ancestor presumably did so also. The living echinoderm classes, with their torsionlike transformation to pentameral symmetry, are far more derived from the bodyplan of any such ancestor than are living hemichordates.

Two forms have been repeatedly hypothesized as echinoderm ancestors: the dipleurula, a hypothetical organism believed to be recapitulated by the dipleurula larval type of the deuterostomes, chiefly suggested by early workers; and the pterobranch hemichordates. These hypotheses have been reviewed by Holland (1988). In order to explain the torsion and loss of many right-side features by echinoderms, supporters of a dipleurula ancestor suppose that the earliest echinoderms were sessile suspension feeders that became attached to the substrate by their right surfaces and therefore suppressed the right hydrocoel and associated feeding organs. As the practice of regarding developmental types as recapitulatory waned, this ancestral dipleurula model lost favor, being replaced by an ancestral pterobranch model, but with torsion still considered to have arisen as an accommodation to attachment. However, the fossil record has now revealed that the earliest echinoderms may not have been attached.

Accordingly, Jefferies (1986) suggested that a pterobranch lineage came to lie down on its right side, but without attachment, to give rise to a suspension-feeding form with an ancestral echinoderm architecture. Jefferies' model lacks an explanation of the advantage to be accrued by a bilateral suspension feeder in becoming

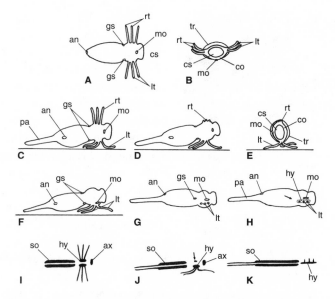

FIG. 11.17 Holland's evolutionary scenario for the origin of laeo-thetism (lying on the left side) during the evolution of the echinoderm bodyplan from the stem deuterostomes. (*A, B*) Symmetrical stem deuterostome, a suspension feeder, (*A*) ventral view and (*B*) ante-rior view. (*C*) Descendant that has become a deposit feeder with left tentacles, ancestrally ventral view. Posterior appendage has evolved to facilitate locomotion. (*D*) Right tentacles are lost, same view. (*E*) Anterior view of bodyplan in D. (*F*) Mouth migrates to ancestral left side; ancestral ventral view. This is the last common ancestor of the Chordata and Echinodermata. (*G*) Ancestral left-side view of bodyplan in F. (*H*) Gill slits lost, ancestral left-side view. (*I–K*) Stages in evolution of coelomic compartments seen in ancestral ven-tral view: (*I*) stem deuterostome such as in A; (*J*) hydrocoel reduced as in D; (*K*) reconstructed compartments in a solutan homalozoan. *an,* anus; *ax,* axocoel; *co,* collar; *cs,* cephalic shield; *gs,* gill slits; *hy,* hydrocoel; *lt,* left tentacle; *mo,* mouth; *pa,* posterior appendage; *rt,* right tentacle; *so,* somatocoel; *tr,* trunk. After Holland 1988.

recumbent; as Holland (1988) points out, such a change would if anything seem to be disadvantageous. Therefore, Holland has modified this scenario by postulating that the bilateral suspension feeder switched to a deposit-feeding mode, and in fact became oriented with the left side down rather than up, feeding on particles on or in the substrate, rather than in the water column (fig. 11.17). Deposit feeding of this sort is in fact quite common among the unattached echinoderms today. Hyper-trophy of the left hydrocoel followed. Early echinoderm evolution is thus viewed as proceeding from a bilateral suspension feeding form, to an asymmetrical deposit feeder lying on its left side, to trimeral or pentameral types. In this model an infau-nal habitat for helicoplacoids appears more plausible, and the question is raised as to whether many of the homalozoans might not have been chiefly deposit feeders.

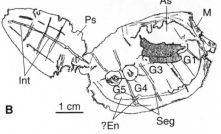

FIG. 11.18 A vetulicolian, *Didazoon*. Interpreted features: *As*, anterior section; *?En*, presumed endostyle; *G1–5*, gills; *Int*, intestine; *M*, mouth; *Ps*, posterior section; *Seg*, segments. From Shu et al. 2001.

Vetulicolia

A group of four fossil genera from the Chengjiang fauna and its near correlative have been assigned to a new fossil phylum of deuterostomes, Vetulicolia (Shu et al. 2001), chiefly because of features that are interpreted as gill slits and an endostyle. The vetulicolian body is bipartite; both parts are segmented in one genus (*Didazoon*, fig. 11.18), but the anterior may not be segmented in another (*Xidazoon*). *Xidazoon* (Shu, Conway Morris, Zhang, Chen, Li, and Han 1999) is columnar to bag-shaped with two circlets of plates surrounding a large anterior mouth. Another genus, *Vetulicola*, has an anteroposterior series of four rigid plates on the anterior body part, which appear to show growth lines, suggesting that they were not exoskeletal in the arthropod manner. Vetucolians are hypothesized to be stem deuterostomes by Shu et al. (2001), but are suggested to be more derived than the echinoderm + hemichordate clade, and possible ancestors of stem chordates, by Gee (2001).

Invertebrate Chordata

Urochordata

Most workers regard the urochordates as a primitive subphylum of the Chordata because they possess a notochord and dorsal neural tube at early stages of their ontogeny (see Barrington 1965; Millar 1966; Goodbody 1982; Jefferies 1986). The adult urochordate bodyplan is sufficiently distinct from those of the other chordates that urochordates are sometimes considered to be a sister to other chordates or to be a phylum of their own. That the notochord is homologous in urochordates and chordates seems entirely plausible.

There are four extant urochordate classes (table 11.2). Two classes are benthic, one being restricted to the deep sea, and two are pelagic. The benthic class Ascidiacea is by far the most diverse of these classes today and has usually been assumed to be the most primitive (Garstang 1928). Early SSU rRNA studies suggested that the pelagic Appendicularia might be more primitive (Wada and Satoh 1994; Turbeville et al. 1994; Swalla et al. 2000), but broader sampling indicates that Appendicularia

TABLE 11.2 Higher taxa of Urochordata (Tunicata). After Brusca and Brusca 1990.

Taxon	Remarks
Class Appendicularia	Larvaceans; pelagic, solitary
Class Ascidiacea	Tunicates; benthic, sessile, solitary or colonial
Class Soberacea	Possibly ascidians; benthic, abyssal, solitary
Class Thaliacea	Salps, etc.; pelagic, solitary or colonial

may be sister to the ascidian order Aplousobranchia, and this clade may be sister to the remaining ascidians (Stach and Turbeville 2002). This still leaves the bodyplan of the stem ancestor of the urochordates in doubt; it could have resembled either appendicularians or ascidians. A member of the other pelagic class, the Thaliacea, may nest within or belong to a sister group of the ascidians (Cameron, Garey, and Swalla 2000; Swalla et al. 2000). This form is a pyrosome, believed to be the most basal group of thaliaceans, and it may have arisen from a sessile ascidian ancestor. There are no sequences as yet from the deep-sea group, which is sometimes also considered to be derived from ascidians. Most morphological and developmental information on Urochordata is on ascidians, which have played important roles in scenarios of chordate origins.

Bodyplan of Ascidiacea. The ascidian body (fig. 11.19) is sac-shaped and surrounded by a tunic that is secreted by the epidermis and is composed of a carbohydrate resembling cellulose. Spicules of calcium carbonate occur in the tunic and in some cases elsewhere in the body; they vary among taxa in morphology and mineralogy (Lowenstam and Weiner 1989). The body wall is lined by bands of longitudinal muscles and by an inner circular muscle layer. The mouth lies at the anterior, upper end of the sac, opening into a buccal cavity that is succeeded posteriorly by an expanded pharynx that takes up perhaps three-quarters or more of the interior of the body. The pharynx is succeeded in turn by a gut that traces a U to end in an anus near the middle of the body dorsal to the pharynx. The pharynx is perforated by numerous openings or stigmata arranged in rows to form a pharyngeal or branchial basket. The edges of the stigmata are ciliated, and ciliary activity produces an inhalant water current through the mouth. This feeding and respiratory current passes through the wall of the branchial basket and is expelled through a dorsal atrial siphon (fig. 11.19). Mucous sheets that collect food particles from the branchial current are produced along the ventral length of the branchial basket by an endostyle. A mucous sheet is conveyed dorsally by cilia along each side of the basket; at the dorsal midline the sheets meet and are twisted into a cord to flow posteriorly to the gut. Gonads lie posterior to the pharynx, lodged in the bend of the intestine or in the body wall, with gonoducts rising dorsally to end near the anus. The atrial excurrent removes fecal material and gonadal products. There are no excretory organs. A nerve ganglion and an associated nerve gland are located

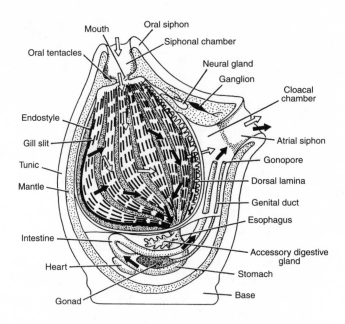

Mouth
Oral siphon
Oral tentacles
Siphonal chamber
Neural gland
Ganglion
Cloacal chamber
Endostyle
Gill slit
Tunic
Mantle
Intestine
Heart
Gonad
Atrial siphon
Gonopore
Dorsal lamina
Genital duct
Esophagus
Accessory digestive gland
Stomach
Base

FIG. 11.19 Bodyplan of an adult ascidian. From Brusca and Brusca 1990.

dorsally between the mouth and the atrial opening. Branches from the ganglion innervate the body wall and viscera, but not the heart.

The main body cavity is a hemocoel. The heart lies ventrally near the base of the intestinal loop, pumping blood through an extensive system of channels and sinuses that serve the viscera, the branchial basket, and the tunic. Curiously, the heart pumps in alternate directions periodically. In adult ascidians small paired epicardial cavities lie on each side of the heart. These cavities arise as evaginations from the posterior pharyngeal wall and may just possibly be coelomic spaces; they are involved in asexual budding and may also function in providing waste storage.

Development in Ascidiacea. The developmental stages of ascidians (Cloney 1982; Katz 1983; Satoh 1994) have furnished the bases for the most famous of speculations as to the origin of the chordates because the larva is tadpolelike, with a notochord (fig. 11.20), and has been imputed to be the vertebrate ancestor.

Ascidian embryogenesis was beautifully described by Conklin (1905), and modern work has been reviewed by Satoh (1994; see also Gerhart and Kirschner 1997). Jeffery and Swalla (1997) review ascidian developmental genetics. All species that have been studied have similar developmental patterns. The urochordates show one of the more extreme examples of mosaic development. Cleavage is radial; fig. 11.21 shows cleavages through the sixty-four-cell stage with blastomeres labeled after Conklin (1905). The oocyte is not visibly regionated, but a polar body forms

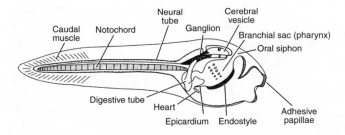

FIG. 11.20 Ascidian tadpole bodyplan. From Brusca and Brusca 1990 after Seeliger.

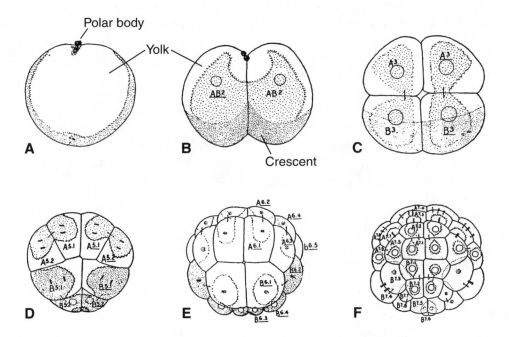

FIG. 11.21 Cleavage in the ascidian *Cynthia partita* (Urochordata); blastomeres named by Conklin's system. (*A*) An uncleaved zygote. (*B*) Two-cell stage; note positioning of "crescent" region. (*C*) Four-cell stage, crescent absent from upper cells, sequestered in lower cells. (*D*) Sixteen-cell stage, vegetal view. (*E*) Thirty-two-cell stage, vegetal view. (*F*) Sixty-four-cell stage, vegetal view. After Conklin 1905.

before fertilization. Following fertilization, the zygote becomes strongly polarized as a mat of actin microfilaments contracts to what becomes the vegetal pole, with visible "crescent" sections, establishing anteroposterior polarity. The first cleavage produces blastomeres that have similar or equal cytoplasmic contents. However, the second cleavage, perpendicular to the first, localizes visible cytoplasmic factors in the B^3 and B^3 blastomeres. At the third cleavage, which is latitudinal, spindles are tilted slightly so that the cleavage planes slant and the dorsal cells are slightly offset toward the anterior, and thus the dorsoventral axis becomes visible. As cleavage unfolds, regulatory factors are segregated among daughter cells, perhaps some being positioned by intracellular migrations. The result is a highly determinate and nearly autonomous system of early development and cell-fate specification. A few cells are induced in the late blastula, and neural tissues are induced via cell-cell interactions in late embryogenesis.

Gastrulation is sometimes by invagination and sometimes by epiboly, the overgrowing cells becoming ectoderm. From the internalized cells, which are endodermal, three main patches of mesoderm arise. A dorsal, longitudinal patch produces the notochord ("urochord"), while paired posterior patches produce the musculature, heart, and some other organs. As the larva develops, a longitudinal ectodermal strip above the notochord cells flattens and rolls up at the lateral edges to close over the top and produce a hollow neural tube, surrounding a dorsal nerve cord that overlies the notochord. The notochord and neural tube elongate posteriorly to form the larval tail, flanked by muscle cells. Meanwhile the gut develops from the archenteron, and the branchial basket differentiates. Paired epicardial cavities pouch from the pharynx, the left lining the body cavity; these are generally reduced in the adult. Ectoderm secretes the tunic, with oral and atrial openings in the head. The tail is postanal. The notochord functions in locomotion; the caudal muscles contract as if to shorten the tail, but that is prevented by the notochord, so instead of shortening, the tail flexes laterally to produce swimming motions. The larval head contains paired adhesive papillae anteriorly, and when the larva settles, they attach to the substrate (fig. 11.22). At this stage the mouth and atrial opening are both dorsal. The tail is resorbed via any of several processes, and a 90° rotation of the mouth and atrial opening ensues, bringing them to what becomes the upper, anterior surface with the atrial opening dorsal to the mouth. All signs of the caudal appendage are now gone.

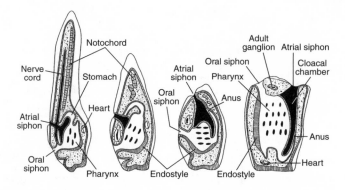

FIG. 11.22 Ascidian tadpole settling and metamorphosis; note the position of the siphons, rotating via differential growth of the body wall so as to open upward in the adult. After Brusca and Brusca 1990.

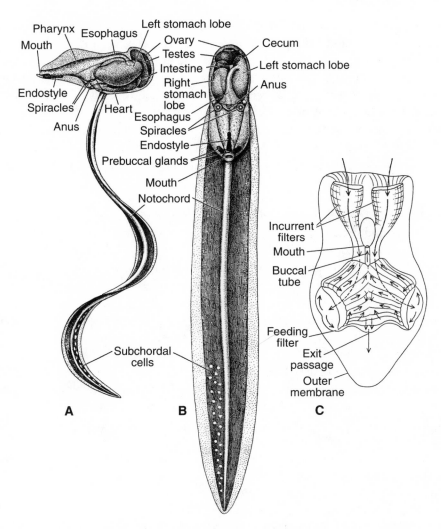

FIG. 11.23 (*A, B*) The appendicularian *Oikopleura*, lateral and ventral views, respectively, of animal removed from its house. (*C*) A house of *Megalocercus*, with position of the animal indicated by the mouth. From Alldredge 1976.

Bodyplan of Appendicularia (Larvacea). If appendicularians are confirmed as basal urochordates, they will have had a much more direct role in chordate evolution than has been understood, and learning more about their biology will become a high priority. Adult appendicularians are generally small (usually <5 mm) and somewhat resemble larval ascidians (fig. 11.23). The digestive tract is U-shaped and dominates the visceral geometry. A gland near the mouth secretes mucus. The mouth opens into a pharynx, which in most species is pierced by ciliated spiracles; the cilia draw water into the oral cavity, which is then expelled through the spiracles and neighboring gill slits. The tail, retained throughout life, is flattened dorsoventrally with a central notochord throughout its length. The nerve cord lies dorsal to the notochord anteriorly but does not extend far along the tail. A mucous structure, the "house" (fig. 11.23C), which lacks celluloselike materials, is secreted around the body by specialized epithelial cells. The house contains filters that catch larger particles; finer material is admitted to the house interior where it is trapped on mucous nets that are ingested periodically. The animal is propelled by caudal waves that force water out an excurrent posterior opening. If the house is damaged or its filters become clogged, it may be jettisoned, a common event, and a new house is rapidly secreted.

Development in Appendicularia. Unfortunately, pelagic forms are difficult to culture and to study and there is relatively little detailed information on appendicularian development. Early observations by Delsman (1910, 1912) have been summarized by Nielsen (1995). Although many details of early embryogenesis are lacking, from what is known it appears to be quite similar to that of ascidians. Gastrulation is by epiboly, and the early notochord, neural tube, and associated structures are similar to ascidians at similar stages. Thus juvenile appendicularians and ascidians resemble one another, but then they diverge toward their adult bodyplans.

Fossil Record of Urochordata. The Lower Cambrian Chengjiang fauna has yielded the earliest well-defined urochordate, *Shankouclava*, which resembles modern aplousobranch ascidians (Chen et al. 2003). Older putative ascidian fossils are from the Upper Cambrian of Nevada (Müller 1977), but as Jefferies (1986) has noted, the evidence for such an assignment is weak. A putative appendicularian fossil from south China (Zhang 1987) is too poorly documented to evaluate. Jaekel (1918) discussed a questionable record of an ascidian from the Permian of Sicily. Possible ascidian spicules are described from the Eocene (Deflandre and Deflandre-Rigaud 1956), and convincing spicules are recorded from the Pleistocene of the Bahamas (Jones 1990). None of these records constrains the origin or radiation of the urochordates. No fossil representatives of the other classes are known.

Urochordate Ancestry. For all intents and purposes the urochordates are not coelomic, nor do their bodies have any clear triregionated architecture that can be homologized with other invertebrate deuterostomes. Furthermore, the well-studied ascidian larva is unique, presenting a particularly knotty problem when considering ancestries. Morphologically, it is chiefly the presence of a notochord and dorsal neural tube in the ascidian larva, and of pharyngotremy in both larva and adult, that have placed urochordates as a deuterostome lineage. As the larva does not feed, yet develops a branchial basket and atrial opening, it seems that adult characters have been accelerated developmentally into the larval stage. Perhaps in coadaptation with such acceleration, the ascidian larva is free for only minutes to hours before settling and rapidly metamorphosing into a feeding form (Cloney 1982). For ascidians the elaboration of the pharynx into a branchial basket suggests an origin in plankton-rich waters, perhaps among shallow rocks or on hardgrounds, sites suggested also by the ascidian habit of cementation to a substrate. It has usually been supposed that the urochordates are descended from coelomic ancestors, in which case it is necessary to account for the lack of a coelom both in the adult and the larva. Coelomic reduction in adult ascidians might be explained as adaptive to their sessile existence as mucociliary feeding baskets. In this case the lack of a larval coelom may be interpreted as another accelerated adult (ascidian) character.

As appendicularians resemble ascidian larvae, it has commonly been assumed that they have descended from ascidians through paedomorphosis. However, the basal position of the appendicularian bodyplan in urochordate phylogeny permitted by SSU rRNA evidence can engender alternative hypotheses. One possibility is that the appendicularians are indeed paedomorphic, but that there is a benthic urochordate paraclade in their ancestry that happens to be extinct but from which the living ascidians branched, somewhat later than did the appendicularians. If appendicularians founded Urochordata, they presumably branched from a stem deuterostome, which is likely to have been a benthic worm. A simple scenario is for a pelagic form to have evolved from this benthic ancestor, with the notochord as an adaptation for swimming and with an expanded branchial region for feeding. If this hypothetical benthic ancestor was coelomate, the coelom has been lost, but if not, there was never a coelomate in urochordate ancestry. Appendicularian and ascidian clades then diverged to produce the adult bodyplans found among living urochordates, the ascidians adopting a sessile life but retaining the larval form of their pelagic ancestor, and the thaliaceans arising from among ascidian (or at least benthic) groups, returning secondarily to a pelagic existence. The early notochord thus becomes primarily an adult character; it is a novel invention but its origin is not made more complicated by whether it was originally larval or adult. If the hypothetical deuterostome ancestor of urochordates was coelomate, the loss of a coelom may have been associated with the very small body size of early pelagic urochordates. Some of the steps in this scenario appear in fig. 11.24.

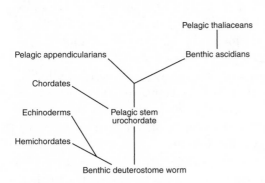

FIG. 11.24 Possible steps in the evolution and diversification of urochordates.

Cephalochordata
Bodyplan of Cephalochordata. The chordate subphylum Cephalochordata contains only two living genera (Barrington 1965; Jefferies 1986). The best-known is *Branchiostoma*, the lancelet (fig. 11.25). The bilateral, laterally compressed, elongate body is

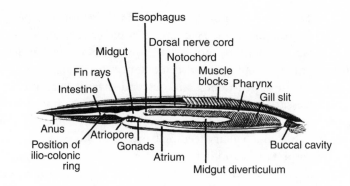

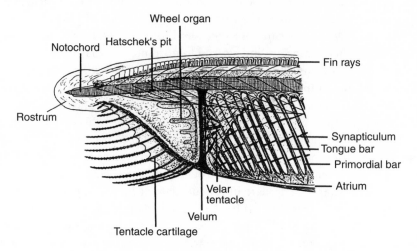

FIG. 11.25 Anatomy of *Branchiostoma*. (*A*) General anatomy from right side. (*B*) Anterior anatomy from left side. From Jefferies 1986, A after Burn, B after Franz.

characterized by a notochord overlaid dorsally by a prominent nerve cord that gives off segmentally arranged roots dorsally and ventrally. The body-wall epithelium is underlaid by a connective tissue layer lined by relatively massive muscle blocks or myotomes that are chevron-shaped (point forward). Anteroventrally the muscles thin, and in their place are found the pharynx, various portions of the digestive tract, and paired and segmented gonads. The pharynx is perforated by gill slits, up to 200 or so, with gill bars that would closely resemble the gill structures of enteropneusts except that they entirely separate the adjoining gill slits rather than stopping short to produce U-shaped slits. Surrounding the pharynx is an atrial cavity that opens via a ventral atriopore about two-thirds of the way toward the posterior. At the base of the pharynx is an endostyle that produces mucus, and at the top is a ciliated epibranchial groove wherein a mucous rope is transported to the gut. The anterior mouth is screened by buccal cirri and velar tentacles, which remove unwanted particles from the feeding stream. Accepted particles are carried to the mouth, a perforation in the velum, by ciliated bands of the so-called wheel organ.

The digestive tract narrows posteriorly to an intestine, and the anus opens short of the posterior end in the ventral midline. Excretion is accomplished by nephridia that supply waste products to a duct opening into the atrium. Although nephridial filtration cells have similarities to protonephridial cells, the nephridial system is connected to the coelom and has been considered to be metanephridial by some workers (see Ruppert 1994).

Much of the interior of cephalochordates is taken up by the atrial space. A coelomic cavity (splanchnic coelom) surrounds the gut and communicates with a narrow ventral cavity below the atrium through extensions in the pharyngeal gill bars. Anteriorly there is another important coelomic cavity (pterygial coelom) below the atrium. Various anterior structures, such as the rostrum, oral cirri, and tentacles, also contain coelomic spaces.

Lancelets have closed circulatory systems but lack a localized heart; blood is pumped chiefly by peristalsis of the major vessels. Paired dorsal vessels overlie the pharynx, joining to run posteriorly as an unpaired aorta; blood flows posteriorly through these dorsal vessels, and the myotomes, notochord, and intestine are supplied by arterial branches. Blood collected from these tissues flows anteriorly in veins to a ventral aorta, and thence through branchial arteries that run through the gill bars, finally returning to the paired dorsal vessels.

Development in Cephalochordata. Cephalochordate development (fig. 11.26) has been carefully reviewed by Jefferies (1986; see also Conklin 1932; Whittaker 1997). Cleavage is radial, and gastrulation is by invagination of a coeloblastula. Mesoderm arises from the archenteron, the notochord from a dorsal strip, and somites, from which myotomes develop, from dorsolateral regions. Each of the first pair of somites to form in this manner becomes crossed by a transverse groove that delimits a future somite. Subsequently, grooves appear serially toward the posterior to delimit a growing series of paired somites; the somites subsequently shift so that the right and left series become offset by one-half somite. The more-anterior somites capture coelomic spaces from the archenteron in the approved enterocoelic manner; more posterior somites are solid when delimited, and their coeloms appear via schizocoely. The coelomic spaces have a variety of fates, some disappearing, others growing (usually asymmetrically) to form the coelomic spaces in adult structures; these spaces are sometimes cut off from each other. Some of the coelomic spaces in the somites become confluent ventrally and join the general splanchnocoel. Anteriorly the archenteron gives off paired diverticula (considered to represent the most anterior somite pair), and these separate from the gut. The right diverticular space enlarges, growing ventrally to become the ventral rostral coelom of the adult. By some accounts (van Wijhe 1914; Franz 1927; Jefferies 1986) another distinctive coelomic space develops from the left second somite, which becomes in part the adult velar coelom. Aside from the fact that they arise from archenteric outpockets, there are no clear developmental homologies of these coelomic spaces with the triregionated coelom of the hemichordates.

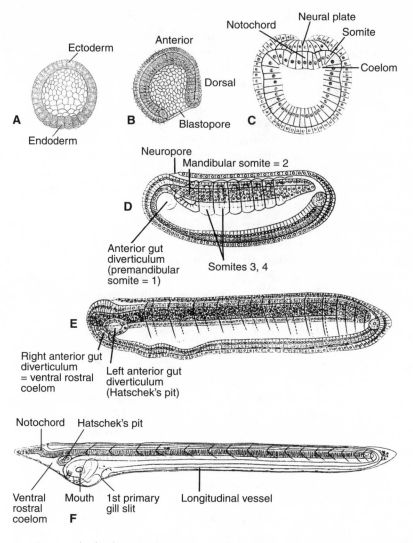

FIG. 11.26 Some early developmental stages in *Branchiostoma*. (A) Coeloblastula. (B) Gastrula. (C) Neural-plate formation, transverse section. (D) Neurula, left side. (E) Late neurula stage, left side. (F) Later larva. From Jefferies 1986, not to scale, after Hatschek.

The early larva is a highly ciliated nonfeeding form that develops quite asymmetrically, becoming essentially bilateral only after metamorphosis. The developmental history of the gill slits is a case in point. The first gill slit arises in the ventral midline of the larva beneath the third somite, and a series of thirteen more slits form behind it. These slits become the left gill slits, and in due course are enclosed by tissue folds from the body wall to form an atrium, still along the ventral midline. Later, at metamorphosis, a second series of eight gill slits forms along the right wall of the pharynx above the primary series. Six of the primary slits close, and the secondary slits elongate transversely, displacing the primary slits to the left; the two series of gill slits are now symmetrically arrayed. Tertiary sets of slits arise after metamorphosis, one set behind each of these two early series. The mouth also arises asymmetrically, on the left side.

Fossil Record of Cephalochordata. A likely Cambrian cephalochordate, not yet thoroughly described, is *Pikaia* (fig. 5.19H) from the Middle Cambrian Burgess Shale (Conway Morris 1979a; Conway Morris, Whittington, Briggs, Hughes, and Bruton 1982). The body outline is lancelet-shaped, and the sides are subdivided by sigmoid features, major V facing forward, that are likely to represent myotomes. A longitudinal bar is interpreted as a notochord, though positioned more dorsally than in modern lancelets. The putative head is not well preserved, but there are indications of oral tentacles and of postoral structures that may mark the sites of gill slits. A fossil from the Permian of South Africa (Oelofsen and Loock 1981) is an undisputed cephalochordate.

Other Possible Invertebrate Chordates

Yunnanozoans are an extinct group of deuterostomes from the Lower Cambrian Chengjiang fauna of Yunnan, China. First described by Hou et al. (1991) as metazoans of unknown affinities, they have been assigned to the Chordata (Chen, Dzik, Edgecombe, Ramsköld, and Zhou 1995; Chen and Li 1997; Chen et al. 1999; Mallatt et al. 2003) and to the Hemichordata (Shu, Conway Morris, and Zhang 1996), and have been considered as possible stem deuterostomes allied to vetulicolians (Shu 2003; Shu et al. 2003). Interpretations of their morphological features are heavily disputed. In fig. 11.27, *Yunnanozoon* is reconstructed as a chordate; it has a series of gills (which may be external) and a dorsal "fin" that appears to be divided into segments by straight sutures. The presence of a notochord now seems unlikely. While literally hundreds of yunnanozoans have been recovered, they are so unusual that their affinities remain uncertain. Perhaps the best bets are that they are either a (nonchordate) branch in chordate ancestry, or stem deuterostomes.

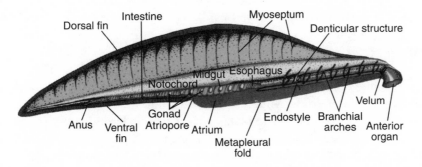

FIG. 11.27 *Yunnanozoon lividum* interpreted as a primitive chordate, Lower Cambrian Chengjiang Formation. Specimens range from about 12 to 40 mm in length. Compare with *Pikaia* in fig. 5.19H. From Chen and Li 1997.

Early Vertebrata

The presence of vertebrates during the Cambrian explosion is documented by fossils from China (see below), indicating that the major synapomorphies of the Vertebrata had evolved by the close of that event. Preeminent among those synapomorphies is the neural crest, a tissue that has been accorded the status of a fourth germ layer (Hall 1998, 1999), which would make the vertebrates tetrablastic. In vertebrate embryos, neural crest cells arise from ectoderm at the neurula stage, from within neural folds. Many cell types derive from neural crest cells, migrating in streams to produce sorts of "endomesenchymal" tissues (see Meier and Packard 1984); they contribute to important organs such as the branchial arches, the brain, and the heart.

Earliest Agnathans

Rather convincing fish fossils have been found in Early Cambrian rocks (Shu, Luo, Conway Morris, Zhang, Hu, Chen, Han, Zhu, and Chen 1999; and see Janvier 1999), the earliest vertebrates known. There are two forms, *Myllokunmingia* and *Haikouichthys,* both from Haikou, Yunnan, China, about 50 km from the classic Chengjiang locality, with which they appear to be roughly correlative (see fig. 5.18). These fish are interpreted as agnathans, related in some way to hagfish and lampreys (fig. 11.28). Both fish have a dorsal fin and probably paired ventrolateral fin folds. The bodies have zigzag myotomes, as do living agnathans. Clear indications of skulls and suggestions of gill skeletons, presumably cartilaginous, provide further evidence of vertebrate affinities, and indicate that the developmental system associated with the neural crest had evolved by this early date.

Euconodonta

The taxon Conodonta was originally based on minute sclerites that are found from the Middle Cambrian to the Triassic, with peaks of diversity in the Ordovician and Devonian (fig. 11.29). They are locally abundant in marine rocks, with over 4,000 described form species, and have proven to be excellent index fossils for correlation. Their biological affinities have long been debated, with suggestions ranging across a wide spectrum of possibilities from algae to vertebrates (see Müller 1981). Conodonts are of fluorapatitic composition (chiefly calcium phosphate), and thus some sort of affinity with chordates has always seemed a possibility. There are two sorts, sometimes ranked as classes: the paraconodonts, known from at least the middle Middle Cambrian to the Middle Ordovician; and the euconodonts (sometimes termed the conodontophorids), known from the middle Late Cambrian to the Late Triassic. Paraconodonts are typically coniform, and, when in sets, the elements tend to resemble each other closely; they have a deep basal cavity. Euconodonts include coniform types but also have elements in a variety of shapes that may occur in sets of up to seven different types. Each element is composed of subunits: a basal body,

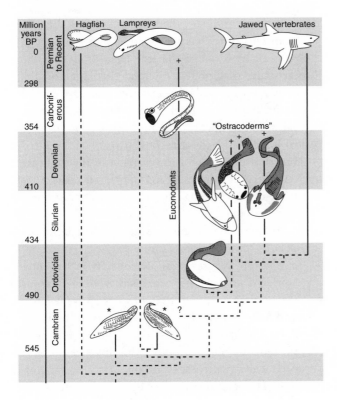

FIG. 11.28 Possible phylogenetic positions of Early Cambrian fish (starred drawings at bottom) and euconodonts, among the vertebrates. From Janvier 1999.

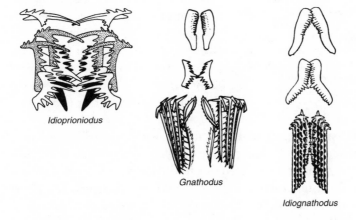

FIG. 11.29 Conodonts. Reconstructions of assemblages of disparate elements that are believed to have formed integrated functional systems. From Rhodes and Austin 1981.

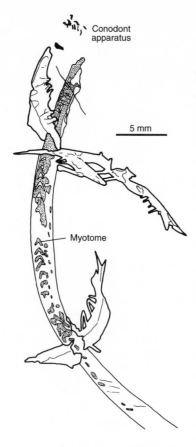

Conodont
apparatus

5 mm

Myotome

FIG. 11.30 Conodont animal from the Carboniferous of Scotland, camera-lucida sketch of a specimen. This locality is known as the shrimp bed, and sure enough there are three fossil shrimp that cross the specimen. Note the position of myotomes and of the conodont apparatus. After Aldridge et al. 1986.

a crown, and so-called white matter. Euconodont histology has been studied by Sansom et al. (1992), who report that the basal body has the structure of globular calcified cartilage and the crown is of lamellar mineralized tissue, while the white matter is clearly a cellular bone. Euconodonts thus furnish the first known occurrence of bone in the fossil record. It has been pointed out that bone might have arisen in invertebrate chordates (Pridmore et al. 1997), but there is no trace of bone in the acraniate cephalochordates, nor even in early fish (see below).

There was a momentary sidetrack in the debate on conodont affinities when sets of euconodont sclerites were discovered in the interior of a soft-bodied organism from the Lower Carboniferous of Montana (Scott 1973; Melton and Scott 1973). For a while it looked as if the euconodont animal had been found, but as it turned out the soft-bodied fossil represented a predator or scavenger of the euconodont animal. The sclerites were residing in the gut (Lindstrom 1974; for a thorough description and evaluation of this fossil see Conway Morris 1990). An actual euconodont animal was finally discovered in an old and neglected collection of fossils from the Lower Carboniferous of Scotland (Briggs et al. 1983). Additional specimens from the Silurian of Wisconsin (Mikulik et al. 1985a, 1985b) from the original Carboniferous locality (Aldridge et al. 1986), and from the Upper Ordovician of South Africa (Gabbott et al. 1995), have now been described. The body of the Scottish euconodont is eellike (fig. 11.30) and was evidently laterally flattened in life, with chevron-shaped myotomes. The head is expanded laterally and contains a bilaterally symmetrical apparatus of conodont sclerites, usually interpreted as a feeding apparatus, with slicing and grinding elements. Indeed, microscopic wear on conodont elements strongly suggests that they were used as teeth (Purnell 1995). Paired, rounded structures at the dorsal margin of the head region are interpreted by Aldridge and Theron (1993) as eye capsules. There is indirect evidence of cranial cartilage (Donoghue et al. 1998). Posterior raylike structures occur in positions occupied by fin rays in many fish (Briggs and Kear 1994). The possible position of a notochord is suggested by two parallel longitudinal lines running much of the body length in the Scottish specimens, and by a stronger horizontal barlike space in the South African specimen; experiments on the preservation of notochords, conducted with decaying lancelets, support this interpretation (Briggs 1992). One noticeable difference between euconodont and craniate morphology is the simple chevron-shaped myotomes of the euconodonts, which resemble cephalochordate myotomes more closely than the zigzag pattern of vertebrates. However, other anatomical evidence is strong that euconodonts are early vertebrates (see Donoghue et al. 1998).

As the presence of cellular bone has been identified only in vertebrates, Sansom et al. (1992) assigned the euconodonts to the Vertebrata unequivocally. A likely position for euconodonts is indicated in fig. 11.28. This assignment has the effect of increasing the known genera of Cambrian and Ordovician vertebrates from about 7 to about 150 (Briggs 1992). Wherever they branched precisely, conodonts would seem to merit recognition as an extinct higher taxon of chordates, perhaps a class in Linnean terms, and students of conodonts are vertebrate paleontologists! Sweet (1988) has argued that conodonts are distinct enough to represent a phylum of their own. Whatever their Linnean ranking, their early radiation and obvious vertebrate affinities invest them with great interest.

Paraconodonta

The paraconodont animal is not known, though it is sometimes suggested that para-conodonts represent an ancestral stock from which the euconodonts descended. The growth lamellae of the paraconodonts resemble the growth pattern of lamellae in the basal unit of euconodonts, and it has been plausibly suggested that the eu-conodont feeding apparatus arose by the addition of a crown to a paraconodont element (see Müller and Nogami 1971; Bengtson 1983). There are intermediates between paraconodonts and euconodonts that are suggestive of this relationship (Szaniawski and Bengtson 1993). Such an innovation might have permitted or at least facilitated the rapid burst of morphological diversification and specialization observed in Ordovician euconodont apparatuses.

Chordate Ancestry

Older Scenarios

The various hypotheses of vertebrate ancestry have been well reviewed by Gee (1996); the debate has been lively and is by no means settled. A famous preevolu-tionary model for the origin of chordate gross anatomy was proposed by the great comparative morphologist Geoffroy Saint-Hilaire (1822). In most nonchordate bi-laterian phyla the chief longitudinal nerve cord is ventral, underlying the gut, while in chordates (both cephalochordates and vertebrates) it is dorsal, overlying the gut. Geoffroy produced a model that rationalized this dorsoventral difference, suggest-ing that the dorsoventral axis had simply been reversed, and that (in modern terms) the vertebrate dorsum and the arthropod ventrum were homologous (he used a lobster as his arthropod example). This notion of a dorsoventral reversal was obscured by phylogenetic hypotheses after evolution became established as a guide to relationships, but was given new life by findings that elements of the developmental systems of arthropods and vertebrates are indeed inverted. A number of authors have concluded that a dorsoventral inversion may well have occurred (e.g., Arendt and Nübler-Jung 1994; and see molecular support in DeRobertis and Sasai 1996 and Cornell and Van Ohlen 2000; though see Jacobs et al. 1998 and Gerhart 2000 for other opinions).

Surely the most elegant hypothesis of the origin of higher chordates is that of Garstang (1894, 1928), who suggested that they arose from ascidian larvae. Garstang postulated that some larvae began to reproduce before metamorphosis, thus deleting the sessile adult stage from the life cycle of that lineage; they then proceeded to evolve into lancelets and fish. In Garstang's scheme an early offshoot of the line leading to hemichordates arose somewhere along the echinoderm branch. In this scenario the ciliated larva of a hemichordate or hemichordate ancestor eventually evolved into a tadpole larva with a notochord, which became the chordate ancestor through paedomorphosis. The adult of this tadpole larva was or became a urochordate. As the ascidian larva is not coelomic, if it was itself the ancestor of the chordates, it implies that the chordate coeloms are not homologous with those of any other phylum.

On another tack, quite a different scheme has been vigorously propounded by Jefferies (1986 and references therein), who proposed that the urochordates, cephalochordates, and vertebrates were each derived separately from different lineages of mitrate carpoids (see Echinodermata above). In this view the ascidian larva is reminiscent of an adult ancestor, but has nothing to do with the direct ancestry of the vertebrates. The carpoids that gave rise to the cephalochordates, urochordates, and vertebrates, in that branching order, were mitrates that are interpreted as somewhat resembling advanced coelomate tadpoles with calcite skeletons. The tadpolelike interpretation of mitrate anatomy is highly contentious; see for example Philip 1979, Bone 1981, Parsley 1999, and David et al. 2000 for a number of pertinent criticisms (early criticisms rejoined by Jefferies [1981, 1986]).

Revised Scenarios

The series of branchings within the deuterostomes that seems most likely on present molecular evidence is depicted in fig. 11.31. This phylogeny essentially precludes a chordate ancestry within echinoderms (or realistically from any of the crown protostome groups that have been suggested in the past), falsifying Jefferies' scheme and the echinoderm part of Garstang's. Attention is thus focused on the last common hemichordate + echinoderm/chordate ancestor, which has descended from the last common protostome/deuterostome ancestor. As both these ancestors are hypothetical, no derived characters are available to separate them. The best that can be done

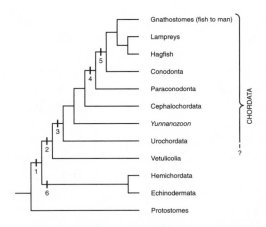

FIG. 11.31 The branching relations among major deuterostome taxa believed to be most likely on present evidence. Some features at the nodes: *1*, hemocoel and/or blood vascular system, distinguishable from the protostome/deuterostome ancestor by the appearance of pharyngotremy; *2*, myotomes; *3*, notochord; *4*, mineralized feeding apparatus; *5*, origin of bone; *6*, tripartite bodyplan, tentacular feeding system.

at present is to compare the radiates and the only known crown bilaterians likely to have preceded these ancestors—acoelomorphs—with the most basally organized of crown deuterostomes, probably the enteropneusts. In light of the molecular evidence in support of dorsoventral inversion in chordates, the origin of their dorsal nerve cord is of major interest.

The nervous system in cnidarians consists of a nerve net, while in ctenophores a similar net is supplemented by cords along the comb rows (chap. 6). In acoels (chap. 6; and see Raikova et al. 2001 and references therein) the nervous system is also a plexus or net, in some forms connected to a number (commonly three to five) of laterally arranged, subequal, longitudinal cords. In enteropneusts the nervous system likewise consists of a nerve plexus, with two cords, ventral and dorsal, which are simply conductors. The findings of Lowe et al. (2003), that the molecules associated with patterning of the central nervous system in chordates are expressed in the diffuse nerve plexus that completely surrounds the body of enteropneusts, provides a basis for a new hypothesis for the evolutionary origins of longitudinal nerve cords. Those authors suggest that the stem ancestor of deuterostomes may have had a diffuse nervous system, which became centralized in a dorsal nerve cord during evolution of basal chordates. Thus the nerve nets found in crown hemichordates and echinoderms have been supplemented by cords during the evolution of those bodyplans. In protostomes, evolution from the diffusely innervated ancestor produced a ventral cord in the major lineages. Nonneurogenic ectoderm arose ventrally in chordates and dorsally in many protostome lineages as central nervous systems evolved, consolidating anteroposterior functional integration in the longitudinal cords.

Therefore, in light of the molecular developmental findings, it is not necessary to postulate dorsoventral inversion in the chordates, but simply the evolution of a primary nerve cord in a dorsal location rather than a ventral one. It is tempting to suggest that the dorsal position of the chordate cord is related to the evolution of swimming, based on the biomechanics of the notochord and its serial muscular system.

Insofar as Garstang's hypothesis of phylogeny within the chordates is concerned, the finding that appendicularians probably branch more basally than ascidians requires a revision of the presumed ancestral chordate. One solution is just to switch the ancestral chordate from an ascidian to an appendicularian larva. In this case there need be little change in the notion of a paedomorphic chordate ancestor. Appendicularians would represent a pelagic descendant of the founding deuterostomes, perhaps sister to the enteropneusts. In an alternative evolutionary model, the adult appendicularian ancestor could have given rise to crown appendicularians and to cephalochordates. This scenario obviates any requirement for a paedomorphic origin of the chordate bodyplan. The history of coelomic cavities in the early deuterostomes is unclear, whichever of the several available scenarios is adopted. The simplest case would be for an enterocoel to have arisen in the early stem deuterostomes, to have been present in the stem hemichordates and stem chordates, and to have been lost in appendicularians before the origin of ascidians.

A draft of the ascidian genome has been described by Dehal et al. (2002). In many cases ascidians contain a single copy of genes that belong to entire gene families in vertebrates, and thus have a simplified version of a basic chordate genome. However, the ascidian genome also has unique attributes that indicate both losses and gains that it does not share with vertebrates. Gains include genes involved in cellulose metabolism, associated with the unique ascidian tunic. Losses include some genes, including some Hox genes, common to protostomes and vertebrates and thus that have been lost somewhere in the lineage leading to crown ascidians. Despite these differences the ascidian genome appears to represent a condition that fits nicely between the last protostome/deuterostome ancestor and the more derived chordates.

PART THREE

Evolution of the Phyla

Theories are excellent servants but very bad masters. THOMAS HENRY HUXLEY

The general outline of the tree of metazoan life seems to be emerging from a combination of traditional and new, chiefly molecular, studies, as reviewed above, but the position of many phyla remains uncertain. Although the imposition of "overarching theories" of evolutionary modes has been criticized here, the model tree that has been adopted as a starting point (fig. 4.17) really depends on the assumption that SSU rRNA data can be used to establish major clades and in some cases to suggest branching patterns within those clades. The information presented in the march through the phyla in part 2 indicates that the major alliances recognized by molecular data are not implausible as clades. However, many features that have been proposed to be informative for determining the relations among phyla have produced topologies that are different, sometimes strikingly different, from the model tree used here.

It has so far proven impossible to trace in the fossil record the pathways of change that led to the bodyplans of any of the phyla; all are cryptogenetic. Probably the ancestors of the phyla did not fossilize readily in large part because they were soft-bodied, and perhaps in many cases small to minute. Although granting that there was an evolutionary buildup of morphological complexity during the Neoproterozoic, the relatively abrupt appearance of new bodyplans just before and during the Cambrian explosion appears on the surface to be real. However, the creation of evolutionary novelty, though slowed, certainly didn't stop with the end of the Cambrian explosion, but has continued into the Phanerozoic, where for durably skeletonized clades at least we have a fairly good fossil record. Therefore it is reasonable to study the origins of those Phanerozoic novelties that occur within clades with good fossil records, to see if their patterns lend any support to the explosive interpretation of the Cambrian appearances, or if they suggest that the Cambrian record is so unique and in some ways biased that the explosion is probably an artifact. As indicated in chapter 12, the Phanerozoic fossil patterns have their explosive aspects also and suggest that the abrupt origin of higher taxa in the Cambrian may be essentially real, given the nature of the record. Chapters 13 and 14 review some current scenarios for the adaptive bases of the origin of Metazoa and the diversification of metazoan bodyplans, in the light of the data presented in previous chapters.

Chapter 13 is concerned chiefly with the prelude to the Cambrian explosion, and chapter 14 with the explosion itself. I have tried to stick as closely to the actual evidence as possible, but there are many features of metazoan history that must be in some sense hypothetical, like the many unknown intermediate forms that are extinct. Those forms certainly existed, but their nature remains uncertain, though clearly open to interpretations that are consistent with the evidence that we do have.

CHAPTER 12

Phanerozoic History of Phyla

To briefly recap what can be said of the relationship between the origins of phyla and the Cambrian explosion, it is certain that the phyla, defined by some key combinations of features that create characteristic stem bodyplans, did not originate precisely when we first find them. The earliest records of phyla must occur some time after they had originated, in rocks representing environments where they were present *and* when conditions were appropriate for their preservation *and* when those rocks are preserved *and* when those rocks are exposed or otherwise available for our exploration. And, I suppose, when we are lucky. So the abruptness and correlation of appearances of many phyla during the Cambrian explosion are due, without a doubt, to the first appearance of appropriately fossiliferous rocks following the origins of the bodyplans.

Attempting to explain the low number of phyla known before the Cambrian explosion is complicated by the fact that the explosion is based partly on mineralized skeletons that became widespread for the first time, appearing independently in many branches of the metazoan tree of life. Some workers have long believed that there was a long hidden history of animals before the Cambrian, when metazoans were simply not very fossilizable, and that many phyla originated long before they appear in the record. Metazoan origins must certainly antedate their earliest records, but whether many of the crown phyla have truly long hidden histories is more uncertain. The bodyplans of most crown phyla can plausibly be interpreted as indicating adaptations to life in benthic environments (part 2 and chap. 13). Sponges are benthic, and anthozoans, the more basal cnidarians, are also benthic. The lobopods that presumably preceded arthropod appendages indicate benthic, not pelagic, locomotion, and arthropods themselves, while quite able to invade pelagic habitats, are most likely to have originated in the benthos. The molluscan foot is a creeping or burrowing organ. The annelid bodyplan, while amenable to pelagic invasion, seems to have evolved as adaptive to burrowing or ambulatory locomotion. Echinoderms certainly appear to be primitively benthic. Many of the paracoelomates are meiofaunal. Thus it is in the benthic fossil record that the basal members of the bulk of the crown phyla should appear. To be sure, if those forms were entirely soft bodied, and especially if they were small, body fossils of those ancestral lineages would be hard to preserve and their absence would be understandable.

If the bodyplans of the crown phyla that appear in the Cambrian explosion were present during the late Neoproterozoic and earliest Lower Cambrian, however, the trace-fossil record should indicate the presence of a diverse benthos of creepers, burrowers, and bioturbators. Traces left by such activities should have been equally fossilizable both before and after the general evolution of durable skeletons. Beginning just before the explosion, trace fossils indicate an increase in the diversity of life modes and in the intensity of faunal-sedimentary interactions that was concurrent and roughly commensurate with increases in the size and diversity of body fossils, both skeletonized and soft-bodied. Before the Cambrian, trace indications of a rich benthos of larger invertebrates are lacking. Trace indications of a fauna of small-bodied creepers and shallow, horizontal burrowers are present in Neoproterozoic rocks, however, suggesting benthic forms in the size range exemplified by paracoelomate bodyplans. Whether any of those forms were stem or crown paracoelomates cannot yet be inferred.

It does not seem likely that miniaturized examples of crown phyla such as arthropods, annelids, and echinoderms abounded on the late Proterozoic seafloor. Rather, the bodyplans of at least some of the phyla that characterize the Cambrian explosion (the stem ancestors) probably arose within, say, several million years of their first fossil appearances. To be sure, their records must be somewhat telescoped to appear so abrupt and nearly contemporaneous, because they follow modest gaps in the known fossil record. We know that the gaps are geologically modest by dint of radiometric dating.

At the level of phyla, the explosion records morphological events but not necessarily cladogenetic ones. It is likely that the branchings that produced some of the clade ancestors of living phyla occurred tens of millions of years before the Cambrian, and may have been completed before the close of the Early Cambrian. However, it is not certain that the stem ancestors of the phyla were all present by then. That is to say, even if the branchings had occurred, the defining morphological elements of all the phyla may not have appeared. Even during the Early Cambrian, some phyla may have been represented by ancestral paracoelomate lineages from which the more complex bodyplans evolved. For example, mollusks were quite small-bodied when they first appeared (chap. 8), and their early skeletonized representatives may have been structurally paracoelomate. Obviously, many of the uncertainties surrounding the timing and extent of the origins of metazoan bodyplans stem from the lack of a good fossil record of soft-bodied forms. If we had abundant Neoproterozoic soft-bodied fossils right across the tree of life, we could reconstruct the prelude and the main events of the Cambrian explosion.

Following the explosion, numbers of clades with mineralized skeleton became quite diverse, and the fossil records of many of them extend throughout the subsequent Phanerozoic Eon, permitting us to trace their histories in some detail over a time span of 520 million years. Thus we may examine the evolutionary patterns that can be documented by the durably skeletonized Phanerozoic clades to see if they can help in reconstructing the more obscure and fragmentary patterns of the Cambrian explosion and its prelude.

Diversification Patterns of Higher Taxa with Mineralized Skeletons Can Be Evaluated by Richnesses and Disparities

There are two aspects of the appearance of higher taxa after the Cambrian that bear directly on evaluating the reality of the explosion and on interpreting the evolutionary modes that are implied. One aspect is the breadth of diversifications within phyla. This breadth is usually expressed by the number of taxa produced at various levels of the Linnean hierarchy, assuming that the taxa are proxies for a hierarchy of morphological disparity. Numerical methods are now being developed to permit the calculation of morphological disparity directly from morphometric data, providing objective information on diversification breadth within clades. The other aspect is the evolutionary rate at which disparity can be achieved, which has usually been expressed as the rate of taxonomic turnover at some (or several) taxonomic level(s). Rates of change in disparity can also be determined by direct morphological measurements when successions of well-preserved fossils are available. This morphometric approach is labor-intensive and has been applied to relatively few clades as yet, but there are nevertheless enough data to evaluate the results of the more subjective taxonomic approaches, which have been applied broadly to many clades, by the more rigorous numerical approaches.

Taxonomic Disparity Commonly Reached High Levels Early in Clade History
A common pattern for higher taxa is that significant taxonomic disparity is often generated very early in the history of a clade, in some cases so early that the fossil record of maximum disparity is close to the first appearance of its members. These early bursts of appearances of novel morphologies were noted even before the acceptance of evolution. However, generalities as to the evolutionary patterns that might be associated with the origins of morphological disparity began to emerge only in the early 1900s (e.g., Jaekel 1902). The first interpretations of these patterns that were firmly rooted in the evolutionary biology of the modern synthesis were by Simpson (1944; see also Rensch 1947). Simpson postulated that the appearance of a new higher taxon denoted the invasion of a new adaptive zone—a region of adaptive space to which access was provided by the novel adaptations that were associated with the novel morphology. The zone may have been empty or simply uninhabited by the invading clade. Once such an interzonal shift (of descendants from ancestors) was accomplished, the opportunity was presented for the evolutionary exploration of related zones and of various subzones and of other, more minor variations within the zones—entrance to the zone provided a breakthrough, so to

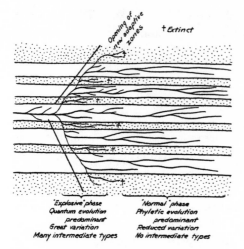

FIG. 12.1 Simpson's figure illustrating "explosive evolution," a rubric for the rapid attainment of morphological disparity, by expansion of a clade into the adaptive zones made accessible by the successful invasion of a novel adaptive zone. Note that forms intermediate between higher taxa are assumed to once have been present, but now are missing because of extinction. Note also that extinct forms are found in the more extreme adaptive (and presumably morphological) outliers. From Simpson 1944.

speak, into an inviting region of adaptive space (fig. 12.1). The process of traversing the interzonal boundaries was termed "quantum evolution." Because the zones were visualized as separated by environmental discontinuities that acted as adaptive barriers, Simpson suggested that the lineages that broached the barriers were poorly adapted during the invasion. Thus, switching zones meant evolving from one sort of adaptive state, in the ancestral zone, through an unstable intermediate state, to a second adaptive state in the new zone. In order successfully to complete such a zonal transition, a lineage might exhibit "preadaptations" to the novel conditions. In the relatively inadaptive situation at the zonal boundary, selection could act upon these fortuitous features to preserve the lineage during adaptation to the new conditions. The more stable adaptive states within the zones were "quanta" between which long-term persistence was not possible. Zonal boundaries could be spatial, temporal, or both.

Crossings of the zonal boundaries were mediated by microevolutionary changes, but occurred rapidly under high selection pressures, so there would be a systematic bias against discovery of the intermediate types as fossils. A new taxon would thus tend to burst on the scene and radiate rapidly into important subtaxa.

Simpson understood the patterns in the fossil record very well indeed, and his suggestion of a quantum evolutionary mode was an attempt to explain the explosive pattern of appearance of evolutionary novelties by invoking what was known of the genetics of evolution. His examples were drawn chiefly from terrestrial vertebrates, and involved characters that were of importance at the familial and ordinal levels, though, as he pointed out, the same patterns could be found among both lower and higher taxa. As indicated in part 2, the phyla that are the most durably skeletonized—the Bryozoa, Brachiopoda, Mollusca, and Echinodermata—certainly do tend to display a burst of diversification at high taxonomic levels soon after they appear in the record. Insofar as we can tell from the Cambrian lagerstätten, at least some of the less easily preservable taxa, particularly arthropods and lobopods, had the same pattern.

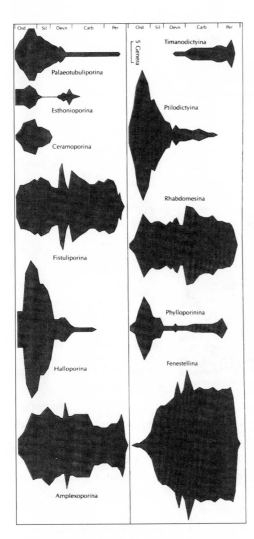

FIG. 12.2 Spindle diagrams showing the diversity of bryozoan suborders through the Paleozoic, measured by the numbers of genera described in each epoch. For time scale see appendix. From Anstey and Pachut 1995.

Bryozoa Radiated at the Ordinal Level while at Low Diversity. The Bryozoa do not appear until the Ordovician, when they are first represented by the durably skeletonized class Stenolaemata. This class is divided into five orders, all of which had appeared by the close of the Early Ordovician (Taylor and Curry 1985), although only thirty-three Early Ordovician genera are known. The other marine class, Gymnolaemata, contains two orders, one of which is nonmineralized, but even this one appears by the Late Ordovician, and it is conceivable that it originated significantly earlier. The final order does not appear until the Mesozoic, after the bryozoans were decimated by Permian and Triassic extinctions. Thus the bryozoans achieved great diversity at high taxonomic levels quite early in their recorded history. Just how strong this trend was, and how far down the taxonomic hierarchy it extended, has been developed in a careful cladistic analysis of Paleozoic stenolaemates by Anstey and Pachut (1995). As some of the usually accepted bryozoan orders may be polyphyletic, these workers focused on the eleven chiefly monophyletic suborders, of which ten appear in the Early Ordovician (fig. 12.2). In fact Timanodictyina, which appears in the Devonian, is suspected of representing a branch of the Ptilodictyina, and if combined therewith, all suborders would have appeared early in the Ordovician. The widths of the spindles in fig. 12.2 represent generic diversity, which is summarized for all suborders in fig. 12.3. Note that the peak generic diversity was attained much later than the first appearances of the suborders. The great morphological disparity implied by the high taxonomic level of the early radiation was clearly not a product of any diffusion of morphology associated with the achievement of high generic diversities.

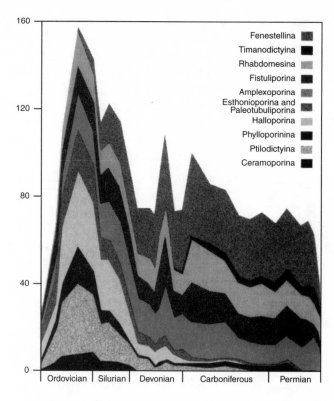

FIG. 12.3 The cumulative diversity of described bryozoan genera through the Paleozoic, partitioned into suborders and binned by epoch. For time scale see appendix. From Anstey and Pachut 1995.

Nine of the suborders radiate significantly in the Lower Ordovician; three of them retain fairly uniform levels of diversity throughout their Paleozoic history, while six are greatly reduced (a couple are eliminated) by mid-Paleozoic extinctions. The remaining suborder increased more gradually in diversity to minor peaks in the mid-Paleozoic and retained a high diversity until the Late Permian. Few Paleozoic genera managed to survive the Permian extinction.

It is quite possible that the Bryozoa originated as a phylum—as a bodyplan—long before we first see their mineralized skeletons in the fossil record. Are we then to postulate that the lack of intermediates between the suborders (which are well-characterized as disparate types; see Anstey and Pachut 1995) is because they are soft-bodied? If so, do we then postulate ten independent origins of bryozoan biomineralization?

Brachiopoda Also Radiated at the Ordinal Level while at Low Diversity. The brachiopods have a pattern of first appearances partly similar to that of the bryozoans at high taxonomic levels (fig. 9.6). Linguliforms and craniforms appear in the Early Cambrian, with five distinctive higher taxa, usually ranked as orders, present by the end of that epoch. Cambrian rhynchonelliforms are represented by three "inarticulate" groups and four "articulate" groups, which are assigned ordinal status.

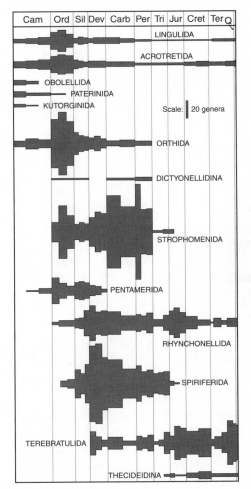

Cam	Ord	Sil	Dev	Carb	Per	Tri	Jur	Cret	Ter Q

LINGULIDA

ACROTRETIDA

OBOLELLIDA

PATERINIDA

KUTORGINIDA

Scale: | 20 genera

ORTHIDA

DICTYONELLIDINA

STROPHOMENIDA

PENTAMERIDA

RHYNCHONELLIDA

SPIRIFERIDA

TEREBRATULIDA

THECIDEIDINA

FIG. 12.4 Spindle diagrams showing the diversity of brachiopod orders through the Phanerozoic, measured by the number of genera described in each epoch. For time scale see appendix. From Williams 1965.

Thus twelve distinctive brachiopod taxa appear during the Cambrian. Diagrams of brachiopod generic diversity through time, partitioned by order, are given in figs. 12.4 and 12.5; these data are based on a somewhat outdated ordinal classification but do not misrepresent the salient features of Phanerozoic brachiopod diversity. It is clear that the brachiopods do display a high taxonomic disparity abruptly when the phylum first appears, and that this occurs during an interval when taxonomic diversity at lower taxonomic levels was quite modest.

The entire marine invertebrate fauna, which enjoyed a modest diversification at lower levels during the Cambrian, underwent perhaps the most important single burst of species diversification in Phanerozoic history during the Ordovician (see below). By that time the brachiopods had already undergone an initial radiation and had already lost some higher taxa. Some new higher brachiopod taxa appeared in the latest Cambrian and Early Ordovician, and they flourished during this Ordovician diversification. Nevertheless, this second diversification of higher taxa was probably unrelated to the general diversification of lower-level taxa insofar as the origin of morphological disparity is concerned. In other words, it need not be inferred that the origins of those higher taxa were driven by high levels of speciation. Three rhynchonelliform orders did arise within the Ordovician diversification, however, and it is certainly possible that their origins were related to ameliorating environmental factors that favored the general enrichment of invertebrate species.

Mollusca Produced Disparate Stem Groups in the Early Cambrian. The mollusks appear in the Early Cambrian and display a variety of disparate forms, representing at least six and possibly nine or more taxa that could be ranked as classes, a disparity

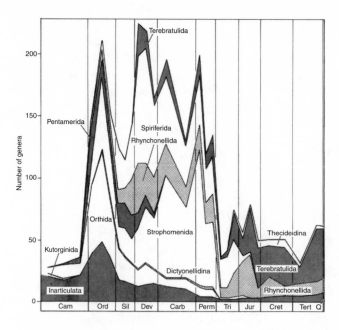

FIG. 12.5 The cumulative diversity of described brachiopod genera through the Phanerozoic, partitioned into orders and binned by epoch. The inarticulate orders are lumped. For time scale see appendix. From Williams 1965.

that is submerged in some schemes of classification (see table 8.2). A question for the mollusks, as for the other durably skeletonized phyla, is whether they radiated from some mineralized ancestor or whether mineralized skeletons evolved independently in at least some of the groups. If the high taxonomic rank accorded these early taxa indicates a corresponding morphological uniqueness, there is certainly an impression that great morphological disparity was acquired quite early in the history of these phyla. An alternative possibility is that the diversification had been under way for a long time among the soft-bodied ancestors of these classes and orders, and that what we are seeing is the sudden acquisition of durable skeletons (a suggestion that has often been made for phyla as a whole). If this is the case, it means that each of these many early-appearing classes or orders evolved a mineralized skeleton in parallel and independently.

Echinodermata Diversified into Disparate Clades from Early Cambrian to Middle Ordovician Time. The argument of parallelism is particularly weak for another durably skeletonized phylum, the Echinodermata, for which the diversity pattern has been reviewed by many workers (e.g., Paul 1977; Sprinkle 1980; Campbell and Marshall 1987). The unique mode of echinoderm skeletonization, with the production of the characteristic stereom skeletal texture throughout the phylum, is difficult to explain as homoplastic (chap. 11). Three nominal echinoderm classes appear first in the Early Cambrian (Atdabanian). Perhaps six more classes appear during the remaining Cambrian. Then seven more appear in the Early Ordovician, and two more in Middle Ordovician time. These appearances are not so temporally concentrated as in the phyla discussed above, but by Middle Ordovician time there are twenty nominal classes of echinoderms recognized by some authorities (table 11.1). And however they are ranked, these taxa differ morphologically in fairly fundamental ways, although they nearly all share the characteristic echinoderm features of endoskeletal plates with stereom texture. Yet the classes do not appear to converge on each other when traced back to their origins—these higher taxa are cryptogenetic, just as are all the metazoan phyla themselves. Smith (1984, 1988a) and Paul and Smith (1984) have analyzed character distributions among these nominal classes and have been able to hypothesize plausible interrelations and pathways of morphological change leading from common ancestors. The point remains, however, that morphological disparity among the earliest classes appears abruptly.

With the echinoderms it is more difficult than with the other phyla to make the argument that forms transitional between classes are missing because they were entirely soft-bodied. If they were, it would required several soft-bodied Early Cambrian echinoderm clades with rather different bodyplans to have given rise independently to more or less distinctive skeletonized groups with plated endoskeletons with stereom structure. So here we have a series of body subplans appearing in the record after mineralized skeletons had evolved within the echinoderms but without fossil intermediates. One explanation proposed for the missing intermediates is that the early echinoderms all had tests of many small plates that were sometimes imbricated and rarely if ever attached to one another. Rather, the plates were born on the wall of a flexible body, so that when the organisms died their skeletons fell into small pieces—certainly a plausible interpretation considering the skeletons that we have. Intermediates may be missing, then, not because they were entirely soft-bodied but because their skeletal elements were scattered and indications of their bodyplans lost, although echinoderm workers are adept at identifying many sorts of isolated plates. However, even generalized echinoderm skeletal "hashes" are not found before the Atdabanian; remains of Tommotian or older echinoderms are simply not known.

Other Durably Skeletonized Phyla Show High Early Disparity, and the Records of Soft-Bodied Forms Are Not Inconsistent with Such a Pattern. Three other phyla—Cnidaria, Arthropoda, and Hemichordata—contain groups with skeletons durable enough to produce fair to good fossil records. For arthropods, a broad range of distinctive stem groups, judged by patterns of appendage types and tagmoses, were present in the Early Cambrian, and the first records of still others are provided by the rich Middle Cambrian lagerstätten of the Burgess Shale and its correlatives. The Phanerozoic history of morphological disparity among arthropods has been the subject of some dispute, but it is generally conceded that the range of Cambrian disparity was at least equal to the range of disparity documented among the living fauna (Briggs et al. 1992; Wills et al. 1994). As the full range of Cambrian disparity is surely not known, it is clear that great disparity was achieved very early in the history of the phylum. As for the Hemichordata, only the pterobranchs and graptolites have somewhat durable parts; these groups first appear in the Middle Cambrian, when they are represented by one and three orders, respectively. Four additional graptolite orders appear in the Ordovician. As pterobranchs are represented by only a single order today, their disparity was clearly greater early in their history, if higher taxonomic diversity is any guide.

The Cnidaria contain a few groups with hard parts—the corals—the chief members of which can be thought of as a subset of anemones with mineralized skeletons. There are a number of little-known corallike fossils from the Cambrian and Ordovician that seem to indicate a mineral skeletonization of a number of distinctive clades of actinarians, and the two well-known and dominant coral orders of the Paleozoic appear in the Ordovician. How well the coral taxa represent the total disparity of anemone architectures, which, it seems, was considerably greater than today's in the Early Paleozoic, is unknown. Here, then, is a case in which multiple independent evolutionary origins of mineral skeletons *does* appear to have occurred within a phylum.

In sum, the taxonomic data of durably skeletonized forms support the notion that, among phyla, early attainment of great morphological disparity is the most common pattern. While we cannot expect the fossil record of soft-bodied phyla to provide an accurate measure of their disparity, the record is not inconsistent with the presumption that those groups have had a similar disparity pattern. When a soft-bodied group happens to have something of an early fossil record, significant disparity is usually found. For example, priapulids and lobopodians are relatively common within soft-bodied lagerstätten of the Early and Middle Cambrian (chap. 7). The Cambrian priapulids are lodged in six genera; commenting on the five that are found in the Burgess Shale, Conway Morris (1977a) notes that each is distinctive enough to warrant assignment to a family of its own. The other Cambrian genus, in the Chengjiang fauna, appears at least as distinctive. There are only three living priapulid families. As for the lobopodians, the known Cambrian forms indicate that they were far more morphologically disparate than the living onychophorans.

Morphometric Disparity Has Been Evaluated Only within Phylum Subclades, Which Sometimes Reach High Levels Early in Clade History

There are as yet no morphometric studies of Phanerozoic disparity for most of the phyla, and none for any of the soft-bodied groups. Yet there are some more limited studies of high quality that permit the generalized account of taxonomic disparity given above to be evaluated by quantitative data.

Within Blastozoa and Crinoidea (Echinodermata) High Morphological Disparity Is Achieved Early. Foote (1992a) has studied the Blastozoa explicitly to determine the patterns of morphological diversity during their early history. He studied species representing over half of the known blastozoan genera, coding them for sixty-

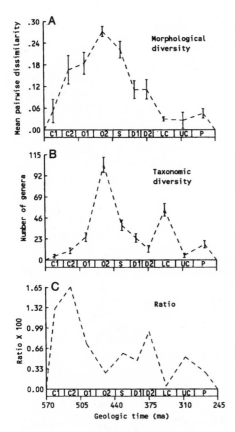

FIG. 12.6 Morphological disparity and taxonomic diversity in Paleozoic blastozoans. (*A*) Mean morphological disparity (dissimilarity) between pairs of species, measured as match/mismatch ratios within a pool of sixty-five characters. (*B*) Taxonomic diversity measured as generic richness. (*C*) The ratio of morphological disparity to taxonomic diversity. In the Cambrian, morphological disparity increased far faster than taxonomic diversity to produce an early burst of disparity; when taxonomic diversity did expand rapidly, in the Ordovician, it was not matched by a proportionate rise in disparity. Note that the estimate of the span of Cambrian time has been significantly reduced since this figure was prepared, so the rises at that time are even steeper than indicated. Error bars indicate one standard error based on bootstrap sampling. C1, C2, time bins of the Cambrian; O1, O2, time bins of the Ordovician; *S*, Silurian; *D*1, *D*2, time bins of the Devonian; *LC*, Lower Carboniferous; *UC*, Upper Carboniferous; *P*, Permian. From Foote 1992a.

five discrete morphological characters, and calculating the average morphological dissimilarity displayed by all pairs of these species in each stratigraphic interval (which ranged from systems to series). The results cannot be taxonomic artifacts, for they are based on the morphology of species. As depicted in fig. 12.6A, morphological diversity peaked during the Middle and Upper Ordovician, when taxonomic diversities also reached their highest levels (fig. 12.6B). However, morphological diversity was quite high in the Cambrian, considering that taxonomic diversity was low; indeed, the ratio of morphological to taxonomic diversity (fig. 12.6C) peaked at well over three times as high in the Cambrian as during the Middle and Upper Ordovician. As Foote points out, this pattern indicates that the morphologies that originated during the initial blastozoan radiations occupied quite a large range in morphospace, even though that morphological range was sparsely occupied. These blastozoan divergences were not driven by simple taxonomic proliferation. Morphologies did expand somewhat further during the subsequent taxonomic diversifications, however, and this later expansion could be interpreted as taxonomically driven, although the additional forms chiefly filled in morphospace, which became more densely occupied.

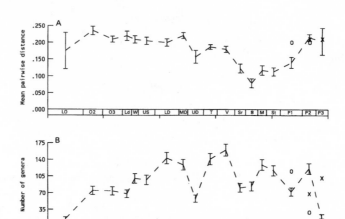

FIG. 12.7 Morphological disparity and taxonomic diversity in Paleozoic crinoids. (A) Mean morphological disparity between pairs of species, measured as a sort of squared Euclidean distance between characters. (B) Taxonomic diversity, measured as generic richness. Error bars indicate one standard error based on bootstrap sampling. Time scale, from Lower Ordovician (LO) to Upper Permian (P3), indicates binning system; see appendix. From Foote 1995.

The morphologies of crinoids have also been subjected to formal quantitative studies (Foote 1994a, 1994b, 1995, 1999), using somewhat similar approaches as in the studies of blastozoans. Cambrian crinoid records are few, so in effect these studies begin with the earliest well-defined indications of crinoid diversity in the Ordovician. As shown in fig. 12.7, the crinoids appeared on the scene with a significant degree of morphological disparity in the Lower Ordovician and reached their maximum disparity in the next youngest interval into which they were binned (essentially the Caradocian). Disparity then remained high through most of the middle Paleozoic but declined in the Late Devonian and Carboniferous. However, crinoid diversity, measured as generic richness, rose from the Lower Ordovician and peaked first in the Lower Devonian, declined sharply in the Upper Devonian, and reached an even higher peak in the late Lower Carboniferous (Viséan). The pattern of disparity parallels the diversity pattern of taxonomic ranks; most higher taxa appear early in crinoid history, and Foote shows that higher taxa tend to occupy distinctive regions of morphospace, and thus disparity arose early. Extinctions of major taxa, such as the diplobathrid camerates in the Viséan, resulted in losses of disparity.

Most post-Paleozoic crinoids belong to the subclass Articulata, which appeared following the end-Permian extinctions that removed most of the Paleozoic taxa. Articulates radiated abruptly early in their histories to produce significant disparity at low species richnesses, thus paralleling in this respect the Paleozoic taxa. However, the post-Paleozoic crinoids do not achieve nearly the morphological disparity of their Paleozoic predecessors.

Thus, in groups of echinoderms that have been studied morphometrically, morphological disparity is achieved early in their fossil records, a result that is consistent with their taxonomic histories and with the inferred history of disparity within the phylum as a whole. However, when the classes of the subphylum Blastozoa, or the orders of the class Crinoidea, are examined individually, they do not all display the same pattern. For example, the blastozoan class Blastoidea and the crinoid orders Camerata and Cladida + Flexibilia show only gradual increases in morphological disparity rather than initial jumps. Thus, although subtaxa may occupy distinctive regions of morphospace and contribute significantly to the disparity of the taxa that contain them, they do not necessarily display a similar pattern internally; subclades of morphologically disparate clades may be morphospatially compact.

Among Mollusca, Gastropoda Shows Larger Early Morphological Disparities while Rostroconcha Shows a Complicated Pattern. Appropriate quantitative studies are available for two class-level molluscan groups, gastropods and rostroconchs. Wagner (1995b) studied disparity among the best-defined Paleozoic gastropod group ("Archeogastropoda") by analysis of the transitions found between morphologically defined sister species pairs. Differences between species were quantified by measuring morphological landmarks on fossils, which were then subject to a principal components analysis, and the resulting eigenvectors treated as transition axes. Euclidean distances were also calculated between species pairs. Both measures of the morphological distance traversed between species pairs proved to be highly congruent (fig. 12.8), and they indicate that, during the time span investigated, the greatest morphological separation between species occurred in the earliest time intervals, while low levels of disparity characterized transitions during the later intervals.

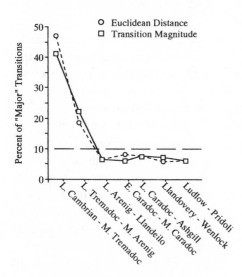

FIG. 12.8 Trends in the evolution of major morphological disparities (the top 10%) in Paleozoic (Late Cambrian through Silurian) gastropods, as measured by (*circles*) Euclidean distance and by (*squares*) transition analysis. For time units see appendix. From Wagner 1995b.

*An interesting feature of the gastropod shell characters that Wagner used is that they can be partitioned into two

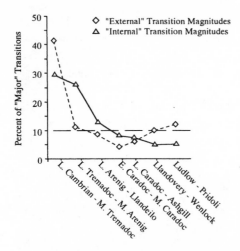

FIG. 12.9 Trends in the evolution of major morphological disparities (the top 10%) in Paleozoic (Late Cambrian through Silurian) gastropods, partitioned into (*diamonds*) features inferred to be associated with relations between the skeletal morphology and the environment, and (*triangles*) those inferred to be associated with soft-part anatomical changes. For time units see appendix. From Wagner 1995b.

major sorts, those that are associated with direct environmental interactions, which Wagner termed "external" characters, and those that are associated with the internal anatomy of the organisms, or "internal" characters. In marine gastropods, there has been extensive convergence, so that snails with quite different internal anatomies share similar ranges of (convergent) shell morphologies. Fig. 12.9 shows the temporal trend in the rate of accumulation of disparity, represented as transition magnitudes, between shell features reflecting external factors and those reflecting internal factors. Early in gastropod history the rate of changes in internal characters lagged the rate of external changes, suggesting that diversification was weighted toward exploitation of ecological opportunities. Later, rates of internal changes equal or exceed external ones, suggesting that the limits to the external shell characters of gastropods had been approached, and that the locus of chief evolutionary change had shifted to internal characters, for which open fitness pathways were more easily discovered. Such a history of change in internal characters produced homoplasies in external ones.

The only other large molluscan taxon that has been studied with respect to disparity is Rostroconcha (Wagner 1997), which does not show the pattern of larger early disparities and modest later ones that is displayed by "archeogastropods." The two rostroconch subclades, ribeirioids (a paraclade) and the derived conocardioids, have different patterns of morphological evolution (fig. 12.10). Ribeirioids had a gradual rise in disparity throughout their history, while disparity decreased through most of conocardioid history. Ribeirioids become extinct about a third of the way through rostroconch history, so the class as a whole does display a marked change from rather high early disparity to rather low later disparity. This is merely due to the survival of the subclade with the lower disparity, however, rather than representing an evolutionary trend within the clade. The differential survival of the more conservative subclade may not be just a random occurrence, but may have macroevolutionary significance (see below).

Trilobita Are Most Disparate Well after Their First Appearance. Trilobites show a complex disparity pattern (Foote 1991, 1992b). In the Cambrian, total disparity

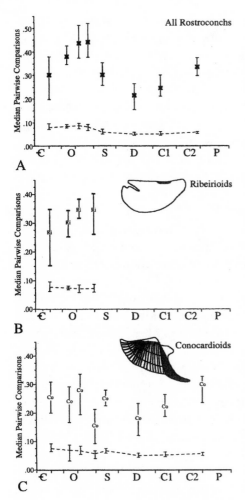

A

B

C

FIG. 12.10 Trends in the evolution of morphological disparity in the Rostroconchia. (A) All rostroconchs; although the pattern is complicated, they are characterized by high early disparities and lower later ones. (B) Ribeirioids; note rising disparity through time. (C) Conocardioids. Overall, disparities are flat through time. Note the effect of the extinction of the disparate ribeirioids on rostroconch disparity as a whole. Error bars are derived from bootstrapping. X, all rostroconchs; Ri, ribeirioids alone; Co, conocardioids alone. From Wagner 1997.

rose from the Lower to the Middle and Upper Cambrian interval as species richness increased, but the disparity per species was higher in the Lower Cambrian. Morphospace was not much expanded by the added species richness in the Middle and Upper Cambrian. Species richness then fell, but disparity rose to a high in the Middle and Late Ordovician, and appears to have fluctuated thereafter. Thus a modest burst of disparity was associated with the first appearance of trilobites, but this was surpassed in their later history. However, morphological disparity was not correlated with species richness. The time interval over which the early disparity was attained is uncertain; it may have been as little as 10 million years.

Within Phyla, Disparity and Diversity Seem to Be Independent

Most of the phyla for which adequate data are available achieved significant diversity at higher taxonomic levels very early in their known histories. This pattern has been taken to imply that morphological diversity was achieved early as well. If the onset of their known histories is not too far displaced from the origin of their bodyplans, this pattern implies that the breadth of morphological disparity that lies at the core of the distinctive bodyplans of phyla can be achieved relatively rapidly, geologically and evolutionarily speaking. Many classes, orders, and suborders appear during the Phanerozoic, within clades that had evolved mineralized skeletons and that had established fossil records before those appearances. Many of these clades are nevertheless cryptogenetic, and great taxonomic disparity among and within their subclades is commonly, though not exclusively, achieved early in their histories.

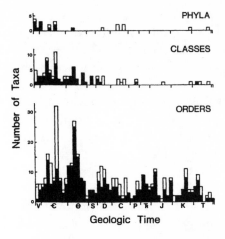

FIG. 12.11 First appearances of phyla, classes, and orders in the marine fossil record; vertebrates are included. The impression is that evolution is operating "from the top down." Since this figure was prepared, redefinition of the Neoproterozoic-Cambrian boundary has moved most of the first appearances of the taxa shown here as Vendian into the Lower Cambrian, but the general pattern remains valid. Scale of geologic time is from Vendian (about 565 Ma) to the Recent; see appendix for time divisions. From Erwin et al. 1987.

Patterns recognized by morphometric studies of disparity agree well with patterns of disparity inferred from the Linnean taxonomy of the same groups—which is not surprising, as Linnean taxonomy is based on associations of morphologically similar forms, in theory within a hierarchy of increasing morphological disparity. So far as can be told, increases in disparity need not be driven by increases in species richness.

This history of the achievement of great disparity early in the history of taxa produces a characteristic pattern of appearance of taxa of different ranks, given that there is some correlation between the earliest appearances of phyla (Valentine 1969, 1995; Erwin et al. 1987). For phyla the time of peak first fossil appearances is the Early Cambrian, when eleven living phyla appear; by the close of the Cambrian, seventeen living phyla are known (fig. 12.11). For classes fossil appearances are somewhat more dispersed, with the times of most first appearances shifting slightly later, but still, by the close of the Cambrian, 58% of the well-defined marine classes with readily fossilizable representatives had appeared. Ordinal first fossil appearances are still more dispersed and shift again toward the present; for example, four times as many invertebrate orders appear first in the Early Paleozoic than in the Mesozoic; the major ordinal modes are in the Cambrian and Ordovician. Inspection of numerous compilations reveals that the times of peak fossil appearances of families are shifted even farther toward the present (see Valentine 1969). Thus the record makes it appear that phyla evolved first; then classes evolved, as major architectural themes based on the phylum bodyplans; then orders evolved, as modifications of class bodyplans; and so on. This record runs counter to what might be expected during the origin of phyla, which would be the divergences of two lineages from common ancestors, at first at the species level only. Then as time passed their differences would become more pronounced, the two lineages becoming as distinctive as are average genera, and then as are average families, then as orders, and so forth. In fact such stages of divergence must have occurred during the history of Metazoa, but we don't see them. We don't even see those stages in the history of marine invertebrate classes (or commonly in the history of orders) that have entirely Phanerozoic records. Yet there must have been a bottom-up but hidden buildup of morphological differences and often of complexity in major taxa that preceded the first appearances of phyla in the Cambrian record. Much of this bottom-up evolution, however, presumably occurred within lineages that had not yet achieved the bodyplans characteristic of the phyla that appeared in Cambrian time. The fossil evidence does suggest that, when those bodyplans were finally assembled, body subplans of significant disparities commonly evolved rapidly.

Macroevolutionary Dynamics of Phyla Run the Gamut from Stability to Volatility

After the phyla appear in the record and undergo their early diversifications, their histories are quite varied. Some phyla increased more or less steadily in diversity right up until today; others have remained at quite low diversities but have nevertheless persisted throughout the Phanerozoic; still others have displayed great volatility in diversity, undergoing rapid diversifications and extinctions. In studying the various dynamics of evolution within and among phyla and other major clades, one is concerned with principles and processes that are roughly analogous with those in population biology, but with taxa rather than individual organisms as the basic units. Much attention has been given to population parameters, and therefore there is a large body of concepts and techniques available for application to the study of clade dynamics. (Strictly speaking, the dynamics are realized in hierarchies within the clades, into which the cladistic tree is transformed—see chap. 1.) Populations and taxa are not necessarily analogous, however, so it is prudent to examine applications that are important to early metazoan evolution to ensure that they are appropriate.

Clade Dynamics Reflect Speciation and Extinction Rates

Speciation produces "individual" species that form "populations" of species in clades. The behavior of species numbers—richness—in clades can be modeled in ways similar to those used for population growth (fig. 12.12). A rise in species numbers within a clade requires an excess of speciation over extinction to produce a realized rate of species diversity increase, and when speciation equals extinction, there will be an equilibrium diversity (see Stanley 1979). Fig. 12.12A shows the rise of species diversity in a clade invading a new adaptive zone when the rate of extinction is diversity-dependent. In the real world, adaptive zones are not homogeneous in factors affecting species numbers, being subdivided into environments that represent potential adaptive subzones for subclades. Furthermore, the world is divided into geographic regions that are somewhat isolated, but are available for colonization, and in which ensuing diversification might exceed that in the original "home" zone. Thus for a given clade originating in a given region, expansion may be quite episodic, and diversification may accelerate after an adaptive breakthrough

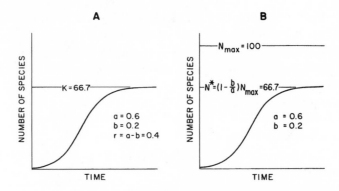

FIG. 12.12 Two models of species diversity increase after invasion of an "empty" adaptive zone. a is the speciation rate, b the extinction rate, r the intrinsic speciation rate minus the extinction rate, N_{max} the number of adaptive opportunities—"niches"—available in the zone, N^* the equilibrium diversity. (A) Extinction is diversity-dependent, and an equilibrium level K is regulated by the diversity itself. (B) Extinction is diversity-independent, and adaptive opportunities replete with resources ("empty niches") are present at equilibrium, but are not exploited because of the intrinsic extinction rate. From Walker and Valentine 1984.

into previously unoccupied subzones, or following the colonization of a distant region by an unusual immigration event, if in each case there is opportunity for the production of additional species in adaptive space. The expansion of a clade may be slowed or ended by a falling speciation rate or a rising extinction rate.

Ecological factors such as living space or trophic resource availability, which can be used up by species, are termed diversity-dependent factors, while those that affect species but are not related to species richness, such as temperature, are termed diversity-independent factors (Valentine 1973b). A plausible hypothesis is that the speciation rate is commonly diversity-dependent, for as adaptive space is taken up by the species in an expanding clade, opportunities for speciation should decline as resources accessible to the clade are used up.

The hypothesis that extinction is also diversity-dependent has been entertained by a large number of workers (e.g., MacArthur 1969; Rosenzweig 1975; Sepkoski 1979). Density-dependent mortality plays a major role in population theory, and density-dependent *local* extinction has an important role in island-biogeographic theory (MacArthur and Wilson 1967). MacArthur (1969), extrapolating to "real" (global) extinction from local extinctions on islands, proposed that such extinctions were density-dependent also. The data for "real" extinctions are found in the fossil record, however, and there is little support, if any, for diversity-dependent extinction there. Most of the data are recorded at higher taxonomic levels than species (chiefly generic and familial). Nevertheless it is clear that high extinction rates within clades do not tend to correlate with their peak diversities; indeed there is a tendency among some clades for high extinction rates to occur early in clade history, when it is unusual for high standing diversities to occur. Furthermore, most clades exhibit strikingly consistent per-taxon extinction rates regardless of major fluctuations in their diversities (Gilinsky and Bambach 1987). There are distinctive differences in per-taxon extinction rates between major clades (Simpson 1944, 1953; Stanley 1979; Van Valen 1985; Valentine 1990), but these differences do not correlate with the relative clade diversities.

Such evidence as exists, then, suggests that extinction is usually independent of diversity, and achieving a dynamic equilibrium diversity, when speciation equals extinction, must be regulated chiefly by speciation (see Gilinsky and Bambach 1987). In this case, so long as there is any extinction, there will be "open" adaptive space (fig. 12.12B), for the inherent speciation rate of a clade can never provide for an instantaneous filling of "empty niches." As the fossil record indicates that extinctions are geologically incessant, extinction is keeping some portion of adaptive space open, in the sense that it is available for occupation by additional species. As the resources represented as open adaptive space have chiefly been released by extinction, there should not be any unusual adaptive barriers to their reexploitation. Exceptions could occur if the capacity of a region to support species is reduced by the factors associated with the extinction, or if all the species that have adaptive access to the resources have become extinct. In general, the proportion of adaptive space that is being kept open can be roughly estimated by the extinction rate divided by the speciation rate (Walker and Valentine 1984).

Macroevolutionary "Competition" Arises from Differential Speciation and Extinction Rates

The taxonomic richness of some phyla has increased and of others has decreased over Phanerozoic time. The question has been posed as to whether some phyla have grown in richness at the expense of others. Questions about the outcome of competition between two species' populations that vie for common resources in ecological time have generated a large literature, while the outcome when two clades occupy the same range of adaptive space in evolutionary time has not been extensively investigated. When two species' populations compete for the same resource in a homogeneous experimental environment, one is usually eliminated (see Pianka 1976 and discussions of ecological competition such as in Ricklefs 1990). The outcomes of experimental competitions vary, with the odds usually favoring the survival of one of the populations but with an element of uncertainty in the outcome. When the environment is heterogeneous, however, both populations may survive. The population that is usually eliminated in homogeneous environments finds refugia within the heterogeneous environment; in natural situations such complexities abound. Experiments are not possible with clades, and observations based on the fossil record are inconclusive.

In theory, a simple stochastic model of two clades inhabiting a region to which they are equally well adapted ecologically results in the clade with the lower equilibrium diversity having the greatest chance of being eliminated (Walker 1984), other things (such as initial diversities) being equal. This is because lower equilibrium diversity implies a lower ratio of extinction to speciation. Therefore, the clade that is more at risk of extinction has either a higher extinction rate, a lower speciation rate, or both, than the clade at less risk. If a clade has low species diversity, however, its chances of being eliminated by a temporary wave of extinction will obviously increase even if it has a low ratio of extinction to speciation.

This simple model becomes very complicated when an adaptive region is treated more realistically, subdivided into subregions among which the speciation rate of clade members varies, and which are separated by adaptive barriers of different values so that their accessibility varies. The outcomes are not only probabilistic, but the probabilities depend partly on such factors as the regime of environmental fluctuation and the diversity (and diversity changes) of each clade in subregions that could contribute invaders. Despite the myriad complications that can be imagined, some clades nevertheless dwindle to extinction relative to others that are stable or increasing. Evidently "winning" is possible because, although extinction is diversity-independent, the degree of resistance or vulnerability is set intrinsically, and clades vary significantly in that respect.

Linnean Taxa May Form Macroevolutionarily Dynamic Units

An evolutionary unit must reproduce and must differ from other units in heritable properties that have differential success in their propagation within the collective that forms the evolving entity. Individuals as evolutionary units, within populations as evolving entities, have these properties. Species also can qualify as evolutionary units within clades as evolving entities. To consider taxa higher than species as units in macroevolution requires that they also possess the appropriate properties, but it has been strongly argued that they do not do so (Wiley 1981; Eldredge and Salthe 1984; Vrba and Eldredge 1984; Levinton 1988). It is true that so long as they are embedded in a tree, or represent units in a classification that has the properties of a tree, taxa do not have dynamic qualities (chap. 1). However, if clades and paraclades are arrayed within a hierarchy, such as the Linnean hierarchy, they may exhibit dynamic behaviors. Higher Linnean taxa may form spatiotemporally delimited "individuals" in the same sense that species do (Ghiselin 1974, 1997; Eldredge 1989; for holophyletic taxa only). Thus transformation of the tree of life to a hierarchy permits the study of macroevolutionary dynamics, although this procedure raises important questions. Is there enough objectivity in the definition of higher taxa, and is there enough similarity among taxa assigned to the same rank, that study of their macroevolutionary patterns is rewarding? Is fresh information on the evolution of life to be gained from such studies?

An objection that has been raised against the use of higher taxa as evolutionary units is that they are composed of lower taxa—species—that can evolve independently of one another (Wiley 1980, 1981). However, interdependence is not among the properties required of evolutionary units, which may be aggregative as well as constitutive. In short, there does not seem to be anything special about higher Linnean taxa, as compared to species, that would prevent them from being macroevolutionary units, provided that they have heritable features than can undergo sorting. It should be just as possible to discuss the macroevolution of a class in terms of its genera as of its species. Evolutionary branching events are associated with apomorphies in daughter species, and these may become autapomorphies among any clade founded by the daughter. The autapomorphies need not be only morphological but may involve physiological, behavioral, developmental, or other features that may or may not be associated with obvious morphological characters, so long as they are heritable.

Sorting Strategies Arise from Selection among Individuals

That sorting does in fact occur among higher taxa is suggested by the history of their diversities. Phanerozoic biotas display more or less continuous compositional changes, not only at low taxonomic levels where turnover rates are relatively rapid over the era, but also among higher taxa whose ranges span nearly or quite all of Phanerozoic time. The waxing and waning of family diversities among major marine taxa are depicted by the spindle diagrams in fig. 12.13, where a number of distinctive patterns can be seen. For example, trilobites reached their peak family diversity in the Cambrian and faded to extinction in the Permian, while bivalves have increased almost steadily in family diversity and seem to be at their peak today. The variety of spindle shapes suggests that they might be the result of random processes. This possibility was investigated by Raup et al. (1973), who modeled clade diversity behaviors resulting from random patterns of diversification and extinction. When the size of the clades (the number of units—lower taxa—of which they were composed) was small, the random clades rivaled the real clades in the richness of distinctive shapes. However, model clades of moderate to large sizes similar to the large natural clades in fig. 12.13 are much less volatile in response to random processes, displaying far fewer and far less abrupt changes than the real clades (Stanley et al. 1981). Such considerations of scale indicate that randomness is unlikely to be a general description of the pattern of Phanerozoic clade diversities.

It has nevertheless proven to be difficult to specify with any confidence those features on which clade sorting occurs, even though each clade is clearly different from all others and thus possesses unique features, or unique combinations of features, on which sorting is imaginable. The literature of paleontology is replete with hypotheses as to why one clade failed while another prospered. There may be more suggestions as to why dinosaurs became extinct than there are types of dinosaurs known to science. There are also abundant speculations on the "key" adaptations that have led to the success of many groups. In a well-known example, Stanley (1968) proposed that the evolution of siphons among the Bivalvia has been chiefly responsible for the successful deployment of that group in infaunal benthic habitats, leading to the great diversity of the class at present. Such scenarios are

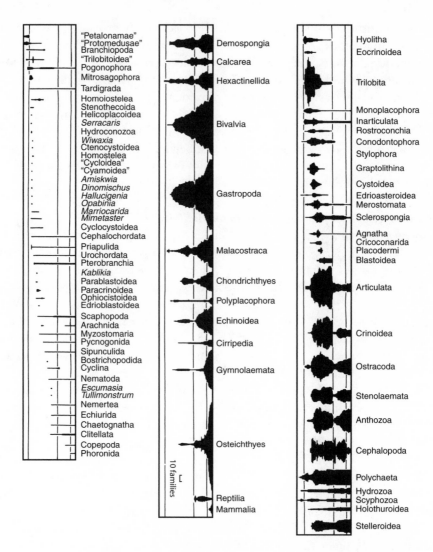

FIG. 12.13 Sepkoski's spindle diagrams (1981) of marine phyla and classes that reflect their family diversities as known from the fossil record, arranged according to their similarities in diversity history. A few assignments are now considered to be unlikely (such as early records of Pogonophora), but the patterns of the major clades are holding up well. Diversities are binned in stages, averaging about 7 million years. Within diagrams, left column is Vendian, next is Paleozoic, next is Mesozoic, and right column is Cenozoic. Arranged by Bambach 1985.

exceedingly difficult to refute even if they happen to be incorrect, as many must be. As higher taxa differ in numerous morphological features and even in basic architectures, picking and choosing among these differences to explain differential success and failure has the earmarks of special pleading and is sometimes criticized as an improper approach (e.g., Forey 1982). A feature of adaptive scenarios of this sort is that by their nature they apply to a particular clade. Clade sorting does occur, however, and is in fact a major feature of life's history, and it is part of the job description of a scientist to find explanations for such processes. It is the limited application and the difficulty in testing of such hypotheses, and not the inappropriateness of proposing them, which are responsible for the many competing adaptive scenarios.

It might be concluded from the foregoing that as each taxon is different, each adaptation is ad hoc, and the history of life may be seen as a series of capricious successes and failures for which there is no overriding rhyme or reason. However, the ad hoc nature of individual adaptations does not preclude the existence of sorting principles, or of general effects that may be shared among particular adaptations. In ecology such general features are sometimes called adaptive strategies (Levins 1968), so I shall call them sorting strategies. It seems reasonable to search for sorting strategies that result in similarities in the evolutionary behavior of clades. If common strategies are found, the underlying adaptations may be sought. Among the behaviors that can be examined in the fossil record are clade-specific regimes of speciation and extinction, of rates of origination and extinction of higher taxa, and of standing diversity levels and trends among taxonomic ranks.

The dynamic features of clades cannot originate as products of natural selection per se; rather, they must be effects of selection for other properties, in the sense of Williams (1966). The propensity of a clade to speciate rapidly might derive, for example, from restrictions on gene flow that grow out of life-history properties, but those properties did not evolve under selection to enhance speciation, but because they contribute to the fitness of the individual organisms that display them. For a hypothetical example, offspring that are concentrated in a given area might be so much more successful than those that migrate that genotypes that favor dispersal are eliminated via selection. But any ensuing increase in speciation rate growing from the increased genetic isolation of local populations is simply a side effect of the adaptive changes. Extinction rates, similarly, may be influenced by the results of selection but are not themselves selected. Williams has put it as follows: "Selection has nothing to do with what is necessary or unnecessary, or what is adequate or inadequate, for continued survival. It deals only with an immediate better-vs.-worse within a system of alternative, and therefore competing, entities. It will act to maximize the mean reproductive performance regardless of the effect on long-term population survival. It is not a mechanism that can anticipate possible extinction and take steps to avoid it" (1966, 31). Neither is sorting a mechanism capable of anticipating a need for speciation and providing it.

Effects or no, when speciation and extinction rates are clade-specific, they and other measures and consequences of clade dynamics can provide a basis for macroevolutionary sorting whenever they are relatively advantageous or disadvantageous. As noted by Eldredge and Gould (1972) and reiterated by numerous workers, sorting on the properties of species can override the selective values that may be important among individuals. In the same vein, sorting on the effects at one taxonomic rank may produce effects at a higher rank, on which an overriding sorting may occur in turn. For example, orders with many families inhabiting a variety of habitats and employing a number of distinctive family-level adaptations can sort differently than orders with many families inhabiting a narrow range of habitats but over broader geographic regions, or than orders with few families, and so on. An extinction so massive that many orders lose all their members might sort on such features, overriding the effects of sorting within lower ranks as well as of natural selection within populations. This phenomenon is hardly surprising, being simply an example of the constraints that higher levels impose upon lower levels within hierarchies.

Clade Histories of Invertebrate Taxa with Mineralized Skeletons Reflect Turnover Dynamics

There Are Hints That the Origin of Marine Clades Is Favored in Shallow Tropical Waters

As many Phanerozoic clades have mineralized skeletons and are likely to begin producing a fossil record once they become at all common, it is possible to examine their first occurrences for clues as to whether there are any regions, climates, or ecological settings in which they originate preferentially. For the Early Paleozoic, it has been documented by a number of workers (Sepkoski and Sheehan 1983; Jablonski et al. 1983; Sepkoski and Miller 1985; Conway Morris 1989) that fossil communities or community complexes appear earliest nearshore, and then spread offshore across the shelf. A similar pattern of onshore origination and later offshore spread is documented among post-Paleozoic faunas as well (see Jablonski et al. 1983; Bottjer and Jablonski 1988). Since the earlier faunas tend to contain the first records of many major clades, their clade pattern would seem to be similar, but ecological aspects of their first occurrences have not been investigated in detail. However, the patterns of origin of higher taxa within some major clades have been tallied for the post-Paleozoic. Crinoid, echinoid, and bryozoan orders have been studied, and all show preferential nearshore originations (Jablonski and Bottjer 1988, 1990; Jablonski et al. 1997). Those post-Paleozoic patterns are not inconsistent with those of the Neoproterozoic and Early Cambrian fossil record, though paleoenvironmental reconstructions are less secure for those times.

Most Paleozoic records are for tropical latitudes, which creates a strong bias in studying climatic aspects of clade originations during that era. The post-Paleozoic record is rich in temperate as well as tropical faunas, however, and Jablonski (1993) studied the origination of marine orders within climate zones for that interval. He found that more orders appear first in tropical waters. Whether this bias is due to the larger number of taxa in the tropics or represents a per-taxon differential between tropical and extratropical environments is not certain. In any event, if a latitudinal diversity gradient was present during post-Paleozoic time, higher taxa would have tended to originate preferentially in shallow tropical settings, which therefore would have been the major source of evolutionary novelty in the marine realm. There is very little climate signal from Neoproterozoic and Early Cambrian faunas, but most of them do seem to have been deposited chiefly in relatively low latitudes (chap. 5).

Turnover Rates Influence Clade "Shape" over Time

Many of the differences in the clade histories that are indicated by the spindle diagrams of fig. 12.13 must be associated with differences in origination and extinction rates. It can be seen that many clades have somewhat similar histories of family diversity. The changing dominance of clades through time is also clearly indicated in the figure. For example, trilobites form the dominant class in Cambrian faunas, although a few other groups, such as the inarticulate brachiopods, also make a showing there. (One important invertebrate group in the Cambrian, the Archaeocyatha, was not included in the figure.) In the post-Cambrian Paleozoic, the invertebrate benthos was dominated by crinoids and by the articulate brachiopods; bryozoans, ostracods, and corals also did well then. And in the Mesozoic and Cenozoic, the gastropod and bivalve mollusks, which had been expanding fairly steadily through time, finally became dominant. A factor-analytic evaluation of the faunal associations found through Phanerozoic time has revealed three major faunas (Sepkoski 1981). Fauna I is the Cambrian fauna; fauna II is the post-Cambrian Paleozoic fauna, which was terminated by the major extinction at the close of the Permian; and fauna III is the Mesozoic-Cenozoic ("Modern") fauna, still represented in the living marine biosphere (fig. 12.14).

There are clear differences in turnover dynamics among higher taxa. A special point of those differences was made by Simpson (1944, 1953), who noted that (among others) mammals and bivalves had very different turnover rates, mammals being significantly faster. A large number of origination- and extinction-rate studies have been made since, especially of marine genera and families within phyla, classes, and orders (e.g., Stanley 1979; Sepkoski 1979; Raup and Marshall 1980; Ward and Signor 1983; Walker 1984; Van Valen 1985; Van Valen and Maiorana 1985; McKinney 1987; Gilinsky and Bambach 1987; Raup and Boyajian 1988; Foote 1988; Holman 1989; Valentine, Tiffney, and Sepkoski 1991). In table 12.1, results from some extinction-rate studies are presented for three taxonomic levels within the five classes that dominate the Phanerozoic marine faunas. Table 12.2 gives the extinction rates for each of three taxonomic levels within these five classes, per taxon per million years, during the periods dominated by each of the major Phanerozoic faunas. The results of studies of other taxa and ranks not included in the table are consistent with the trends that are shown. This consistency holds even though the fossil record is different for each taxon and each rank, and data sets and methodologies of rate estimations vary somewhat from worker to

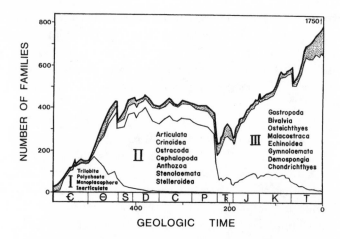

FIG. 12.14 The classic three Phanerozoic faunas identified by Sepkoski through a factor analysis of the geologic ranges of marine families. *I,* the "Cambrian fauna"; *II,* the "Paleozoic fauna"; *III,* the "Modern fauna." Note that elements of the "Paleozoic" and "Modern" fauna were present from the earliest Cambrian, but came to dominance in subsequent periods. The sharp drops in diversity represent mass-extinction episodes, particularly noticeable in the Cambrian fauna near the end of the Ordovician, and in the Paleozoic fauna at the close of the Permian. For names of periods see appendix. From Sepkoski 1981.

worker. The most commonly used marine data sets from which such rates have been calculated are those of Sepkoski (1982 and supplements for families; 2002 for genera).

The rates in tables 12.1 and 12.2 are simply the number of taxa that have gone extinct in a given time period (i.e., that are unknown thereafter), per number of taxa present, divided by the length of the time period in millions of years. Thus the taxa are binned within the times of interest, and represent the literal historical record (as preserved, collected, and interpreted by taxonomists) for those times. Determining extinction rates from the number of taxa that disappear from subsequent bins can produce a special sort of artifact, which is among artifacts that Raup (1979) christened the pull of the Recent. As one approaches the Recent, an increasingly high percentage of taxa in the bins are still alive. As the living fauna is better known than the fossil fauna for a given bin, it was presumed that taxa that would otherwise disappear from the record and be counted as extinct are (correctly) tallied as being present. This lowers the calculated extinction rate as one approaches the present. One remedy for this problem is to cull the living taxa from the sample. Evaluations of a very large database of generic extinction rates with and without culling show essentially no difference (Raup and Boyajian 1988), however, and the trend can be seen in extinct groups as well as living. Furthermore, it is not true that the

TABLE 12.1 Average extinction rates of orders, families, and genera of some classes of Phanerozoic marine invertebrates, per taxon per million years. Data for families and genera from Valentine, Tiffney, and Sepkoski 1991.

	Articulate				
	Trilobites	Crinoids	Brachiopods	Gastropods	Bivalves
Order	0.002	0.001	0.001	<0.001	<0.001
Family	0.019	0.013	0.010	0.003	0.003
Genus	0.079	0.043	0.052	0.015	0.016

TABLE 12.2 Average extinction rates of orders, families, and genera of some classes of Phanerozoic marine invertebrates, per taxon per million years, binned by fauna (I, II, or III). Asterisk indicates data uncertain (because of taxonomic uncertainty) or not available; dash indicates class is not certainly known to be present. Data for families and genera from Valentine, Tiffney, and Sepkoski 1991.

	Fauna I	Fauna II	Fauna III
Orders			
Trilobites	0.017	0.004	—
Crinoids	*	0.003	0.002
Articulate brachiopods	0.000	0.002	0.002
Gastropods	0.000	0.000	<0.001
Bivalves	*	<0.001	<0.001
Families			
Trilobites	0.028	0.016	—
Crinoids	*	0.021	0.008
Articulate brachiopods	0.017	0.016	0.009
Gastropods	*	0.004	0.003
Bivalves	*	0.005	0.002
Genera			
Trilobites	0.110	0.044	—
Crinoids	*	0.051	0.024
Articulate brachiopods	*	0.049	0.006
Gastropods	*	0.021	0.014
Bivalves	*	0.021	0.015

recent fauna extends the ranges of a large percentage of genera that would otherwise be missing from the record; only about 5 percent of ranges are so extended (Jablonski, Roy, Valentine, Price, and Anderson, unpublished data). Finally, the taxa in tables 12.1 and 12.2 retain their extinction-rate distinctiveness at all levels. Trilobite genera, families, and orders each have higher extinction rates than corresponding levels among the crinoids and so forth, and the relative magnitudes of the rate changes from level to level is comparable among all the taxa. This has been the general case found in rate studies. Such a pattern suggests, incidentally, that there is some consistency in the recognition of taxonomic levels. As rates slow at higher levels, distinctions between fast and slow clades could disappear or could be greatly increased, as an artifact, if a lower level of one clade were compared with a higher level of another clade. But as the differences in extinction rates are clear, and maintain a roughly similar relationship from genera to orders, it appears that any such artifacts are not causing the differences. Within the classes that have dominated Phanerozoic marine ecosystems, the taxonomic levels appear to be operationally similar for dynamic studies, though of course they are not identical. Congruence between extinction-rate trends at different taxonomic levels suggests that it may be possible to use the fossil records of taxa such as families or genera, which are more complete, as a proxy from which to estimate the relative extinction dynamics of clades at the species level, where the record is much poorer.

Within each class in table 12.1, the extinction rate tends to be lower for each higher taxonomic level. A falling dynamic with rising level is required of aggregative hierarchies, or more correctly, the dynamics of a higher level can be equal to or slower than those of a contained lower level, but never faster. If extinctions occurred within higher-level taxa that have few subtaxa, leaving larger taxa with their lower-level subtaxa unscathed, it is possible for extinction rates to appear higher at upper taxonomic levels. However, the taxa listed in table 12.1 have the expected trends of extinction from level to level. The extinction rates of orders are commonly lower than those of families by an order of magnitude, and extinction rates of families are lower than those of genera by factors of around four or five. In addition to a general lowering of extinction rate with increasing taxonomic rank, there is also a lowering of extinction rates from clades that are dominant earlier in Phanerozoic time to those that are dominant later. Trilobites have the highest extinction rates, articulate brachiopods and crinoids have intermediate rates, and gastropods and bivalves have the lowest. The Early Cambrian archaeocyathids, which are not figured, may have had the highest extinction rate of all.

The same trend of highest extinction in fauna I dominants and lowest extinction in fauna III dominants is evident in table 12.2. Furthermore, within these dominant groups, there is a general trend for extinction rates to slow over time; for example, crinoid and brachiopod families and genera have intermediate extinction rates in fauna II, where they dominate, and lower extinction rates in fauna III, when gastropods and bivalves dominate. Thus one might expect that there has been an overall decrease in Phanerozoic marine extinction rates. Fig. 12.15 shows the extinction rates of all marine families through Phanerozoic time; among the class of "background" extinctions, which accounts for something over 95% of marine species extinctions, rates have definitely slowed. As Sepkoski pointed out, the decreases in extinction intensity are at least partly owing to the replacement of higher-turnover clades (high extinction and speciation rates) by lower-turnover clades through time. Note that bivalves and gastropods, the dominant fauna III clades, are present in fauna I, but at a low diversity. These groups increase steadily in diversity through the duration of fauna II and come to dominance in fauna III; even during the intense Permian-Triassic extinctions their diversities are affected only in a minor way. This is evidently macroevolutionary sorting on extinction rate; those groups with lower rates tend to become dominant as those with higher rates wane or even disappear. There is no a priori reason, though, that

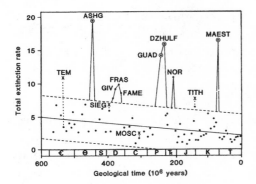

FIG. 12.15 Phanerozoic extinction rates of marine families. Five episodes of unusually intense extinction have been recognized as "mass-extinction" events; the solid line is a regression on the remaining "background extinction" intensities, with a 95% confidence interval indicated by dashed lines. *TEM* is Templetonian, early Middle Cambrian in Australia; *SIE* is Siegenian, middle Early Devonian in Germany; the other time periods can be found in the appendix. Modified from Raup and Sepkoski 1982.

extinction must always govern clade history; speciation could do so as well, if speciation rates were rapid enough to more than offset extinctions. Among terrestrial plants, there is some evidence that major taxa with high extinction rates (and even higher speciation rates—very high turnover clades) have succeeded slower-turnover taxa during plant history (Valentine, Tiffney, and Sepkoski 1991). Among marine invertebrate clades, however, extinction resistance seems to have been the clue to long-term success.

The histories of the diversities of some of the clades with distinctive spindle shapes in fig. 12.13 can be interpreted with relative turnover rates in mind. The shapes of clades with low fossilization potentials are likely to be affected by taphonomic artifacts and are difficult to interpret. The shapes of spindles among the durably skeletonized forms, however, suggest a number of generalities. Steadily increasing spindle widths such as those of the gastropods and bivalves indicate sorting strategies in which their speciation rates have nearly always exceeded extinction rates when averaged over the time intervals (about 7 million years) in which they are binned. The measured tempo of their increasing diversity suggests low levels of extinction and speciation, as are in fact documented in tables 12.1 and 12.2, and only a very slight advantage of speciation over extinction.

Some spindles, such as those of the cephalopods and the Paleozoic crinoids, display great volatility with generally high levels of diversity; this pattern indicates sorting strategies with both high extinction and high speciation, for if either rate falters, diversity levels are quickly affected (geologically speaking). Indeed, both the Paleozoic crinoids and the ammonoids among the cephalopods (largely responsible for their volatility) had high levels of both extinction and speciation and finally became extinct when speciation failed to respond adequately during episodes of high extinction. It is plausible that these clades lasted as long as they did because they were able to attain high diversities rapidly under favorable conditions.

A few spindles, such as those of the inarticulate brachiopods, the scaphopods, and the polyplacophorans, show low diversities throughout their histories but have managed to endure. One would predict that these low-diversity groups have very low extinction and speciation, and that their low-turnover sorting strategies are responsible for their long-term successes—they seem relatively immune to waves of extinction, but do not diversify rapidly during times of general clade expansions either. Certainly, inarticulate brachiopods are famous for their low turnover rates; scaphopod and chiton turnovers seem never to have been studied.

A number of the spindles display different characteristics during different time periods, and usually these reflect the replacement of one major subtaxon by another. For example, the crinoids of fauna III are not so volatile as are their predecessors in fauna II, which belonged to orders that became extinct. Similarly, the corals (Anthozoa) of fauna III are less volatile than the corals of fauna II, and they represent entirely different orders. Again, the less volatile clades replace the more volatile ones, as Sepkoski noted. As discussed earlier (chap. 5), the fossil record is incomplete and biased, but the manner in which these turnover statistics hang together in a coherent fashion suggests they represent real signals of macroevolutionary dynamics.

The Early History of Phyla Is Consistent with the Evolutionary Patterns Shown Following the Cambrian

There is no obvious reason that the events and behaviors that can be inferred among the various classes and orders during Phanerozoic time do not reflect evolutionary processes and principles that are applicable to phyla as well, although there are certainly no guarantees in making such extrapolations. A most obvious evolutionary characteristic of classes and orders is that they are nearly all cryptogenetic, just as are phyla, and they tend to appear early in the history of their phyla or classes, respectively. The abrupt appearance of disparate body types without a record of intermediate forms, even after these taxa had acquired durable skeletons, suggests that significant morphological change may evolve very rapidly, at least when it represents variations (major modifications in some cases) on a basic architectural theme. The implication is that the abrupt appearance of many phyla during the Tommotion-Atdabanian interval could mirror an actual rapid radiation of body-plans, just before or during that time period.

Among the clades that make their appearance during the explosion are some that have slow turnover dynamics, such as the inarticulate brachiopods, and some that have rapid dynamics, such as the trilobites and the archaeocyathids. The clades with potentially rapid speciation potentials were, naturally enough, the ones that quickly came to dominate the benthic associations, at least the skeletonized fraction of them; the slower clades persevered, however, some until the present day. As high extinction rates correlate with high speciation potentials, the high-turnover clades have very volatile diversities. Archaeocyathids were soon extinct; trilobites continued fairly successfully through the Lower Paleozoic but then were hit by major extinctions, and thereafter acted as a "slow" clade until their extinction at the close of the Paleozoic. The ability of "fast" clades to radiate explosively, both during the Early Cambrian and, especially at lower taxonomic levels, during the Ordovician, suggests that there were empty or at least underutilized adaptive zones present or potentiated, opened to exploitation by the novel morphologies (and perhaps physiologies) of these clades. A broad generalization, with many exceptions, would be that at lower taxonomic levels the Early Cambrian radiation involved chiefly benthic detritus feeders, suspension feeders utilizing bacteria, and their predators, while the Ordovician radiation was particularly enriched by many clades of suspension feeders that chiefly ate protistans and, probably, larvae (see Signor and Vermeij 1994). The early diversity patterns do not suggest a biosphere replete with previous occupants that were preempting much of the ecospace available to the radiating clades. The diversification history is consistent with the notion of an important episode of bodyplan origination in the late Neoproterozoic and Early Cambrian.

The correlation of taxonomic turnover dynamics with the Phanerozoic history of clade diversities suggests that key morphological adaptations, such as are commonly held to underpin the success of phyla or of their major subtaxa, have a less important place in the overall history of life than they are sometimes accorded. It seems true enough that the innovations that mark morphologically novel clades are chiefly responsible for the ability of those clades successfully to invade new adaptive zones (and depending on how one structures the theory, it may be said that the new adaptations create the zones). The association of partitioned coelomic skeletons with parapodia, the evolution of jointed appendages, the encasement of a lophophore within a bivalved shell—such features are clearly related to the assumption of ways of life that would not otherwise be possible, and at least in the first two cases provided a basis for significant expansions of primitive ecospaces. So the novel attributes in new bodyplans and their modifications are probably involved in breakthroughs into new adaptive zones and subzones, and become characteristic features—synapomorphies—of higher taxa. As such features are naturally widespread among members of the successful group, they are commonly called key adaptations and their prevalence is sometimes taken as evidence of their efficacy in promoting diversity and longevity to a clade. But such a conclusion may be a case of Whig history. It seems more probable that, although putative key features may account for access, and therefore for early diversification in novel ecospace, they do not generally account for the ultimate success of clades. It can be argued that the success or failure of clades is usually due to their dynamic properties.

It might also be argued that the turnover dynamics are themselves related to bodyplans, which confer differential extinction resistance. This notion is very difficult to demonstrate, however. Take the trilobites; it is plausible that their dominance in benthic fossil communities of the Early Paleozoic is owing to high speciation rates, despite their high extinction levels, and that their endurance at low species diversities in the Middle to Late Paleozoic is due to the presence or emergence of a few lineages with low extinction levels, albeit also with low speciation rates. Evidently a significant range of turnover regimes, characterizing different clade behaviors, was possible in clades with trilobite bodyplans. Although trilobite clades with either extreme of fast or slow turnover regimes had their morphological apomorphies, there is nothing to suggest that their turnover characteristics were morphologically based. Indeed, most taxa display a significant range of turnover dynamics within their subtaxa, with "normal" extinction rates varying up to an order of magnitude (Jablonski 1995), although the modal turnovers are distinctive, as expressed by the extinction rates in tables 12.1 and 12.2. Within taxa, extinction frequencies are skewed, with short-lived taxa being the more common (Jablonski 1995).

There have been many attempts to discover the factors that make some species or clades more vulnerable than others to extinction (see Jablonski 1986b, 1994, 1995) and some attempts to understand why one species or clade would be more likely to speciate than others (see Stanley 1979, 1990). Generally, factors that are usually identified as promoting species longevity (during times of low "background" extinction) include tolerance to variations in habitat parameters (Jackson 1974), high dispersal capabilities, and wide species ranges. For clades, longevity may be correlated with position in the trophic pyramid (the lower the better), species richness, and breadth of adaptive zones. For the most part, these factors are due to biological aspects of species or clades that are at least partly independent of bodyplans. Jablonski (1986a) has shown that factors that tend to protect species or clades from extinction at background extinction intensities appear to be different from those that are effective at high extinction intensities. For example, species richness of clades or the breadth of species ranges seem to have no effect on extinction intensities during mass extinctions, but genera with wider geographic ranges survive some mass extinctions better (Jablonski 1986b). Thus clades that prosper at low extinction intensities can be relatively vulnerable during a severe extinction event, even though they have lasted for hundreds of millions of years. The properties that confer extinction resistance during mass and background times are not antithetical, however, so that a clade may have both sets of properties and be relatively well off under either extinction regime—Gastropoda and Bivalvia, for example, seem to do relatively well in either case. At any rate, the extinction resistances evidenced by the more successful clades appear to be related to biological factors such as species diversity, geographic and ecological range, fecundity, and larval dispersal capability, rather than to their bodyplans per se.

Is the Number of Phyla Related to the Gross Heterogeneity of the Marine Environment?

Controls on the diversity of species are subjects of continuing debate, and similar questions may be asked of bodyplans. To put the question in taxonomic terms, why are there only 35 (or so) phyla? Why not 17, 70, or 140? Just before and during the

Cambrian explosion, phylum-level bodyplans were appearing in the fossil record at a geologically rapid clip, possibly at an average rate of one per million years, yet arguably there have been no phyla (at best only a few) founded after those times. One suggestion has been that the lack of new phyla after the Cambrian is entirely an artifact of the tree of life; that as the main branches have themselves branched, the many features that characterized the main branches are naturally inherited by the new branches, which we therefore simply define as classes, or as some subsidiary taxa, rather than as new phyla. There is certainly an argument to be made as to how distinctive bodyplans must be to qualify as phyla. The phyla were not recognized because they had all evolved at an early date (which wasn't known), however, but because of their morphological differences—the judgments were not made with reference to the tree of life, but with reference to bioarchitectural disparities. An incidental point about phyla is that in most cases their bodyplans have not evolved one from another but from last common ancestors with different bodyplans. When it is discovered that a group accorded phylum status has in fact evolved as a branch of an existing phylum, as in the case of the pentastomids, that group is commonly assigned a subsidiary status with the parent phylum. Other nominal phyla may have to be broken up (such as the brachiopods and flatworms) or be subsumed within other phyla (such as the echiuroids within the Annelida) if some of the molecular evidence holds and if one wishes to avoid paraphyly of living phyla.

Other explanations for the number of phyla can be divided into three sorts (see for example Erwin 1994, 1999; Valentine 1995; Foote 1999). Two hypotheses call upon internal features: constraints that are imposed by developmental processes and constraints that are imposed by structural limitations. A third relies on external features, essentially on ecological factors. The main developmental-constraint hypotheses are based chiefly on the "top down" pattern of appearance of bodyplans, suggesting that developmental flexibility within the genomes of phyla may have been progressively constrained (Erwin 1999). Developmental pathways were visualized as being canalized to provide the basic architecture of the group, with major variations on the developmental theme, such as gave rise to classes, becoming constrained in turn (Valentine 1986; Valentine and Erwin 1987; Erwin 1999). The early arguments for developmental constraints on bodyplan origins were made before the basic architecture of gene regulation was understood, and now seem less likely to serve as a general explanation for the bias toward an early origin of nearly all bodyplans. The last common protostome/deuterostome ancestor had a wide array of regulatory genes, seemingly all of the important ones involved in body patterning.

The hypothesis of structural constraint implies either that all of the major design modifications possible on a given bodyplan have been explored, or at least that surviving designs do not permit further important modifications. These propositions are difficult to assess. Numerous constraints limit the functional range of any given morphology, but are there constraints that prevent the evolution of novel forms through intermediate steps? Some clades have produced unusual design variants that lack even the defining morphological features of their phyla—such as carnivorous sponges (chap. 6).

The ecological hypotheses are generally couched in terms of the availability of adaptive space. One sort of model of adaptive space requires the presence of broad, unoccupied adaptive zones to permit new bodyplans to become established; somewhat narrower zones permit new subplans; still narrower zones permit new variants of these; and so on (Valentine 1980). Computer simulations based on such a model can produce a hierarchy of taxa similar to hierarchies found in nature (Valentine and Walker 1986). An implication of these models is that, when some region of adaptive space is already occupied, novelties produced to reinvade that space will usually be less well adapted than are the inhabitants, and will lose out in any sorting. Thus the Cambrian radiations may have produced enough metazoan types to occupy the broad range of available marine environments, thereby foreclosing the production of additional bodyplans.

It seems that ecological constraints should play the more important role in the diversification of lower taxa, while developmental constraints should play the more important role in the diversification of novel morphologies and therefore of higher taxa. To the extent that morphological change is driven by speciation, ecology might be the more important constraining factor. When morphology changes in response to major adaptive opportunities, though, constraints might be more likely to arise from development. The pattern of early evolution of novelties, at low diversities, suggests that developmental radiations were not simply driven by massive taxonomic diversification. However, the Phanerozoic pattern of decreasing novelty through time may not be related directly to developmental constraints but may simply grow from ecological propinquity. If a major adaptive zone has been occupied but there is a potential for further expansion of metazoans into some subregion of that zone where they have not yet penetrated, the invaders are more likely to be drawn from among the occupants of the major adaptive zone than from some entirely separate adaptive pool. The lineages most likely to have evolutionary access to an uninhabited adaptive region are simply those that have established themselves "nearby"; new bodyplans are no longer required to exploit untapped resources (Erwin et al. 1987). If bodyplans are evolutionary solutions to life in distinctive adaptive zones, the diversity of bodyplans should be dictated by the structure of the environment, though the precise morphological solutions may certainly be constrained by developmental factors. How heterogeneous are the conditions for life? How strong are the barriers within the environmental mosaic? It seems likely that the pattern of environmental domains, as perceived by evolving lineages, determined how many bodyplans were enough. In this formulation it was not active evolutionary constraints that impeded the creation of novel bodyplans so much as it was evolutionary indifference—enough was enough.

The Late Neoproterozoic and Early Cambrian Pattern of Appearances Is Consistent with Patterns Found throughout the Phanerozoic

The abrupt appearance in the fossil record of major marine clades is the rule during Phanerozoic time, and it is common for these clades to be represented by numbers of morphologically disparate subclades at the time of their appearance or shortly thereafter, geologically speaking. As most major clades appear relatively early in metazoan history, this pattern of abrupt appearances is mostly recorded in early Phanerozoic rocks. Evaluating the Cambrian appearances of phyla is complicated by the fact that their ancestral lineages were soft-bodied. When a phylum has become skeletonized, however, it would seem that the course of its subsequent diversification into body subplans could be recorded by fossils. In fact such fossils are usually not known; the lineages linking invertebrate classes, and most orders, have generally not been identified as fossils. Thus even those lineages whose individuals are presumably of average preservability fail to produce fossil evidence of the origins of major morphological novelties. The pattern of the appearance of phyla during and after the Cambrian explosion is consistent with this general Phanerozoic pattern. With due regard for both the smearing and the truncating effects arising from processes of accumulation and preservation of fossils, the explosion can be interpreted as reflecting an actual interval of rapid morphological evolution without special pleading.

The numbers of distinctive animal architectures that have evolved are significantly higher than the numbers that have survived. A search for the factors that seem to have promoted diversity and longevity among the survivors during the Phanerozoic suggests that sorting on clade dynamics was a major contributor. An obvious factor in the success of a clade is that it maintains at least as high an origination rate as its extinction rate. In practice this seems to have meant that clades with low volatilities have been more successful, other things being equal. Perhaps this is because the history of Earth's environment contains many episodes of more or less rapid change, ranging from mild to quite severe in intensity. These changes produce extinctions that sweep through the biota. Low-volatility clades are best able to weather these episodes and to maintain fairly high diversities. Clades that survive but that have been reduced to low diversities are more liable to be extinguished by the next unfavorable change if it occurs before they can recover. However, clades of extremely low volatility—that have very low origination rates but are extremely resistant to extinction—may survive indefinitely even at low diversities.

Early in their history, clades with novel morphologies are not very rich in taxa, but as the fossil record tells us, they commonly produce numbers of quite disparate branches, which may be attributed to their having broken through into a "new" region or zone of adaptive space. The disparate branches are themselves presumably adapted to unique environmental subzones, to which their novel morphologies and/or physiologies have afforded access, and are of low diversities early in their histories. This seems to be a fair description of the faunal situation of Early and Middle Cambrian time. As ripples to waves of extinction occurred, interplays between the inherent dynamics of the clades and the patterns and intensities of the extinctions produced a variety of winners and losers. With many distinctive clades of low diversity present, the extinction rate of what would be considered higher taxa on the basis of their morphological disparities, perhaps up to and including phyla, was much higher in the Cambrian than later in the Phanerozoic. As the original definitions of phyla are based on distinctive crown taxa, it is not really very useful at this point to argue about the ranks of those early, distinctive, extinct clades. It is extremely useful to know that they existed, however, that the patterns of abrupt origination of novel bodyplans and subplans are likely to have been real and pervasive, and that their fates were perfectly ordinary as clades go.

CHAPTER 13

Metazoan Evolution during the Prelude to the Cambrian Explosion

The multiplicity of topologies that have been suggested as representing the phylogeny of metazoans indicates just how little predictive power there is in the distribution of morphological and developmental characteristics among the phyla. Given almost any particular metazoan tree of life, it is possible to imagine evolutionary events that would produce it. By the same token, it is difficult to falsify any given topology, although it's usually easy to imagine evolutionary events that would not produce it. Because there are clearly many homoplasies among the phyla no matter what topology is assumed, any hypothesized evolutionary event almost always seems implausible from the standpoint of some character or other. It can be hard to choose between alternative implausibilities.

One of the advantages of molecular phylogenetic data is that they permit one to choose a topology without regard to the morphology. Because molecular-based topologies produced with different exemplars, different mixes of taxa, or different algorithms commonly differ among themselves, however, they must be evaluated further by some other criteria, and all the data that are available are those contradictory data from morphology and development, and from the spotty fossil record. One combs these data for congruences. This brings to mind a crack of Richard Levins's about truth being the overlap of independent lies.

The hypothesis of metazoan phylogeny that is being used here (fig. 4.17) relies heavily on molecular data but has been modified by evidence from the more traditional sources, permitting the retention of some alliances in topologies that are not indicated with much confidence by molecules. Although the resulting tree implies a series of evolutionary events that seem quite plausible from the standpoint of some important characters, at least to my mind, it is inevitably contradicted by other characters. In this chapter and chapter 14, the evolutionary histories of major morphological and developmental features reviewed in part 2 are reconstructed partly from evidence supplied by the tree, and some of the contradictions are also evaluated in the light of what is known of the metazoan genome and the fossil record.

Metazoan Multicellularity Evolved from Protistan Pluricellularity

Protistans Set the Stage

Many features critical to metazoan evolution clearly evolved in premetazoan lineages. Unicellular eukaryotes must have inherited many of their features from still more basal types of cell organization, and they certainly also "invented" many of the features that were inherited in turn by metazoans. The eukaryotic cytoskeleton, composed of molecules such as tubulin and actin that are unknown in prokaryote clades, is an example of a feature involved with the evolution of protistans. Possible prokaryotic antecedents to those molecules have been identified structurally, but the sequences are quite dissimilar (see Doolittle 1995). The origins of the cytoskeleton may lie in the development of an underpinning of the membranes of earlier cells by a skeletal cage. This event may have been associated with the rise of food ingestion via endocytosis. Early cytoskeletal molecules must also have been associated with the origin of eukaryotic chromosomes and their separation during mitosis (Cavalier-Smith 1987; Doolittle 1995). The ability to ingest particles should have preceded the acquisition of endosymbionts and their eventual recruitment as eukaryotic organelles. Thus a hypothesis can be formed that polarizes the origins of some important eukaryotic features: the cytoskeleton preceded symbiotic organelles and therefore eukaryotes as we know them today. Many of the regulatory motifs found in metazoan genes, including those of important body-patterning genes such as the homeobox, must have arisen within protistans, for they are found in more than one multicellular clade, although their functions in unicellular organisms are unknown. It is likely that a choanoflagellate sort of cell founded the metazoans, but the relations of choanoflagellates to the parasitic mesomycetozoans and to fungi are not well established. There is little genetic information on the choanoflagellates. Knowledge of the choanoflagellate genome is obviously of critical importance in establishing which features of the metazoan genome are derived.

Attempts to date radiations among the Protista have not led to closely constrained dates, as molecular-clock dates have been inconsistent (chap. 4), but there is some agreement that there was an important radiation between about 1,000 and 1,250 Ma (see Sogin 1989, 1994; Knoll 1992b; Butterfield 2000). Presumably the last common ancestor of plants, fungi, and animals lived during that radiation. A protistan radiation within that interval is consistent with the known fossil record of metazoans, which would not have originated for some undetermined interval of time after the split from the last common fungal + metazoan ancestor, and is also consistent with the fungal fossil record (Berbee and Taylor 1993; Redecker et al. 2000). Butterfield (2000) has proposed that the evolution of sex may have been a key factor contributing to the timing of the protistan radiations that produced branches that included the clade ancestors of multicellular forms.

A Novel Bodyplan involving Differentiated Cell Types Founded the Kingdom Metazoa

If the origin of metazoans is defined as the onset of multicellularity in their common ancestor, the first metazoan bodyplan could have involved just two cell morphotypes, though more are possible. The ancestral lineage would presumably have evolved from a colonial form. Protistan colonies ancestral to metazoans may have become organized into associations of dissimilar cell-cycle phases before actually achieving status as multicellular organisms. Many protistans rely on environmental signals to switch between cell phases, such as reproductive and vegetative phases. But for evolution from two or more morphological phases in the life cycle of a cell to two or more actual cell morphotypes, there must have been a significant gain in independence from environmental cues, to produce reliably the cell types required by the bodyplan of a multicellular organism. Cell-cell signaling would be a plausible first step in capturing morphogenesis from the cellular level. Thus functions such as are now filled by ligand receptors, molecular signaling pathways, and transcription factors would be required as tissues evolved. King and Carroll (2001) and King et al. (2003) have identified a tyrosine kinase gene in choanoflagellates. It is possible that genes that mediated the organization of colonial cellular architectures (cytopatterning genes, so to speak) and colonial organization were the genes originally employed to mediate the morphology of evolving cell types.

The protistan-metazoan developmental transition presumably occurred during the evolution of early sponges, which were surely benthic, from choanoflagellates or their allies. A scenario that derives sponges from a colonial benthic form with cell types that have phenotypic phases for feeding (protochoanocytes) and reproduction (protoarchaeocytes) fits the facts well. The evolution of the sponge grade involved many important innovations, including multicellular bodies, with extracellular matrix to provide cohesion and support; differentiated cell types; and tissues and organs composed of characteristic cell-type associations. Such novelties imply the evolution of regulative genes in signaling cascades controlling a pattern of gene expression that reliably provided an assemblage of specific cell types. Many sponge cell types are not terminally differentiated; nevertheless they are not simply phases in a unicellular ontogeny but parts of a multicellular bodyplan that is mediated by key developmental genes. In a sense these pluripotent cells have it both ways: they are members of a distinctive bodyplan but may vary in response to ecological challenges.

Another innovation associated with the evolution of sponges is the metazoan egg. In sponges, choanocytes may transform into oocytes (and spermatocytes), although processes of oogenesis vary among living sponge groups (Simpson 1984). The sponge egg is highly specialized, a large cell built up partly through the activity of nurse or follicle cells (which are also sometimes recruited from choanocytes). Many of the advantages suggested for the origin of large eggs, such as enhanced cell survivability (Kerszberg and Wolpert 1998) and provision of larger targets for sperm (e.g., Levitan 1998) have been proposed to explain the origin of mating types in haploid/diploid protistans. For metazoans an important advantage of large eggs would seem to be that they permit cleavage. A large egg stuffed with organelles and abundant cytoplasmic materials is able to cleave into successively smaller cells simply by partitioning its contents. So the metazoan egg may have evolved as a cleavage system.

One advantage of cleavage is that it permits rapid early development. As blastomeres do not have to grow, a multicellular embryo can develop essentially at the speed required to position, duplicate, and cleave cycles of mitotic spindles—it can all be over in minutes, if necessary. A second advantage is in the role of cleavage in the patterning of cellular differentiation. The establishment of cleavage provides a convenient opportunity to establish commitment among differentiating cell lines. Indeed, if all blastomeres were to be identical, cleavage would be difficult to arrange, for it would require eggs to be isotropic, or at least to have similar properties with respect to cleavage planes. This does not seem to be an impossible requirement, but it is probably never met. Recall that both autonomous and conditional specification of blastomeres is associated with the pattern of cleavage planes. The developmental pattern of autonomously differentiating blastomeres depends on the relation between cleavage-plane position and factors that are positioned or activated either maternally (as in arthropods) or by events associated with fertilization (as in nematodes). And the pattern of differentiation by conditional specification depends upon the relative positions of neighboring blastomeres as produced by cleavage.

It seems appropriate that eggs and cleaving embryos are among the earliest known metazoan body fossils (fig. 5.9; Xiao et al. 1998; Xiao and Knoll 2000). As these fossils confirm the evolution of metazoan-style eggs probably near 600 Ma, they indicate the presence then of multicellular adults with differentiated cell types. The fossil eggs are large, and the embryos are therefore interpreted as representing direct developers, which unfortunately means that the cleavage pattern cannot be confidently compared with the highly conserved indirect cleavage patterns characteristic of some major metazoan alliances. The possible affinities of the Neoproterozoic embryos thus include sponges, other prebilaterians, and in fact primitive members of any of the bilaterian alliances as well. The possibilities even extend to the enigmatic Vendobionta (chap. 6).

Many of the features that characterize metazoan structure and development are already established in sponges. Sponge cells are linked to each other by aggregation factor (AF), a cell adhesion factor that bonds to receptors in plasma membranes. Although the mechanisms are not completely elucidated, aggregation factor mediates in some way the famous sorting out of sponge cell types that follows cell disaggregation (Humphries et al. 1977). The extracellular matrix of sponges contains several collagen types, including the type (IV) that is associated with basal laminae in higher metazoans. Also as in higher metazoans, extracellular fibronectins occur;

these molecules are involved in the linkage of extracellular matrix to cell membranes. Integrins are also known in sponges; these form a class of membrane receptors that bind fibronectins and other extracellular molecules and span cell membranes to link with proteins that connect to microfilaments; thus even sponges have a system to link extracellular matrix to the cytoskeleton. Sponge cell membranes also contain receptors, such as tyrosine kinases, that lead to signal transduction paths in higher metazoans, and presumably do in sponges as well. Evidence for other typical eumetazoan molecules and molecular systems in sponges (G proteins have been identified, for example) is reviewed by Müller (1998) and extended by Suga et al. (1999) and Ono et al. (1999), who develop evidence of extensive gene duplication in sponges before the branching that led to diploblastic forms.

In short, although sponges do not generally behave developmentally as the more derived phyla do, they display many basic features of metazoan development and evidently use molecular mechanisms that are essentially universal across the eumetazoans today. The processes that pattern sponge morphology have not been described; however, homeobox genes have been identified in sponges, though no Hox-type genes have been confirmed (Seimiya et al. 1994; Seimiya et al. 1996; Coutinho et al. 1994; Degnan et al. 1995; and especially Manuel and Le Parco 2000). As these regulatory genes belong to common gene classes found in higher metazoans, where they are involved in specifying cell identities and positions (chap. 3), it is a reasonable supposition that they function similarly in sponges.

Although a number of authors have espoused a planktonic ancestry for metazoans (see Jägersten 1972; Nielsen 1995, 1998), sponges are epibenthic, and thus it is almost certain that all living metazoans share a common benthic ancestor. As epibenthic animals must cope with the biomechanical requirements of life at and just above the seafloor, it is easy to imagine that the origin of collagenous extracellular matrix was associated with structural adaptation to benthic life modes. The original differentiated cell functions in adult metazoans are likely to have been feeding (rather than locomotion) and reproduction. The finding that hexactinellids are at the base of crown Porifera, if confirmed, probably indicates that there were earlier, cellular sponges or spongelike organisms that are now extinct. The evolving protistan lineage that led to Metazoa also appears to be extinct, so we are left with a sponge model for the earliest metazoans, which is probably close to the truth. The simplest pathway to Metazoa is for benthic choanoflagellates (perhaps stem forms) to have given rise to the earliest, and benthic, sponges, some time around 600 Ma.

The Complexity of Metazoan Bodies Led to the Emergence of a Hierarchical Somatic Organization and to Hierarchical, Scale-Free Networks within the Developmental Genome

In a unicellular organism that reproduces by cell division, "development" is essentially coeval with the reproductive phase, as new cell "bodies" are patterned that duplicate the parental one. The populations consist of clones, perhaps with interclonal genetic variation among related clone arrays. Genetic variability in unicellular cytopatterning is thus unlike the genetic variability in development found within sexual metazoan populations. There seems to be little information on the architecture of the part of the genome concerned with cytopatterning within protistans. However, there is evidence that plants, fungi, and animals all have similar developmental genomic architectures, although the genes used for developmental functions are commonly not the same. It would therefore seem possible that there are features of protistan genomes that form templates for the multicellular genome architectural systems, or at least in some way predispose their evolution along similar lines. Another possibility, which does not exclude the previous one, is that the pathway to evolving an entity that will function in complex ways leads inevitably to the sort of genomic architectures that we find in common among multicellular kingdoms.

The key to complex biological architectures is surely the requirement that they function—that they be alive. One can imagine a very large pile of rubble that would be more complex than any living organism, for if made large enough its minimum description could be far larger. But the rubble doesn't do anything; it is complex but inert, unorganized, and nonfunctioning. To produce a complexly functioning entity requires organization (McShea 1991). Simon suggested (1962; see chap. 1) that the only architecture of organized complexity may be a hierarchy. The finding that metabolic functions may be organized as modular, scale-free networks (chap. 3; see Ravasz et al. 2002 and references therein) suggests that it is such an architecture that transforms trees of gene-expression events into functioning somatic hierarchies.

A speculative scenario of the origins of the architectures of metazoan complexity can be based on these observations. As multicellular organisms arose from protistans, the evolution of differentiated cell types, which are somatic modules, was accompanied by the evolution of regulatory modules. At first the regulatory system captured distinctive cell phases to form cell types, with the gene(s) responsible for the capture operating as a regulatory node but not as players in the morphologies or in the physiological processes themselves. At least one of the cell types must have functioned as a stem cell. For the earliest sponges a reasonable speculation is that the original cell types were the choanocytes that formed the tissues in a feeding chamber and archaeocyte-like cells that differentiated into supporting pinacocyte-like tissues. The spatial patterning of the tissues became regulated by the first "key developmental genes," producing a genetic map that patterned the morphology, and, aided by the evolution of extracellular matrix, the first metazoan bodyplan appeared.

As cells became more specialized or extended their functions (e.g., by producing mineralized spicules), the gene(s) regulating those novelties came under the aegis of (usually) earlier regulatory nodes, whose domains of influence they subdivided or extended. New cell types could arise by at least two primary routes. One route involves the capture of physiological variant phenotypes, as mentioned. A second route involves the evolution of a new type from a parental type by specialization of a clade of daughter cells. The new clade might involve the acquisition of new alleles, the expression of new gene combinations, or the silencing of some parentally expressed genes. Such a route requires that the daughter cells be regulated differentially, and it appears that this was often accomplished by using different enhancer signals to regulate different cell clades produced by a single parental cell. The use of a new transcriptional/translational sequence does not solve the problem of maintaining both daughter cell types in the same module—it creates two regulatory gene nodes, but there must still be a gene upstream to pattern them. The genes lying at the nodes that regulated evolving modules (often being subdivided into submodules) evolved by increasing the number of transcriptional signals recognized by their enhancer systems. The translated binding sequences of these genes were nearly or quite stabilized because of their multiple usages, but through enhancer multiplication the genes became the centers of expression clusters that formed modularities, corresponding closely to the morphological/physiological modularities of the evolving organism. Indeed, it must have been selection on the somatic aspects of the organisms, to meet adaptive challenges or opportunities, that permitted enhancer mutations to increase fitness.

It is possible to speculate that most of the early cell types were founded by capture of physiological variants and the concomitant creation of stem cells; for example, in sponges the line between cell phases and differentiations is not always clear. However, it seems likely that many cell types arising later were products of the isolation and subsequent specialization of cell clades that did not previously exist as cell phases. Once modularity set in at tissue and organ levels, homologous cells in different modules were free to evolve independently, under the aegis of selector genes that helped regulate the positional identity of morphological entities—organs or modules, such as segments or other major body parts, that included originally similar tissues and, commonly, organs. The signaling pathways in these evolving systems presumably assumed a modular architecture.

Diploblastic Somatic Architecture Evolved from Sponges

Diploblastic Bodyplans Employ Epithelia

Early sponges must have had choanocytes, but they have disappeared from diploblastic phyla. Diploblastic forms (and their larvae) do possess flagellated cells, however, as do sponge larvae. It is possible, then, that the last common ancestor of sponges and diploblastic forms passed through an early ontogenetic stage at which flagellated cells were present, but at which no choanocytes were yet differentiated, and that in the branch(es) leading to eumetazoan phyla the choanocyte was dropped from the repertoire of differentiation. This could be described as a progenetic shift of reproduction into premetamorphosis, producing an adult with exterior flagellated cells surrounding or partly surrounding larger interior cells as in tissue-level metazoans. Such an organism is usually termed a planuloid, resembling as it does the planula larva of cnidarians.

One of the ways to account for the abrupt appearance of a distinctive bodyplan from a different but distinctive ancestral bodyplan is to postulate that the line of descent was carried for a while by a juvenile or larval form. If reproduction devolves upon an early ontogenetic stage, then the former adult characters of that lineage are hardly under the aegis of selection and can be abandoned. A new sort of adult bodyplan is then free to evolve. Such a heterochronic (progenetic) shift in the timing of reproduction provides a way for lineages to escape from their adult specializations (Hardy 1954), and provides evolutionists with an interesting way of explaining away jumps between entire grades. A number of authors have postulated a planuloid stage in early metazoan history, sometimes as the founding metazoan, sometimes as a progenetic sponge larva; a history of these views with appropriate references is given by Salvini-Plawen (1978).

If a diploblastic bodyplan evolved from a planuloid ancestor, there are two important steps that had to be taken. Originally the planula must have been at a spongelike level of organization rather than a tissue level, so that epithelial tissue sheets had yet to appear. Furthermore, planulae lack the principal adult feeding structures of sponges, although nutriment can be taken up by cells other than choanocytes. A shift to a feeding system without choanocytes may not have been a difficult step. It seems plausible that the loss of choanocytes was associated with the evolution of digestive tissue. The evolution of digestive cells, for example from phagocytic archaeocyte-type cells, is not hard to imagine, but this probably occurred before dispensing with choanocytes altogether. Therefore, digestive surfaces that employed macromeres and the products of the last micromeres to be produced (cells that were to become endoderm) evolved in organisms that possessed choanocytes, with the eventual integration of those cells into a digestive tissue that rendered choanocytic feeding unnecessary at some point (Ivanov 1973; Salvini-Plawen 1978). The evolving endoderm was probably disposed ventrally, was used to digest food items on the substrate, and then became internalized to provide a greater digestive surface in association with body-size increases. The internalization of digestive tissue produces gastrulation. These forms, then, no longer had a sponge bodyplan, and although they need not have had more cell types than sponges, they were organized quite differently.

Some modern sponges have replaced their choanocyte-based feeding system through the evolution of other feeding methods, as exemplified by the discovery of carnivory in the Cladorhizidae, a group whose demosponge affinities are indicated by their siliceous spicular types (Vacelet and Boury-Esnault 1995). These forms trap minute (<1 mm) crustaceans on hook-shaped spicules that coat filamentous extensions of the epithelium; the crustaceans are then overgrown by migrating cells and digested. There are no traces of choanocytes or of intake pores or oscula in these sponges. In effect, these forms have abandoned the entire pumping system that is the basis for the sponge bodyplan. Vacelet and Boury-Esnault (1995) suggest that this trophic shift is adaptive to environments that are poor in the usual sponge food items. Thus there is some basis to hypothesize that the Neoproterozoic sponges that gave rise to diploblastic animals shifted from feeding on suspended food items to feeding on aggregates of unicellular organisms or detritus, and that pumping systems with choanocytes thus became obsolete in those forms. There is actually no need to invoke progenesis to eliminate choanocytes from a bodyplan, although the origin of epithelial tissues is nicely explained by a planuloid scenario.

Ctenophora appears to have branched earliest among the living diploblastic clades. The diploblastic bodyplan may have evolved from a planuloid form, although evolution of eumetazoan-style tissues in an adult sponge lineage cannot be ruled out. In either of these cases, the diploblastic architecture originated in the benthos. Ctenophora, however, obviously invaded and thrived in the pelagic environment, implying that there are important stem diploblastic forms that are missing from the record, unless they are to be found among the vendobionts. Perhaps ctenophores are descended from a stem group that branched to give rise to cnidarians as well. If the Anthozoa are the most primitive of living cnidarians, as suggested by molecular phylogenies, crown Cnidaria originated in the benthos. The origins of some of the rather sophisticated regulatory systems that characterize metazoans may lie in selection for increasingly integrated tissues, leading to the epithelial diploblastic construction, with specialized organs placed definitively within an established symmetry. The increasing bodyplan complexity must have required many signaling pathways, some of which were evidently recruited for multiple usages.

The Developmental Genomes of Prebilaterians Are Foreshadowed in Their Protistan Ancestors

As some binding motifs used in developmental regulation, such as homeobox sequences, are shared in such distant relatives as metazoans and plants, they are presumably widespread among protistans as well, though obviously mediating different functions. The similarity between developmental genes is greater still in metazoans and fungi. Nevertheless, comparative study of the annotated genomes of yeasts and metazoans (*Caenorhabditis*, *Drosophila*, and, in one study, *Homo*) reveals that genes involved in metazoan development, especially those involved in cell-cell and cell-substrate interactions, are far less well represented in yeast, with some types of genes being absent altogether (Rubin et al. 2000; V. Wood et al. 2002). However, among choanoflagellates, the genomic similarity with metazoans is even closer (King and Carroll 2001; King et al. 2003).

The protistan nature of choanoflagellates is indicated not only by their unicellular condition but also by their distinctly protistan mitochondrial genome, which is far less compact (intergenic regions are long even for protistans) and contains 1.5

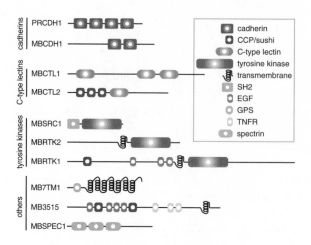

FIG. 13.1 Some protein domains (*box, upper right*) in common between choanoflagellates and metazoans, not known in any other eukaryotic clades. Multidomain proteins from choanoflagellates shown on left. *MB*, protein predicted from *Monosiga*; *PR*, protein predicted from a *Proterospongia*-like form. In metazoans, cadherins are cell-adhesion molecules; the C-lectins (these contain carbohydrate recognition domains) mediate cell adhesion and signaling; tyrosine kinases are signaling molecules; and the "others" include domains from molecules used in signaling and in cytoskeletal structure. From King et al. 2003.

to 3 times as many genes as metazoan mitochondria (Burger et al. 2003). Nevertheless, study of only a small fraction of the choanoflagellate nuclear genome in *Monosiga* and a *Proterospongia*-like species has identified a number of genes whose homologues are found in metazoans, but that are unknown in any other eukaryotes (King et al. 2003). Strikingly, some of these genes are, in metazoans, employed in activities required of multicellular bodyplans, but not of unicellular ones; they include multiple genes coding for tyrosine kinases that form parts of important signaling pathways in metazoans (chap. 3), other components of such tyrosine signaling pathways, and multidomain proteins that are used in cell adhesion (fig. 13.1). The functioning of these genes in unicellular organisms that do not have a developmental repertoire is still largely unknown, though experiments by King et al. (2003) suggest that tyrosine kinase signaling may be involved in regulation of the cell cycle, perhaps in response to nutrient availability. It will be fascinating to learn if fibroblast growth factor–type molecules, and others associated with the manufacture of extracellular matrix, also have ancestors in choanoflagellates.

Sponges, then, surely inherited molecules of tyrosine kinases, molecules with homeobox and cell adhesion domains, and doubtless many other molecules that became important for multicellularity, from their protistan ancestors. It was using this molecular toolkit as a basis that the sponge stem ancestor invented cell differentiation. From this perspective it is not surprising that the prebilaterian phyla contain many types of developmental genes that are widespread in bilaterians (see review in Wilkins 2002, chap. 13). Hox-type genes are the best-studied class (e.g., Schierwater et al. 1991; Kuhn et al. 1996; Finnerty and Martindale 1997, 1999; Schierwater and Kuhn 1998; Finnerty 1998). Hox and ParaHox clusters are present in cnidarians, and it is possible that Hox-type clusters mediated seriation in vendobionts (chap. 6). Other regulatory genes may have been present as clusters. Pollard and Holland (2000) have identified four arrays of homeobox genes that appear to be quite ancient, and have proposed a model of their early evolution. It is suggested that the ProtoHox gene cluster was originally a chromosomal neighbor of another homeobox gene cluster, the NKL (*Natural Killer*–like) cluster, and that these genes were also linked to an array of still other homeobox genes. Presumably these linked gene clusters arose by tandem duplications of their ancestral genes. This situation implies that the linked genes were likely to have been present before the Hox-ParaHox cluster duplication, and thus were present at least in stem diploblasts. *Eve*-type genes are known in cnidarians and ctenophores (Finnerty et al. 1996; Finnerty and Martindale 1997). *Pax* genes, widely employed in neural and sense organ development in bilaterians, have been studied in the embryos of anthozoan cnidarians (Catmull et al. 1998). One of these genes (*Pax-Cam*) contains residues that are characteristic of several bilaterian *Pax* genes, so that this prebilaterian gene may have given rise to an array of bilaterian developmental regulators. *Empty spiracles (ems)*, an anterior gap gene expressed in the head in *Drosophila*, is represented by a homologue that is expressed proximally to the mouth in hydrozoan polyps (Mokady et al. 1998). The development of muscles in hydrozoan cnidarians is regulated by genes (including *Brachyury* and *Snail*) that are important in mesoderm patterning in bilaterians (Spring et al. 2002). There are growing numbers of other examples of the likely presence in diploblastic lineages of genes that are developmentally important in bilaterians.

If such key regulatory genes and gene clusters give some indication of the evolution of the metazoan gene regulatory systems involved in bodyplan evolution, the basic system of patterning a body through differentiation and placement of cell types was largely in place within the metazoan genome before the living diploblastic phyla emerged and was enhanced separately in descendant clades.

The Nature of Early Bilateria Is Widely Debated

The Prebilaterian/Bilaterian Gap Is Wide

Living bilaterians appear to be monophyletic and to have descended from a stem diploblastic ancestor, but there are no living taxa, and no body fossils, that are known to represent morphological intermediates between the diploblasts and bilaterians. The two grades of organization are separated by a wide morphological gap. There is also a molecular gap; the SSU rRNA phylogenetic trees indicate that prebilaterian and bilaterian rRNA gene sequences are separated by a long branch (see Collins 1998), though acoelomorphs must shorten it. This gap suggests either geologically rapid reorganization of the rRNA sequence or a lengthy period of evolution that is unrepresented by living forms. The organisms that evolved across these gaps are evidently extinct and may be missing from the body-fossil record.

If those forms left fossil remains that we know of, they are probably represented among Neoproterozoic fossils, possibly only as traces.

The Earliest Bilaterians May Descend from Stem Diploblastic Larvae

Interpretations of the evolutionary steps leading to bilaterians are tied up with scenarios on the origins of the major crown bilaterian taxa. Some of those scenarios are mentioned in chapter 4. Reviews of many early notions of bilaterian origins include those of Clark (1964), Salvini-Plawen (1978), and Willmer (1990). If the major features of molecular phylogenies are correct, some of those interpretations have now been falsified, at least in the form in which they were first presented. There are, however, several current hypotheses that are consistent with molecular phylogenies, or that can be modified to accommodate these new data, and some of those contain elements of the early hypotheses. Some hypotheses derive bilaterians from pelagic adults, derived in turn from the earliest metazoans, which are also assumed to be pelagic in some versions. Contrasting hypotheses derive bilaterians from benthic ancestors. In one such hypothesis the ancestor is a modular sponge; in others it is a planuloid of diploblastic ancestry.

The Trochaea Hypothesis Proposes Evolution in the Plankton. One scenario is to derive bilaterians from adult planktonic forms, a notion that has been most explicitly worked out in the Trochaea hypothesis of Nielsen and Nørrevang (1985). This is a highly recapitulatory hypothesis that assumes the evolution of a series of increasingly complex pelagic forms, adults modeled in part after embryonic and larval stages in metazoan development. In this scenario an actual adult blastula stage, Blastaea, evolves through bodyplans termed Gastraea, Trochaea, and Protornaea to a Tornaea (fig. 13.2). These forms are all visualized as having been adult holopelagic organisms, feeding, reproducing, and evolving as plankters. Organisms at the gastraean grade produced a branch that colonized the benthic environment, evolving into Cnidaria, while a branch that remained as a holopelagic form evolved into Trochaea. A branch of trochaeans also colonized the benthos, and it is within these that the first bilaterian bodyplan evolved, with an anteroposterior axis appropriate to a creeping form. As the adult stage evolved, the trochaean bodyplan was retained as a planktonic larval stage, producing a biphasic bilaterian life cycle. Further evolution within each phase resulted in the ancestral protostomian bodyplan, termed Gastroneuron, with a trochophorelike planktotrophic larval phase.

In the plankton a branch of trochaeans continued to evolve as holopelagic organisms, passing through the bodyplan of Protornaea to Tornaea. Tornaeans then also colonized the seafloor, where adults became the ancestors of Deuterostomia; the founder of this branch is termed Notoneuron, which, it is postulated, had a tornaria-like (i.e., enteropneust larva–like) planktotrophic larval phase descended from Tornaea (fig. 13.2). The evolution of features within the holoplanktonic adults is worked out so that Gastroneuron and Notoneuron display different

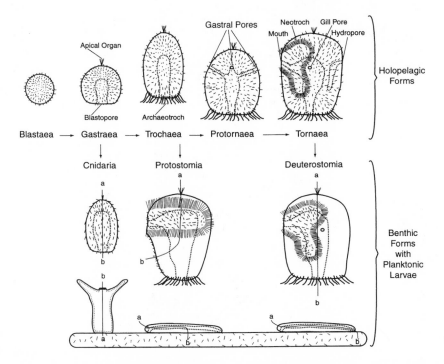

FIG. 13.2 Aspects of the Trochaea hypothesis. The evolution of holopelagic lineages resulted in the upper series of organisms, from Blastaea to Tornaea. Gastraea-grade organisms descended to the benthos and produced Cnidaria; Trochaea-grade organisms produced benthic Protostomia, retaining planktonic larva of the trochophore type; and Tornaea-grade organisms produced benthic deuterostomes, retaining planktonic larvae of the dipleurula type. *a–b*, apical-blastopore axis. From Collins and Valentine 2001 after Nielsen 1995.

organizations that led to the distinctive protostome and deuterostome characters (see Nielsen 1995).

In the Trochaea hypothesis, the last common protostome/deuterostome ancestor is a holoplanktonic trochaean. Nielsen (1995) has divided Protostomia into two sister branches, Spiralia and Aschelminthes, which are visualized as having diverged from a benthic Gastroneuron. Both Spiralia and Aschelminthes include mixtures of forms that are separated into Ecdysozoa and Lophotrochozoa on 18S rRNA evidence (fig. 13.3). Further, in Nielsen's scheme, Deuterostomia contains a number of phyla that are now considered to be protostomes. Although these changes present difficulties with the evolutionary pathways associated with the Trochaea hypothesis, the overriding notion is that much of the evolution of the differences in organization between major metazoan clades occurred in ancestral holopelagic forms, wholly feeding and breeding as plankters. It should be possible to modify this general approach so as to account for the newer phylogenetic data. There is no evidence that the putative holopelagic forms ever existed as independent

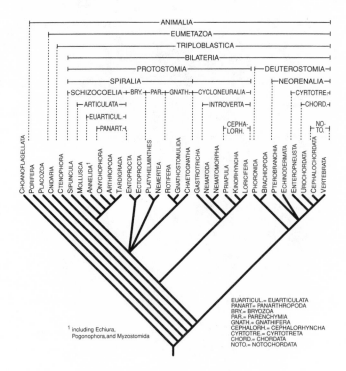

FIG. 13.3 The phylogenetic scheme of Nielsen 1995, which was interpreted in terms of the Trochaea hypothesis.

organisms, however; their reconstruction relies entirely upon assuming an extreme recapitulatory scenario (reminiscent of Haeckel 1866).

The Set-Aside Hypothesis Proposes Evolution via Deferred Complexity. Another hypothesis that invokes larval-style bodyplans as ancestral to complex metazoans was proposed by Davidson et al. (1995; see also Peterson et al. 1997; Arenas-Mena et al. 1998). In some metazoans, many adult organs are not derived from cells within larval organs, but rather from pluripotent cells sequestered during larval life, set aside as primordia from which adult structures form, such as the imaginal disks of insects. Davidson et al. proposed that such a set-aside system was an adaptation to the growing complexity of adult bodyplans. An evolutionary scenario was suggested in which the bodyplans that are now represented by the planktotrophic larvae of many deuterostomes and eutrochozoans were characteristic of adult bodyplans of early branches of Bilateria. As more complex bodyplans evolved, the cells that were used for adult bodyplans were not employed in the early developmental stages but were set aside then, and their fates specified during or after a metamorphosis. As with the Trochaea hypothesis, this scenario requires a highly recapitulatory life history.

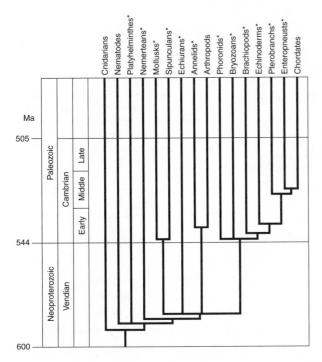

FIG. 13.4 The phylogenetic scheme of Davidson et al. 1995. Asterisks indicate phyla that have planktonic larvae that are believed to be descendants of a minute early bilaterian that resembled modern marine larvae. Geologic divisions not to scale.

Peterson et al. (1997) suggested that the rotifers, which are direct developers and lack set-aside cells, may broadly represent the less complex, ancient, adult body type from which complex bilaterians evolved, ancestral to the protostome/deuterostome ancestor and preceding the evolution of indirect development in bilaterians. This suggestion was supported by a phylogenetic tree that is at odds with the SSU rRNA tree (fig. 13.4). The molecular tree suggests that the rotifers and their paracoelomate allies arose, not only after the protostome-deuterostome split, but after the split that separated ecdysozoans and lophotrochozoans.

The set-aside hypothesis was subsequently modified and brought into accord with some molecular phylogenetic and clock data (Peterson et al. 2000). In this formulation metazoans are assumed to have originated over 1,500 Ma, and the origin of set-aside cells is placed at some later date during the origin of Bilateria (fig. 13.5). The Hox cluster is inferred to have been assembled after the origin of adult development via set-aside cells but before the evolution of the protostome/deuterostome ancestor; the bilaterian lineages that predated the protostome/deuterostome ancestor are presumably extinct. However, the evidence that the Hox and ParaHox clusters originated before the evolution of crown cnidarians (Finnerty and Martindale 1999) indicates that the "modern" patterning apparatus was in place long before the origin of the early bilaterians in which set-aside cells are inferred to have arisen. The earliest Hox-type gene clusters were clearly not used to pattern the sorts of bodyplans found in adults of complex crown Bilateria. Furthermore, it is clear that

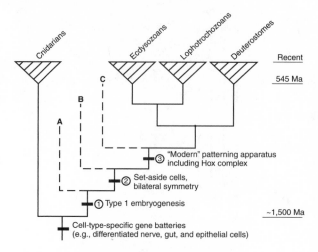

FIG. 13.5 The set-aside hypothesis of Peterson et al. 2000. The origin of Hox cluster patterning and similar regulatory systems is suggested as occurring after the origin of bilaterians that sequestered cells for development of adult bodyplans. Evidence now available indicates that formation of the Hox complex, and of many opposite developmentally important gene arrays, preceded the origin of Cnidaria.

the protostome/deuterostome ancestor had a cluster of at least seven Hox genes and probably more (de Rosa et al. 1999). Those genes were without doubt used in patterning the adult bodyplan, and their activities predated the evolution of set-aside cells. Evidently any larval sequestrations of classic Hox-gene activities found in higher Bilateria is owing to the evolution of larval bodyplans that are intercalated between gastrulation and the adult stages, and may employ other patterning systems (Valentine and Collins 2000).

The Colonial Hypothesis Proposes Evolution via Rounds of Individuation. Dewel (2000) has suggested that the ancestral bilaterian bodyplan arose through the evolution of a colonial diploblastic bodyplan into a modular triploblastic one. She has hypothesized that the sort of complexity that arises from the association of choanoflagellate cells to form colonies, which then became individuated as multicellular sponges, was continued in an association of functional sponge modules, which became individuated to form pennatulacean-like Cnidaria on one hand and ancestral Bilateria on the other (fig. 13.6). The ancestral bilaterian modules were serially arranged, producing a segmented bodyplan. The early bilaterians were thus not particularly small organisms, and as they evolved they became quite complex, so that the protostome/deuterostome ancestor was at a rather advanced coelomate grade. This ancestor was regionated into a head, segmented trunk, and tail, and had gill slits. Blocks of innervated mesoderm occurred within the segments, as did gonads with associated gonocoels, gonoducts, other serially repeated coeloms, nephridial organs, and elements of a circulatory system. Still other, unsegmented coelomic compartments, a brain, and a contractile "heart," were also present.

    With such a complex protostome/deuterostome ancestor, it is necessary to postulate the simplification of many metazoan bodyplans, a requirement that is reminiscent of the archicoelomate hypotheses that derive segmented coelomates directly

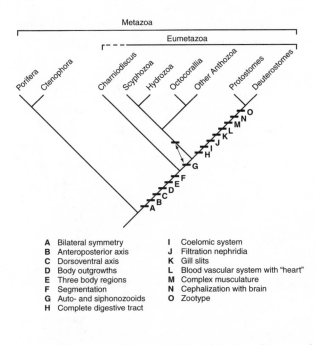

FIG. 13.6 The colonial hypothesis of Dewel 2000. Porifera are inferred to have originated from colonial ancestors (not shown). Eumetazoans then arose from modular sponges; cnidarians are considered to be segmented, and cnidarian colonies gave rise to complex bilaterians before the divergence of protostomes and deuterostomes.

A Bilateral symmetry
B Anteroposterior axis
C Dorsoventral axis
D Body outgrowths
E Three body regions
F Segmentation
G Auto- and siphonozooids
H Complete digestive tract
I Coelomic system
J Filtration nephridia
K Gill slits
L Blood vascular system with "heart"
M Complex musculature
N Cephalization with brain
O Zootype

TABLE 13.1 Crown phyla listed by Dewel (2000); those in bold are suggested as possibly having simplified bodyplans.

Porifera	**Nematomorpha**	**Sipuncula**
Ctenophora	**Kinorhyncha**	**Echiura**
Cnidaria	**Loricifera**	Annelida
Placozoa	Onychophora	Brachiopoda
Myxozoa	**Tardigrada**	**Bryozoa**
Gnathostomulida	Porifera	**Phoronida**
Rotifera	Arthropoda	Chaetognatha
Acanthocephala	**Mesozoa**	Hemichordata (**Pterobranchia**)
Cycliophora	**Platyhelminthes**	Echinodermata
Gastrotricha	**Nemertea**	Chordata
Nematoda	Mollusca	
Priapulida	**Entoprocta**	

from Cnidaria (chap. 4; see fig. 4.5). Dewel has listed phyla that may have been simplified (table 13.1); they include twenty (or twenty and a half) of the twenty-nine bilaterian phyla she recognizes. As she points out, the simpler phyla are generally either meiofaunal and therefore small-bodied, or are parasitic. In fact, the earliest known members of some of the "unsimplified" phyla, such as Mollusca and Brachiopoda, are also as small as many of those listed as simplified, and might well be added to the list, for it is unlikely that such small-bodied forms would require the complex anatomical features associated with their crown relatives. There is, however, no trace (pun intended) of the large-bodied ancestral bilaterians required by Dewel's hypothesis. As Budd and Jensen (2000) emphasize, there is little chance that larger Neoproterozoic trace fossils would not have been produced by such bilaterians, or would have escaped our attention, given that many small traces are preserved.

Much of Dewel's hypothesis rests on the assumption that similar morphological features found in complex bilaterians are homologues. For example, coelomic cavities and segmentation are considered to be synapomorphies of protostomes and deuterostomes, and therefore properties of their last common ancestor. However, the homology of those features seems unlikely (chap. 4); they are solutions to particular biomechanical problems in elongate organisms that pursue certain modes of life, and their details are consistent with independent evolutionary histories. Dewel argues the homology of complex morphological features from the standpoint of their developmental genetics as well, pointing out that similar and often homologous developmental genes control similar features—limbs, eyes, nerves—in both protostome and deuterostome taxa. While these homologues indicate the presence of the genes or of some paralogs in a common ancestor, they do not necessarily certify the homology of the features whose development they mediate (chap. 3). In these cases it is more likely that the genes mediated the development of features in the last common ancestors that were precursors to the complex organs in the descendants: eye spots instead of eyes (e.g., Tomarev et al. 1997), body-wall protuberances instead of limbs (e.g., Panganiban et al. 1997), and diffuse nerve nets instead of condensed nerve cords (e.g., Gerhart 2000). Finally, Dewel believes that the large number of developmental genes inferred to be present in early bilaterians (chap. 3; Gerhart and Kirschner 1997; Knoll and Carroll 1999) indicates that complex organs and organ systems were present. However, present evidence favors the idea that the developmental genome was evolved early in metazoan history to permit cell differentiation and bodyplan patterning in relatively simple multicellular organisms, perhaps as early as the ctenophore-cnidarian common ancestor. Although metazoan bodyplans were relatively simple at first, the marvelous, flexible structure of metazoan genomic regulation permitted the eventual rise of the rich diversity of body types that characterize the Phanerozoic Eon.

The Planuloid Hypothesis Proposes Benthic Evolution with Complexity Increases within the Somatic Hierarchy. For proponents of the earlier phylogenies that placed the flatworms at the base of Bilateria, the evolution of a benthic protoflatworm from an elongate larva—a planuloid somewhat like that postulated as ancestral to diploblasts—seemed to be a reasonable hypothesis (see Salvini-Plawen 1978). For a creeping benthic worm, subsequent evolution of anteroposterior differentiation and mesodermal body-wall muscles makes good adaptive sense. Although rhabditophoran flatworms are now assigned to Protostomia on the basis of SSU rRNA data (chap. 8), bilaterian ancestry in a planuloid organism remains a viable scenario (fig. 13.7). The benthic adult planuloid is essentially a paracoelomate. As a very common evolutionary pattern is for stem groups to radiate and then to be

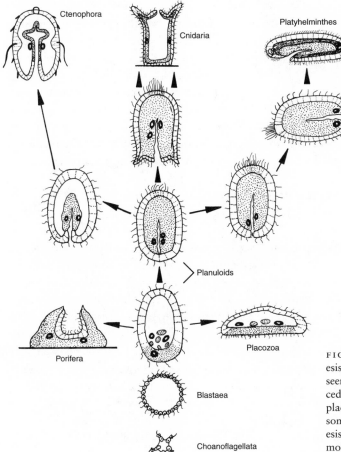

FIG. 13.7 A planuloid hypothesis of bilaterian origins. It now seems likely that the sponges preceded the planuloid, and that placozoans arose later, but with some modifications this hypothesis can be made consistent with molecular data. From Willmer 1990 after Salvini-Plawen 1978.

sorted by subsequent events (chap. 12), it seems likely that the early bilaterians radiated into many habitats, with morphologies appropriate to those varied modes of life.

As body-fossil evidence is largely lacking, the nature of a protostome/deuterostome ancestor must be reconstructed chiefly from features among crown bilaterians. Such reconstructions create a sort of bilaterian that would easily evolve from a planuloid founder. Most living paracoelomates have bodyplans that involve about fifteen to twenty cell morphotypes, and it is plausible that the protostome/deuterostome ancestor would be at about this grade. Judging from the attributes of living paracoelomates, a likely protostome/deuterostome ancestor, if arising from a planuloid, would be a small soft-bodied worm, benthic and either epifaunal or meiofaunal, probably with protonephridia, and with a complete gut. There is likely to have been a fluid space that was topologically on the site of the blastocoel or, alternatively, a tissue mass between body-wall muscles and gut. This worm would have moved via a mucociliary locomotory tract, feeding on organic detritus and/or small organisms. In common with other Neoproterozoic paracoelo-mates, the body would have been patterned by homologues of the key developmental genes found in living members of highly derived metazoan taxa. The hypothesis that the last protostome/deuterostome common ancestor was a paracoelomate has gar-nered significant support from the evidence that acoelomorph flatworms may well branch basally to that ancestor (chap. 6). Acoelomorphs would thus be in the phy-logenetic position formerly occupied by stem bilaterians, although as acoelomorphs are extant, the effect would be to significantly expand the bilaterian crown.

A Benthic Hypothesis Can Explain Both Fossil and Molecular Data and Is Not Incompatible with Developmental Patterns

I believe that the bulk of available evidence favors the hypothesis that the origin and radiation of bilaterian bodyplans occurred almost entirely in the benthos, and prob-ably in shallow water, and that once Metazoa had appeared, evolution proceeded among individual organisms (rather than involving the individuation of colonies). An evolutionary model based on these assumptions obviously draws on the plan-uloid hypothesis and on many other sources, and is not inconsistent with some aspects of the Trochaea, set-aside, and colonial hypotheses, but contrasts sharply with other aspects, and with the main thrusts, of those hypotheses. Inevitably, such a hypothesis involves much speculation. Fig. 13.8 shows the main aspects of this model. My purpose in this section is simply to show that a benthic history of origin is feasible for most metazoan bodyplans and is quite parsimonious in the general sense, and not that any particular history is confirmed by available data. Based on the molecular evidence of their phylogeny, Metazoa can be grouped into several major alliances: a prebilaterian alliance, which is paraphyletic; an early bilaterian alliance that preceded the last common protostome/deuterostome ancestor, which has been quasi-hypothetical but is probably represented among living organisms by acoelomorphs, and is probably paraphyletic; the three bilaterian alliances rep-resented by crown phyla, Deuterostomia, Ecdysozoa, and Lophotrochozoa, and possibly a fourth, Platyzoa. Additionally, there are some living phyla that cannot be placed within these alliances by SSU rRNA evidence, at least some of which may lie outside any of these groups. Crown Bilateria are largely treated in the next chapter.

Molecular Norms of Metazoan Development Were Established
in Prebilaterian Genomes
There are no outstanding problems to accepting a benthic origin for diploblastic forms, probably first represented as body fossils by the vendobionts. The early ap-pearance of so many aspects of metazoan developmental systems in relatively simple crown bodyplans suggests that most of the major gene types appeared early, and

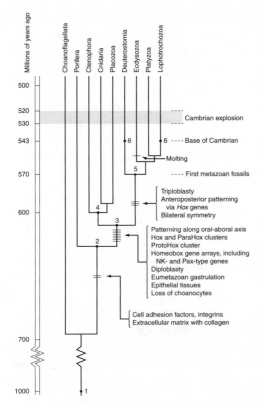

FIG. 13.8 A hypothesis of the evolution of some major features in the early diversification of Metazoa. *1*, Origin of stem Choanoflagellata, perhaps near 1,000 Ma. *2*, Origin of stem Porifera, perhaps between 600 and 700 Ma; basic elements of multicellular development evolved. *3*, Origin of stem diploblasts and of the clade ancestor of bilaterians, which was not yet triploblastic, perhaps between 600 and 650 Ma; basic patterning system of eumetazoans evolved. *4*, Origin of earliest crown diploblasts. *5*, Origin of triploblasty and of the clade ancestor of most crown metazoan phyla, perhaps between 570 and 600 Ma. The bodyplans of most crown phyla are not present until latest Neoproterozoic and earliest Cambrian time, perhaps between 525 and 550 Ma, and they may be represented by stem lineages for some time. *6*, Ramping up of body sizes and origins of eucoeloms, and perhaps the origins of ciliated planktotrophic larvae. The timing of most of these events is not well constrained.

that even vendobiont development may have involved patterning by transcription factors and *cis*-regulatory promoter systems (Valentine 2001). Subsequent increases in morphological complexity were met in large part, though not exclusively, by increases in numbers of gene-expression events, through gene duplications, through the eventual divergence of paralogs, and especially through the addition of enhancer modules. Certainly, the early bilaterians possessed Hox and ParaHox clusters, which mediated patterning of their bodyplans.

The Early Bilaterians Presumably Radiated during the Prelude to Produce a Diverse Paracoelomate Fauna

The pattern of a strong early radiation of basic bodyplans into disparate types, followed by a rising diversity of variations on those types, is so common when we can actually trace macroevolution in the fossil record (chap. 12) that it seems likely that such a pattern was present during many of the early bilaterian diversifications. Accordingly, the diversification of stem bilaterians was probably concentrated relatively early in their history, and was probably quite broad. That the genomes of stem bilaterians may have been enlarged through gene duplications, producing, for example, larger Hox clusters, is suggested by the relatively large Hox cluster inferred for the genome of the protostome/deuterostome ancestor. The extent of genomic differences that appeared within the stem bilaterian alliance will never be known. However, it should be possible to infer the sorts of genomic differences that arose among the clade ancestors of surviving bilaterians. Judging from what evidence exists to date, the differences were not very great, for even after the independent rise of larger and more complex bodyplans in crown phyla, their genomes are more characterized by differences in gene organization than in gene content. Critical parts of the regulatory machinery—such as signaling pathways, the binding sequences of transcriptional regulators, and processing of transcripts and of translated proteins—did not vary much. The combinatorial flexibility with which this machinery could meet adaptive challenges, by producing selectively advantageous variations in gene usages, permitted it to be highly conserved. Even so, there have certainly been additions, modifications, and deletions to regulatory mechanics, associated with important developmental evolution. If the survival of early clades was mediated by their dynamics, then the lineages to which the clade ancestors of crown phyla belonged should have been preferentially sorted as being among the more successful, relatively slow turnover groups among stem bilaterians.

A possible time for a radiation of these forms is in the range between 550 and 650 Ma. Among those forms, and possibly a rather early product of this radiation, was the protostome/deuterostome ancestor. The radiation that proceeded from this ancestor probably produced the clade ancestors of at least many of the crown paracoelomates, which would thus be monophyletic, though some may be paraphyletic with respect to more complex crown phyla. In this case crown paracoelomate bodyplans need not be particularly reduced morphologically, though some, such as nematodes, evidently are.

Among the early bilaterians, benthic creeping was probably based on mucociliary locomotory systems. Mucociliary creeping (chap. 2) is a common locomotory method among very small members of many bilaterian phyla, including Gastrotricha, Platyhelminthes, Nemertea, Mollusca, and Annelida (Fransen 1980; Brusca and Brusca 1990). Mucociliary locomotion is also known to occur in the prebilaterian Placozoa (Grell 1982), Ctenophora (Emson and Whitfield 1991), and Cnidaria (Collins et al. 2000). In metazoans with mucociliary systems, the cilia operate within a two-layered system (Sleigh et al. 1988; Beninger et al. 1997). The layers consist of a more solid mucous layer on the outside of the mucous stream, away from the tissue and in contact with the substrate, and a more watery layer on the inside of the stream, next to the tissue, within which the cilia can beat. The ciliary motion propels the bodies across the mucous substrate. At larger body sizes, mucociliary locomotion came to be supplemented and eventually replaced by peristaltic or introvert mechanics involving body-wall muscles (Clark 1964).

If acoelomorph flatworms are indeed basal to the protostome/deuterostome ancestor (chap. 6), then we have the exciting prospect of being able to study a living representative of what would otherwise be the stem bilaterians. It is not yet clear where the acoelomorph bodyplan would have branched within the early bilaterian tree. This bodyplan might be quite primitive with respect to a more complex protostome/deuterostome ancestor, or it might represent a bodyplan close to that of that

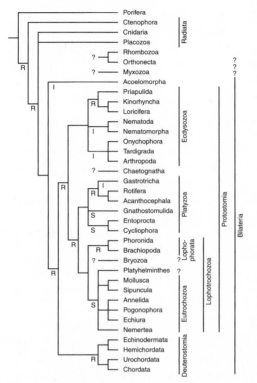

FIG. 13.9 The distribution of cleavage types in the model tree. Cleavages are those of indirect developers when present. *R*, radial; *S*, spiral; *I*, idiosyncratic. After Valentine 1997.

ancestor, or it may even have been simplified from a more complex early bilaterian bodyplan. My guess is that it is simpler than the protostome/deuterostome ancestor, but not by much, and that the last ecdysozoan/lophotrochozoan common ancestor was of a similar architectural grade. We shall see. Acoelomorph genomes may hold important, perhaps critical, evidence as to the course of the early evolution of developmental systems in Bilateria.

Cleavages and Larval Modes Are Related in Extant Bilaterians, and Suggest Models for the Fauna of the Prelude

Fig. 13.9 depicts the distribution of cleavage types reviewed in chapters 6–11. There is a wide range of developmental modes in marine invertebrates (see Gilbert and Raunio 1997 for morphological aspects, and Jablonski and Lutz 1983 for a review of ecological aspects). Three common categories of developmental mode (chap. 2) seem to be related to cleavages: direct developers (species that develop within egg capsules and emerge as juveniles, and generally have yolky eggs to provide nutrition during early development; all phyla contain at least some direct-developing species); planktotrophs (indirect developers that feed as larval stages in the plankton and have little or no yolk); and lecithotrophs (indirect developers that have free-living larval stages but do not feed, and which may have little to much yolk). In indirect development, cell specifications and differentiations in early embryogenesis are chiefly directed toward producing a larval rather than an adult bodyplan, although morphological features useful only to adults often make an appearance in larval stages (a presumed heterochrony termed adultation by Jägersten 1972). The two classic cleavage patterns, spiral and radial, are displayed chiefly in indirect-developing species.

Hexactinellid sponges have a gastrulalike embryo (chap. 6), and the cleavage pattern is spiral for three cleavages, and then radial. The cleavage pattern that was first associated with gastrulation in Eumetazoa was evidently not spiral, but biradial, as in Ctenophora, or radial, as in Cnidaria. Indirect-developing members of

Deuterostomia show radial cleavage, while indirect-developing members of Lophotrochozoa may be either radial (as the lophophorates) or spiral (as most eutrochozoans). In clades with both indirect and direct development, the direct-developing species usually do not cleave like their indirect-developing allies, in either a radial or a spiral pattern. The egg capsules of direct developers often contain much yolk, up to half their volume. The blastomere geometries must accommodate to this yolk. As a result, the cleavage patterns are modified from the symmetrical geometries of indirect developers. In cases that have been investigated, changes from indirect to direct development have concomitant changes in gene expressions and cell fates (e.g., Wray and Raff 1989, 1990 for echinoids; Swalla et al. 1993 for ascidians; see also Ferkowicz et al. 1998; Raff et al. 1999). Emlett and Hoegh-Guldberg (1997) removed yolk from eggs of a direct-developing echinoid and observed their development; cleavages did not revert to an ancestral radial pattern but proceeded as in untreated eggs of that species, though the resulting juveniles were smaller. These studies indicate that, during the evolution of those direct developers, not only are the cell geometries mechanically modified from their ancestral patterns, but novel patterns of cell lineage specification have also evolved to produce their developmentally unique early ontogenies.

Indirect development is conspicuously absent in solitary paracoelomates today. However, cnidarians have indirect-developing larvae, which are lecithotrophic. Unlike direct developers, lecithotrophs commonly show the spiral or radial cleavage patterns of their planktotrophic allies. For example, basal molluscan clades, which have lecithotrophic larvae whose eggs have relatively little yolk, show classic spiral cleavage. Most crinoids, basal among crown echinoderms, are lecithotrophic and have classic radial cleavage. In echinoids, however, a more derived echinoderm group, the eggs of lecithotrophs show modified cleavages (Wray and Bely 1994 and references therein).

Larval Modes and Therefore Cleavage Patterns Are Related to Environmental Conditions

The common, reasonable explanation for the adaptive utility of indirect development is that it provides for the dispersal of a population and for gene flow between populations once they are dispersed. While free-living larvae complicate the final development of the adult bodyplan, the situation is somewhat ameliorated by larval adaptations that enhance success during the planktonic dispersal phase. Planktotrophic larvae are estimated to be present in 70% of living benthic invertebrate species. Some feeding larvae are demersal, feeding near or at the seafloor and developing rapidly into juveniles, and thus are ecologically distinct from planktotrophs. Finally, significant numbers of marine invertebrates are direct developers, hatching from eggs as juveniles, although some of these pass through larvalike stages in the egg. There are additional complications; for example, some larvae are retained (brooded) by the parent and released as advanced larvae or juveniles;

some of these are direct developers, some are lecithotrophic, and a few are even planktotrophic.

There is a well-known pattern to the proportion of indirect to direct developers today; perhaps 50% or more of the species in high-latitude faunas are direct developers, but this proportion decreases equatorward to around 10% to 20% (Thorson 1936; Jablonski and Lutz 1983). There is a parallel trend in the average length of free-living larval stages; the duration of the average larval stage becomes longer in lower latitudes, where planktotrophy predominates. The poleward shortening or elimination of larval stages is interpreted as resulting from a trade-off between fecundity and mortality rates (Vance 1973; Christiansen and Fenchel 1979), which may be summarized as follows. In general, species with larvae that can feed themselves require less maternal energy (such as yolk) per egg than direct developers or some nonfeeding larval forms—and thus planktotrophic species can produce more eggs per female. In regions where conditions are favorable for planktonic existence, species with planktotrophic larvae are favored because of this relatively high reproductive potential. However, in regions where conditions in the water column are inclement, at least seasonally as in high-latitude winters, the mortality of planktonic larvae can become so great that there is more reproductive benefit, in terms of survival, in fewer nonplanktotrophic offspring, among which mortality is lower.

Strathmann (1978) has postulated that the evolution of developmental modes is largely one-way, and that while direct developers have commonly evolved from indirect developers, the opposite pathway is quite rare. Judging from the present distribution of embryonic types in echinoderms, evolutionary shifts from indirect to direct developmental modes have been quite commonplace, occurring repeatedly in a wide variety of lineages. As an example, Hart et al. (1997) established a phylogeny for twelve species of starfish in two allied genera by mitochondrial gene comparisons. Assuming that the phylogeny is correct, planktotrophic larvae had been lost four times, three of these being changes to benthic development with relatively large yolky eggs (fig. 13.10). None of these lineages appeared to have changed from nonplanktotrophy to planktotrophy. One reason that planktotrophy does not simply disappear because of the preponderance of shifting to direct developers is that planktotrophs are less prone to extinction, other things being equal. The fossil record indicates that indirect-developing species last between two and three times longer than direct developers (Jablonski and Lutz 1983; Jablonski 1986a); these data are based on larger-bodied phyla, however.

Although changes from direct to indirect development may be rare, changes from lecithotrophy to planktotrophy may be quite common. They are documented best in classic spiralians (Haszprunar et al. 1995; Rouse 2000a, 2000b; Hickman 1999). Rouse (2000a) presented a phylogenetic tree of annelid families (fig. 13.11), along with other spiralians shown at the phylum level, each branch coded for larval feeding. The independent evolution of planktotrophy from lecithotrophic larvae is indicated in nine cases. Multiple switches from planktotrophy to lecithotrophy are

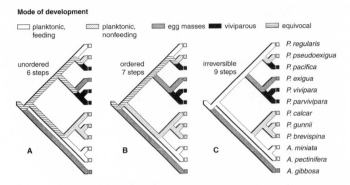

FIG. 13.10 Developmental modes in twelve species of starfish, arranged in a phylogeny established by mitochondrial gene comparisons, showing alternative hypotheses of the minimum number of steps involved in changing from an ancestral feeding type to the present ones. If the ancestor was a planktonic feeding form, as is likely from other considerations, that feeding mode was lost four times and never regained. (A) Direction of change unordered. (B) Direction of change ordered from planktonic feeding to viviparous brooding, but order is reversible. (C) Like B but order is irreversible. A., *Asterina*; P., *Patiliella*. From Hart et al. 1997.

recorded in other phyla as well. For example, among the Mollusca, larval feeding appears to have been acquired independently by Bivalvia and Gastropoda, each of which evolved a larval velum (Hickman 1999). The repeated evolution of these similar feeding larvae suggests that the intervening stages may have been lecithotrophic and shared preconditions that channeled modifications in parallel directions when larval feeding evolved. Evidently, among various spiralians, when planktotrophy becomes disadvantageous and lineages become lecithotrophic, they are able to revert to planktotrophy later under appropriate conditions.

Cleavage Patterns among Living Clades May Reflect the History of Their Larval Modes

From these data on developmental modes it is possible to derive a speculative history of the evolution of cleavage types that is fairly consistent with the distribution of cleavages in the model tree (fig. 13.9). Perhaps the earliest eumetazoan cleavages were radial, simply in preparation for development of a radially symmetrical bodyplan. When free or in capsules that permit cleavage in any direction, radially cleaving eggs produce spherical blastulae, which could well have been characteristic of early eumetazoans. Ctenophores are direct developers, and the bell-shaped bodyplan begins to be laid down as the blastula develops, presaged by some tilting of spindle axes to produce a biradial cleavage pattern, presumably apomorphic. In indirect-developing cnidarians, coelogastrulae arising by invagination are easily

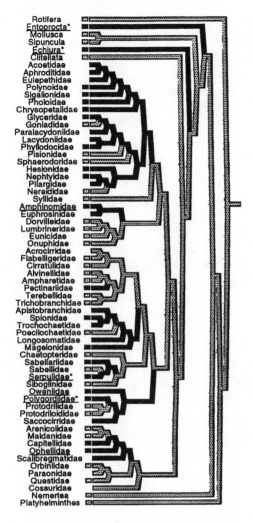

FIG. 13.11 Gain and loss of larval feeding found in a phylogenetic analysis of annelid families, together with some outgroups. This particular tree, created in a parsimony analysis, minimizes convergence and maximizes reversals in larval feeding. Black branches indicate planktotrophy, stippled branches lecithotrophy. Planktotrophy has arisen nine times, nonplanktotrophy eight times. Underlined taxa have opposed-band feeding; those with asterisks have strictly defined trochophores (see Nielsen 1995). From Rouse 2000a.

(i.e., without topological complications) elongated into planula larvae, and indeed solid gastrulae make equally simple stages in the development of planulae.

The small body sizes of early bilaterians, and therefore relatively low fecundities, suggest that their larval stages were unlikely to have been planktotrophic (see Chaffee and Lindberg 1986). However, a nonfeeding planktonic larva, similar to those found in most cnidarians, would seem to be plausible for those hypothetical early paracoelomates. In such a larva a plesiomorphic radial cleavage might be retained. Indeed, Costello and Hanley (1976) postulated that radial cleavage might be primitive for bilaterians because an early cleavage plane can divide the developing embryo cleanly into mirror-image halves. In any event there is no reason to suspect that early bilaterians were not radially cleaving. If their larvae were chiefly lecithotrophic, radial cleavage might have dominated among a diversifying paracoelomate fauna. In this event radial cleavage may have been present in the protostome/deuterostome and ecdysozoan/lophotrochozoan clade ancestors.

If this scenario is correct, it is necessary to account for the fact that all crown vermiform paracoelomates, including acoelomorphs, are direct developers. There are two points to be made in this regard. First, the adaptive value of early larvae must be judged in terms of their contemporary communities and not by today's standard, and it is possible that predation and other pressures on planktonic populations were lower in the Neoproterozoic. Perhaps it was not a great adaptive risk for small-bodied forms to reproduce via planktonic larvae, and to obtain the advantages therefrom.

Neoproterozoic seas probably did have a planktonic flora rich in prokaryotes and protistans, however, and therefore a planktonic fauna rich in heterotrophic protistans, and perhaps with some metazoan predators. There is scanty evidence as yet by which to judge the intensity of planktonic predation at that time.

A second point is that small-bodied paracoelomates with little internal space for eggs would be the least likely of metazoans to retain planktonic larvae over the course of the Phanerozoic, given the history of episodic inclement conditions implied by the many marine extinction peaks. Several of the extinction events may have removed over 60%, and one well over 80%, of the fauna (chap. 12). The loss of planktonic lineages during such times must have been significant, and the complete loss of planktonic larvae within very small bodied phyla is quite easily imagined. The direct developmental systems that were selected or that evolved during inclement times should commonly have involved a concomitant increase in yolk and the evolution of a new pattern of cleavage and of blastomere fate. The unique cleavages of some crown paracoelomate clades may have evolved from blastulae that were modified in accommodating to yolk, with cell-cell signaling systems that depended on the modified positions of the blastomeres. The unusual positions of the early blastomeres in nematodes, for example, are required for proper signaling (chap. 7). Other unusual blastomere configurations, as in many marine arthropod groups and in gastrotrichs, may also imply constraints imposed by signaling systems. If the spindle axes of blastomeres become tilted to produce cell geometries that conform to the presence of yolk bodies, a spirallike pattern would evolve. Acoelomorphs may have such a history.

Although other aspects of developmental and morphological attributes of living bilaterian phyla probably originated before the Cambrian explosion, in Early Cambrian time or even before, they are best discussed when reviewing the early appearances of crown phyla in the fossil record. Those events define the explosion, and are discussed in the following chapter.

Ectoderm, Endoderm, and Endomesoderm Are Probably Homologous throughout the Eumetazoa

Gastrulation is found in all Eumetazoa, and although there is a variety of mechanisms by which it is achieved, there is every reason to assume that all have been inherited from the earliest diploblastic forms, and thus are homologues or at least transformational homologues. If this assumption is correct, then ectodermal and endodermal germ layers are homologous throughout Eumetazoa. Mesodermal tissues also display a variety of developmental patterns. If acoelomorphs are indeed basal to Bilateria, the earliest known mesoderm is endomesoderm arising from third duet macromeres (Henry et al. 2000).

The expression patterns of some of the transcription factors implicated in endomesoderm determination, including the T-box gene *Brachyury*, have been described in several taxa (Peterson et al. 1999; Holland 2000; Technau 2001; and references therein). In Cnidaria, *Brachyury* homologue is expressed around the blastopore in an anthozoan, and in endodermal epithelium surrounding the mouth of adult *Hydra. Brachyury* homologues are also expressed around the blastopore of several protostomes and deuterostomes, and later in their larval hindgut and foregut, and later still in splanchnic endomesoderm, despite the different cell-line origins of endomesoderm in classic spiralians (from 4d) and deuterostomes (delaminating from the archenteron). As with transcription factors generally, *Brachyury* genes are expressed in other contexts as well. It appears that *Brachyury* genes have been involved with the organization of germ layers in Bilateria, functions that may have arisen from being important transcriptional regulators in diploblastic animals. Those genes must have played important roles in the origins of bilaterian organization. It will be fascinating to learn how early germ lines are determined by gene regulatory activities in acoelomorphs.

Crown Paracoelomate Bodyplans Largely Represent a Radiation of Small-Bodied Protostomes

The complexity increases found in the evolution of many crown bilaterian clades seem to be associated with body-size increases (chap. 14). Free-living clades that have evidently remained small-bodied from Neoproterozoic or Early Cambrian times—the paracoelomate phyla—have tended to retain a level of bodyplan complexity characterized by about fifteen to twenty or so cell types (table 2.3). The diverging evolutionary pathways of these forms have nevertheless led to many apomorphies. The nerve net presumably present in early paracoelomates is usually consolidated into longitudinal cords, variously connected by rings or lateral commissures. For example, the longitudinal nerve cords are multiple in loriciferans, lateral in gastrotrichs, ventral, lateral, and dorsal in kinorhynchs (with ventral being largest), ventrolateral in most rotifers, and dorsal in some rotifers, while priapulids have nerve nets and a ventral cord, with a pharyngeal nerve ring and incomplete lateral extensions around the body. Entoprocts have a ventral ganglion but no longitudinal cords. These arrangements doubtless reflect varied solutions to accommodate neurogenic patterns to the morphological changes accompanying adaptation to disparate life habits and environments.

Metazoan Complexity Increased before the Cambrian Explosion, Perhaps Chiefly during the Early Cambrian

In fig. 13.12 the estimated numbers of cell types required to produce the basic bodyplans (the stem ancestors) of each phylum are arranged on a model tree. These numbers are based on accounts of cell morphology and contents, usually as visualized by transmission electron microscopy, and do not include cell-type differences

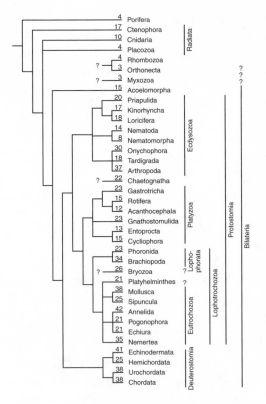

FIG. 13.12 Estimated numbers of cell types required to produce the basic (stem) bodyplan of each metazoan phylum, placed on the model phylogenetic tree. The actual number of cell types present in the hypothetical stem ancestor of each phylum may well have been somewhat more than the estimate, which is for a stripped-down animal with only essential features. Highly specialized gland and sensory cells are not tallied unless closely associated with the functional anatomy of the bodyplan. Chordata is estimated from cephalochordates, Urochordata from ascidians, Hemichordata from pterobranchs, Arthropoda from anostracans, and Mollusca from amphineurans. Data sources are mentioned in chapters 6–11.

that are identified only by molecular criteria. These data played no part in construction of the tree. As discussed in chapter 2, cell-type numbers might be taken as a rough index of the relative complexity of the bodyplans. As cell-type numbers may increase or decrease along any given branch, it is difficult to use such numbers as a guide to the complexity of bodyplans at ancestral nodes. Still, the distribution of cell-type numbers on the tree in fig. 13.12 permits some speculative interpretation of complexity trends within Metazoa. Most prebilaterian stem ancestors are estimated to have had few cell morphotypes; the complexity of the ctenophores is assumed to be derived. Bodyplans at early bilaterian nodes, as of the protostome/deuterostome ancestor and the stem ancestors of the major protostome alliances, may have had in the range of fifteen to twenty cell morphotypes. Evolution within each of those clades produced complex bodyplans with about thirty to forty or so cell morphotypes—the arthropods, brachiopods, mollusks, annelids, and echinoderms—that appeared during the Cambrian explosion. Although it is perfectly possible for relatively simple animals to produce mineralized skeletons, as witness corals (or for that matter, unicellular protistans such as foraminifers), the living bilaterian bodyplans that have mineralized skeletons are believed to have been founded by stem ancestors with at least about thirty to thirty-five cell morphotypes.

The early history of complexity within metazoans, then, appears to read something like the following. Multicellularity arose probably before 600 Ma, perhaps only tens of millions of years earlier. The early metazoans may have had two or three cell morphotypes and are likely to have been benthic. Bodyplans of the vendobionts required tissue sheets, which evidently evolved sometime before or near 570 Ma; vendobionts were probably of diploblastic grade, but whether they were organized like the surviving radiates is uncertain, to say the least. Their complexity cannot be estimated except to speculate that they may have been less complex than living radiates. Diploblastic organisms that were organized like living radiates, consisting of forms with perhaps seven to ten cell morphotypes, probably also arose before or near 570 Ma. Meanwhile, benthic bilaterians with mesodermal tissues also appeared, probably establishing forms with fifteen or more cell morphotypes, to judge by living paracoelomates, that were patterned by a large suite of developmental genes. Many of the early bilaterians may have been hemocoelic from the first, in the sense that they retained a fluid-filled blastocoel to complement muscles associated with the body wall that functioned chiefly in locomotion. Circumpharyngeal muscles evolved in forms of this grade to aid in feeding, and in some cases muscle fibers extended along the gut. As these forms attained larger body sizes, schizocoelic organ coeloms may have evolved to serve as renal and gonadal ducts and other spaces. Nerve nets became consolidated into cords as the need for integrated control of differentiated and iterated tissues and organs increased. Some of these forms were capable of disturbing the bottom sediments, and their traces entered the fossil record.

During the later Neoproterozoic and Early Cambrian, complexity must have increased in some stem lineages within all of the major bilaterian alliances, to bodyplans with perhaps thirty to thirty-five or more cell morphotypes. Some of the bodyplans of these forms, still soft-bodied and largely unknown from this late Neoproterozoic interval, can be inferred from Cambrian fossils or, with far less certainty, by speculating from their positions within the metazoan tree. For example, lobopodians are likely to have evolved within the ecdysozoan clade before the Cambrian explosion. No fossils have been assigned to lobopodians during that time, but there is reason to suspect, based on their position in the tree and on the Cambrian records of their allies, that they were present. Any such hypothetical lobopodians were presumably soft-bodied.

As the early metazoans evolved, the patterning system of the zootype arose (chap. 3; Slack et al. 1993), a genetic "map" that regulated the positions of cell types, and of the tissues and organs they compose. The zootype thus should include the genes in major developmental pathways with their signaling systems and transcription factors. The anteroposterior and dorsoventral conservation of gene expressions, characterizing Bilateria and foreshadowed in radiates, formed a developmental framework or scaffolding, a three-dimensional map within which morphological innovations could be located.

The Hox gene cluster, which mediates anteroposterior regional identities in bilaterian bodyplans (chap. 3), contains unique combinations of genes in each phylum (except arthropods and onychophorans) but tends to be stable within phyla. In Nematoda, the living phylum of paracoelomate grade for which Hox function is known, Hox genes mediate anteroposterior positioning of individual cell types. Thus, as paracoelomate bodyplans diversified, Hox clusters must have been quite directly involved in their morphological evolution, suffering gene losses or acquiring new paralogs as appropriate. As more complex bodyplans arose, leading to the Cambrian explosion, Hox genes came to control the expression of the downstream cascades of genes, cascades that were evolved during the rising complexity of antero-posteriorly arrayed structures. Evolutionary changes in these structures were then met by changing patterns or associations of downstream gene expressions, some-times involving changes in Hox gene domains, but the Hox clusters themselves were highly conserved. Today, homologous Hox genes occur in the same order in such morphologically disparate forms as limbless priapulids and flatworms, annelids with parapodia, and arthropods with legs.

CHAPTER 14

Metazoan Evolution during the Cambrian Explosion and Its Aftermath

Independent Trends in Body-Size Increases Produced the Major Bilaterian Alliances

The Degree of Disparity among Neoproterozoic Paracoelomates Is Entirely Conjectural

As emphasized repeatedly, we do not see the last common ancestors of crown phyla in the fossil record, and it is possible that the clade ancestors of all crown phyla were present by the close of the Neoproterozoic. It is doubtful, however, that the bodyplans of all crown phyla—the stem ancestors—had appeared by that time, though stem members of prebilaterians, vermiform paracoelomate phyla, some limbed ecdysozoans, small mollusks, and perhaps other stem forms, might well have been present. Between the close of the Neoproterozoic, with the appearance of a distinctly larger trace fauna, and the appearance of the relatively abundant fauna that produced small shelly fossils and that marks the beginning of the Cambrian explosion, is a period of between 13 and 23 million years, probably closer to the lower figure. It was evidently during this interval that the disparate bodyplans of many crown phyla evolved, although most were represented by stem groups. The presence of stem groups is testimony only to the later ravages of extinctions on branches of the phylogenetic tree, however, and does not certify that the crown ancestors, which branched later, represent any special advances or any superior attributes, though some may. The stem ancestors of crown phyla are assumed to have arisen from among a diverse paracoelomate fauna that was variously adapted to the many distinctive habitats across the floor of late Neoproterozoic and Early Cambrian seas. Just how much morphological disparity may have existed among those forms is unclear. It does not seem possible to identify clear synapomorphic features among the founders of major alliances of crown phyla, with the exception that ecdysozoan ancestors molted. The last protostome/deuterostome common ancestor was surely among those paracoelomates (for a review see Erwin and Davidson 2002).

Cambrian Selection for Body-Size Increases Involved Regulation of Cell-Division Cycles

The advantages of larger body sizes became significant among many benthic bilaterian lineages during the Early Cambrian, after many tens of millions of years of bilaterian evolution at relatively small body sizes. Although the reasons for those advantages are still uncertain, some things may be inferred about changes in the developmental processes that underpinned them. As for body size itself, it may be increased by cell enlargement or by cell proliferation. However, most metazoan cells are of roughly the same size, about 10 to 20 μm in diameter, and there is not a regular change in cell sizes from smaller-bodied to larger-bodied invertebrate phyla. Evolving bigger bodyplans has chiefly involved cell proliferation.

Control of mitosis has received much attention, partly owing to its involvement in cancer, but involves a large complex of molecular processes that are still not completely understood. A schematic cell-division cycle involves two resting stages, G1 and G2 (gaps), which precede and follow DNA replication stage S (synthesis) and cell-division stage M (mitosis). Entry to and exit from the stages is probably regulated by a variety of mechanisms (see discussion in Wilkins 2002), including degradation, through activities of the molecule ubiquitin, of the cyclins or kinases that set up mitosis. Increasing cell numbers would probably involve changing the checks on mitosis. Changes in growth at the cellular level could be systemic, or could be regional or quite local, depending upon the range of activity of the regulatory signals and of patterns of quenching. Growth differentials would certainly occur as body sizes increased, for example, to accommodate the disproportionate demands on two-dimensional surfaces that service three-dimensional organs. As by the Cambrian the stem groups of phyla had largely or entirely separated, the size increases would have occurred independently in most or all phyla. Thus it will not be surprising if details of the regulation of cell-division cycles vary among phyla.

The Homology of Body Cavities across Bilateria Is Unlikely

Primary Body Cavities—Pseudocoels and Hemocoels—Have Been Lost in Some Lineages but Are Main Body Spaces in Others
The presence early in the metazoan fossil record of horizontal, infaunal traces suggests the presence of some sort of tissue or fluid skeleton to facilitate peristaltic locomotion. Pseudocoels would have been more efficient than tissue skeletons at transmitting locomotory forces, and it seems likely that they were present in some of the early bilaterians. Unfortunately, we aren't sure as yet just which phylogenetic branch points were present before and which after the appearance of traces. It is not certain that all primary body cavities are homologous; although by definition they all form on the site of the blastocoel, they may have originated more than once, from among different lineages that lacked them. It is conceivable that the clade ancestors of the major metazoan alliances lacked primary body cavities (aside from embryonic blastocoels); Acoelomorpha, Rhabditophora, Catenulida, Gnathostomulida, Gastrotricha, Entoprocta, and Cycliophora are structurally acoelomate, and any primary body cavities in their ancestries have been lost or occluded. Primary body cavities even within some of the coelomates (as Sipuncula, Bryozoa,

and Echinodermata) have been lost or greatly reduced. Some protostome taxa have fairly capacious pseudocoels and hemocoels, and indeed the main body spaces in ecdysozoans are primary body cavities.

Secondary Body Cavities—Coeloms—Are Highly Functional and Are Not Likely to Be Homologous across Bilateria

Coelomic spaces, classed as either enterocoels or schizocoels, are among the characteristic features of early development that have phylogenetically interesting distributions. Their origins are presumably associated with body-size increases, and while it is not clear that the earliest coeloms originated during the size increases associated with the earliest Cambrian, it is likely that coelomic spaces were elaborated within many lineages during that time. The possibilities of enterocoely are quite constrained by the origin of mesoderm (chap. 2). When mesoderm originates from cells such as 4d and its products, there is no opportunity to capture space within mesoderm from the archenteron; any cavity inside the mesoderm must have a schizocoelous origin. By contrast, when mesoderm arises from endodermal cells along the archenteron, enterocoely is possible, but schizocoely is also possible. Therefore it is expected that the classic spiralians with 4d mesentoblasts must be entirely schizocoelous, which they are, while radialians may be either schizocoelous or enterocoelous, which indeed they are (Valentine 1997; Lüter 2000).

The origin of coelomic space from the enteron would seem most appropriate for producing perivisceral coeloms—eucoeloms—while for organ coeloms, in tissues developing away from the archenteron, schizocoely would seem more likely. It is certainly not topologically impossible that enterocoelous space could migrate to form organ coeloms even in organisms that lack perivisceral coeloms, and given the many twists and turns of development, that possibility can hardly be discounted. Still, the more complex phyla that lack perivisceral coeloms, such as the arthropods and their allies, seem all to have organ coeloms that are schizocoels. These are the forms with chiefly unique cleavages, and in general they do not seem to derive mesoderm from archenteric endoderm. If the first metazoan coelomic spaces were organ coeloms, which is quite likely, they may well have been gonocoels formed by schizocoely, which would become the earliest sort of coelom formation, but not necessarily ancestral in the sense that those spaces are homologous with other coeloms, even with other schizocoels.

As coeloms such as eucoeloms and their extensions into tentacles or introverts are likely to have been primitively associated with muscular systems of locomotion or feeding, they probably arose within mesoderm because they were evolving as fluid skeletons within muscular tissues. As fluids they are perfectly deformable yet can transmit forces generated by muscular contractions that can be antagonized by other sets of muscles. Such an enclosed space functions as a hydrostatic skeletomuscular system and should be superior to a system that transmits muscular forces within a primary body cavity that is simply bounded by somatic musculature. By specializing as the fluid portion of the muscular system, the eucoelom may become regionalized into hydrostatic skeletal elements, while the hemal system may specialize in circulatory and respiratory functions.

The assumption of the monophyly of coelomic spaces and their implied derived origins and states were once major principles used to relate the phyla, and are still under consideration today (see Budd and Jensen 2000). There has always been some skepticism on such points, however, with notions of the polyphyly and reduction of coeloms playing important roles in many alternative phylogenies. In the phylogenetic hypothesis used here, coeloms are polyphyletic. Few characters are simpler than fluid-filled cavities, and it is not difficult to visualize them as evolving many times for a number of purposes. The character of coelomic linings, epithelia with common cell orientations, has been cited as indicating that coeloms are monophyletic, but they vary much more than was realized before they were studied widely by transmission electron microscopy. Furthermore, there may well be a large component of self-organization involved in the evolution of coelomic linings; such features may have evolved repeatedly to create homoplasies.

The function to which any given eucoelomic space was first devoted and the bodyplan in which it first arose remain speculative. An assumed primitiveness of enterocoely is sometimes tied to the theory that the enterocoels are descended from cnidarian or cnidarian-like enteric lobes (chap. 4). However, it seems that, if mesodermal tissues are recruited from the archenteron lining, the simultaneous infolding of a fluid-filled space within the elaborating mesoderm is a simple developmental trick, and an elaborate recapitulatory scenario is not needed to explain it. When packets or strings of mesodermal cells are produced, whether from archenteric or 4d cells, eucoeloms filled by expressed fluids are formed by schizocoely. In at least some unregionated bodyplans that do not require bypassing of septa by a blood vascular system, eucoelomic fluids have evidently acquired physiological functions generally associated with the hemal system. Perhaps the coelomic space that might be considered a eucoelom, which is most likely to be derived from an organ coelom, is that of the cephalopod mollusks, which may be an expanded gonocoel. The other crown molluscan classes, although possessing gonadal and pericardial organ coeloms with adjunct ducts or kidneys, use hemal fluid or sea water in skeletomuscular functions. The places in the model tree where it seems most likely that a eucoelomic architecture has been reduced and essentially lost in a free-living phylum are on the urochordate branch and within the polyphyletic "archiannelid" assemblage (chap. 8).

Systems Associated with Body Cavities, Such as Blood Vascular and Nephridial Systems, May Be Homoplastic

Other features that form quite basic design elements in metazoans may be homoplasies in some bodyplans. A speculative example is provided by the blood vascular system, absent in paracoelomates but a basic feature in large deuterostomes, ecdysozoans, and lophotrochozoans. It would seem from this broad taxonomic distribution that a blood vascular system would have been a feature of the last common protostome/deuterostome ancestor. Yet recent developmental studies suggest that this need not be the case. Mukouyama et al. (2002) find that, in mice embryos, the pathways of developing arteries are regulated by molecular signals from nerves. Such a mechanism thus underlies the development of a branching circulatory system that serves tissues innervated by a branching nervous system. If this mechanism of vascular patterning is ancient (the speculative part), it may have first been expressed when body sizes increased near the Proterozoic-Cambrian boundary, entailing the evolution of circulatory systems to oxygenate internal tissues. Advanced paracoelomates, with nerve strands at least partly condensed into cords, could integrate neuronal and vascular systems developmentally with such a patterning system. The stem ancestors of the major bilaterian alliances probably separated well back in Neoproterozoic time, while large-bodied bilaterians did not appear in any numbers until about the onset of the Cambrian. Therefore it seems that blood vascular systems may have evolved independently several times in various large-bodied clades, using their neural architectures as a template.

Another example of organ systems that are likely to represent homoplasies among some phyla is provided by invertebrate excretory organs, the nephridia. Nephridia are absent in all of the phyla believed to be prebilaterians and in the exceedingly modified parasitic groups of uncertain affinities such as the mesozoans. Nephridia are also lacking in some minute bilaterian phyla, for which it is assumed that excretory requirements can be met by diffusion. Nevertheless, nephridia first evolved in early bilaterians. The simpler types are termed protonephridia, which are ectodermal derivatives consisting of a single terminal cell that functions for filtration, associated with an excretory duct opening (sometimes indirectly) to the exterior. The terminal cells are of two chief types, solenocytes (perhaps the more primitive, with one or two flagella) and flame cells (with numerous cilia). According to Ruppert and Smith (1988) the characterizing feature of protonephridia is that filtration is mediated by flagellar activity. Protonephridia are chiefly found in organisms that are acoelomate or pseudocoelomate, including the larvae of some large coelomate forms, but are also found in some coelomates that nearly or quite lack circulatory systems (fig. 14.1). More complicated nephridial types are termed metanephridia, in which filtration is by muscular forcing, the filtrate entering a coelomic cavity. Metanephridia correlate well with possession of a coelom, which may be perivisceral or an organ coelom (fig. 14.2). If metanephridia are homologous, their distribution has important phylogenetic consequences—organisms bearing metanephridia must form a clade (Goodrich 1945).

However, Bartolomaeus and Ax (1992) have reviewed bilaterian nephridia and concluded that, while protonephridia are likely to be homologous across the bilaterian phyla that have them, metanephridia are likely to have arisen more than once. Phoronid metanephridia develop from larval protonephridia by in situ growth of

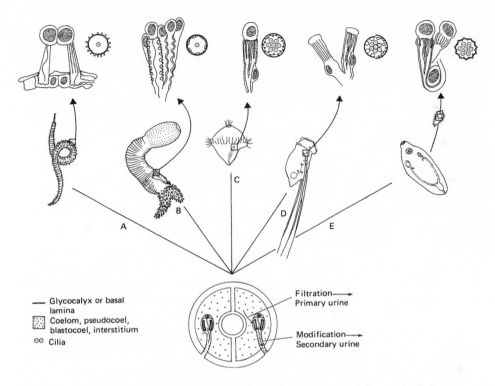

FIG. 14.1 Protonephridial systems in several phyla. The basic relations of protonephridial structures to the body cavities is shown at the bottom of the figure. A and B have solenocytes as terminal cells; C–E have flame cells. (A) Annelid, *Glycera*. (B) Priapulida, *Priapulus*. (C) Annelid larva, *Pomatoceros*. (D) Rotifera, *Asplanchna*. (E) Platyhelminthes larva, *Fasciola*. From Ruppert and Smith 1988.

a funnel, while annelid metanephridia develop de novo from a solid progenitor field following metamorphosis. Such distinctive developmental histories are taken to indicate independent origins. If this is the case, then nephridial evolution correlates nicely with a possible independent origin of coelomic spaces in lophophorates and eutrochozoans. The prediction from these data is that, if the model tree is correct, metanephridia have also evolved independently in both ecdysozoans and in deuterostomes. Those organs, which qualify as metanephridia because of how they function, are in fact rather varied in their structural details (fig. 14.2).

Nephridia are absent in Acoelomorpha, Chaetognatha, Gnathostomulida, Tardigrada, and Nematomorpha; it seems possible that protonephridia were once present in the ancestry of some of these groups but have been lost. Some Acanthocephala have protonephridia, though most do not. Some free-living marine nematodes are provided with specialized cells, termed renette cells, with tubes that lead into pores, which may be chiefly osmoregulatory but perhaps have an excretory function as well. Other nematodes have highly derived tubule systems believed to be excretory. Ascidian urochordates evidently store excretory products in paired

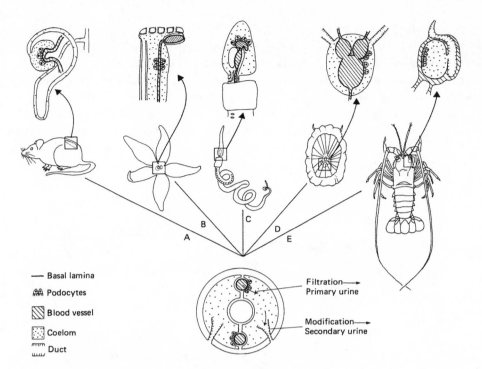

FIG. 14.2 Metanephridial systems in several phyla; the basic relations of metanephridial structures to the body cavities are shown at the bottom of the figure. (*A*) Chordata, mouse. (*B*) Echinodermata, asteroid. (*C*) Hemichordata, enteropneust. (*D*) Mollusca, gastropod, kidney. (*E*) Arthropoda, crustacean, antennal gland. From Ruppert and Smith 1988.

renal sacs that lie on either side of the heart and are associated with the epicardial organs; otherwise they seem to have no excretory organs. Malpighian tubules and the other derived excretory systems of terrestrial organisms do not bear on the relations among phyla.

Still other characters that are widespread among metazoans, but that do not seem to be arrayed within clades, may also represent homoplasies. The origins of these features were not necessarily associated with body-size increases. Examples include multiciliate cell types and cell-junction types. Some of these are tabulated and reviewed by Willmer (1990).

Body-Size Increases Are Consistent with the Early Cambrian Evolution of Planktotrophy and Divergences in Early Development

Arendt et al. (2001 and references therein) have shown that key genes involved in the development of larval features in a protostome (an annelid species) are homologous with genes expressed in similar patterns in some larval deuterostomes (particularly the genes *Brachyury,* expressed in larval fore- and hindguts, and *orthodenticle,* expressed in ciliary bands). They conclude that those similarities indicate a common ancestral larva for protostomes and deuterostomes, a ciliated form with a tube-shaped gut. This ancestral larva is a feeding form in their evolutionary model, but the data are not inconsistent with the notion that the small-bodied paracoelomates of the late Neoproterozoic had lecithotrophic larvae. Constraints against the deployment of relatively long lived planktotrophic larvae would presumably have been removed as larger body sizes evolved. Higher levels of fecundity would have been permitted, and whatever advantages favor planktotrophy now, a strategy so common in today's oceans, would presumably have been at work in the Early Cambrian seas as well. In this event the evolution of planktotrophy would have proceeded independently among different lineages in which larger, complex body plans were emerging.

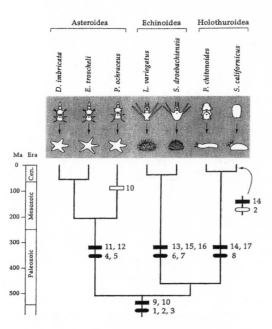

Vagaries in patterns of gene expression in larvae appear to be quite similar to those described in adults (chap. 3). Lowe et al. (2002) studied patterns of expression of *orthodenticle* and *distal-less* in echinoderm larvae (fig. 14.3). Expression in some developing features in different classes is often quite similar, but is also quite distinctive in some cases. Each of these genes has been recruited for use in the development of novel features that are derived both within and between some classes. Further, expression events for homologous developmental features have been lost differentially within and between some classes. Interestingly, in holothuroids, *orthodenticle* expression was found both in feeding and nonfeeding larval ciliary bands, although the intensity of expression was different (less in the

FIG. 14.3 Gains (solid bars) and losses (open bars) in larval morphology and in larval expression of *distal-less* (Dlx) and *orthodenticle* (Otx) found among some echinoderm taxa. Morphological features (ovals): *1*, podia; *2*, feeding larvae; *3*, adult rudiment on left larval side; *4*, brachiolar complex; *5*, subtrochal cells; *6*, larval skeleton; *7*, adult rudiment within vestibular invagination; *8*, anterior torsion of rudiment. Genetic features (rectangles): *9*, Dlx expression in podia; *10*, Dlx expression in hydrocoel; *11*, Dlx expression in brachiolar complex; *12*, Dlx expression in subtrochal cells; *13*, Dlx expression in vestibular ectoderm; *14*, Otx expression in ciliary band; *15*, Dlx expression in scattered cells in ciliary band; *16*, Otx expression in early ectoderm; *17*, Dlx expression throughout ciliated ectoderm. Distinctive expression patterns are commonly associated with distinctive morphological change. From left to right: *D.*, *Dermasterias*; *E.*, *Evasterias*; *P.*, *Pisaster*; *L.*, *Lytechinus*; *S.*, *Strongylocentrotus*; *P. Psolus*; *S.*, *Stichopus*. From Lowe et al. 2002.

nonfeeding type). The recruitment of homologous regulatory genes to mediate the development of analogous structures, which are in some cases transformed homologues, has been common in adults (e.g., limbs and eyes in arthropods and chordates), and appears to be prevalent among larvae as well. Until changes in the roles of gene expressions that are associated with morphological and developmental novelties can be traced and understood, it seems premature to reconstruct ancestral larval forms (or adult forms for that matter) solely from the expression patterns of regulatory genes.

Having said that, it is possible to reconcile the use of similar genes in similar features in independently evolved planktotrophic larvae. Acoelomorph flatworms appear to be stem bilaterians, and if so, Neoproterozoic paracoelomates included forms with blind guts. If such a form possessed a lecithotrophic larva, it would not be surprising if larval foregut development was mediated by *Brachyury*, thus being expressed around the single gut orifice, with *distal-less* associated with ciliary patterning. Subsequent evolutionary trends within diversifying bilaterians would have produced adult and therefore larval forms with complete guts, with *Brachyury* expressed around the larval mouth and the anus and thus mediating development of both gut openings. In this scenario it so happened that there were at least two separate inventions of larval planktotrophy, one involving protostomy and the other deuterostomy, associated with independent body-size increases. Both of those larval types would have used ciliary tracts mediated by *distal-less* in feeding, in swimming, and in maintaining orientation. Differences between the larvae of the protostome and deuterostome crown ancestors would thus represent synapomorphies of those clades, with feeding being a functional homoplasy, while similarities in the use of key developmental genes would represent plesiomorphies.

There Are Similarities in the Gross Morphological Adaptations of Some Phyla in the Separate Alliances

The adult bodyplans of the crown ancestor of each major bilaterian alliance, and by implication the stem ancestors as well, are generally agreed to be designed for benthic life. This is clearly recognized in the Trochaea hypothesis and is consistent with the colonial, set-aside, and planuloid hypotheses as well. The presumably rapid radiations associated with the Cambrian explosion must have built upon bodyplan differences among stem groups and further established the significant bodyplan disparities that form the basis of the recognition of phyla. During the early radiation of metazoans that were durably skeletonized, or for which we have a fortuitous fossil record, we find distinctive but short-lived forms that include stem taxa of the crown alliances. By analogy with Phanerozoic patterns, these short-lived stem forms may have belonged to higher-turnover branches within their clades.

Although their radiations were entirely independent, there are some rather generalized developmental and morphological convergences or parallelisms among bilaterian alliances. For example, within the lophotrochozoans and deuterostomes, feeding larvae that are not modified juveniles (i.e., trochophores and dipleurulae) have evidently been intercalated between embryonic and juvenile stages. Within each of the alliances is a segmented form—Cephalochordata, Arthropoda, and Annelida—at about the same level of complexity. Each bilaterian alliance also contains fluid-filled wormlike forms of rather similar complexities—Enteropneusta, Priapulida, and Echiura. Deuterostomia and Lophotrochozoa each contain soft-bodied coelomate forms that feed by ciliated tentacles—Pterobranchia and Phoronida—which are also of similar complexities. The lack of ciliated larvae and ciliated suspension-feeding systems among Ecdysozoa is easily explained: the molting habit essentially precludes the evolution of ectodermal ciliary tracts. To be sure, there are body types that are unique to a given alliance: the chordates, for example, are unique and unparalleled in the morphological complexity of their derived members. Nevertheless, the rough sorts of architectural similarities found among some phyla in clearly separate alliances suggest that there were some general parallelisms in the sorts of adaptations that were worked out independently as their bodyplans evolved, near or during the Cambrian explosion. Biomechanical solutions to life on the seafloor have been organized around only a few thematic designs, such as that of a hollow-bodied worm with a hydrostatic skeleton. The themes were sometimes employed as the central features of several bodyplans that arose independently. The thematic organizations entailed lesser, secondary biomechanical requirements and had physiological consequences that were also met independently, commonly in similar ways in groups that were only distantly related.

Indeed, as time went on, the morphospace within each of the expanding phyla become ever more densely populated by diversifying lineages, and the more easily discovered morphological solutions to adaptive problems were produced independently in increasing numbers of lineages (Wagner 2000). Barring a new adaptive breakthrough, the character states that were more easily available to a given bodyplan or subplan were soon used up. This exhaustion of morphological character states, as Wagner has put it, led to increasing amounts of homoplasy over time and an iteration of many similar morphologies through time within clades. Perhaps this trend can be analogized to the development of stock phrases or ritual statements within languages—as in love letters, or in the Miranda warning. Thus essentially from the first, metazoan morphology has had its stereotypical aspects, large thematic ones for bodyplans, smaller ones for body subplans, and minor ones for lower taxa, all complicating the lives of morphological systematists.

Deuterostomes Radiated in Two Major Clades, One of Which Evolved a Notochord, and the Rest Is History

The founders of the deuterostome clade of course had exactly the same bodyplan as the founders of the protostome clade. If this hypothetical form was indeed a paracoelomate, then the deuterostome coelom arose independently of the protostome one(s). It seems likely that evolution of the deuterostome eucoelom was associated with hydroskeletal functions that supported locomotion. The triregionation of primitive deuterostomes is surely related to the provision of functional independence in different body regions. The protocoel is the earliest coelomic space in developing enteropneust larvae (see review in Nielsen 1995), though whether this reflects an evolutionary precedence is unknown. One branch of the deuterostomes remained benthic, and while those phyla, Echinodermata and Hemichordata, are interesting, they do not stand out as complex groups. The other branch took to the water column and developed swimming behaviors based on a notochord, radiating into the clades of Chordata, some of which are spectacularly successful and quite complex, and probably in some cases (as the ascidians) they recolonized the benthos.

Pharyngotremy is one of the more important deuterostome synapomorphies. In pterobranchs there is a hint that pharyngeal slits are related to the intake of large volumes of water in feeding currents; slits are not present in the minute *Rhabdopleura*. Assuming that early hemichordates were small and fed by ciliary currents, significant increases in body size with the concomitant increases in feeding requirements could involve the evolution of increasingly strong feeding streams, leading to a water surplus at the mouth. A speculative possibility is that, in a small-bodied ancestral group, lateral oral grooves were developed to permit water (and perhaps rejected particles) to escape at the mouth, perhaps aided by lateral ciliary tracts in the buccal and pharyngeal regions. As body size increased, these grooves became deeper reentrants, eventually closing anteriorly to preserve support for the mouth region, and creating pharyngotremy. The resulting pharyngeal currents were subsequently exploited for respiration, unnecessary at small body sizes but important for larger forms, such as enteropneusts, in which slits are multiplied. The development of gill slits, such an ordinary sort of morphological adaptation among the many spectacular inventions of the Early Cambrian, led to the vertebrate jaw. Many of the early echinoderm clades appear to have had some form of pores or slits for the escape of respiratory currents, but crown echinoderms do not.

Because of the rise of vertebrates, the origin of the notochord in invertebrates has special evolutionary significance. Notochord-based locomotion represented something new in the way of locomotory mechanics, and the groups that possess it, Urochordata and Chordata, are certainly most easily visualized as having been founded by pelagic rather than benthic forms. In segmented invertebrates the body compartments are semi-independent during locomotion, and body flexibility is produced at the sites where the segments or their tagmata join. In the invertebrate chordates the presence of a springy longitudinal rod permitted a new sort of flexibility in a continuously graded body motion along the anteroposterior axis, integrated so as to unite the locomotory system. To be sure, there are individual muscle bundles or myotomes, required to produce waves of flexure along the notochord, but because of the notochord this locomotory architecture does not produce segmental responses in the way segmentation does in arthropods and annelids. The most obvious function of a primitive notochord system is for swimming, although the system is also used for wriggling through loose sand (Webb and Hill 1958). The dorsal development of a coordinating central nervous system seems likely to have been coadaptive with the evolution of a dorsal notochord system, furnishing the chief basis of locomotion. The early deuterostomes do not appear to be exceptional invertebrates, although each group has its own peculiarities and interest. Even the cephalochordates, though they had evolved a unique bodyplan with unusual locomotory mechanics, are associated with so many other invertebrates with novel features that they do not stand out as exceptional, although they certainly have lots of Hox genes. There seems to be no reason that, when first evolved, invertebrate chordate bodyplans would have been identified as foreshadowing the great complexity to be found in modern vertebrate bodies.

Adaptive Trends within Ecdysozoa Implicate the Molting Habit as Providing Both Major Opportunities and Constraints

In contrast with chordates, major protostome phyla appear to have been primitively benthic, and their locomotory techniques involved creeping, wriggling, or ambulating on or within the seafloor. At least in larger-bodied protostome phyla, coordinating functions are associated with a ventral position of the main anteroposterior nerve cord(s), perhaps evolved in coadaptation with their essentially ventral locomotory mechanics.

At first glance the molting habit does not seem like a very good candidate for an important synapomorphy for a major metazoan alliance, but as it characterizes Ecdysozoa (Aguinaldo et al. 1997) and therefore most living species of animals, it is reasonable to believe that it may have some basic biological consequences. Molting implies the presence of a cuticle, which, though it may be flexible, can expand very little or not at all and thus cannot accommodate much growth. Presumably some of the advantages that would underlie the evolution of such a cuticle are biomechanical, in helping to control body shape and providing some attachment for body-wall muscles. Relatively thin cuticles may support muscle attachments at minute or small body sizes, but as body size increases, cuticular stiffness must increase proportionately to support similar muscular functions. Longitudinal body-wall muscles predominate in living paracoelomate ecdysozoans. It is possible that ecdysozoan trunk musculature was originally involved in a complex of feeding and locomotory functions, supporting a hydrostatic system associated with a buccal pumping apparatus. As mucociliary locomotion at small body sizes was eventually replaced with locomotory modes based instead on body-wall muscles, in some cases antagonized by pseudocoels, this more powerful locomotory mechanism entailed a stiffer cuticle. Most vermiform ecdysozoans use some sorts of cuticular anchors during creeping or burrowing. Once formed, such structures cannot be enlarged during growth, but if the integument is molted, they may be replaced by larger structures, functionally appropriate to larger body sizes, when new cuticle is secreted (Valentine and Collins 2000). In this scenario, molting originated to assist anchoring in a protoecdysozoan that crept or burrowed via peristaltic waves or perhaps used an introvert—that is, did not employ a mucociliary locomotory system. The priapulans, which employ peristalsis in burrowing, are largely restricted to very soft sediments, and it may be that the ecdysozoan cuticle has discouraged evolution of efficient peristaltic burrowing in firm sediments.

Whatever the origin of stiffened cuticles and molting habits, it turned out that the acquisition of molting would be a key feature in the evolution of most of the world's animal species. Evidently a radiation of paracoelomate molting forms occurred, eventually producing lobopodians with their locomotory limbs and finally, with sclerotization of the cuticle, producing arthropods. Some of the various organs that appeared in larger lineages, needed to accommodate physiological requirements at larger sizes, involved intramesodermal spaces for their operation, or for communication with the exterior or with each other, and organ coeloms became part of the bodyplans of these forms. Hydrostatic requirements, however, were mediated chiefly by hemocoels, possibly homologues of the primary body cavities of earlier paracoelomates.

Several phyla of paracoelomate grade survive within Ecdysozoa, though they do not seem to form a clade, and one (Nematoda) may be secondarily reduced. The appearance of limbs within Ecdysozoa can be linked to increasing locomotory efficiency, probably in larger creeping organisms. The ancillary locomotory anchoring devices, which include spines, flanges, scales, annulations, various rugae, and body-wall protuberances, occur widely in ecdysozoans. A passively employed series of outpocketing anchors in the body wall could become more efficient if body-wall musculature were to become involved in actively manipulating those protuberances to provide some locomotory impulse. On some sorts of substrates, such as hardgrounds, epifaunal creepers (peristaltic or perhaps pedal creepers) would be able to reduce drag on their ventral surfaces by lengthening such protuberances and strengthening the associated muscular systems. Continued evolution of such organs could eventually raise the ventral surface entirely off the substrate, completing a transfer of locomotory activity from the body wall to what have become limbs (Valentine 1999). In a nonsclerotized or lightly sclerotized hemocoelic ecdysozoan, those limbs would be lobopods, becoming jointed if a more rigid exoskeleton evolved, as occurred in the arthropod bodyplan.

The great diversity achieved by the arthropods seems most likely to be a direct consequence of their bodyplans; their segments provide an architecture that is capable of great morphological variety by elaboration or suppression of features on individual segmental modules, in many combinations. Not all segmented organisms display quite such a morphological bounty, however; lobopods and annelids are cases in point. It is therefore suggested that it is the sclerotized exoskeleton, in combination with the modular architecture to be sure, that provides the key to the morphological diversity of arthropods. The exoskeleton can be perfectly molded to even rococo elaborations. Furthermore, the possible varieties and combinations of structures are multiplied further because the appendages, segmented (jointed) and modular themselves, are also capable of such modifications. Thus the seemingly limitless morphological variations on the arthropod theme are based on the modular bodies and appendages as supported by the sclerotized exoskeletons. The small body size of terrestrial arthropods, and the resulting complexity of the habitats at their disposal, may well represent another key factor in their diversity on land.

The diversity of early segmented ecdysozoans that was achieved during the Cambrian explosion is striking (Jacobs 1990), and as they include many stem groups with extinct bodyplans and bodyplan variations, they pose challenges both for systematics and evolution. Again, the possession of modular exoskeletal elements permitted numerous modifications to the patterns of tagmosis and of appendage number, morphology, and placement, features that are diagnostic of major clades today. The evolutionary inventiveness was surely aided by the modular nature of the regulatory genome (Jacobs 1990; Valentine and Hamilton 1997), based as it is on key nodes in the trees of gene-expression events. The correspondence between morphological and genetic modular architectures is likely to be responsible for the variety of body design found among the more complex Cambrian ecdysozoans.

In marine representatives of Ecdysozoa, the larvae appear to have evolved from instars of direct developers, and most have nonclassical cleavage patterns. These cleavages are sometimes characterized as "modified" spiral or radial, depending on the author (although cleavage in Priapulida is almost always reported as radial). Among the arthropods, some taxa show radial cleavages, some show a spirallike pattern (although the blastomeres do not have the classic spiralian fates), and some show unique patterns. Perhaps the plesiomorphic cleavage was lost in Ecdysozoa, presumably when development shifted from indirect to direct modes, with the evolution of yolkier eggs. As the resulting cleavage patterns vary widely among ecdysozoan clades, the shifts must have occurred after the various clade ancestors had diverged. Whether the advent of molting and concomitant suppression of motile ectodermal cilia in ecdysozoan adults was a factor in the loss of ciliated larvae is uncertain. At any rate, when planktonic larvae were again favored, ciliated larvae did not evolve, but juvenile stages were modified for planktonic life instead.

Lophotrochozoan Phylogeny Is Problematic, Suggesting a Rapid Radiation from the Crown Ancestor

In Lophotrochozoa the relations among phyla whose bodyplans appear during the Cambrian explosion are even more uncertain than in the other major alliances. Although flatworms have been considered to be basal to this clade on morphological grounds, giving rise to mollusks for example, little support for this relationship has been forthcoming from SSU rRNA studies. Indeed, rhabditophorans lie outside the SSU rRNA node on which the concept of Lophotrochozoa was founded (Halanych

THE CAMBRIAN EXPLOSION AND ITS AFTERMATH 511

et al. 1995). As rhabditophoran cleavage is in classic spiralian quartets and their endomesoderm is derived from blastomere 4d, it is hard to accept that they are not closely related to the classic spiralians. Why classic spiral cleavage might have arisen is uncertain. Spiral packing is more compact than radial, and it has been suggested that this close packing provides an advantage. Thus it might be hypothesized that spiral packing originated as one possible response to crowding of blastomeres by yolk within the egg capsule, presumably within a lecithotrophic lineage. Another possible value of spiral cleavage is that it places each of the blastomeres in contact with more cells. In early embryos in which cell-cell signaling is dominant, communication with more cells could conceivably be advantageous. Costello and Henley (1976) imply that embryos with spiral cleavage can better develop into trochophore larvae than can those with radial cleavage. The blastulae resulting from cleavages are commonly coeloblastulae in both radially and spirally cleaving phyla, however, the major exceptions being among the more derived clades of these phyla. That spiral cleavage is particularly adaptive to the trochophore bodyplan is not at all obvious.

There may well be quite a number of "empty" nodes within the eutrochozoan tree, some of which were occupied by missing bodyplans. If a flatworm bodyplan is ancestral to the mollusks, then the stem ancestor of mollusks was probably a creeping form. The eucoelomate eutrochozoans, on the other hand, may be descended from a burrowing form, perhaps a sister of the molluscan ancestor. The hypothesized history of complexity is somewhat similar along both these postulated creeping and burrowing branches. Acoelomate and pseudocoelomate forms at about the fifteen to twenty cell-morphotype level, which have left fossil traces, evidently gave rise during the late Neoproterozoic and Early Cambrian to forms with thirty or forty cell morphotypes. Some of these forms became skeletonized and contributed to the mid–Early Cambrian explosion, while others were soft-bodied or at least not durably skeletonized and if not represented in lagerstätten are unknown among the body fossils of the Cambrian.

The aplacophorans, basal living molluscan classes, are vermiform; they may have descended from creeping paracoelomates. Although they are somewhat elongate bilaterians, aplacophorans do not display the sort of coordinated seriation among organs that characterize segmented groups. Some molluscan clades do have strongly seriated features, such as chitons and especially tergomyan monoplacophorans, but this architecture was not developed in a pattern of distinctive anteroposterior modules analogous to either arthropods or annelids. Rather, molluscan modules are characterized by differential expressions, some essentially hypertrophied, and by differential rotations or other displacements (chap. 8), to produce the different bodyplan themes (see Ponder and Lindberg 1997).

Annelids, classic spiralians with trochophores, must share a common spiralian ancestor with mollusks, but have quite a different pattern of bodyplan evolution. The most influential discussion of annelid bodyplan origins has been by Clark

(1962, 1964). In annelids that burrow actively and continuously, such as burrowing predators, some mechanism restricts the flow of fluid between coelomic compartments; the septa are entire or are so constructed as to retard or prevent intercompartmental exchange. In burrowing annelids that are fairly sedentary, such as those that form a burrow and live in it as a semipermanent domicile—commonly as suspension feeders—many of the septa are reduced and much or all of the trunk operates as a single mechanical unit. The implication is that an undivided trunk coelom is sufficient to permit burrowing but is less efficient than a compartmentalized one if burrowing is an everyday activity. If the annelid eucoelom and its segmentation evolved as a burrowing system, the earliest annelids may have been closely segmented and either lacked parapodia or had small parapodia as auxiliary gripping devices. Perhaps, then, the earliest annelids were oligochaete-like in structure (Clark 1969; Fauchald 1975). Another possibility is that segmentation evolved in support of epifaunal locomotion, with parapodia evolved to supplement or replace peristalsis (see Mettam 1971, 1985). The parapodia can have evolved from body-wall protuberances that helped grip the substrate during locomotion, somewhat analogous to the evolutionary pathway envisioned for lobopod limbs.

The brachiopods and phoronids seem unequivocally associated with the eutrochozoans by SSU rRNA data, yet they differ in fundamental developmental features from classic spiralians and have some characters otherwise found only in Deuterostomia (chap. 9). One solution to this enigma is to consider these phyla as branching basally to the classic spiralians and thereby sharing plesiomorphic developmental features with the deuterostomes that are lost in eutrochozoans (Valentine 1997). The protostome/deuterostome ancestor and the ecdysozoan/lophotrochozoan ancestor would thus have been radial cleavers and have had mesoderm derived from archenteric cells. In this event, when eucoelomic spaces evolved in larger-bodied forms in both deuterostomes and lophophorates, they were enterocoels. The planktotrophic larvae of brachiopods and phoronids display adult characters and seem highly derived.

A vermiform lophophorate is commonly proposed as ancestral to these phyla. Such an ancestral form, presumably lacking some of the specialized features of living lophophorate anatomies, might require twenty-five to thirty cell morphotypes. Some protolophophorate branches might have invaded hardgrounds, where epifaunal brachiopods may have originated and radiated, attaching to rocks or to shells, while other branches, such as the phoronids, burrowed. The planktotrophic larvae of these phyla are not classic trochophores and may have descended from lecithotrophic larvae of paracoelomate ancestors, becoming planktotrophic independently of the classic spiralian larvae.

Bryozoa are allied to protostomes in SSU rRNA trees but do not appear to be most closely related to brachiopods and phoronids on molecular evidence, despite having radial cleavage. Morphologically, their lack of respiratory, excretory, and possibly circulatory systems are presumably derived features that can be ascribed to their minute body size, while their lophophore suggests an alliance with brachiopods and phoronids, although it has unique characteristics.

The bodyplans of lophotrochozoan phyla are not much more diverse than those of Ecdysozoa, but uncertainties about the relations between the disparate developmental patterns, and the extent of the clade, make it difficult to formulate hypotheses for plausible evolutionary pathways on developmental or morphological bases. As SSU rRNA data do not resolve these problems, I shall speculate that the general historical pattern inferred for deuterostomes and ecdysozoans may underlie lophotrochozoans as well. That is, lophotrochozoans arose from their last common ancestor with ecdysozoans or from a stem group within their branch. Their clade ancestor was a small-bodied paracoelomate, probably radially cleaving and with planktonic but nonfeeding larvae. One branch evolved into the phoronid + brachiopod group, another branch, probably including flatworms, acquired spiral cleavage and radiated into Eutrochozoa. If this is correct, eucoelomic architectures arose twice within this alliance.

The Remaining Protostomes Are Still a Troublesome Group

In general, systematists have had little success in placing the paracoelomate phyla within the tree of life. The former aschelminth assemblage of phyla has been characterized variously as a "ragbag group" (Inglis 1985), "odd bedfellows" (Ruppert 1991b), and an "unholy alliance" (Willmer 1990), testifying to a certain frustration with these difficult forms. Although the molting paracoelomate phyla can be assigned to Ecdysozoa, and the colonial Entoprocta and Bryozoa to Lophotrochozoa, the remaining groups are still in somewhat of a phylogenetic limbo. For consistency with the general scheme of evolution adopted here, it is speculated that the nonecdysozoan paracoelomates arose from (probably several) stem groups of the protostome clade that excludes Ecdysozoa. Considering the nature of the evidence, this speculation is not falsified by what is now known of their relationships, and is consistent with such trace evidence as is known to exist in the fossil record. This assemblage includes most of the Platyzoa of Cavalier-Smith (1998) and Giribet et al. (2000).

The Cambrian Explosion Produced Widespread Homoplasy: A Summary

During Early Cambrian times, increasingly elaborate anatomical structures must have been required to meet the demands of the larger body sizes that were evolving in many lineages. Neoproterozoic bilaterians were largely creepers and shallow burrowers on or just below the seafloor. Many of the Cambrian forms included vertical burrowers and ambulators. These changes in body size and style imply the evolution of new design elements such as circulatory and excretory systems, specialized respiratory surfaces, and enhanced locomotory modes. The well-known principles of physiology and biomechanics that are associated with body-size

differences were involved in size and complexity increases within the evolving lineages, independently producing many similar morphological solutions. At the same
time the radiation of the larger forms into disparate habitats engendered novel morphological features in response to the opening of new ecological opportunities. One
result was the organization of a number of bodyplans whose living branches are
recognized as phyla.

As many of the physiological and ecological problems faced by the early members of those clades were solved in similar ways, there is a sort of common aspect to
bilaterian anatomies. This commonality has been reflected in the way in which their
features were named and described in comparative literature. It seemed likely that
those features were homologues—after all, they were similar evolutionary products
with analogous functions—and the same descriptive terms were applied across different clades. In many cases similar features have indeed evolved as modifications
of common ancestral features and represent true homologues. However, metazoan
anatomical terminology has in some instances effectively masked true differences
in evolutionary histories among organs and other morphological structures. The
status of many cases as either homoplasies or homologies has yet to be sorted out.
An important result has been that the application of cladistic methodologies to de-
ciphering the phylogeny of higher taxa from morphological characters suffers from
mistaken character definitions, an ongoing problem (chap. 4; see Jenner 2001).
The task of assembling early taxa into a Linnean hierarchy suffers from the same
difficulties.

Much Evolution of the Developmental Genome Occurred in the Service of Bodyplan Originations: A Summary

The rise of new and more complex bodyplans involved the deployment of body-
design elements, such as tissues and organs, into new geometries, as well as involving
the invention of novel types and combinations of elements. The varied bodyplans
postulated among Neoproterozoic paracoelomates evolved as their lineages de-
ployed across environmental gradients or discontinuities. The major bilaterian patterning genes were presumably involved directly in mediating these morphological
changes (chap. 13), resulting in the evolution of distinctive developmental gene
combinations, as among Hox clusters. As complex bodyplans arose, new developmental pathways were elaborated in service of novel features, but many patterning
genes and gene clusters were conserved as upstream regulators. As these changes
proceeded, some morphological elements retained a topological integrity that helps
us trace their evolution within some assemblages of phyla—germ layers and their
principal derivatives, for example. Other elements, such as serial muscle blocks,
organs such as nephridia, and so forth, once assembled within a stem lineage, can
be repeated and modified as bodyplans are diversified. The elements, underlaid by

suites of regulatory circuits tied together by key upstream regulators, can be expressed as modules. It seems clear that genetic elements within modules can be recruited separately and perhaps shuffled as new modules evolve. The modules appear to be arranged in a rough hierarchy, which, as in other biological hierarchies, emerged with the evolution of increasing complexity, both morphological and genomic. Thus module recruitment and shuffling may occur in supermodules and in submodules and sub-submodules, and so forth, within the hierarchy.

The recruitment of modules is made possible by the evolutionary flexibility of the mechanics of *cis*-regulation, operating within a system that is, in Simon's phrase (1962), semidecomposable. Even though the evolutionary gene tree may be obscured by a network of developmental signals, there must be nodes at which at least near-independence of downstream gene-expression cascades permits the duplication, transfer, or silencing of a developmental module. The ability of the modules to maintain a significant degree of independence has thus been critical to body-plan evolution. The independence is achieved because, although many regulatory genes have many expression events and thus are responsible for many developmental functions, the individual expression events, mediated by a *cis*-regulatory signal, are not necessarily pleiotropic when scaled to the level at which they occur—they do not necessarily affect more than one function (see Stern 2000 for an excellent exploration of this situation; see also Wilkins 2002). Hence, even developmental variants with relatively large effects may be scrutinized by selection for their own unique contributions to fitness, without the confounding difficulties of deleterious pleiotropies, so long as their effects are confined within a module. To be sure, modules at higher levels of the hierarchy are more inclusive and, one would expect, should be less likely to contribute to fitness when altered, though one can imagine that, say, some high-level heterochronic or heterotopic changes might be successful in promoting the invasion of new environments.

Among the most puzzling feature of metazoan macroevolution is the relatively abrupt appearances of higher taxa, as evidenced by the appearance of phyla during the Cambrian explosion itself, and by class- and order-level miniexplosions, followed by a tailing off of the appearance of evolutionary novelty within those clades (chap. 12). The two leading explanations are that constraints are imposed ecologically, by the "filling" of adaptive zones, or are imposed developmentally, by the rise of complex interconnections within genomes that prohibit or at least inhibit major changes. Evolutionarily, the developmental arguments boil down to the question of whether selection tends to favor a continuing integration of developmental genomic modules or tends to favor their independence. One can imagine advantages for either case, but the documented persistence of at least semi-independence among modules within some genomes on a half-billion-year time scale suggests to me that independence is commonly selectively advantageous. Whether such an advantage may stem from selection within populations or from sorting at higher levels is unclear, considering the underexplored qualities of the architectures—trees, hierarchies, and networks—that underlie the complexity of metazoan genomes and morphologies.

Why Are Problems of Early Metazoan Evolution So Hard?

If Only It Were Just the Data

Trying to arrive at plausible interpretations for the early history of Metazoa is certainly a lot of fun, but it is also very frustrating. At the end of the day one is reasonably certain of the correctness of one's conclusions for only a limited range of the evolutionary problems. For nearly every facet of early metazoan history there is an array of hypotheses that cannot be definitively falsified by the available data. It therefore might be hoped that, as more data become available, fewer hypotheses will be left standing and, eventually, a strongly supported consensus will emerge. Hopes such as these imply that the chief problems are associated with a lack of data—half a billion years and more of earth and life history have obscured those distant events, but continued work with modern techniques, and with such improved techniques as will be developed in the near future, will soon provide much additional information. Our ability to interpret the Cambrian explosion and its prelude has certainly increased spectacularly over the last decades, and important further advances are clearly within the reach of the evidence and techniques that are available today. More research—more paleontology, more evo-devo biology, more molecular phylogenetics—and we shall know lots more.

There are gaps in the available evidentiary materials that can never be filled, however, and therefore events within the gaps that cannot be established from direct observations. For example, only crown taxa are available for study by molecular developmental techniques. It has become quite clear that nearly all the phyla were founded by extinct, stem and prestem taxa that were not members of other crown phyla, and within which their bodyplans were assembled. The key evolutionary changes in genome organization that were associated with the rise of the various bodyplans are simply not available to be traced through series of ancestral-to-descendant conditions. If those gaps are to be filled, we must rely upon some sorts of first principles of genome evolution that constrain the evolutionary routes between the developmental systems that can actually be studied—those within the available living animals.

It might be hoped that paleontologists would discover the remains of the extinct stem groups, and thus at least close some of the gaps with morphological evidence. If this can be done, it may also provide some clues to the underlying evolution of the genome. Many examples of stem groups have indeed been described, throwing much light on patterns of the early metazoan radiations within phyla. For the most part, however, it is not clear that the stem groups discovered to date narrow the gaps between phyla. Many of the early groups are of obscure affinities, and perhaps some will eventually yield information on the nature of "gap taxa." Furthermore, there is a hopeful sign in the continuing discovery of fossil lagerstätten containing interesting fossils from critical late Neoproterozoic and Early Cambrian times. Just within the last few years there are reports of new finds around the margins of Neoproterozoic western North America, in Alberta, British Columbia, Nevada, and California. But for the present we must soldier on with the gaps, and it seems unlikely that many will be filled, considering the incompleteness of the fossil record, especially with respect to the sorts of organisms—small, soft-bodied, and perhaps rare—that are likely to have characterized many prestem and stem taxa.

Perhaps the molecular phylogeneticists are in the best position to cross some of those gaps, at least for crown taxa. In phylogenetic systematics, interpolation is the basis of the methodology. The techniques are set up to recognize the sorts of sequence changes that occur during gaps—the time that has elapsed since the life of the last common ancestor of the taxa under study. To be sure, differential evolutionary rates, reversals of sequence changes, base substitution biases, and other phenomena must be dealt with in interpreting patterns of sequences of genes or gene products. But there are lots of genes, and lots of crown species, and molecular phylogenetics may be the one case in which more data, which are potentially available in large quantities, will soon resolve a number of the present uncertainties. This would seem to be an extremely promising prospect, as establishing the nature of a pregap ancestor and its postgap descendants are basic requirements in the reconstruction of an evolutionary pathway connecting them. Many of the pathways that have been hypothesized in the literature are clearly not between actual ancestor-descendant pairs, and uncertainties associated with the phylogenetic tree at present permit many hypotheses to stand unfalsified. Eliminate those uncertainties, and many falsifications will ensue. But this approach helps only indirectly with stem taxa.

The Incredible Richness of Choice

Gaps in the data are not the only devils that make paleobiology difficult. There is an ultimate devil that will preclude any very complete knowledge of life's history; it lurks in the epistemology of paleobiological inference (see for example Ruse 1973, chap. 5; Elsasser 1987; Richards 1992; Boyd and Richerson 1992; these authors do not necessarily agree on other points). Many of the processes in evolution involve combinatorial systems that can generate enormous numbers of alternative outcomes, only a relative few of which can be realized. Some processes are very small scale, such as those that rely on stochastic events such as mutations that establish one new molecule or one new useful molecular developmental pathway rather than another—either of which could be preserved by selection. We cannot predict the character or sequence of the next mutations. A larger-scale feature is the nature of the changing adaptive landscape that directs fitness pathways, which is unknown and probably unknowable for living organisms in nature (Lande 1986), much less for extinct ones. The course of evolutionary change across unknown landscapes is unpredictable. Near the largest end of this scale, mass extinctions sort taxa and bite deeply within phylogenetic trees, pruning away major branches and drastically thinning others, and only the survivors can contribute to the future biota. Part of survival in such circumstances is surely stochastic—being in the right place at the right time during a meteorite impact, for example. But without knowing where and how the next large meteorite will strike, we cannot predict survival.

Once something happens at any of these scales, it becomes possible to explain it if there is adequate information—one can sometimes reconstruct which "choice" was made. Historical gaps remove all signals from the record of gap events, however, leaving us only with their consequences and with inadequate information—there is no way of being sure which of myriad possible events occurred and which of alternative processes operated during gap intervals, to shape what we observe afterward. In other words, in evolution there is not just one set of "necessary" conditions for an observed outcome, but huge numbers of alternative sets of conditions, which occur at each level in the evolutionary process, from molecules to clades, and different combinations of which can lead to a sufficient explanation in appropriate circumstances. Clearly there are some ultimate limits to our ability to explain the past. Of course, we are very far from approaching those limits as yet for any level of evolutionary history concerned with the origin of phyla. Plenty of work remains within all of the many fields that contribute to understanding the history of life. At least in interpreting the past we are far better off than in forecasting the future.

Closing Thoughts

How to explain the Cambrian explosion? In this book I have been concerned almost entirely with a side of early metazoan evolutionary history that involves morphological evolution, and ecological relations chiefly among organisms. There is another side on which I have been nearly silent, that of relations to the physical environment. Incomplete though they are, we nevertheless have lots of signals from the morphological side of evolution—the fossils themselves, aided by our knowledge of relationships within the living fauna and of biological processes in today's world—that can inform plausible hypotheses about what went on within the fauna in those early times. But physical environmental factors certainly must have been supremely important, as natural selection acts through differential fitnesses that are responding chiefly to environmental factors, many of which are physical. True, there are "internal" factors that affect fitness, and constraints arising from internal or constructional factors may well have helped determine what morphologies did not evolve, but the ones that we see should have depended on their environmental adaptations and histories (for a thoughtful general discussion see Fisher 1985). It is logical for workers to search for environmental triggers that could produce the explosive events inferred for the Early Cambrian, and a great many have been suggested, as indicated by the selection of hypotheses reviewed in chapter 5.

Paleoenvironmental signals that can be related to the origins of bodyplans seem exceedingly difficult to discover. The signals from late Neoproterozoic and Cambrian rocks are difficult to interpret, and none of them as yet confirm that any extraordinary environmental events were causally associated with the Cambrian explosion. Therefore, the emphasis in this book has been on those factors that seem to be absolutely required to produce the fossil record that we find. Given the well-known processes of evolution that involve population genetics, most of the additional requisite factors concern the evolution of development. This situation suggests a sort of null hypothesis: the early evolutionary history of Metazoa, including the Cambrian explosion, can be explained simply as the result of an interplay of normal evolutionary processes (including "evo-devo") with physical environmental changes that were within the norm of geologic events. Specifically, this hypothesis asserts that there were no unique sorts of changes in the physical environment that triggered the Cambrian explosion. The widespread independent changes observed within metazoan clades, and changes in organisms belonging to other kingdoms, must be accounted for by "ordinary" biological processes acting internally or as "everyday" biologically mediated ecological factors.

Many of the evolutionary events during Neoproterozoic and Early Cambrian times seem explicable in terms of the general biological upheaval associated with the explosion—events that may have been consequences of the explosion and not independent responses to some physical trigger. For example, a diversification of planktonic algae, which evolved more elaborate tests near the Neoproterozoic-Cambrian boundary interval (acritarchs; see Butterfield 2001), might well reflect a response to increased predation from a burgeoning invertebrate fauna that came to include zooplankters. Similarly, the Early Cambrian acquisition of agglutinated tests by benthic protistans (foraminifers; see Lipps 1992) may have been a response to competition and predation from expanding invertebrate clades. The independent evolution of skeletons among some invertebrate lineages may have had two major biological causes (chap. 2), representing on one hand adaptations to increasing predation, perhaps chiefly the case with early mollusks, and on the other representing biomechanical responses to increasing complexity and body size, perhaps chiefly the case with early arthropods. Whatever the original impetus for skeletonization, many lineages that evolved rigid skeletons found adaptive pathways to the evolution of larger bodies, and increasing complexities must commonly have followed.

Physical environmental fluctuations are certainly incessant, and provide the basic challenges that drive natural selection over macroevolutionary time scales. Those challenges do not necessarily require unique or extreme conditions, however. So long as lineages are evolving across a changing adaptive landscape, some may closely approach adaptive thresholds that can then be crossed during an environmental fluctuation that is well within the norm, geologically speaking, and thus they can come to occupy a new or previously empty region of adaptive space. Secular paleoenvironmental trends that may last over long time spans also present adaptive challenges, and might also produce adaptive thresholds within the evolving biota. Changes that are extreme, so great or abrupt as to create large waves of extinction, appear often to extirpate the occupants of some regions of adaptive space. Adaptive regions that have been occupied once, and so have proven to be accessible to invasion by evolving lineages, may be reopened for exploitation by invaders that originate within a newly evolved suite of taxa. Thus even mass extinctions play a significant though indirect role in the creative side of macroevolution, although we could certainly get by without any more. A severe extinction of Neoproterozoic lineages has been postulated, but it is not clear that it involved a single extinction wave, or that it was not simply a result of the rise of the diverse benthic communities themselves, rather than being a factor that contributed to the explosion. I do not believe that the null hypothesis has been falsified as yet, though falsification may well be forthcoming from continuing studies of early paleoenvironmental conditions.

The early metazoan lineages evolved within a mosaic of marine environments, in which the mosaic units were bounded by challenging adaptive thresholds, some sharp and some rather gradational, each with a distinctive assortment of resources. The relatively simple early clades "invented" adaptations that permitted entry into a variety of adaptive regions: onto rocks; into the substrates—into mud, or sand, or gravel; up into the water column; offshore into deeper waters; and into larger body sizes and greater complexities. We see the colonization of such marine adaptive zones, probably multiple colonizations, recorded in the bodyplans of metazoan phyla. Today an adaptive opportunity can be quickly taken up by any of numerous adaptive types. During most of the Neoproterozoic, however, adaptive types were relatively few, presumably essentially restricted to prebilaterians and paracoelomates, and novel morphological innovations involving the evolution of the developmental system were required to exploit the many untapped resources in unoccupied adaptive zones. Those evolutionary inventions led relatively quickly to the establishment of a fauna that included a fair diversity of larger invertebrates interacting within complex communities. In place of Darwin's tangled bank, we find a Cambrian seafloor.

Appendix: The Geologic Time Scale

The absolute ages of boundaries between geologic time units are under constant investigation and are subject to revision (see chap. 5). Continuing advances in radiometric dating and improvements in correlation have steadily reduced the magnitude of possible errors, however, and dates from even the very early Phanerozoic are thought to be accurate to within less than half a percent. See the table overleaf, adapted from Wood 1999.

Era	Period		Epoch	Stage	Age Ma
Cenozoic	Quaternary		Holocene		
			Pleistocene		
	Tertiary	Neogene	Pliocene	Placenzian	1.8
				Zanclian	5.2
			Miocene	Messinian	
				Tortonian	
				Serravalian	
				Langhian	
				Burdigalian	
				Aquitanian	23.8
		Paleogene	Oligocene	Chattian	
				Rupelian	33.5
			Eocene	Priabonian	
				Bartonian	
				Lutetian	
				Ypresian	55.6
			Paleocene	Thanetian	
				Danian	65.0
Mesozoic	Cretaceous		Late	Maastrichtian	
				Campanian	
				Santonian	
				Coniacian	
				Turonian	89
				Cenomanian	98.9
			Early	Albian	
				Aptian	
				Barremian	
				Hauterivian	127
				Valanginian	
				Berriasian	144
	Jurassic		Late	Tithonian	
				Kimmeridgian	
				Oxfordian	160
			Middle	Callovian	
				Bathonian	
				Bajocian	
				Aalenian	180
			Early	Toarcian	
				Pliensbachian	
				Sinemurian	
				Hettangian	206
	Triassic		Late	Rhaetian	
				Norian	
				Carnian	228
			Middle	Ladinian	
				Anisian	242
			Scythian	Spathian	
				Nammalian	
				Griesbachian	

Era	Period	Epoch	Stage	Age Ma
Paleozoic	Permian	Tatarian	Changxingian	251
			Djulfian	
		Guadalupian	Capitanian	
			Wordian	
			Roadian	
		Early	Leonardian	
			Artinskian	
			Sakmarian	
			Asselian	290
	Carboniferous	Pennsylvanian	Gzelian	
			Kazimovian	
			Moscovian	
			Bashkirian	
		Mississipian	Serpukhovian	
			Viséan	
			Tournaisian	353.7
	Devonian	Late	Famennian	
			Frasnian	
		Middle	Givetian	
			Eifelian	
		Early	Emsian	
			Pragian	
			Lochkovian	408.5
	Silurian	Pridoli		
		Ludlow		
		Wenlock		
		Llandovery		439
	Ordovician	Ashgill		
		Caradoc		
		Llandeilo		
		Llanvirn		
		Arenig		
		Tremadoc		500
	Cambrian	Late	Franconian	
		Middle		520
		Early	Botomian	
			Atdabanian	
			Tommotian	531
			Nemakit-Daldynian	
				543
	Vendian			565
	Proterozoic			2500
	Archaean			3600

Glossary

The glossary includes terms that are deemed most likely to be unfamiliar to readers with more specialized backgrounds. In some cases, when terms have been used more broadly or in more than one way, sometimes in different fields, the usage in this book is indicated, hopefully to forestall confusion. However, definitions are notoriously slippery, and the need for brevity precludes lengthy qualification of these terms; please forgive the resulting imprecision.

acoelomate A bilaterian animal that lacks both pseudocoel and coelom, and thus has no body cavity.

aggregative hierarchy A nested hierarchy in which discrete, independent entities are assembled into entities on the next higher level.

analogues Similar morphological or functional features in different taxa that have not evolved from a common ancestral feature but are independently derived.

animal pole The apex of an egg that produces ectoderm and associated structures.

animal-vegetal axis The line connecting the animal and vegetal poles of an egg or early embryo.

apodeme An infold of an exoskeleton to which muscles are attached (chiefly in arthropods).

apomorphic feature A character found within a clade that is uniquely derived with respect to other clades, and can therefore be used as a defining feature.

archenteron The gut formed by gastrulation during embryogenesis.

autonomous specification The method of establishment of cell lineages whose identities depend on internal constituents only.

basal membrane A layer of extracellular matrix on which epithelial tissues rest.

benthos The animals living on the seafloor (epibenthos) or within the bottom sediment (infauna).

blastocoel Cavity within the blastula-stage early embryo, filled with fluid or gel.

blastomere Cell produced by cleavage of the egg.

blastopore Opening into the archenteron.

blastula Early embryo, formed of blastomeres.

blood vascular system The circulatory system of blood vessels and sinuses in triploblastic organisms. The blood spaces arise on the site of the blastocoel, the primary body cavity, between mesoderm and either endoderm or ectoderm, and may penetrate tissues.

body cavity A fluid-filled internal space, such as a pseudocoel or coelom, that may communicate to the exterior only through special ducts (i.e., not a digestive space).

body fossil An indication of the morphology or anatomy of a fossil organism; includes mineralized skeletons, molds or casts, and tissue petrifactions, and may be quite fragmentary.

Burgess Shale fauna A rich marine fauna from the middle Middle Cambrian of British Columbia, Canada, in which the soft-part anatomy of many taxa is exceptionally preserved.

Cambrian explosion The earliest appearance of fossils of many living phyla and of rich invertebrate faunas, between about 515 and 530 Ma.

cassette A set of functionally related genes that may be employed as a bloc in development, as genes forming a signaling pathway.

cell morphotype A cell type classed by morphological characters, by either light or electron microscopy.

cell types Cells that are developmentally similar or identical.

cell junctions Specialized molecules or molecular structures that connect cells to each other or to extracellular matrix.

Chengjiang fauna A rich marine fauna of middle Lower Cambrian age from Yunnan, China, in which the soft-part anatomy is exceptionally preserved.

choanocyte A flagellated collar cell type found in sponges.

clade A group that includes a founding common ancestor and all of its descendants.

clade ancestor The founding ancestral taxon of a clade.

cladogram A diagram that specifies the relations among taxa based on the distribution of homologous characters. Cladograms have the form of trees, but are not phylogenetic trees.

cleavage Cell divisions of an egg without cell growth, thus dividing the egg into ever smaller cells (blastomeres).

coelom A fluid-filled body cavity that is surrounded by mesoderm and usually lined by a cellular tissue (peritoneum); it may communicate with the exterior via specialized ducts or organs.

coeloblastula A hollow blastula, which thus contains an open blastocoel.

commissure A transverse nerve that joins other nerves or ganglia.

conditional specification The method of establishment of cell lineages whose identities depend at least in part on interactions with other cells.

constitutive hierarchy A nested hierarchy in which entities on one level form physical parts of entities on the next higher level.

crown ancestor The last common ancestor of all members of a living clade.

crown taxon The last common ancestor of members of a living clade and all of its descendants, living or extinct.

cryptogenetic A clade or taxon, the ancestry of which cannot be traced from fossil evidence.

cytoskeleton Scaffolding within a eukaryotic cell consisting of proteinaceous filaments, providing support, intracellular structure, and motor functions.

demersal Of motile animals that may swim but whose movements and/or distributions are closely associated with the seafloor.

desmosome Cell junction between epithelial cells.

diploblastic A grade of construction consisting of only two tissue layers, endoderm and ectoderm, and organs derived therefrom, separated by a gelatinous extracellular matrix.

disparity The degree of morphological differentiation among taxa, within groups. A similar concept may be used with other features, such as genomes.

downstream (genes or gene expressions) See **upstream**

ectoderm Tissue arising as a germ layer in eumetazoans that gives rise to epidermis, the nervous system, and some associated features.

endoderm Tissue arising as a germ layer in eumetazoans that gives rise to the gut and other features chiefly associated with digestion.

endostyle A ciliated groove in the pharynx of ascidians that produces a mucous string to convey food to the esophagus.

enterocoel A coelomic cavity that arises in development as an outpocketing of the larval gut; characteristic of lower deuterostomes and some lophophorates.

epithelium A tissue formed of a sheet of similarly oriented cells that are connected in some fashion, usually by occluding junctions. Many epithelia are underlaid by a basal membrane and line body surfaces, internally or externally.

eucoelomate Possessing an adult coelomic cavity that surrounds the gut.

Eumetazoa Metazoans that have ectodermal and endodermal germ layers; evidently includes all extant phyla except the sponges.

extracellular matrix A collagenous substance that forms the ground mass of connective tissues and the basal membranes of epithelia; it contains a variety of fibers and other macromolecules.

fossil completeness The fraction of an original living fauna or flora at a given taxonomic level that is now represented by fossils.

gastrula Early embryonic stage following the blastula, in which endoderm (and often prospective mesoderm) has become internalized within an ectodermal layer.

gastrulation Production of a gastrula from a blastula.

genome The entire DNA content of a cell.

germ layer Region of the early embryo that will give rise to ectoderm, endoderm, or mesoderm.

gonocoel A coelomic (i.e., intramesodermal) space containing gonads; in some early hypotheses of coelom evolution, thought to have given rise to paired coelomic cavities.

hemocoel A blood-filled space on the site of the blastocoel, and thus a primary body cavity; used herein for such spaces that function as fluid skeletons.

heterochrony An evolutionary change in the relative timing of the development of a feature.

heterotopy An evolutionary change in the spatial relations of a feature.

holopelagic Having an entire life cycle in the water column; usually said of plankton.

holophyletic A group that is monophyletic, that is, descended from a founding common ancestor, and that includes all descendants of that founder.

homeobox A DNA motif found on many regulatory genes that codes for a region (the homeodomain) on its protein product that binds to *cis*-regulatory portions of target genes, affecting their transcription.

homeodomain A DNA-binding sequence present in many transcription factors.

homologues Similar features in different taxa that have evolved from a similar ancestral feature and that owe their similarity to this common inheritance.

homoplasy A character that is not a synapomorphy, that is, that has evolved independently within the taxa under consideration.

Hox gene A regulatory gene that includes a homeobox motif and acts as a transcription factor. Hox genes occur in linked clusters that pattern metazoan bodies, and are known in all phyla except sponges and ctenophores; they function chiefly in ectoderm and mesoderm.

hypermorphosis A type of heterochrony in which ontogeny is extended beyond the ancestral condition; usually involves increasing body size or complexity.

idiosyncratic cleavage A pattern of cleavage and, usually, cell fate that is unique to a taxon (and thus is neither spiral nor radial).

infauna The fauna living within the bottom sediments.

introvert An anterior proboscis that invaginates in order to be shortened.

lagerstätte A fossil deposit in which preservation is extraordinary by normal paleontological standards; commonly applied to fossil assemblages in which soft-part anatomies are preserved.

late Neoproterozoic An informal subdivision of geologic time that ranges from 543 to 650 Ma.

lecithotrophic Larvae that swim but do not feed.

lorica Case of a cell or body part; may consist of plates or ribs, and may be organic or mineralized.

Ma Million years ago.

macroevolution Used here for evolutionary processes that do not involve changes in the frequency of structural genes, and includes gene regulatory evolution and patterns and rates of speciation and extinction.

meiofauna Usually, the fraction of marine species that passes through a 1,000-μm screen but is retained on a 42-μm screen (smaller sizes are nanofauna).

mesoderm Tissue arising as a germ layer that gives rise to muscles including the heart, as well as to blood, kidneys, and endoskeletons.

mesoglea A collagenous layer of extracellular material that lies between ectoderm and endoderm in Cnidaria and Ctenophora (the "jelly" of jellyfish).

messenger RNA (mRNA) A molecule transcribed from DNA that specifies the sequence of amino acids in a protein.

metanephridium See **nephridium**

microtubule organizing center An intracellular replicating system to produce structures employing tubulin, which include cilia and flagella, the axons of neurons, and mitotic spindles.

molecular sequence In DNA or RNA molecules, the sequence of bases; in proteins, the sequence of amino acids.

monophyletic A group that is descended from a founding common ancestor; it may be paraphyletic or holophyletic.

monotypic In Linnean taxonomy, a taxon that contains only a single member that is classed within the next lower category; for example, a genus with a single species is monotypic.

morphocline A gradient in a morphological feature in which the end members are connected by intermediate steps. The gradient may be in space or in time, or simply composed of a series proposed for heuristic purposes.

morphospace An imaginary space, usually multidimensional, whose axes represent morphological parameters, so that organisms may be compared and contrasted with respect to those features.

mRNA See **messenger RNA**

myoepithelium An epithelial tissue, the cells of which contain myofilaments in basal processes.

Neoproterozoic An era of geologic time that ranges from about 1,000 to about 543 Ma.

neoteny A variety of paedomorphosis wherein features of early developmental stages appear in the adult due to retardation of somatic development.

nephridium Excretory organ commonly communicating to the exterior via a tubule or duct.

neurula The vertebrate embryo following gastrulation, when the neural tube is forming.

node In phylogenetics, a branch point in a phylogenetic tree or in a cladogram.

notochord A rod-shaped mesodermal derivative that forms a skeletal or preskeletal median dorsal structure in chordates, lying beneath the dorsal nerve cord.

ortholog A gene that is directly inherited from a gene in ancestral forms (without duplication).

paedomorphosis The retention of features of early developmental stages in the adult.

paraclade A monophyletic group that is paraphyletic with respect to some descendant taxa.

paracoelomate A small-bodied bilaterian that is either acoelomate or pseudocoelomate in construction (i.e., lacks a coelom).

ParaHox gene A regulatory gene with a homeobox motif, arising in a cluster as paralogs of a small Hox gene cluster. ParaHox genes are probably present in all phyla except sponges and ctenophores; they function in endoderm (but not exclusively).

paralogs Homologous genes that are related because of their duplication in an ancestral gene.

paraphyletic A group of taxa that is monophyletic but not holophyletic; that is, there are descendant taxa that are not members of the paraphyletic group.

pelagic Of waters above the seafloor, including waters of the shelf and open ocean.

peramorphosis The evolutionary addition of novel features to the end of adult development.

pharyngotremy The perforation of the pharynx by gill slits.

phenotype The physical traits of an organism, including morphology and physiology.

phylogenetic tree A diagram that indicates the pattern of descent among taxa; most such diagrams are hypotheses inferred from morphological or molecular data.

plankton The biota that live suspended in the water column.

planktotrophic Of larvae that swim and feed in the plankton.

plesiomorphic feature A character that is ancestral to a clade and thus is shared with taxa that branched more basally and cannot be used as a defining feature of the clade.

pluripotent cell A cell that can differentiate into any of a number of somatic cell types.

point d'appui The point of contact of a body-wall section or a limb with the substrate during locomotion.

polyphyletic Groups of organisms that include taxa that do not share a common ancestor.

polythetic A class that is defined by possession of a number of features in common, none of which is necessary, and any of which may be found in other classes.

positional structure An architecture in which the positions of elements are fixed by their histories, such as individuals in a family tree or species in a phylogenetic tree.

protonephridium See **nephridium**

pseudocoel A fluid-filled body cavity that lies between mesoderm and either ectoderm or endoderm.

quenching Repression of the activity of a gene in some but not all cells.

radial cleavage Pattern of blastomeres that results when spindles, and therefore cleavage planes, are oriented either parallel to or at right angles to the pole of the egg.

radiates Members of the diploblastic phyla Ctenophora and Cnidaria, which have quasi-radial symmetries.

resolution interval (stratigraphic) The length of time used to evaluate the completeness of a stratigraphic section.

rRNA The RNA in ribosomes, composed in metazoans of a large subunit with three RNAs and a small subunit with one RNA.

schizocoel A coelomic cavity arising by splitting within mesodermal tissue.

sclerite An element such as a spine, blade, plate, or sheath that combines with other sclerites to form a skeleton; commonly used for elements of arthropod and coeloscleritophoran exoskeletons.

scleritome A skeleton composed of sclerites; scleritome assemblages are often made up of morphologically disparate sclerites that have distinctive functions.

segmentation The correlated repetition of a suite of similar morphological features along the anteroposterior body axis.

selector genes Genes that regulate the identity of structures that develop within their domains of expression.

seriation The repetition of a morphological feature along the anteroposterior body axis.

seta (pl. setae) Chitinous bristles, characteristic of polychaete annelids but found in several phyla, commonly used as ancillary locomotory devices and in feeding.

sibling species Two biological species that are so similar morphologically that nonmorphological criteria are required to define or identify them.

sister species or taxa Groups that have descended directly (without intervening groups) from the node representing their last common ancestor.

small-subunit rRNA An RNA subunit of the ribosome, with a sedimentation coefficient of 18S for eukaryotes and 16S for prokaryotes.

somatic muscles The muscle system lying along the body wall, and thus between coelom and ectoderm in coelomates.

spiral cleavage Pattern of blastomeres resulting when spindles and therefore cleavage planes are tilted with respect to the polar axis of the egg.

splanchnic muscles The muscle system lying along the gut, and thus between coelom and endoderm in coelomates.

SSU rRNA See **small-subunit rRNA**

stem ancestor The last common ancestor of a clade (which may include stem and/or crown groups), which first exhibits defining morphological features of the clade.

stem cell A cell that can divide to produce one identical, undifferentiated daughter and one differentiated daughter.

stem taxon An extinct group that belongs to a living clade but that branched more basally than the last common ancestor of living members of the clade (i.e., than the crown ancestor).

stereoblastula An early embryo formed of blastomeres that is solidly constructed, without a blastocoel.

stereom Network texture of a mineralized endoskeleton that is permeated by tissue in life, characteristic of Echinodermata.

stratigraphic completeness The fraction of a given interval of geologic time that is represented by sediments in a stratigraphic section.

stratotype A horizon within a selected section of rock chosen to represent the instant that forms the beginning of a unit of geologic time, such as the Cambrian period.

synapomorphic feature A character that is shared between taxa, uniting them within a monophyletic group.

syncytium A unicellular organism or a cell or tissue that contains multiple nuclei but is not partitioned by cell membranes.

tagmosis Anteroposterior regionation of a segmented bodyplan; segments are grouped into morphologically distinctive regions called tagmata.

taxon (pl. taxa) A group of organisms recognized as a formal systematic unit in classification, either as a member of a ranked hierarchy, or as a branch of a phylogenetic tree.

taxonomic richness The number of taxa at a given level that are present in a biotic sample, which may range in size from a single locality to a global sample.

time-averaged Applied to associations of fossils that have been assembled over a period of geologic time and reflect more than one biotic situation.

total group A group of organisms that includes all stem and crown taxa of a clade.

totipotent cell An animal cell that has the ability to differentiate into any cell type normally produced by its genome; normal oocytes are totipotent, for example.

trace fossil The record of the activity of an organism, such as a trail, burrow, or footprint, that does not indicate body morphology per se.

transcription factor A protein molecule that binds to the regulatory region of a gene, affecting its expression.

tree See **phylogenetic tree** or **positional structure**

triploblastic A grade of construction consisting of three tissue layers, ectoderm, endoderm, and mesoderm, and organs derived therefrom; characteristic of bilaterians.

unconformity A surface within or beneath a sedimentary sequence that represents a period of erosion or nondeposition, and thus a gap in the temporal record.

upstream (genes or gene expressions) Of genes, the end at which transcription begins is upstream; of gene expressions, earlier transcription events are upstream.

vegetal pole The apex of an egg that produces endoderm and associated structures.

Vendian System of Neoproterozoic rocks on the Russian Platform that underlies Cambrian formations; also applied to the fauna recovered from those rocks, and sometimes to presumably coeval faunas elsewhere.

Vendobionta Fossil organisms, chiefly of late Neoproterozoic age, that are constructed of modules suggesting a quiltlike constructional plan; they may represent stem diploblasts.

vicariance For an event, the splitting of a taxon into two or more geographically disjunct divisions by the appearance of some physical barrier(s).

References

Abele, L. G., W. Kim, and B. E. Felgenhauer. 1989. Molecular evidence for inclusion of the phylum Pentastomida in the Crustacea. Mol. Biol. Evol. 6:685–691.

Adams, M. D., et al. 2000. The genome sequence of *Drosophila melanogaster.* Science 287:2185–2195.

Agassiz, J. L. R. 1857. Essay on Classification. Contributions to the Natural History of the United States, vol. 1, pt. 1. Little Brown, Boston.

Aguinaldo, A. M. A., J. M. Turbeville, L. S. Linford, M. C. Rivera, J. R. Garey, R. A. Raff, and J. A. Lake. 1997. Evidence for a clade of nematodes, arthropods, and other moulting animals. Nature 387:489–493.

Akam, M. 1987. The molecular basis for metameric pattern in the *Drosophila* embryo. Development 101:1–22.

Akam, M., P. Holland, P. Ingham, and G. Wray, eds. 1994. The Evolution of Developmental Mechanisms. Development 1994 suppl.

Åkesson, B. 1958. A study of the nervous system of the Sipunculoideae with some remarks on the development of the two species *Phascolion strombi* Montagu and *Golfingia minuta* Keferstein. Undersokningar over Oresund (Lund) 38:1–249.

Aksarina, N. P. 1968. Probivalvia: A new class of ancient mollusks. Pp. 77–86 in New Data on Geology and Natural Resources of Western Siberia, vol. 3. Tomsk, USSR. In Russian; not seen.

Alberch, P., S. J. Gould, G. F. Oster, and D. B. Wake. 1979. Size and shape in ontogeny and phylogeny. Paleobiology 5:296–317.

Albert, R., H. Jeong, and A.-L. Barabási. 2000. Error and attack tolerance of complex networks. Nature 406:378–382.

Alberts, B., D. Bray, J. Lewis, M. Raff, K. Roberts, and J. Watson. 1989. Molecular Biology of the Cell. 2nd ed. Garland, New York.

Aldridge, R. J., D. E. G. Briggs, E. N. K. Clarkson, and M. P. Smith. 1986. The affinities of conodonts: New evidence from the Carboniferous of Edinburgh, Scotland. Lethaia 19:279–291.

Aldridge, R. J., and J. N. Theron. 1993. Conodonts with preserved soft tissue from a new Upper Ordovician *Konservat-Lagerstätte*. J. Micropaleontol. 12:113–117.

Alessandrello, A., G. Pinna, and G. Teruzzi. 1988. Land planarian locomotion trail from the Lower Permian of Lombardian pre-Alps. Atti Soc. Ital. Sci. Nat. Mus. Civ. Stor. Nat. Milano 129:139–145.

Alldredge, A. 1976. Appendicularians. Sci. Am. 235:94–102.

Allison, P. A., and D. E. G. Briggs, eds. 1991. Taphonomy: Releasing the Data Locked in the Fossil Record. Plenum, New York.

Almond, J. E. 1985. The Silurian-Devonian fossil record of the Myriapoda. Philos. Trans. R. Soc. London B 309:227–238.

Amores, A., A. Force, Y.-L. Yan, L. Joly, C. Amemiya, A. Fritz, R. K. Ho, J. Langeland, V. Prince, Y.-L. Wang, M. Westerfield, M. Ekker, and J. H. Postlethwaite. 1998. Zebrafish *hox* clusters and vertebrate genome evolution. Science 282:1711–1714.

Anbar, A. D., and A. H. Knoll. 2002. Proterozoic ocean chemistry and evolution: A bioinorganic bridge? Science 297:1137–1142.

Anderson, D. T. 1969. On the embryology of the cirripede crustaceans *Tetraclita rosea* (Krauss), *Tetraclita purpurascens* (Wood), *Chthamalus antennatus* (Darwin), and *Chamaesipho columna* (Spengler) and some considerations of crustacean phylogenetic relationships. Philos. Trans. R. Soc. London B 256:183–235.

———. 1973. Embryology and Phylogeny in Annelids and Arthropods. Pergamon, Oxford.

———. 1979. Embryos, fate maps, and the phylogeny of arthropods. Pp. 59–105 in A. P. Gupta, ed., Arthropod Phylogeny. Van Nostrand Reinhold, New York.

———. 1982. Embryology. Pp. 1–41 in L. G. Abele, ed., The Biology of Crustacea, vol. 2. Academic Press, New York.

Andreyeva, O. N. 1987. The Cambrian articulate brachiopods. Paleontol. J. 4:27–33.

Anstey, R. L., and J. L. Pachut. 1995. Phylogeny, diversity history, and speciation in Paleozoic bryozoans. Pp. 239–284 in D. H. Erwin and R. L. Anstey, eds., New Approaches to Studying Speciation in the Fossil Record. Columbia Univ. Press, New York.

Antequera, F., and A. Bird. 1993. Number of CpG islands and genes in human and mouse. Proc. Nat. Acad. Sci. USA 90:11995–11999.

Aparicio, S., et al. 2002. Whole-genome shotgun assembly and analysis of the genome of *Fugu rubripes*. Science 297:1301–1310.

Appel, T. A. 1987. The Cuvier-Geoffroy Debate: French Biology in the Decades before Darwin. Oxford Univ. Press, Oxford.

Arduini, P., G. Pinna, and G. Terruzi. 1981. *Megaderaion sinemuriense* n. g. n. sp.: A new fossil enteropneust of the Sinemurian of Osteno in Lombardy. Atti Soc. Ital. Sci. Nat. Mus. Civ. Stor. Nat. Milano 122:104–108.

———. 1983. *Eophasma jurasicum* n. g. n. sp.: A new fossil nematode of the Sinemurian of Osteno in Lombardy. Atti. Soc. Ital. Sci. Nat. Mus. Civ. Stor. Nat. Milano 124:61–64.

Arenas-Mena, C., P. Martinez, R. A. Cameron, and E. H. Davidson. 1998. Expression of the *Hox* gene complex in the indirect development of a sea urchin. Proc. Nat. Acad. Sci. USA 95:13062–13067.

Arendt, D., and K. Nübler-Jung. 1994. Inversion of dorsoventral axis? Nature 371:26.

Arendt, D., U. Technau, and J. Wittbrodt. 2001. Evolution of the bilaterian larval foregut. Nature 409:81–85.

Arnone, M., and E. H. Davidson. 1997. The hardwiring of development: Organization and function of genomic regulatory systems. Development 124:1851–1864.

Ausich, W. I., and I. E. Babcock. 1998. The phylogenetic position of *Echmatocrinus brachiatus*, a probable octocoral from the Burgess Shale. Palaeontology 41:193–202.

———. 2000. *Echmatocrinus*: A Burgess Shale animal reconsidered. Lethaia 33:92–94.

Avise, J. C. 1994. Molecular Markers, Natural History, and Evolution. Chapman and Hall, New York.

Avise, J. C., and W. S. Nelson. 1995. Reply to a commentary on the use of sequence data for phylogeny reconstruction by S. J. Hacket, C. S. Griffiths, J. M. Bates, and N. K. Klein. Mol. Phylogenet. Evol. 4:353–356.

Ax, P. 1965. Die Gnathostomulida: Eine ratselhafte Wurmgruppe aus dem Meeressand. Abh. Math. Naturwiss. Kl. Akad. Wiss. Lit. Mainz 8:1–32.

———. 1985. The position of the Gnathostomulida and Platyhelminthes in the phylogenetic system of the Bilateria. Pp. 168–180 in S. Conway Morris, J. D. George, R. Gibson, and H. M. Platt, eds., The Origins and Relationships of Lower Invertebrates. Clarendon Press, Oxford.

———. 1987. The Phylogenetic System: The Systematization of Organisms on the Basis of Their Phylogenesis. Wiley, New York.

Ayala, F. J. 1997. Vagaries of the molecular clock. Proc. Nat. Acad. Sci. USA 94:7776–7783.

Ayala, F. J., A. Rzhetsky, and F. J. Ayala. 1998. Origin of the metazoan phyla: Molecular clocks confirm paleontological estimates. Proc. Nat. Acad. Sci. USA 95:606–611.

Babcock, L. E., W. Zhang, and S. A. Leslie. 2001. The Chengjiang biota: Record of the Early Cambrian diversification of life and clues to exceptional preservation of fossils. GSA Today 11 (2):3–9.

Baguñà, J., S. Carranza, J. Paps, I. Ruiz-Trillo, and M. Riutort. 2001. Molecular taxonomy and phylogeny of the Tricladida. Pp. 49–56 in D. T. J. Littlewood and R. A. Bray, eds., Interrelationships of the Platyhelminthes. Taylor and Francis, London.

Baguñà, J., I. Ruiz-Trillo, J. Paps, M. Loukota, C. Ribera, U. Jondelius, and M. Riutort. 2001. The first bilaterian organisms: Simple or complex? New molecular evidence. Int. J. Dev. Biol. 45:S133–S134.

Baker, A. N., F. W. E. Rowe, and H. E. S. Clark. 1986. A new class of Echinodermata from New Zealand. Nature 321:862–864.

Bakke, T. 1980. Embryonic and post-embryonic development in the Pogonophora. Zool. Jb. Abt. Anat. 103:276–284.

Balavoine, G. 1997. The early emergence of platyhelminths is contradicted by the agreement between 18S rRNA and Hox genes data. C. R. Acad. Sci. 320:83–94.

Baldauf, S. L., and J. D. Palmer. 1993. Animals and fungi are each other's closest relatives: Congruent evidence from multiple proteins. Proc. Nat. Acad. Sci. USA 90:11558–11562.

Balinski, A. 2001. Embryonic shells of Devonian linguloid brachiopods. Pp. 91–101 in C. H. C. Brunton, L. R. M. Cocks, and S. L. Long, eds., Brachiopods Past and Present. Taylor and Francis, London.

Balsam, W. L., and S. Vogel. 1973. Water movement in archaeocyathids: Evidence and implications of passive flow in models. J. Paleontol. 47:979–984.

Bambach, R. K. 1977. Species richness in marine benthic habitats through the Phanerozoic. Paleobiology 3:152–167.

———. 1985. Classes and adaptive variety: The ecology of diversification in marine faunas through the Phanerozoic. Pp. 191–253 in J. W. Valentine, ed., Phanerozoic Diversity Patterns: Profiles in Macroevolution. Princeton Univ. Press, Princeton, NJ.

Barabási, A.-L., and R. Albert. 1999. Emergence of scaling in random networks. Science 286:509–512.

Bardack, D. 1997. Wormlike animals: Enteropneusta (acorn worms). Pp. 89–92 in D. W. Shabica and A. A. Hay, eds., Richardson's Guide to the Fossil Fauna of Mazon Creek. Northeastern Illinois Univ., Chicago.

Barfod, G. H., F. Albarede, A. H. Knoll, S. Xiao, P. Telouk, R. Frei, and J. Baker. 2002. New Lu-Hf and Pb-Pb age constraints on the earliest animal fossils. Earth Planet. Sci. Lett. 201:203–212.

Barnes, R. D. 1987. Invertebrate Zoology. 5th ed. Saunders, Philadelphia.

Barnes, R. S. K., ed. 1984. A Synoptic Classification of Living Organisms. Sinauer, Sunderland, MA.

Barnes, R. S. K., P. Calow, and P. J. W. Olive. 1993. The Invertebrates: A New Synthesis. 2nd ed. Blackwell Scientific, Oxford.

Barrell, J. 1917. Rhythms and the measurements of geologic time. Geol. Soc. Am. Bull. 28:745–904.

Barrier, M., R. H. Robichaux, and M. D. Purugganan. 2001. Accelerated regulatory gene evolution in an adaptive radiation. Proc. Nat. Acad. Sci. USA 98:10208–10213.

Barrington, E. J. W. 1965. The Biology of Hemichordata and Protochordata. Oliver and Boyd, Edinburgh.

Bartolomaeus, T. 1993. Die Leibeshöhlenverhältnisse und Verwandtschaftsbeziehungen der Spiralia. Verh. Dtsch. Zool. Ges. 86:1–42.

———. 1994. On the ultrastructure of the coelomic lining in the Annelida, Sipuncula, and Echiura. Microfauna Marina 9:171–220.

Bartolomaeus, T., and P. Ax. 1992. Protonephridia and metanephridia: Their relation within the Bilateria. Z. Zool. Syst. Evolutionsforsch. 30:21–45.

Bateson, W. 1894. Materials for the Study of Variation, Treated with Especial Regard to Discontinuity in the Origin of Species. Johns Hopkins Univ. Press, Baltimore.

Bayer, F. M., and H. B. Owre. 1968. The Free-Living Lower Invertebrates. Macmillan, New York.

Beauchamp, P. de. 1910. Sur la presence d'un hemocoele chez *Dinophilus*. Soc. Zool. Fr. Bull. 35:18–25.

———. 1929. Le développement des gastrotriches (note préliminaire). Bull. Soc. Zool. Fr. 54:549–558.

Beckner, M. 1959. The Biological Way of Thought. Columbia Univ. Press, New York.

Bedini, C., and F. Papi. 1974. Fine structure of the turbellarian epidermis. Pp. 108–147 in N. W. Riser and M. P. Morse, eds., Biology of the Turbellaria. McGraw-Hill, New York.

Beklemishev, W. N. 1969. Principles of Comparative Anatomy of Invertebrates. 2 vols. Univ. Chicago Press, Chicago.

Bell, B. M. 1980. Edrioasteroidea and Edrioblastoidea. Pp. 158–174 in T. W. Broadhead and J. A. Waters, eds., Echinoderms. Studies in Geology 3. Univ. Tennessee, Department of Geological Science, Knoxville.

Bengtson, S. 1983. The early history of the Conodonta. Fossils and Strata 15:5–19.

———. 1992a. The cap-shaped Cambrian fossil *Maikhanella* and the relationship between coeloscleritophorans and molluscs. Lethaia 25:401–420.

————. 1992b. Proterozoic and earliest Cambrian skeletal metazoans. Pp. 1017–1033 in J. W. Schopf and C. Klein, eds., The Proterozoic Biosphere: A Multidisciplinary Study. Cambridge Univ. Press, Cambridge.

Bengtson, S., and S. Conway Morris. 1984. A comparative study of Lower Cambrian *Halkieria* and Middle Cambrian *Wiwaxia*. Lethaia 17:307–329.

Bengtson, S., S. Conway Morris, P. A. Jell, and B. Runnegar. 1990. Early Cambrian fossils from South Australia. Association of Australasian Palaeontologists, Brisbane, Mem. 9:1–364.

Bengtson, S., and V. V. Missarzhevsky. 1981. Coeloscleritophora: A major group of enigmatic Cambrian metazoans. U.S. Geol. Surv. Open-File Rep. 81-743:19–21.

Bengtson, S., and A. Urbanek. 1986. *Rhabdotubus:* A Middle Cambrian rhabdopleurid hemichordate. Lethaia 19:293–308.

Bengtson, S., and Y. Zhao. 1997. Fossilized metazoan embryos from the earliest Cambrian. Science 277:1645–1648.

Beninger, P. G., J. W. Lynn, T. H. Dietz, and H. Silverman. 1997. Mucociliary transport in living tissue: The two-layer model confirmed in the mussel *Mytilus edulis* L. Biol. Bull. 193:4–7.

Benito, J. 1982. Hemichordata. Pp. 819–821 in S. P. Parker, ed., Synopsis and Classification of Living Organisms, vol. 2. McGraw-Hill, New York.

Benito, J., and F. Pardos. 1997. Hemichordata. Pp. 15–101 in F. W. Harrison and E. E. Ruppert, eds., Microscopic Anatomy of Invertebrates, vol. 15, Hemichordata, Chaetognatha, and the Invertebrate Chordates. Wiley-Liss, New York.

Benton, M. J. 1993. Late Triassic extinctions and the origin of dinosaurs. Science 260:767–770.

Benus, A. P. 1988. Sedimentological context of a deep-water Ediacaran fauna (Mistaken Point Formation, Avalon Zone, eastern Newfoundland). Pp. 8–9 in E. Landing, G. M. Narbonne, and P. Myrow, eds., Trace Fossils, Small Shelly Fossils, and the Precambrian-Cambrian Boundary. N.Y. State Mus. Bull. 463.

Berbee, M. L., and J. W. Taylor. 1993. Dating the evolutionary radiations of the true fungi. Can. J. Bot. 71:1114–1121.

Bereiter-Hahn, J., A. G. Matoltsy, and K. S. Richards. 1984. Biology of the Integument. Vol. 1. Invertebrates. Springer-Verlag, Berlin.

Berg, G. 1985. *Annulonemertes* gen. nov.: A new segmented hoplonemertean. Pp. 200–209 in S. Conway Morris, J. D. George, R. Gibson, and H. M. Platt, eds., The Origins and Relationships of Lower Invertebrates. Clarendon Press, Oxford.

Bergquist, P. R. 1978. Sponges. Univ. California Press, Berkeley and Los Angeles.

————. 1985. Poriferan relationships. Pp. 14–27 in S. Conway Morris, J. D. George, R. Gibson, and H. M. Platt, eds., The Origins and Relationships of Lower Invertebrates. Clarendon Press, Oxford.

Bergström, J. 1973. Organization, life, and systematics of trilobites. Fossils and Strata 2:1–69.

————. 1986. *Opabinia* and *Anomalocaris:* Unique Cambrian "arthropods." Lethaia 19:241–246.

————. 1989. The origin of animal phyla and the new phylum Procoelomata. Lethaia 22:259–269.

———. 1991. Metazoan evolution around the Precambrian-Cambrian transition. Pp. 25–34 in A. M. Simonetta and S. Conway Morris, eds., The Early Evolution of Metazoa and the Significance of Problematic Taxa. Cambridge Univ. Press, Cambridge.

Bergström, J., and X. Hou. 1998. Chengjiang arthropods and their bearing on early arthropod evolution. Pp. 151–184 in G. D. Edgecombe, ed., Arthropod Fossils and Phylogeny. Columbia Univ. Press, New York.

Berkner, C. V., and L. C. Marshall. 1964. The history of growth of oxygen in the earth's atmosphere. Pp. 102–106 in C. J. Brancuzio and A. G. W. Cameron, eds., The Origin and Evolution of Atmospheres and Oceans. Wiley, New York.

Berney, C., J. Pawlowski, and L. Zaninetti. 2000. Elongation factor 1-alpha sequences do not support an early divergence of the Acoela. Mol. Biol. Evol. 17:1032–1039.

Berry, W. B. N. 1987. Phylum Hemichordata (including Graptolithina). Pp. 612–635 in R. S. Boardman, A. H. Cheetham, and A. J. Rowell, eds., Fossil Invertebrates. Blackwell Scientific, Palo Alto, CA.

Bertolani, R. 1990. Tardigrada. Pp. 49–60 in K. G. Adiyodi and R. G. Adiyodi, eds., Reproductive Biology of Invertebrates, vol. 4, pt. B. Wiley, Chichester.

Binford, M. W. 1982. Ecological history of Lake Valencia, Venezuela: Interpretation of animal microfossils and some chemical, physical, and geological features. Ecol. Monogr. 52:307–337.

Bird, A. 1995. Gene number, noise reduction, and biological complexity. Trends Genet. 11:94–100.

Bischoff, G. C. O. 1990. Phosphatised soft parts in *Costipelagiella* n. sp. A (Mollusca, Pelagiellidae) from the Cambrian of New South Wales, Australia. Senckenb. Lethaea 70:51–67.

Blaxter, M. L., P. DeLey, J. R. Garey, L. X. Liu, P. Scheldeman, A. Vierstraete, J. R. Vanfleteren, L. Y. Mackey, M. Dorris, L. M. Frisse, J. T. Vida, and W. K. Thomas. 1998. A molecular evolutionary framework for the phylum Nematoda. Nature 392:71–75.

Bleidorn, C., A. Schmidt-Rhaesa, and J. R. Garey. 2002. Systematic relationships of Nematomorpha based on molecular and morphological data. Invert. Biol. 121:357–364.

Boaden, P. J. S. 1985. Why is a gastrotrich? Pp. 248–260 in S. Conway Morris, J. D. George, R. Gibson, and H. M. Platt, eds., The Origins and Relationships of Lower Invertebrates. Clarendon Press, Oxford.

Boardman, R. S., A. H. Cheetham, and P. L. Cook. 1983. Introduction to the Bryozoa. Pp. 3–48 in R. A. Robison, ed., Treatise on Invertebrate Paleontology, pt. G, Bryozoa (Revised). Geological Society of America, Boulder, CO; Univ. Kansas Press, Lawrence.

Boardman, R. S., A. H. Cheetham, and A. J. Rowell, eds. 1987. Fossil Invertebrates. Blackwell Scientific, Palo Alto, CA.

Bock, G. R., and G. Cardew, eds. 1999. Homology. Novartis Foundation Symposium 222. Wiley, Chichester.

Bode, H. R. 2001. The role of Hox genes in axial patterning in hydra. Am. Zool. 41:621–628.

Boero, F., C. Gravili, P. Pagliara, S. Piraino, J. Bouillon, and V. Schmid. 1998. The cnidarian premises of metazoan evolution: From triploblasty, to coelom formation, to metamery. Ital. J. Zool. 65:5–9.

Bone, Q. 1981. The neotenic origin of chordates. Atti Convegni Lincei 49:465–486.

Bone, Q., H. Kapp, and A. C. Pierrot-Buls. 1991. Introduction and relationships of the group. Pp. 1–4 in Q. Bone, H. Kapp, and A. C. Pierrot-Buls, eds., The Biology of Chaetognaths. Oxford Univ. Press, Oxford.

Bonner, J. T. 1965. Size and Cycle. Princeton Univ. Press, Princeton, NJ.

———. 1988. The Evolution of Complexity by Means of Natural Selection. Princeton Univ. Press, Princeton, NJ.

Boore, J. L., and W. M. Brown. 2000. Mitochondrial genomes of *Galathealinum*, *Helobdella*, and *Platynereis*: Sequence and gene arrangement comparisons indicate that Pogonophora is not a phylum and Annelida and Arthropoda are not sister taxa. Mol. Biol. Evol. 17:87–106.

Boore, J. L., T. M. Collins, D. Stanton, L. L. Daehler, and W. M. Brown. 1995. Deducing the pattern of arthropod phylogeny from mitochondrial DNA rearrangements. Nature 376:163–165.

Borchiellini, C., N. Boury-Esnault, J. Vacelet, and Y. Le Parco. 1998. Phylogenetic analysis of the Hsp70 sequences reveals the monophyly of Metazoa and specific phylogenetic relationships between animals and fungi. Mol. Biol. Evol. 15:647–655.

Boring, L. 1989. Cell-cell interactions determine the dorsoventral axis in embryos of an equally cleaving opisthobranch mollusc. Dev. Biol. 136:239–253.

Borojevic, R. 1970. Différentiation cellulaire dans l'embryogenèse et le morphogenèse chez les spongiaires. Symp. Zool. Soc. London 25:467–490.

Bottjer, D. J., and D. Jablonski. 1988. Paleoenvironmental patterns in the evolution of post-Paleozoic benthic marine invertebrates. Palaios 3:540–560.

Boury-Esnault, N., S. Efremova, C. Bézac, and J. Vacelet. 1999. Reproduction of a hexactinellid sponge: First description of gastrulation by cellular delamination in the Porifera. Invertebr. Reprod. Dev. 35:187–201.

Boury-Esnault, N., and J. Vacelet. 1994. Preliminary studies on the organization and development of a hexactinellid sponge from a Mediterranean cave, *Oopsacas minuta*. Pp. 407–415 in R. W. M. van Soest, T. M. G. van Kempen, and J.-C. Braekman, eds., Sponges in Time and Space. Balkema, Rotterdam.

Bowring, S. A., J. P. Grotzinger, C. E. Isachsen, A. H. Knoll, S. M. Pelechaty, and P. Kolosov. 1993. Calibrating rates of Early Cambrian evolution. Science 261:1293–1298.

Boyd, R., and P. J. Richerson. 1992. How microevolutionary processes give rise to history. Pp. 179–209 in M. H. Nitecki and D. V. Nitecki, eds., History and Evolution. State Univ. New York Press, Albany.

Boyer, B. C. 1971. Regulative development in a spiralian embryo as shown by cell deletion experiments on the acoel, *Childia*. J. Exp. Zool. 176:97– 106.

Boyer, B. C., J. Q. Henry, and M. Q. Martindale. 1996a. Dual origins of mesoderm in a basal spiralian: Cell lineage analyses in the polyclad turbellarian *Hoploplana inquilina*. Dev. Biol. 179:329–338.

———. 1996b. Modified spiral cleavage: The duet cleavage pattern and early blastomere fates in the acoel turbellarian *Neochildia fusca*. Biol. Bull. 191:285–286.

Brasier, M. D. 1979. The Cambrian radiation event. Pp. 103–159 in M. R. House, ed., The Origin of Major Invertebrate Groups. Academic Press, New York.

Brasier, M. D., and D. McIlroy. 1998. *Neonereites uniserialis* from c. 600 Ma year old rocks in western Scotland and the emergence of animals. J. Geol. Soc. London 155:5–12.

Brasier, M. D., G. Shields, V. Kuleshov, and E. A. Zhegallo. 1996. Integrated chemo- and biostratigraphic calibration of early animal evolution: Neoproterozoic–Early Cambrian of southwest Mongolia. Geol. Mag. 133:445–485.

Breimer, A., and D. B. Macurda, Jr. 1972. The phylogeny of the fissiculate blastoids. Verh. K. Ned. Akad. Wet., eerste Reeks 26:1–390.

Bremer, K. 1988. The limits of amino-acid sequence data in angiosperm phylogenetic reconstruction. Evolution 42:795–803.

Bresciani, J. 1991. Nematomorpha. Pp. 197–218 in F. W. Harrison and E. E. Ruppert, eds., Microscopic Anatomy of Invertebrates, vol. 4, Aschelminthes. Wiley-Liss, New York.

Bridge, D., C. W. Cunningham, R. DeSalle, and L. W. Buss. 1995. Class-level relationships in the phylum Cnidaria: Molecular and morphological evidence. Mol. Biol. Evol. 12:679–689.

Bridge, D., C. W. Cunningham, B. Schierwater, R. DeSalle, and L. W. Buss. 1992. Class-level relationships in the phylum Cnidaria: Evidence from mitochondrial genome structure. Proc. Nat. Acad. Sci. USA 89:8750–8753.

Brien, P. 1953. Étude sur les Phylactolaemates. Ann. Soc. R. Zool. Belg. 84:301–444.

Briggs, D. E. G. 1976. The arthropod Branchiocaris, n. gen., Middle Cambrian, Burgess Shale, British Columbia. Geol. Surv. Can. Bull. 264:1–29.

———. 1978. The morphology, mode of life, and affinities of Canadaspis perfecta (Crustacea: Phyllocarida), Middle Cambrian, Burgess Shale, British Columbia. Philos. Trans. R. Soc. London B 281:439–487.

———. 1983. Affinities and early evolution of the Crustacea: The evidence of the Cambrian fossils. Pp. 1–22 in F. R. Schram, ed., Crustacean Phylogeny. Balkema, Rotterdam.

———. 1991. Extraordinary fossils. Am. Sci. 79:130–141.

———. 1992. Conodonts: A major extinct group added to the vertebrates. Science 256:1285–1286.

Briggs, D. E. G., E. Clarkson, and R. J. Aldridge. 1983. The conodont animal. Lethaia 16:1–14.

Briggs, D. E. G., and D. Collins. 1988. A Middle Cambrian chelicerate from Mount Stephen, British Columbia. Palaeontology 31:779–798.

Briggs, D. E. G., D. H. Erwin, and F. J. Collier. 1994. The Fossils of the Burgess Shale. Smithsonian Institution Press, Washington, DC.

Briggs, D. E. G., and R. A. Fortey. 1989. The early radiation and relationships of the major arthropod groups. Science 246:241–243.

Briggs, D. E. G., R. A. Fortey, and M. A. Wills. 1992. Morphological disparity in the Cambrian. Science 256:1670–1673.

Briggs, D. E. G., and A. J. Kear. 1994. Decay of Branchiostoma: Implications for soft-tissue preservation in conodonts and other primitive chordates. Lethaia 26:275–287.

Briggs, D. E. G., and C. Neddin. 1997. The taphonomy and affinities of the problematic fossil Myoscolex from the Lower Cambrian Emu Bay Shale of South Australia. J. Paleontol. 71:22–32.

Briggs, D. E. G., and H. B. Whittington. 1985. Modes of life of arthropods from the Burgess Shale, British Columbia. Trans. R. Soc. Edinburgh 76:149–160.

Britten, R. J., and E. H. Davidson. 1969. Gene regulation for higher cells: A theory. Science 165:349–357.

———. 1971. Repetitive and non-repetitive DNA sequences and a speculation on the origins of evolutionary novelty. Q. Rev. Biol. 46:111–138.

Bromham, L., A. Rambaut, R. Fortey, A. Cooper, and D. Penny. 1998. Testing the Cambrian explosion hypothesis by using a molecular dating technique. Proc. Nat. Acad. Sci. USA 95:12386–12389.

Bromley, R. G. 1990. Trace Fossils: Biology and Taphonomy. Unwin Hyman, London.

Brooke, N. M., J. Garcia-Fernàndez, and P. W. H. Holland. 1998. The ParaHox gene cluster is an evolutionary sister of the Hox gene cluster. Nature 392:920–922.

Brown, F. A., Jr., ed. 1950. Selected Invertebrate Types. Wiley and Sons, New York.

Brown, W. M., E. M. Prager, A. Wang, and A. C. Wilson. 1982. Mitochondrial DNA sequences of primates: Tempo and mode of evolution. J. Mol. Evol. 18:224–239.

Brusca, G. J. 1975. General Patterns of Invertebrate Development. Mad River Press, Eureka, CA.

Brusca, R. C., and G. J. Brusca. 1990. Invertebrates. Sinauer, Sunderland, MA.

Bruton, D. L. 1991. Beach and laboratory experiments with the jellyfish Aurelia and remarks on some fossil "medusoid" traces. Pp. 125–129 in A. M. Simonetta and S. Conway Morris, eds., The Early Evolution of Metazoa and the Significance of Problematic Taxa. Cambridge Univ. Press, Cambridge.

Budd, G. E. 1993. A Cambrian gilled lobopod from Greenland. Nature 364:709–711.

———. 1996. The morphology of Opabinia regalis and the reconstruction of the arthropod stem-group. Lethaia 29:1–14.

———. 1997. Stem group arthropods from the Lower Cambrian Sirius Passet fauna of North Greenland. Pp. 125–138 in R. A. Fortey and R. H. Thomas, eds., Arthropod Relationships. Chapman and Hall, London.

———. 1998. Arthropod body-plan evolution in the Cambrian with an example from anomalocaridid muscle. Lethaia 31:197–210.

———. 1999. The morphology and phylogenetic significance of Kerygmachela kierkegaardi Budd (Buen Formation, Lower Cambrian, N Greenland). Trans. R. Soc. Edinburgh Earth Sci. 89:249–290.

Budd, G. E., and S. Jensen. 2000. A critical reappraisal of the fossil record of the bilaterian phyla. Biol. Rev. 75:253–295.

Bullock, T. H. 1965. Chaetognatha, Pogonophora, Hemichordata, and Chordata Tunicata. Pp. 1560–1592 in T. H. Bullock and G. A. Horridge, eds., Structure and Function in the Nervous Systems of Invertebrates. Freeman and Co., San Francisco.

Bullock, T. H., and G. A. Horridge, eds. 1965. Structure and Function in the Nervous Systems of Invertebrates. 2 vols. Freeman and Co., San Francisco.

Bullock, W. L. 1969. Morphological features as tools and pitfalls in acanthocephalan systematics. Pp. 9–24 in G. D. Schmidt, ed., Problems in Systematics of Parasites. University Park Press, Baltimore.

Bulman, O. M. B. 1970. Graptolithina. In C. Teichert, ed., Treatise on Invertebrate Paleontology pt. 5, 2nd ed. Geological Society of America, Boulder, CO; Univ. Kansas Press, Lawrence.

Burdon-Jones, C. 1952. Development and biology of the larva of Saccoglossus horsti (Enteropneusta). Philos. Trans. R. Soc. London B 236:553–590.

Burger, G., L. Forget, Y. Zhu, M. W. Gray, and B. F. Lang. 2003. Unique mitochondrial genome architecture in unicellular relatives of animals. Proc. Nat. Acad. Sci. USA 100:892–897.

Bushdid, P. B., D. M. Brantley, F. E. Yull, G. L. Blaeuer, L. H. Hoffman, L. Niswander, and L. D. Kerr. 1998. Inhibition of NF-$_K$B activity results in disruption of the apical ectodermal ridge and aberrant limb morphogenesis. Nature 392:615–618.

Buss, L. W. 1987. The Evolution of Individuality. Princeton Univ. Press, Princeton, NJ.

Buss, L. W., and A. Seilacher. 1994. The phylum Vendobionta: A sister group of the Eumetazoa? Paleobiology 20:1–4.

Butterfield, N. J. 1990. A reassessment of the enigmatic Burgess Shale fossil *Wiwaxia corrugata* (Matthew) and its relationship to the polychaete *Canadia spinosa* Walcott. Paleobiology 16:287–303.

———. 1997. Plankton ecology and the Proterozoic-Phanerozoic transition. Paleobiology 23:247–262.

———. 2000. *Bangiomorpha pubescens* n. gen., n. sp.: Implications for the evolution of sex, multicellularity, and the Mesoproterozoic/Neoproterozoic radiation of eukaryotes. Paleobiology 26:386–404.

———. 2001. Ecology and evolution of Cambrian plankton. Pp. 200–216 in A. Yu. Zhuravlev and R. Riding, eds., The Ecology of the Cambrian Radiation. Columbia Univ. Press, New York.

———. 2002. *Leanchoilia* guts and the interpretation of three- dimensional structures in Burgess Shale–type fossils. Paleobiology 28:155–171.

Butterfield, N. J., and C. J. Nicholas. 1996. Burgess Shale–type preservation of both non-mineralizing and "shelly" Cambrian organisms from the Mackenzie Mountains, northwestern Canada. J. Paleontol. 70:893–899.

Caldwell, W. 1882. Preliminary notes on the structure, development, and affinities of *Phoronis*. Proc. R. Soc. London 34:371–383.

Cameron, C. B., J. R. Garey, and B. J. Swalla. 2000. Evolution of the chordate body plan: New insights from phylogenetic analyses of deuterostome phyla. Proc. Nat. Acad. Sci. USA 97:4469–4474.

Cameron, R. A., and E. H. Davidson. 1991. Cell type specification during sea urchin development. Trends Genet. 7:212–218.

Cameron, R. A., G. Mahairas, J. P. Rast, P. Martinez, T. R. Biondi, S. Swartzell, J. C. Wallace, A. J. Poustka, B. T. Livingston, G. A. Wray, C. A. Ettensohn, H. Lehrach, R. J. Britten, E. H. Davidson, and L. Hood. 2000. A sea urchin genome project: Sequence scan, virtual map, and additional resources. Proc. Nat. Acad. Sci. USA 97:9514–9518.

Campbell, K. S. W., and C. R. Marshall. 1987. Rates of evolution among Paleozoic echinoderms. Pp. 61–100 in K. S. W. Campbell and M. F. Day, eds., Rates of Evolution. Allen and Unwin, London.

Campos, A., M. P. Cummings, J. L. Reyes, and J. P. Laclette. 1998. Phylogenetic relationships of platyhelminthes based on 18S ribosomal gene sequences. Mol. Phylogenet. Evol. 10:1–10.

Canfield, D. E. 1998. A new model for Proterozoic ocean chemistry. Nature 396:450–453.

Canfield, D. E., and A. Teske. 1996. Late Proterozoic rise in atmospheric oxygen concentration inferred from phylogenetic and sulphur-isotope studies. Nature 382:127–132.

Carle, K. J., and E. E. Ruppert. 1983. Comparative ultrastructure of the bryozoan funiculus: A blood vessel homologue. Z. Zool. Syst. Evolutionsforsch. 21:181–193.

Carlson, S. J. 1991. Phylogenetic relationships among brachiopod higher taxa. Pp. 3–10 in D. I. Mackinnon, D. E. Le, and J. D. Campbell, eds., Brachiopods through Time. Balkema, Rotterdam.

———. 1995. Phylogenetic relationships among extant brachiopods. Cladistics 11:131–197.

Carmena, A., S. Giselbrecht, J. Harrison, F. Jiménez, and A. M. Michelson. 1998. Combinatorial signaling codes for the progressive determination of cell fate in the *Drosophila* embryonic mesoderm. Genes Dev. 12:3910–3922.

Carranza, S., J. Baguñà, and M. Riutort. 1997. Are the Platyhelminthes a monophyletic primitive group? An assessment using 18S rDNA sequences. Mol. Biol. Evol. 14:485–497.

Carranza, S., D. T. J. Littlewood, K. A. Clough, I. Ruiz-Trillo, J. Baguñà, and M. Riutort. 1998. A robust molecular phylogeny of the Tricladida (Platyhelminthes: Seriata) with a discussion on morphological synapomorphies. Proc. R. Soc. London 263:631–640.

Carroll, S. B., J. K. Grenier, and S. D. Weatherbee. 2001. From DNA to Diversity: Molecular Genetics and the Evolution of Animal Design. Blackwell Scientific, Malden, MA.

Carter, G. S. 1954. On Hadzi's interpretation of animal phylogeny. Syst. Zool. 3:163–167.

Casares, F., and R. S. Mann. 1998. Control of antennal versus leg development in *Drosophila*. Nature 392:723–726.

Caster, K. E. 1967. Homoiostelea. Pp. 581–623 in R. C. Moore, ed., Treatise on Invertebrate Paleontology, pt. S, Echinoderms. Geological Society of America, Boulder, CO; Univ. Kansas Press, Lawrence.

Catmull, J., D. C. Hayward, N. E. McIntyre, J. S. Reece-Hoyes, R. Mastro, P. Callaerts, E. E. Ball, and D. J. Miller. 1998. *Pax-6* origins: Implications from the structure of two coral *Pax* genes. Dev. Genes Evol. 208:352–356.

Caullery, M., and A. Lavellée. 1908. La fécondation et le développement de l'oeuf des orthonectides: 1, *Rhopalura ophiocomae*. Arch. Zool. Exp. Gen., ser. 4, 8:421–469.

Cavalier-Smith, T. 1987. The origin of eukaryote and archaebacterial cells. Ann. N.Y. Acad. Sci. 503:17–34.

———. 1998. A revised six-kingdom system of life. Biol. Rev. 73:203–266.

Cavalier-Smith, T., M. T. E. P. Allsopp, E. E. Chao, N. Boury-Esnault, and J. Vacelet. 1996. Sponge phylogeny, animal monophyly, and the origin of the nervous system: 18S rRNA evidence. Can. J. Zool. 74:2031–2045.

C. elegans Sequencing Consortium. 1998. Genome sequence of the nematode *C. elegans*: A platform for investigating biology. Science 282:2012–2018.

Chaffee, C., and D. R. Lindberg. 1986. Larval biology of Early Cambrian molluscs: The implications of small body size. Bull. Mar. Sci. 39:536–549.

Chapman, D. M. 1974. Cnidarian histology. Pp. 2–92 in L. Muscatine and H. M. Lenhoff, eds., Coelenterate Biology, Reviews, and New Perspectives. Academic Press, New York.

Chapman, G. 1958. The hydrostatic skeleton in invertebrates. Biol. Rev. 33:338–371.

Chen, J. Y., J. Dzik, G. D. Edgecombe, L. Ramsköld, and G. Zhou. 1995. A possible Early Cambrian chordate. Nature 377:720–722.

Chen, J. Y., and B.-D. Erdtmann. 1991. Lower Cambrian lagerstätte from Chengjiang, Yunnan, China: Insights for reconstructing early metazoan life. Pp. 57–76 in A. M. Simonetta and S. Conway Morris, eds., The Early Evolution of Metazoa and the Significance of Problematic Taxa. Cambridge Univ. Press, Cambridge.

Chen, J. Y., X. Hou, and H. Z. Lu. 1989a. Early Cambrian hock glass-like rare sea animal *Dinomischus* (Entoprocta) and its ecological features. Acta Paleontol. Sin. 28:58–71. In Chinese; English summary.

———. 1989b. Lower Cambrian leptomitids (Demospongea), Chengjiang, Yunnan. Acta Paleontol. Sin. 28:17–27. In Chinese; English summary.

Chen, J. Y., and D. Y. Huang. 2002. A possible Lower Cambrian chaetognath (arrow worm). Science 298:187.

Chen, J. Y., D. Y. Huang, and C. W. Li. 1999. An early Cambrian craniate-like chordate. Nature 402:518–522.

Chen, J. Y., D. Y. Huang, Q. Q. Peng, H. M. Chi, X. Q. Wang, and M. Feng. 2003. The first tunicate from the Early Cambrian of south China. Proc. Nat. Acad. Sci. USA 100:8314–8318.

Chen, J. Y., and C. Li. 1997. Early Cambrian chordate from Chengjiang, China. Bull. Nat. Mus. Natl. Sci. (Taipei) 10:257–274.

Chen, J. Y., and M. Lindström. 1991. Lower Cambrian non-mineralized fauna from Chengjiang, Yunnan, China. Geol. Foeren. Stockh. Foerh. 113:79–81.

Chen, J. Y., P. Oliveri, C. W. Li, G. Q. Zhou, F. Gao, J. W. Hagadorn, K. J. Peterson, and E. H. Davidson. 2000. Precambrian animal diversity: Putative phosphatized embryos from the Doushantuo Formation of China. Proc. Nat. Acad. Sci. USA 97:4457–4462.

Chen, J. Y., and G. Zhou. 1997. Biology of the Chengjiang fauna. Bull. Nat. Mus. Natl. Sci. (Taipei) 10:11–105.

Chen, J. Y., G. Zhou, and L. Ramsköld. 1995a. The Cambrian lobopodian *Microdictyon sinicum*. Bull. Nat. Mus. Natl. Sci. 5:1–93.

———. 1995b. A new Early Cambrian onychophoran-like animal, *Paucipodia* gen nov., from the Chengjiang fauna, China. Trans. R. Soc. Edinburgh Earth Sci. 85:275–282.

Chervitz, S. A., L. Aravind, G. Sherlock, C. A. Bell, E. V. Koonin, S. S. Dwight, M. A. Harris, K. Dolinski, S. Mohr, T. Smith, S. Weng, J. M. Cherry, and D. Bostein. 1998. Comparison of the complete protein sets of worm and yeast: Orthology and divergence. Science 282:2022–2028.

Chia, F.-S., and R. Koss. 1994. Asteroidea. Pp. 169–245 in F. W. Harrison and F.-S. Chia, eds., Microscopic Anatomy of Invertebrates, vol. 14, Echinodermata. Wiley-Liss, New York.

Chitwood, B. G., and M. G. Chitwood. 1974. Introduction to Nematology. University Park Press, Baltimore.

Christiansen, F. B., and T. M. Fenchel. 1979. Evolution of marine invertebrate reproductive patterns. Theor. Popul. Biol. 16:267–282.

Cisne, J. L. 1975. Anatomy of *Triarthrus* and the relationships of the Trilobita. Fossils and Strata 4:45–63.

Clark, R. B. 1962. On the structure and functions of polychaete septa. Proc. Zool. Soc. London 138:543–578.

———. 1964. Dynamics in Metazoan Evolution. Clarendon Press, Oxford.

———. 1969. Systematics and phylogeny: Annelida, Echiura, Sipuncula. Pp. 1–68 in M. Florkin and B. T. Scheer, eds., Chemical Zoology, vol 4. Academic Press, New York.

Clarke, K. U. 1979. Visceral anatomy and arthropod phylogeny. Pp. 467–549 in A. P. Gupta, ed., Arthropod Phylogeny. Van Nostrand Reinhold, New York.

Clément, P. 1985. The relationships of rotifers. Pp. 224–247 in S. Conway Morris, J. D. George, R. Gibson, and H. M. Platt, eds., The Origins and Relationships of Lower Invertebrates. Clarendon Press, Oxford.

Clément, P., and E. Wurdak. 1991. Rotifera. Pp. 219–297 in F. W. Harrison and E. E. Ruppert, eds., Microscopic Anatomy of Invertebrates, vol. 4, Aschelminthes. Wiley-Liss, New York.

Cloney, R. A. 1982. Ascidian larvae and the events of metamorphosis. Am. Zool. 22:817–826.

Coe, W. R. 1943. Biology of the nemerteans of the Atlantic coast of North America. Trans. Conn. Acad. Arts Sci. 35:129–328.

Cohen, B. L. 2000. Monophyly of brachiopods and phoronids: Reconciliation of molecular evidence with Linnaean classification (the subphylum Phoroniformea nov.). Proc. R. Soc. London B 267:225–231.

Cohen, B. L., and A. Gawthrop. 1997. The brachiopod genome. Pp. 189–211 in R. L. Kaesler, ed., Treatise on Invertebrate Paleontology, pt. H, Brachiopods (Revised), vol. 1. Geological Society of America, Boulder, CO; Univ. Kansas Press, Lawrence.

Cohen, B. L., A. Gawthrop, and T. Cavalier-Smith. 1998. Molecular phylogeny of brachiopods and phoronids based on nuclear-encoded small subunit ribosomal RNA gene sequences. Philos. Trans. R. Soc. London B 353:2039–2061.

Cohen, B. L., S. Stark, A. Gawthrop, M. E. Burke, and C. W. Thayer. 1998. Comparison of articulate brachiopod nuclear and mitochondrial gene trees leads to a clade-based redefinition of protostomes (Protostomozoa) and deuterostomes (Deuterostomozoa). Proc. R. Soc. London B 265:473–482.

Collins, A. G. 1998. Evaluating multiple alternative hypotheses for the origin of Bilateria: An analysis of 18S molecular evidence. Proc. Nat. Acad. Sci. USA 95:15458–15463.

———. 2002. Phylogeny of Medusozoa and the evolution of cnidarian life cycles. J. Evol. Biol. 15:418–432.

Collins, A. G., J. W. Lipps, and J. W. Valentine. 2000. Modern mucociliary creeping trails and the body plans of Neoproterozoic trace-makers. Paleobiology 26:47–55.

Collins, A. G., and J. W. Valentine. 2001. Defining phyla: Evolutionary pathways to metazoan body plans. Evol. Dev. 3:432–442.

Collins, D. H. 1996. The "evolution" of Anomalocaris and its classification in the arthropod class Dinocarida (nov.) and order Radiodonta (nov). J. Paleontol. 70:280–293.

Collins, D. H., D. E. G. Briggs, and S. Conway Morris. 1983. New Burgess Shale fossil sites reveal Middle Cambrian faunal complex. Science 222:163–167.

Compston, W., I. S. Williams, J. L. Kirschvink, Z. Zichao, and G. Ma. 1992. Zircon U-Pb ages for the Early Cambrian time scale. J. Geol. Soc. London 149:171–184.

Conklin, E. G. 1902. The embryology of a brachiopod, Terebratulina septentrionalis Couthouy. Proc. Am. Philos. Soc. 41:41–76.

———. 1905. The organization and cell lineage of the ascidian egg. J. Acad. Nat. Sci. (Phil.) 13:1–119.

———. 1932. The embryology of *Amphioxus*. J. Morphol. 54:69–151.

Conway Morris, S. 1977a. Fossil priapulid worms. Spec. Pap. Palaeontol. 20.

———. 1977b. A new metazoan from the Cambrian Burgess Shale, British Columbia. Palaeontology 20:623–640.

———. 1977c. A redescription of the Middle Cambrian worm *Amiskwia sagittiformis* Walcott from the Burgess Shale of British Columbia. Palaeontol. Z. 51:271–287.

———. 1979a. The Burgess Shale (Middle Cambrian) fauna. Ann. Rev. Ecol. Syst. 10:327–349.

———. 1979b. Middle Cambrian polychaetes from the Burgess Shale of British Columbia. Philos. Trans. R. Soc. London B 285:227–274.

———. 1981. Parasites and the fossil record. Parasitology 82:489–509.

———. 1985a. The Middle Cambrian metazoan *Wiwaxia corrugata* (Matthew) from the Burgess Shale and *Ogygopsis* Shale, British Columbia, Canada. Philos. Trans. R. Soc. London B 307:507–586.

———. 1985b. Non-skeletalized lower invertebrate fossils: A review. Pp. 343–359 in S. Conway Morris, J. D. George, R. Gibson, and H. M. Platt, eds., The Origins and Relationships of Lower Invertebrates. Clarendon Press, Oxford.

———. 1989. The persistence of Burgess Shale–type faunas: Implications for the evolution of deeper-water faunas. Trans. R. Soc. Edinburgh Earth Sci. 80:271–283.

———. 1990. *Typhloesus wellsi* (Melton and Scott, 1973): A bizarre metazoan from the Carboniferous of Montana, USA. Philos. Trans. R. Soc. London B 327:595–624.

———. 1992a. Burgess Shale–type faunas in the context of the "Cambrian explosion": A review. J. Geol. Soc. London 149:631–636.

———. 1992b. Ediacaran survivors. P. 69 in S. Lidgard and P. R. Crane, eds., Fifth North American Paleontological Convention, Abstracts and Program. Paleontological Society Special Publication 6.

———. 1993a. Ediacaran-like fossils in Cambrian Burgess Shale–type faunas of North America. Palaeontology 36:593–635.

———. 1993b. The fossil record and the early evolution of the Metazoa. Nature 361:219–225.

———. 1997. The cuticular structure of the 495-Myr-old type species of the fossil worm *Palaeoscolex*, *P. piscatorum* (?Priapulida). Zool. J. Linn. Soc. 119:69–82.

———. 1998. The Crucible of Creation. Oxford Univ. Press, Oxford.

Conway Morris, S., and D. H. Collins. 1996. Middle Cambrian ctenophores from the Stephen Formation, British Columbia, Canada. Philos. Trans. R. Soc. London B 351:279–308.

Conway Morris, S., and D. W. T. Crompton. 1982. The origins and evolution of the Acanthocephala. Biol. Rev. 57:85–115.

Conway Morris, S., and J. S. Peel. 1995. Articulated halkieriids from the Lower Cambrian of north Greenland and their role in early protostome evolution. Philos. Trans. R. Soc. London B 347:305–358.

Conway Morris, S., J. S. Peel, A. K. Higgins, N. J. Soper, and N. C. Davis. 1987. A Burgess Shale–like fauna from the Lower Cambrian of north Greenland. Nature 326:181–183.

Conway Morris, S., R. K. Pickerell, and T. L. Harland. 1982. A possible annelid from the Trenton Limestone (Ordovician) of Quebec, with a review of fossil oligochaetes and other annulate worms. Can. J. Earth Sci. 19:2150–2157.

Conway Morris, S., and R. A. Robison. 1986. Middle Cambrian priapulids and other soft-bodied fossils from Utah and Spain. Univ. Kans. Paleontol. Contrib. Pap. 117:1–22.

———. 1988. More soft-bodied animals and algae from the Middle Cambrian of Utah and British Columbia. Univ. Kans. Paleontol. Contrib. Pap. 122:1–48.

Conway Morris, S., H. B. Whittington, D. E. G. Briggs, C. P. Hughes, and D. L. Bruton. 1982. Atlas of the Burgess Shale. Palaeontological Assoc., Nottingham, U.K.

Cooper, J. A., R. J. F. Jenkins, W. Compston, and I. S. Williams. 1992. Ion-probe zircon dating of a mid–Early Cambrian tuff in South Australia. J. Geol. Soc. London 149:185–192.

Cooper, K. W. 1964. The first fossil tardigrade: *Beorn leggi* Cooper from Cretaceous amber. Psyche 71:41–48.

Cornell, R. A., and F. Van Ohlen. 2000. Vnd/nkx, ind/gsx, and msh/msx: Conserved regulators of dorsoventral neural patterning? Curr. Opin. Neurobiol. 10:63–71.

Corsetti, F. A., S. M. Awramik, and D. Pierce. 2003. A complex microbiota from snowball Earth times: Microfossils from the Neoproterozoic Kingston Peak Formation, Death Valley, USA. Proc. Nat. Acad. Sci. USA 100:4399–4404.

Costello, D. P., and C. Henley. 1976. Spiralian development: A perspective. Am. Zool. 16:277–291.

Coutinho, C. C., S. Vissers, and G. Van de Vyver. 1994. Evidence of homeobox genes in the freshwater sponge *Ephydatia fluviatilis*. Pp. 385–388 in R. W. M. van Soest, T. M. G. van Kempen, and J.-C. Braekman, eds., Sponges in Time and Space. Balkema, Rotterdam.

Couzin, J. 2002. Small RNAs make a big splash. Science 298:2296–2297.

Cowie, J. W., and M. D. Brasier, eds. 1989. The Precambrian-Cambrian Boundary. Clarendon Press, Oxford.

Cox, L. R. 1960. General characteristics of Gastropoda. Pp. 84–169 in R. C. Moore, ed., Treatise on Invertebrate Paleontology, pt. 1, Mollusca, vol. 1. Geological Society of America, Boulder, CO; Univ. Kansas Press, Lawrence.

Crick, F. H. C. 1958. On protein synthesis. Symp. Soc. Exp. Biol. 12:138–163.

Crimes, T. P. 1989. Trace fossils. Pp. 166–185 in J. W. Cowie and M. D. Brasier, eds., The Precambrian-Cambrian Boundary. Clarendon Press, Oxford.

Crimes, T. P., and M. L. Droser. 1992. Trace fossils and bioturbation: The other fossil record. Ann. Rev. Ecol. Syst. 23:339–360.

Crompton, D. W. T., and B. B. Nickol, eds. 1985. Biology of the Acanthocephala. Cambridge Univ. Press, Cambridge.

Crowther, R. R., and R. B. Rickards. 1977. Cortical bandages and the graptolite zooid. Geol. Palaeontol. 11:9–46.

Crozier, W. J. 1918. On the method of progression in polyclads. Proc. Nat. Acad. Sci. USA 4:379–381.

Cuénot, L. 1949. Les Tardigrades. Pp. 39–59 in P. P. Grassé, ed., Traité de Zoologie, vol. 6. Masson, Paris.

Cutler, E. B. 1994. The Sipuncula: Their Systematics, Biology, and Evolution. Cornell Univ. Press, Ithaca, NY.

Cuvier, G. 1812. Sur un nouveau rapprochement à établir entre les classes qui composent le règne animal. Ann. Mus. Hist. Nat. 19:73–84.

Dahl, E. 1984. The subclass Phyllocarida (Crustacea) and the status of some early fossils: A neontological view. Vidensk. Medd. Dan. Naturhist. Foren. 145:61–76.

Dales, R. P. 1967. Annelids. 2nd ed. Hutchinson, London.

Dalziel, I. W. D. 1997. Neoproterozoic-Paleozoic geography and tectonics: Review, hypothesis, environmental speculations. Geol. Soc. Am. Bull. 109:16–42.

Darwin, C. 1844. Observations on the structure and propagation of the genus *Sagitta*. Ann. Mag. Nat. Hist. 13:1–6.

———. 1859. On the Origin of Species by Means of Natural Selection. John Murray, London.

Davenport, M. P., C. Blass, and P. Eggleston. 2000. Characterization of the Hox gene cluster in the malaria vector mosquito, *Anopheles gambiae*. Evol. Dev. 2:326–339.

David, B., B. Lefebvre, R. Mooi, and R. Parsley. 2000. Are homalozoans echinoderms? An answer from the extraxial-axial theory. Paleobiology 25:529–555.

Davidson, E. H. 1986. Gene Activity in Early Development. 3rd ed. Academic Press, New York.

———. 1990. How embryos work: A comparative view of diverse modes of cell fate specification. Development 108:365–389.

———. 1991. Spatial mechanisms of gene regulation in metazoan embryos. Development 113:1–26.

———. 1993. Later embryogenesis: Regulatory circuitry in morphogenetic fields. Development 118:665–690.

———. 2001. Genomic Regulatory Systems: Development and Evolution. Academic Press, San Diego.

Davidson, E. H., K. J. Peterson, and R. A. Cameron. 1995. Origin of bilaterian body plans: Evolution of developmental regulatory mechanisms. Science 270:1319–1325.

Debrenne, F. 1983. Archaeocyathids: Morphology and affinity. Pp. 178–190 in T. W. Broadhead, ed., Sponges and Spongiomorphs. Studies in Geol. 7. Univ. Tennessee, Department of Geological Science, Knoxville.

Debrenne, F., and A. Rozanov. 1983. Paleogeographic and stratigraphic distribution of regular Archaeocyatha (Lower Cambrian fossils). Geobios 16:727–736.

Debrenne, F., A. Rozanov, and G. F. Webers. 1984. Upper Cambrian Archaeocyatha from Antarctica. Geol. Mag. 121:291–299.

Debrenne, F., and J. Vacelet. 1983. Archaeocyatha: Is the sponge model consistent with their structural organization? P. 4 in Abstracts of the Fourth International Symposium on Fossil Cnidaria, Washington, DC.

Deflandre, G., and M. Deflandre-Rigaud. 1956. *Micrascidites* manip. nov.: Sclerites de Didemnidés (Ascidies, Tuniciers) fossiles du Lutétien du Bassin parisien et du Balcombian d'Australie. C. R. Somm. Seances Soc. Geol. Fr., 47–49.

Degnan, B. M., S. M. Degnan, A. Giusti, and D. E. Morse. 1995. A *hox/hom* homeobox gene in sponges. Gene 115:175–177.

Dehal, P., et al. 2002. The draft genome of *Ciona intestinalis:* Insights into chordate and vertebrate origins. Science 298:2157–2167.

Delsman, H. C. 1910. Beiträge zur Entwicklungsgeschichte von *Oikopleura dioica.* Verh. Rijksinst. Onderz. Zee 3 (2):3–24.

———. 1912. Weitere Beobachtungen über die Entwicklung von *Oikopleura dioica.* Tijdschr. Ned. Dierkd. Ver., 2nd ser., 12:14–205.

Dennell, R. 1960. Integument and exoskeleton. Pp. 449–472 in T. H. Waterman, ed., The Physiology of Crustacea, vol. 1. Academic Press, New York.

Denny, M. W. 1984. Mechanical properties of pedal mucus and their consequences for gastropod structure and performance. Am. Zool. 24:23–36.

de Queiroz, K., and J. Gauthier. 1990. Phylogeny as a central principle in taxonomy: Phylogenetic definitions of taxon names. Syst. Zool. 39:307–322.

———. 1992. Phylogenetic taxonomy. Ann. Rev. Ecol. Syst. 23:449–480.

DeRobertis, E. M., and Y. Sasai. 1996. A common plan for dorsoventral patterning in Bilateria. Nature 380:37–40.

de Rosa, R., J. K. Grenier, T. Andreeva, C. E. Cook, A. Adoutte, M. Akam, S. B. Carroll, and G. Balavoine. 1999. *Hox* genes in brachiopods and priapulids and protostome evolution. Nature 399:772–776.

Derstler, K. L. 1981. Morphological diversity of early Cambrian echinoderms. Papers for the Second International Symposium on the Cambrian System. U.S. Geol. Surv. Open-File Rep. 81–743.

Dewel, R. A. 2000. Colonial origin for Eumetazoa: Major morphological transitions and the origin of bilaterian complexity. J. Morphol. 243:35–74.

Dewel, R. A., and W. C. Dewel. 1996. The brain of *Echiniscus viridissimus* Peterfi, 1956 (Heterotardigrada): A key to understanding the phylogenetic position of tardigrades and the evolution of the arthropod head. Zool. J. Linn. Soc. 116: 35–49.

Dewel, R. A., W. C. Dewel, and F. K. McKinney. 2001. Diversification of the Metazoa: Ediacarans, colonies, and the origin of eumetazoan complexity by nested modularity. Hist. Biol. 15:93–118.

Dewel, R. A., D. R. Nelson, and W. C. Dewel. 1993. Tardigrada. Pp. 143–183 in F. W. Harrison and M. E. Rice, eds., Microscopic Anatomy of Invertebrates, vol. 12, Onychophora, Chilopoda, and Lesser Protostomata. Wiley-Liss, New York.

Dilly, P. N. 1985. The habitat and behaviour of *Cephalodiscus gracilis* (Pterobranchia, Hemichordata) from Bermuda. J. Zool., ser. A 207:223–239.

———. 1986. Modern pterobranchs: Observations on their behaviour and tube building. Pp. 261–269 in C. P. Hughes, R. B. Rickards, and A. J. Chapman, eds., Palaeoecology and Biostratigraphy of Graptolites. Geol. Soc. London, Special Publication 20.

———. 1993. *Cephalodiscus graptolitoides* sp. nov.: A probable extant graptolite. J. Zool. 228:69–78.

Dobzhansky, T. 1937. Genetics and the Origin of Species. Columbia Univ. Press, New York.

Donoghue, M. A., M. A. Purnell, and R. J. Aldridge. 1998. Conodont anatomy, chordate phylogeny, and vertebrate classification. Lethaia 31:211–219.

Doolitle, R. F. 1995. The origins and evolution of eukaryotic proteins. Philos. Trans. R. Soc. London B 349:235–240.

Droser, M. L., and D. J. Bottjer. 1988a. Trends in depth and extent of bioturbation in Cambrian carbonate marine environments, western United States. Geology 16:233–236.

———. 1988b. Trends in extent and depth of Early Paleozoic bioturbation in the Great Basin (California, Nevada, and Utah). Pp. 123–135 in D. L. Weide and M. L. Faber, eds., This Extended Land: Geological Journeys in the Southern Basin and Range. Geological Society of America, Cordilleran Section, Field Trip Guidebook.

———. 1993. Trends and patterns of Phanerozoic ichnofabrics. Ann. Rev. Earth Planet. Sci. 21:205–225.

Droser, M. L., J. G. Gehling, and S. Jensen. 1999. When the worm turned: Concordance of Early Cambrian ichnofabric and trace-fossil record in siliciclastic rocks of South Australia. Geology 27:625–629.

Droser, M. L., S. Jensen, and J. G. Gehling. 2002. Trace fossils and substrates of the terminal Proterozoic–Cambrian transition: Implications for the record of early bilaterians and sediment mixing. Proc. Nat. Acad. Sci. USA 99:12572–12576.

Dunagan, T. T., and D. M. Miller. 1991. Acanthocephala. Pp. 299–332 in F. W. Harrison and E. E. Ruppert, eds., Microscopic Anatomy of Invertebrates, vol. 4, Aschelminthes. Wiley-Liss, New York.

Durden, C. J. 1969. Gnathostomulida: Is there a fossil record? Science 164:855.

Durham, J. W. 1967. Notes on the Helicoplacoidea and early echinoderms. J. Paleontol. 41:97–102.

———. 1993. Observations on the Early Cambrian helicoplacoid echinoderms. J. Paleontol. 67:590–604.

Durham, J. W., and K. E. Caster. 1966. Helicoplacoids. Pp. 131–136 in R. C. Moore, ed., Treatise on Invertebrate Paleontology, pt. U, vol. 1. Geological Society of America, Boulder, CO; Univ. Kansas Press, Lawrence.

Dzik, J. 1986. Turrilepadida and other Machaeridia. Pp. 116–134 in A. Hoffman and M. H. Nitecki, eds., Problematic Fossil Taxa. Oxford Univ. Press, New York.

Dzik, J., and G. Krumbiegel. 1989. The oldest "onychophoran" *Xenusion:* A link connecting phyla? Lethaia 22:169–181.

Dzik, J., and K. Lendzion. 1988. The oldest arthropods of the East European Platform. Lethaia 21:29–38.

Edelman, G. M., and F. S. Jones. 1995. Developmental control of N-CAM expression by Hox and Pax gene products. Philos. Trans. R. Soc. London B 349:305–312.

Edlinger, K. 1991. The mechanical constraints in mollusc constructions: The functions of the shell, the musculature, and the connective tissue. Pp. 359–374 in N. Schmidt-Kittler and K. Vogel, eds., Constructional Morphology and Evolution. Springer-Verlag, Berlin.

Edwards, A. W. F. 1996. The origin and early development of the method of minimum evolution for the reconstruction of phylogenetic trees. Syst. Biol. 45:79–91.

Eernisse, D. J., J. S. Albert, and F. E. Anderson. 1992. Annelida and Arthropoda are not sister taxa: A phylogenetic analysis of spiralian metazoan morphology. Syst. Biol. 41:305–330.

Eerola, T. T. 2001. Climate change at the Neoproterozoic-Cambrian transition. Pp. 90–106 in A. Yu. Zhuravlev and R. Riding, eds., The Ecology of the Cambrian Radiation. Columbia Univ. Press, New York.

Efron, B., E. Halloran, and S. Holmes. 1996. Bootstrap confidence levels for phylogenetic trees. Proc. Nat. Acad. Sci. USA 93:7085–7090.

Eggers, F. 1924. Zur Bewegungsphysiologie der Nemertinen: 1, *Emplectonema*. Z. Vgl. Physiol. 1:579–589.

Ehlers, U. 1985. Phylogenetic relationships within the Platyhelminthes. Pp. 143–158 in S. Conway Morris, J. D. George, R. Gibson, and H. M. Platt, eds., The Origins and Relationships of Lower Invertebrates. Clarendon Press, Oxford.

———. 1992. On the fine structure of *Paratomella rubra* Riger and Ott (Acoela) and the position of the taxon *Paratomella* Dorjes in a phylogenetic system of the Acoelomorpha (Platyhelminthes). Microfauna Marina 7:265–293.

Eibye-Jacobsen, D., and C. Nielsen. 1996. The rearticulation of annelids. Zool. Scr. 25:275–282.

Elder, H. Y. 1973. Direct peristaltic progression and the functional significance of the dermal connective tissues during burrowing in the polychaete *Polyphysia crassa*. J. Exp. Biol. 58:637–655.

———. 1980. Peristaltic mechanisms. Pp. 71–92 in H. Y. Elder and E. R. Trueman, eds., Aspects of Animal Movement. Cambridge Univ. Press, Cambridge.

Elder, H. Y., and R. D. Hunter. 1980. Burrowing of *Priapulus caudatus* and the significance of the direct peristaltic wave. J. Zool., ser. A, 191:333–351.

Eldredge, N. 1985. Unfinished Synthesis: Biological Hierarchies and Modern Evolutionary Thought. Oxford Univ. Press, New York.

———. 1989. Macroevolutionary Dynamics: Species, Niches, and Adaptive Peaks. McGraw-Hill, New York.

Eldredge, N., and S. J. Gould. 1972. Punctuated equilibria: An alternative to phyletic gradualism. Pp. 82–115 in T. J. M. Schopf, ed., Models in Paleobiology. Freeman, Cooper, San Francisco.

Eldredge, N., and S. N. Salthe. 1984. Hierarchy and evolution. Oxford Surv. Evol. Biol. 1:184–208.

Elsasser, W. M. 1987. Reflections on a Theory of Organisms. Éditions Orbis, Frelighsburg, Canada.

Emig, C. C. 1977. Un nouvel embranchement: Les Lophophorates. Bull. Soc. Zool. Fr. 102:341–344.

———. 1979. British and Other Phoronids. Academic Press, London.

Emig, C. C., and R. Siewing. 1975. The epistome of *Phoronis psammophila* (Phoronida). Zool. Anz. 194:47–54.

Emlet, R. B., and O. Hoegh-Guldberg. 1997. Effects of egg size on postlarval performance: Experimental evidence from a sea urchin. Evolution 51:141–152.

Emschermann, P. 1982. Les Kamptozoaires: État actuel de nos connaissances sur leur anatomie, leur développement, leur biologie, et leur position phylogénétique. Bull. Soc. Zool. Fr. 107:317–344.

Emson, R. H., and P. J. Whitfield. 1991. Behavioural and ultrastructural studies on the sedentary platyctenean ctenophore *Yallicula nubiformis*. Hydrobiologia 216/217:27–33.

Engel, J., V. P. Efimov, and P. Maurer. 1994. Domain organizations of extracellular matrix proteins and their evolution. Pp. 35–42 in M. Akam, P. Holland, P. Ingham, and G. Wray, eds., The Evolution of Developmental Mechanisms. Company of Biologists, Cambridge.

Erwin, D. H. 1994. Early introduction of major morphological innovations. Acta Palaeontol. Pol. 38:281–294.

———. 1999. The origin of bodyplans. Am. Zool. 39:617–629.

Erwin, D. H., and E. H. Davidson. 2002. The last common bilaterian ancestor. Development 129:3021–3032.

Erwin, D. H., and J. W. Valentine. 1984. "Hopeful monsters," transposons, and metazoan radiation. Proc. Nat. Acad. Sci. USA 81:5482–5483.

Erwin, D. H., J. W. Valentine, and J. J. Sepkoski, Jr. 1987. A comparative study of diversification events: The early Paleozoic versus the Mesozoic. Evolution 41:1177–1186.

Farmer, J. D., G. Vidal, M. Moczydlowska, H. Strauss, P. Ahlberg, and S. Siedlecka. 1992. Ediacaran fossils from the Innerelv Member (late Proterozoic) of the Tanafjorden area, northeastern Finnmark. Geol. Mag. 129:181–195.

Fauchald, K. 1975. Polychaete phylogeny: A problem in protostome evolution. Syst. Zool. 23:493–506.

Fautin, D. G., and R. N. Mariscal. 1991. Cnidaria: Anthozoa. Pp. 267–358 in F. W. Harrison and J. A. Westfall, eds., Microscopic Anatomy of Invertebrates, vol. 2. Wiley-Liss, New York.

Fawcett, D. W. 1994. A Textbook of Histology. 12th ed. Chapman and Hall, New York.

Fay, R. O. 1967. Blastoids: Introduction. Pp. 298–300 in R. C. Moore, ed., Treatise on Invertebrate Paleontology, pt. S, Echinodermata, vol. 1. Geological Society of America, Boulder, CO; Univ. Kansas Press, Lawrence.

Featherstone, D. E., and K. Broadie. 2002. Wrestling with pleiotropy: Genomic and topological analysis of the yeast gene expression network. BioEssays 24:267–274.

Fedonkin, M. A. 1985a. Non-skeletal fauna of the Vendian: Promorphological analysis. Pp. 10–69 in B. S. Sokolov and A. B. Ivanovich, eds., The Vendian System, vol. l. Nauka, Moscow. In Russian.

———. 1985b. Paleoichnology of Vendian Metazoa. Pp. 112–117 in B. S. Sokolov and A. B. Ivanovich, eds., The Vendian System, vol. 1. Nauka, Moscow. In Russian.

———. 1985c. Systematic descriptions of Vendian Metazoa. Pp. 70–106 in B. S. Sokolov and A. B. Ivanovich, eds., The Vendian System, vol. 1. Nauka, Moscow. In Russian.

———. 1987. Non-skeletal fauna of the Vendian and its place in the evolution of metazoans. Tr. Paleontol. Inst. Akad. Nauk SSR 226, Moscow. In Russian.

———. 1992. Vendian faunas and the early evolution of Metazoa. Pp. 87–129 in J. H. Lipps and P. W. Signor, eds., Origin and Early Evolution of the Metazoa. Plenum, New York.

———. 1994. Vendian body fossils and trace fossils. Pp. 370–388 in S. Bengtson, ed., Early Life on Earth. Columbia Univ. Press, New York.

———. 1998. Metameric features in the Vendian metazoans. Ital. J. Zool. 65:11–17.

Fedonkin, M. A., and B. M. Waggoner. 1997. The late Precambrian fossil *Kimberella* is a mollusc-like bilaterian organism. Nature 388:868–871.

Fell, P. E. 1974. Porifera. Pp. 51–132 in A. C. Giese and J. S. Pearse, eds., Reproduction of Marine Invertebrates, vol. 1. Academic Press, New York.

Felsenstein, J. 1978. Cases in which parsimony or compatibility methods will be positively misleading. Syst. Zool. 27:401–410.

————. 1985. Confidence limits on phylogenies: An approach using the bootstrap. Evolution 39:783–791.

————. 1988. Phylogenies from molecular sequences: Inference and reliability. Ann. Rev. Genet. 22:521–565.

Ferkowicz, M. J., M. C. Stander, and R. A. Raff. 1998. Phylogenetic relationships and developmental expression of three sea urchin *Wnt* genes. Mol. Biol. Evol. 15:809–819.

Ferreira, L. F., A. Araújo, and A. N. Duarte. 1993. Nematode larvae in fossilized animal coprolites from Lower and Middle Pleistocene sites, central Italy. J. Parasitol. 79:440–442.

Ferrier, D. E. K., and P. W. H. Holland. 2001. Sipunculan ParaHox genes. Evol. Dev. 3:263–270.

Ferrier, D. E. K., C. Minguillón, P. W. H. Holland, and J. Garcia-Fernàndez. 2000. The amphioxus Hox cluster: Deuterostome posterior flexibility and *Hox14*. Evol. Dev. 2:284–293.

Field, K. G., G. J. Olsen, D. J. Lane, S. J. Giovannoni, M. T. Ghiselin, E. C. Raff, N. R. Pace, and R. A. Raff. 1988. Molecular phylogeny of the animal kingdom. Science 239:748–753.

Finnerty, J. R. 1998. Homeoboxes in sea anemones and other nonbilaterian animals: Implications for the evolution of the Hox cluster and the zootype. Curr. Top. Dev. Biol. 40:211–254.

Finnerty, J. R., and M. Q. Martindale. 1997. Homeoboxes in sea anemones (Cnidaria: Anthozoa): A PCR-based survey of *Nematostella vectensis* and *Metridium senile*. Biol. Bull. 193:62–76.

————. 1999. Ancient origins of axial patterning genes: Hox genes and ParaHox genes in the Cnidaria. Evol. Dev. 1:16–23.

Finnerty, J. R., V. A. Master, S. Irvine, M. Kourakis, S. Warriner, and M. Q. Martindale. 1996. Homeobox genes in the Ctenophora: Identification of *paired-type* and *Hox* homologues in the atentaculate ctenophore *Beroe ovata*. Mol. Mar. Biol. Biotechnol. 5:249–258.

Fisher, D. C. 1985. Evolutionary morphology: Beyond the analogous, the anecdotal, and the ad hoc. Paleobiology 11:120–138.

————. 1992. Stratigraphic parsimony. Pp. 124–129 in W. P. Maddison and D. R. Maddison, MacClade: Analysis of Phylogeny and Character Evolution, Version 3. Sinauer, Sunderland, MA.

————. 1994. Stratocladistics: Morphological and temporal patterns and their relation to phylogenetic process. Pp. 133–171 in L. Grande and O. Rieppel, eds., Interpreting the Hierarchy of Nature: From Systematic Patterns to Evolutionary Process Theories. Academic Press, New York.

Fitch, W. M. 1970. Distinguishing homologous from analogous proteins. Syst. Zool. 19:99–113.

————. 1971. Toward defining the course of evolution: Minimal change for a specific tree topology. Syst. Zool. 20:406–416.

Fjerdingstad, E. 1961. The ultrastructure of choanocyte collars in *Spongilla lacustris* (L.). Z. Zellforsch. 53:645–657.

Flessa, K. W., and M. Kowalewski. 1994. Shell survival and time-averaging in nearshore and shelf environments: Estimates from the radiocarbon literature. Lethaia 27:153–165.

Fletcher, T. P., and D. H. Collins. 1998. The Middle Cambrian Burgess Shale and its relationship to the Stephen Formation in the southern Canadian Rocky Mountains. Can. J. Earth Sci. 35:413–436.

Foote, M. 1988. Survivorship analysis of Cambrian and Ordovician trilobites. Paleobiology 14:258–271.

——. 1991. Morphological patterns of diversification: Examples from trilobites. Palaeontology 34:461–485.

——. 1992a. Paleozoic record of morphological diversity in blastozoan echinoderms. Proc. Nat. Acad. Sci. USA 89:7325–7329.

——. 1992b. Rarefaction analysis of morphological and taxonomic diversity. Paleobiology 18:1–16.

——. 1994a. Morphological disparity in Ordovician-Devonian crinoids and the early saturation of morphological space. Paleobiology 20:320–344.

——. 1994b. Morphology of Ordovician-Devonian crinoids. Contrib. Mus. Paleontol. Univ. Mich. 29:1–39.

——. 1995. Morphology of Carboniferous and Permian crinoids. Contrib. Mus. Paleontol. Univ. Mich. 29:135–184.

——. 1997. Estimating taxonomic durations and preservation probability. Paleobiology 23:278–300.

——. 1999. Morphological diversity in the evolutionary radiation of Paleozoic and post-Paleozoic crinoids. Paleobiology 25, Memoir, suppl. to no. 2: 1–115.

Foote, M., J. P. Hunter, D. M. Janis, and J. J. Sepkoski, Jr. 1999. Evolutionary and preservational constraints on origins of biologic groups: Divergence times of eutherian mammals. Science 283:1310–1313.

Foote, M., and D. M. Raup. 1996. Fossil preservation and the stratigraphic ranges of taxa. Paleobiology 22:121–140.

Foote, M., and J. J. Sepkoski, Jr. 1999. Absolute measures of the completeness of the fossil record. Nature 398:415–417.

Force, A., M. Lynch, F. B. Pickett, A. Amores, Y.-L. Yan, and J. Postlethwait. 1999. Preservation of duplicate genes by complementary, degenerate mutations. Genetics 151:1531–1545.

Forey, P. L. 1982. Neontological analysis versus palaeontological stories. Pp. 119–157 in K. A. Joysey and A. E. Friday, eds., Problems of Phylogenetic Reconstruction. Academic Press, London.

Fortey, R. A. 2001. The Cambrian explosion exploded? Science 203:438–439.

Fortey, R. A., D. E. G. Briggs, and M. A. Wills. 1996. The Cambrian evolutionary "explosion": Decoupling cladogenesis from morphological disparity. Biol. J. Linn. Soc. 57:13–33.

Fortey, R. A., and A. Seilacher. 1997. The trace fossil *Cruziana semiplicata* and the trilobite that made it. Lethaia 30:105–112.

Fox, D. L., D. C. Fisher, and L. R. Leighton. 1999. Reconstructing phylogeny with and without temporal data. Science 284:1816–1819.

Franc, J. M. 1972. Activités des rosettes ciliées et leurs supports ultrastructuraux chez les Cténaires. Z. Zellforsch. 130:527–544.

Fransen, M. E. 1980. Ultrastructure of coelomic organization in annelids: l, Archiannelida and other small polychaetes. Zoomorphologie 95:235–249.

———. 1988. Coelomic and vascular systems. Pp. 199–213 in W. Westheide and C. O. Hermans, eds., The Ultrastructure of Polychaeta. Microfauna Marina 4. Fischer, Stuttgart.

Franz, V. 1927. Morphologie der Akranier. Ergeb. Anat. Entwicklungsgesch. 27:464–692.

Frasch, M., and M. Levine. 1987. Complementary patterns of *even-skipped* and *fushi tarazu* expression involve their differential regulation by a common set of segmentation genes in *Drosophila*. Genes Dev. 1:981–995.

Freeman, G. 1983. Experimental studies on embryogenesis in hydrozoans (Trachylina and Siphonophora) with direct development. Biol. Bull. 165:591–618.

Freeman, G., and J. W. Lundelius. 1992. Evolutionary implications of the mode of D quadrant specification in coelomates with spiral cleavage. J. Evol. Biol. 5:205–247.

———. 1999. Changes in the timing of mantle formation and larval life history traits in linguliform and craniform brachiopods. Lethaia 32:197–217.

Frey, R. W. 1975. The realm of ichnology. Pp. 13–38 in R. W. Frey, ed., The Study of Trace Fossils. Springer-Verlag, New York.

Friedlander, T. P., J. C. Regier, and C. Mitter. 1994. Phylogenetic information content of five nuclear gene sequences in animals: Initial assessment of character sets from concordance and divergence studies. Syst. Biol. 43:511–525.

Friedrich, M., and D. Tautz. 1997. An episode of change of rDNA nucleotide substitution rate has occurred during the emergence of the insect order Diptera. Mol. Biol. Evol. 14:644–653.

Funch, P. 1996. The chordoid larva of *Symbion pandora* (Cycliophora) is a modified trochophore. J. Morphol. 230:231–263.

Funch, P., and R. M. Kristensen. 1995. Cycliophora is a new phylum with affinities to Entoprocta and Ectoprocta. Nature 378:711–714.

———. 1997. Cycliophora. Pp. 400–474 in F. W. Harrison and R. M. Woollacott, eds., Microscopic Anatomy of Invertebrates, vol. 13, Lophophorates, Entoprocta, and Cycliophora. Wiley-Liss, New York.

Furuya, H., K. Tsuneki, and Y. Koshida. 1996. The cell lineages of two types of embryo and a hermaphroditic gonad in dicyemid mesozoans. Dev. Growth Differ. 38:453–463.

Gabbott, S. E., R. J. Aldridge, and J. N. Theron. 1995. A giant conodont with preserved muscle tissue from the Upper Ordovician of South Africa. Nature 374:800–803.

Galant, R., and S. B. Carroll. 2002. Evolution of a transcriptional repression domain in an insect Hox protein. Nature 415:910–913.

Galliot, B., and D. Miller. 2000. Origin of anterior patterning: How old is our head? Trends Genet. 16:1–5.

Garcia-Bellido Capdevila, D., and S. Conway Morris. 1999. New fossil worms from the Lower Cambrian of the Kinzers Formation, Pennsylvania, with some comments on Burgess Shale–type preservation. J. Paleontol. 73:394–402.

Gardiner, S. L. 1992. Polychaeta: General organization, integument, musculature, coelom, and vascular system. Pp. 19–52 in F. W. Harrison and S. L. Gardiner, eds., Microscopic Anatomy of Invertebrates, vol. 7, Annelida. Wiley-Liss, New York.

Garey, J. R., M. Krotec, D. R. Nelson, and J. Brooks. 1996. Molecular anaysis supports a tardigrade-arthropod association. Invertebr. Biol. 115:79–88.

Garey, J. R., T. J. Near, M. R. Nonnemacher, and S. A. Nadler. 1996. Molecular evidence for Acanthocephala as a subtaxon of Rotifera. J. Mol. Evol. 43:287–292.

Garrone, R., T. L. Simpson, and J. Pottu-Boumendil. 1981. Ultrastructure and deposition of silica in sponges. Pp. 495–525 in T. L. Simpson and B. E. Volcani, eds., Silicon and Siliceous Structures in Biological Systems. Springer-Verlag, New York.

Garstang, W. 1894. Preliminary note on a new theory of the phylogeny of the Chordata. Zool. Anz. 17:122.

———. 1928. The morphology of the Tunicata and its bearing on the phylogeny of the Chordata. Q. J. Microsc. Soc. 72:51–187.

Gee, H. 1996. Before the Backbone. Chapman and Hall, London.

———. 2001. On being vetulicolian. Nature 414:407–409.

Gehling, J. G. 1987. Earliest known echinoderm: A new Ediacaran fossil from the Pound Subgroup of South Australia. Alcheringa 11:337–345.

———. 1991. The case for Ediacaran fossil roots to the metazoan tree. Geol. Soc. India Mem. 20:181–224.

———. 1994. The earliest known poriferans: Bath tub sponges from the Ediacara fauna of South Australia. PaleoBios 14 suppl.: 6–7.

———. 1999. Microbial mats in terminal Proterozoic siliciclastics: Ediacaran death masks. Palaios 14:40–57.

Gehling, J. G., and J. K. Rigby. 1996. Long expected sponges from the Neoproterozoic Ediacara fauna of South Australia. J. Paleontol. 70:185–195.

Geoffroy Saint-Hilaire, É. 1822. Considérations générale sur la vertèbre. Mem. Mus. Hist. Nat. 9:89–119.

Gerhart, J. 2000. Inversion of the chordate body axis: Are there alternatives? Proc. Nat. Acad. Sci. USA 97:4445–4448.

Gerhart, J., and M. Kirschner. 1997. Cells, Embryos, and Evolution. Blackwell Scientific, Malden, MA.

Gershwin, L. 1999. Clonal and population variation in jellyfish symmetry. J. Mar. Biol. Assoc. U.K. 79:993–1000.

Ghiselin, M. T. 1974. A radical solution to the species problem. Syst. Zool. 23:536–544.

———. 1988. The origin of molluscs in the light of molecular evidence. Oxford Surv. Evol. Biol. 5:66–95.

———. 1997. Metaphysics and the Origin of Species. State Univ. New York Press, Albany.

Gibson, R. 1972. Nemerteans. Hutchinson Univ. Library, London.

———. 1982. Nemertea. Pp. 823–846 in S. P. Parker, ed., Synopsis and Classification of Living Organisms, vol. 1. McGraw-Hill, New York.

Gilbert, S. F. 1997. Developmental Biology. 5th ed. Sinauer, Sunderland, MA.

———. 2000. Developmental Biology. 6th ed. Sinauer, Sunderland, MA.

Gilbert, S. F., and A. M. Raunio, eds. 1997. Embryology: Constructing the Organism. Sinauer, Sunderland, MA.

Gilbert, W. 1987. The exon theory of genes. Cold Spring Harb. Symp. Quant. Biol. 52:901–905.

Gilinsky, N. L., and R. K. Bambach. 1987. Asymmetrical patterns of origination and extinction in higher taxa. Paleobiology 13:427–445.

Gilmour, T. H. J. 1979. Feeding in pterobranch hemichordates and the evolution of gill slits. Can. J. Zool. 57:1136–1142.

———. 1982. Feeding in tornaria larvae and the development of gill slits in enteropneust hemichordates. Can. J. Zool. 60:3010–3020.

Gingerich, P. D. 1979. The stratophenetic approach to phylogeny reconstruction in vertebrate paleontology. Pp. 41–77 in J. Cracraft and N. Eldredge, eds., Phylogenetic Analysis and Paleontology. Columbia Univ. Press, New York.

Giribet, G., S. Carranza, J. Baguñà, M. Riutort, and C. Ribera. 1996. First molecular evidence for the existence of a Tardigrada + Arthropoda clade. Mol. Biol. Evol. 13:76–84.

Giribet, G., D. L. Distel, M. Polz, W. Sterrer, and W. C. Wheeler. 2000. Triploblastic relationships with emphasis on the acoelomates and the position of Gnathostomulida, Cycliophora, Platyhelminthes, and Chaetognatha: A combined approach of 18S rDNA sequences and morphology. Syst. Biol. 49:539–562.

Giribet, G., G. D. Edgecombe, and W. C. Wheeler. 2001. Arthropod phylogeny based on eight molecular loci and morphology. Nature 413:157–161.

Giribet, G., and C. Ribera. 1998. The position of arthropods in the animal kingdom: A search for a reliable outgroup for internal arthropod phylogeny. Mol. Phylogenet. Evol. 9:481–488.

Giribet, G., and W. C. Wheeler. 1999. The position of arthropods in the animal kingdom: Ecdysozoa, islands, trees, and the "parsimony ratchet." Mol. Phylogenet. Evol. 13:619–623.

Gislén, T. 1940. Investigations on the ecology of Echiurus. Lunds Univ. Arsskrift., N.F. 2, 36 (10):1–36.

Glaessner, M. F. 1969. Trace fossils from the Precambrian and basal Cambrian. Lethaia 2:369–393.

———. 1979a. An echiurid worm from the late Precambrian. Lethaia 12:121–124.

———. 1979b. Lower Cambrian Crustacea and annelid worms from Kangaroo Island, South Australia. Alcheringa 3:21–31.

———. 1979c. Precambrian. Pp. A79–A118 in R. A. Robison and C. Teichert, eds., Treatise on Invertebrate Paleontology, pt. A, Introduction. Geological Society of America, Boulder, CO; Univ. Kansas Press, Lawrence.

———. 1984. The Dawn of Animal Life. Cambridge Univ. Press, Cambridge.

Glaessner, M. F., and M. Wade. 1966. The Late Precambrian fossils from Ediacara, South Australia. Palaeontology 9:599–628.

Golvan, Y. J. 1958. Le phylum des Acanthocephala: Premier note: Sa place dans l'échelle zoologique. Ann. Parasitol. Hum. Comp. 33:538–602.

González-Crespo, S., and M. Levine. 1994. Related target enhancers for Dorsal and NF-$_K$B signaling pathways. Science 264:255–258.

Goodbody, I. 1982. Tunicata. Pp. 823–829 in S. P. Parker, ed., Synopsis and Classification of Living Organisms, vol. 2. McGraw-Hill, New York.

Goodman, M., M. M. Miyamoto, and J. Czelusniak. 1987. Pattern and process in vertebrate phylogeny revealed by coevolution of molecules and morphologies. Pp. 141–176 in C. Patterson, ed., Molecules and Morphology in Evolution: Conflict or Compromise? Cambridge Univ. Press, Cambridge.

Goodrich, E. S. 1945. The study of nephridia and genital ducts since 1905. Q. J. Microsc. Sci. 86:113–392.

Goremykin, V. V., S. Hansmann, and W. F. Martin. 1997. Evolutionary analysis of 58 proteins encoded in six completely sequenced chloroplast genomes: Revised molecular estimates of two seed plant divergence times. Plant Syst. Evol. 206:337–351.

Goto, T. P., P. Macdonald, and T. Maniatis. 1989. Early and late periodic patterns of *even-skipped* expression are controlled by distinct regulatory elements that respond to different spatial cues. Cell 57:413–422.

Götting, K.-J. 1980. Origin and relationships of the Mollusca. Z. Zool. Syst. Evolutionsforsch. 18:24–27.

Gould, S. J. 1977. Ontogeny and Phylogeny. Belknap Press, Cambridge, MA.

———. 1989. Wonderful Life: The Burgess Shale and the Nature of History. Norton, New York.

Grant, S. F. W. 1990. Shell structure and distribution of *Cloudina,* a potential index fossil for the terminal Proterozoic. Am. J. Sci. 290-A:261–294.

Gray, S., H. Cai, S. Barolo, and M. Levine. 1995. Transcriptional repression in the *Drosophila* embryo. Philos. Trans. R. Soc. London B 349:257–262.

Green, C. R. 1984. Intercellular junctions. Pp. 5–16 in J. Bereiter-Hahn, A. G. Matoltsy, and K. S. Richards, eds., Biology of the Integument: 1, Invertebrates. Springer-Verlag, Berlin.

Green, C. R., and P. R. Bergquist. 1982. Phylogenetic relationships within the Invertebrata in relation to the structure of septate junctions and the development of "occluding" junctional types. J. Cell. Sci. 53:279–305.

Gregg, J. R. 1954. The Language of Taxonomy. Columbia Univ. Press, New York.

Grell, K. G. 1982. Placozoa. P. 639 in S. P. Parker, ed., Synopsis and Classification of Living Organisms, vol. 1. McGraw-Hill, New York.

Grell, K. G., and A. Ruthmann. 1991. Placozoa. Pp. 13–27 in F. W. Harrison and J. A. Westfall, eds., Microscopic Anatomy of Invertebrates, vol. 2, Placozoa, Porifera, Cnidaria, and Ctenophora. Wiley-Liss, New York.

Grobben, K. 1908. Die systematische Einteilung des Tierreiches. Verh. Zool. Bot. Gesell. Wien 58:491–511.

Grotzinger, J. P., S. A. Bowring, B. Saylor, and A. J. Kaufman. 1995. New biostratigraphic and geochronologic constraints on early animal evolution. Science 270:598–604.

Gu, X. 1998. Early metazoan divergence was about 830 million years ago. J. Mol. Evol. 47:369–371.

Gutmann, W. F. 1981. Relationships between invertebrate phyla based on functional-mechanical analysis of the hydrostatic skeleton. Am. Zool. 21:63–81.

Gutmann, W. F., H. Zorn, and K. Vogel. 1978. Brachiopods: Biomechanical interdependences governing their origin and phylogeny. Science 199:890–893.

Haas, W. 1981. Evolution of calcareous hardparts in primitive molluscs. Malacologia 21:403–418.

Hadži, J. 1953. An attempt to reconstruct the system of animal classification. Syst. Zool. 2:145–154.

———. 1963. The Evolution of the Metazoa. Pergamon Press, Oxford.

Haeckel, E. 1866. Generelle Morphologie der Organismen, vol. 2. Georg Reimer, Berlin.

Halanych, K. M. 1995. The phylogenetic position of the pterobranch hemichordates based on 18S rDNA sequence data. Mol. Phylogenet. Evol. 4:72–76.

———. 1996. Testing hypotheses of chaetognath origins: Long branches revealed by 18S ribosomal DNA. Syst. Biol. 45:223–246.

Halanych, K. M., J. D. Bacheller, A. M. A. Aguinaldo, S. M. Liva, D. M. Hillis, and J. A. Lake. 1995. Evidence from 18S ribosomal DNA that the lophophorates are protostome animals. Science 267:1641–1643.

Halfon, M. S., A. Carmena, S. Gisselbrecht, C. M. Sackerson, F. Jiménez, M. K. Baylies, and A. M. Michelson. 2000. Ras pathway specificity is determined by the integration of multiple signal-activated and tissue-restricted transcription factors. Cell 103: 63–74.

Hall, B. K. 1994. Homology: The Hierarchical Basis of Comparative Biology. Academic Press, San Diego.

———. 1998. Germ layers and the germ-layer theory revisited: Primary and secondary germ layers, neural crest as a fourth germ layer, homology, demise of the germ-layer theory. Evol. Biol. 30:121–186.

———. 1999. The Neural Crest in Development and Evolution. Springer-Verlag, New York.

Hallam, A. 1981. Facies Interpretation and the Stratigraphic Record. Freeman, Oxford.

Hand, C. 1963. The early worm: A planula. Pp. 33–39 in E. C. Dougherty, ed., The Lower Metazoa. Univ. California Press, Berkeley and Los Angeles.

Hanelt, B., D. Van Schyndel, C. M. Adema, L. A. Lewis, and E. S. Loker. 1996. The phylogenetic position of *Rhopalura ophiocomae* (Orthonectida) based on 18S ribosomal DNA sequence analysis. Mol. Biol. Evol. 13:1187–1191.

Hanson, E. D. 1977. The Origin and Early Evolution of Animals. Wesleyan Univ. Press, Middletown, CT.

Hantzschel, W. 1975. Trace Fossils and Problematica: Treatise on Invertebrate Paleontology, pt. W, suppl. 1. Geological Society of America, Boulder, CO; Univ. Kansas, Lawrence.

Harbison, G. R. 1985. On the classification and evolution of the Ctenophora. Pp. 78–100 in S. Conway Morris, J. D. George, R. Gibson, and H. M. Platt, eds., The Origins and Relationships of Lower Invertebrates. Clarendon Press, Oxford.

Hardy, A. C. 1954. Escape from specialization. Pp. 122–142 in J. S. Huxley, A. C. Hardy, and E. B. Ford, eds., Evolution as a Process. Allen and Unwin, London.

Harrington, H. J., and R. C. Moore. 1955. Kansas Pennsylvanian and other jellyfishes. Kans. Geol. Surv. Bull. 114, pt. 5:153–163.

Harrison, F. W., et al., eds. 1991–1999. Microscopic Anatomy of Invertebrates. Vols. 1–15. Wiley-Liss, New York.

Hart, M., M. Byrne, and M. J. Smith. 1997. Molecular phylogenetic analysis of life-history evolution in asterinid starfish. Evolution 51:1848–1861.

Hartman, O. 1954. Pogonophora Johansson, 1938. Syst. Zool. 3:183–185.

Hartman, W. D. 1963. A critique of the enterocele theory. Pp. 55–77 in E. C. Dougherty, ed., The Lower Metazoa. Univ. California Press, Berkeley and Los Angeles.

Haszprunar, G. 1992. The first mollusks: Small animals. Boll. Zool. 59:1–16.

———. 1993. *Sententia:* The Archaeogastropoda: A clade, a grade, or what else? Bull. Am. Malacol. Union 10:165–177.

———. 1996a. The Mollusca: Coelomate turbellarians or mesenchymate annelids? Pp. 1–28 in J. Taylor, ed., Origin and Evolutionary Radiation of the Mollusca. Oxford Univ. Press, Oxford.

———. 1996b. Platyhelminthes and Platyhelminthomorpha: Paraphyletic taxa. J. Zool. Syst. Evol. Res. 34:41–48.

———. 2000. Is the Aplacophora monophyletic? A cladistic point of view. Am. Malacol. Bull. 125:115–130.

Haszprunar, G., L. v. Salvini-Plawen, and R. M. Rieger. 1995. Larval planktotrophy: A primitive trait in the Bilateria? Acta Zool. 76:141–154.

Haszprunar, G., and K. Schaefer. 1997. Monoplacophora. Pp. 415–457 in F. W. Harrison and A. J. Kohn, eds., Microscopic Anatomy of Invertebrates, vol. 6B, Mollusca 2. Wiley-Liss, New York.

Hatschek, B. 1888–1891. Lehrbuch der Zoologie. Fischer, Jena.

Haymon, R. M., R. A. Koski, and C. Sinclair. 1984. Fossils of hydrothermal vent worms from Cretaceous sulfide ores of the Somali ophiolites, Oman. Science 223:1407–1409.

He, J., and P. Furmanski. 1995. Sequence specificity and transcriptional activation in the binding of lactoferrin to DNA. Nature 373:721–724.

Hedges, S. B., P. H. Parker, C. G. Sibley, and S. Kumar. 1996. Continental breakup and the ordinal diversification of birds and mammals. Nature 381:226–229.

Hendricks, L., Y. Van de Peer, M. Van Herck, J. Neefs, and R. DeWachter. 1990. The ribosomal RNA sequence of the sea anemone *Anemonia sulcata* and its evolutionary position among other eukaryotes. Fed. Eur. Biochem. Soc. Lett. 269:445–449.

Hendy, M. D., and D. Penny. 1989. A framework for the quantitative study of evolutionary trees. Syst. Zool. 38:297–309.

Hennig, W. 1950. Grundzuge einer Theorie der phylogenetischen Systematik. Deutsche Zentralverlag, Berlin.

———. 1966. Phylogenetic Systematics. Univ. Illinois Press, Urbana.

Henry, J. J., and M. Q. Martindale. 1997. Nemerteans, the ribbon worms. Pp. 151–166 in S. F. Gilbert and A. M. Raunio, eds., Embryology. Sinauer, Sunderland, MA.

———. 1999. Conservation and innovation in spiralian development. Hydrobiologia 402:255–265.

Henry, J. J., G. A. Wray, and R. A. Raff. 1990. The dorsoventral axis is specified prior to first cleavage in the direct developing sea urchin *Heliocidaris erythrogramma*. Development 110:875–884.

Henry, J. Q., M. Q. Martindale, and B. C. Boyer. 1995. Axial specification in a basal member of the spiralian clade: Lineage relationships of the first four cells in the larval body plan in the polyclad turbellarian *Hoploplana inquilina*. Biol. Bull. 189:194–195.

———. 2000. The unique developmental program of the acoel flatworm, *Neochildia fusca*. Dev. Biol. 220:285–295.

Hernandez-Nicaise, M.-L. 1991. Ctenophora. Pp. 359–418 in F. W. Harrison and J. A. Westfall, eds., Microscopic Anatomy of Invertebrates, vol. 2, Placozoa, Porifera, Cnidaria, and Ctenophora. Wiley-Liss, New York.

Herr, R. A., L. Ajello, J. W. Taylor, S. N. Arseculeratne, and L. Mendoza. 1999. Phylogenetic analysis of *Rhinosporidium seeberi*'s 18S small-subunit ribosomal DNA groups this pathogen among members of the protoctistan Mesomycetozoa clade. J. Clin. Microbiol. 37:2750–2754.

Hessling, R., and W. Westheide. 2002. Are Echiura derived from a segmented ancestor? Immunohistochemical analysis of the nervous system in developmental stages of *Bonellia viridis*. J. Morphol. 252:100–113.

Hibberd, D. J. 1975. Observations on the ultrastructure of the choanoflagellate *Codosiga botrytis* (Ehr.) Saville-Kent with special reference to the flagellar apparatus. J. Cell Sci. 17:191–219.

Hickman, C. S. 1999. Larvae in invertebrate development and evolution. Pp. 21–59 in B. K. Hall and M. H. Wake, eds., The Origin and Evolution of Larval Forms. Academic Press, San Diego.

Higgins, R. P. 1982. Kinorhyncha. Pp. 874–877 in S. P. Parker, ed., Synopsis and Classification of Living Organisms, vol. 1. McGraw-Hill, New York.

Higgins, R. P., and R. M. Kristensen. 1986. New Loricifera from southeastern United States coastal waters. Smithson. Contrib. Zool. 438:1–70.

Hill, D. 1972. Archaeocyatha. Pp. 1–158 in C. Teichert, ed., Treatise on Invertebrate Paleontology, pt. E, 2nd ed. Geological Society of America, Boulder CO; Univ. Kansas Press, Lawrence.

Hillis, D. M. 1994. Homology in molecular biology. Pp. 339–368 in B. K. Hall, ed., Homology: The Hierarchical Basis of Comparative Biology. Academic Press, New York.

Hine, P. M. 1980. A review of some species of *Myxidium* Butschli, 1882 (Myxosporea) from eels (*Anguilla* spp.). J. Protozool. 27:260–267.

Hinegardner, R., and J. Engleberg. 1983. Biological complexity. J. Theor. Biol. 104:7–20.

Hinz, I., P. Kraft, M. Mergl, and K. J. Müller. 1990. The problematic *Hadimopanella, Kaimenella, Milaculum,* and *Utahphospha* identified as sclerites of Paleoscolecida. Lethaia 23:217–221.

Hochberg, F. G., Jr. 1982. The "kidneys" of cephalopods: A unique habitat for parasites. Malacologia 23:121–134.

Hoffman, P. F., A. J. Kaufman, G. P. Halverson, and D. P. Shrag. 1998. A Neoproterozoic snowball Earth. Science 281:1342–1346.

Holland, L. Z. 2000. Body-plan evolution in the Bilateria: Early antero-posterior patterning and the deuterostome-protostome dichotomy. Curr. Opin. Genet. Dev. 10:434–442.

Holland, N. D. 1988. The meaning of developmental asymmetry for echinoderm evolution: A new interpretation. Pp. 13–25 in C. R. C. Paul and A. B. Smith, eds., Echinoderm Phylogeny and Evolutionary Biology. Clarendon Press, Oxford.

Holland, P. W. H., and J. Garcia-Fernàndez. 1996. *Hox* genes and chordate evolution. Dev. Biol. 173:382–395.

Holland, P. W. H., A. M. Hacker, and N. A. Williams. 1991. A molecular analysis of the phylogenetic affinities of *Saccoglossus cambrensis* Brambell and Cole (Hemichordata). Philos. Trans. R. Soc. London B 332:185–189.

Holman, E. W. 1989. Some evolutionary correlates of higher taxa. Paleobiology 13:357–363.

Holmer, L. E., L. E. Popov, M. G. Bassett, and J. Laurie. 1995. Phylogenetic analysis and ordinal classification of the Brachiopoda. Palaeontology 38:713–741.

Hörstadius, S. 1939. The mechanics of sea urchin development, studied by operative methods. Biol. Rev. 14:132–179.

Hou, X. 1987a. Early Cambrian large bivalved arthropods from Chengjiang, eastern Yunnan. Acta Paleontol. Sin. 26:286–298. In Chinese; English summary.

———. 1987b. Three new large arthropods from Lower Cambrian Chengjiang, eastern Yunnan. Acta Paleontol. Sin. 26:272–285. In Chinese; English summary.

———. 1987c. Two new arthropods from Lower Cambrian, Chengjiang, eastern Yunnan. Acta Paleontol. Sin. 26:236–256. In Chinese; English summary.

———. 1999. New rare bivalved arthropods from the Lower Cambrian Chengjiang fauna, Yunnan, China. J. Paleontol. 73:102–116.

Hou, X., and J. Bergström. 1994. Palaeoscolecid worms may be nematomorphs rather than annelids. Lethaia 27:11–17.

———. 1995. Cambrian lobopodians: Ancestors of extant onychophorans? Zool. J. Linn. Soc. 114:3–19.

———. 1997. Arthropods of the Lower Cambrian Chengjiang fauna, southwest China. Fossils and Strata 45:1–116.

Hou, X., J. Bergström, and P. Ahlberg. 1995. *Anomalocaris* and other large animals in the Lower Cambrian Chengjiang fauna of southwest China. Geol. Foeren. Stockh. Foerh. 117:163–183.

Hou, X., J. Bergström, H. Wang, X. Feng, and A. Chen. 1999. The Chengjiang Fauna. Yunnan Science and Technology Press, Yunnan, China. In Chinese; short English summary.

Hou, X., and J. Chen. 1989a. Early Cambrian arthropod-annelid intermediate sea animal, *Luolishania* gen. nov. from Chengjiang, Yunnan. Acta Paleontol. Sin. 28:207–213. In Chinese; English summary.

———. 1989b. Early Cambrian tentacled worm-like animals (*Fascivermis* gen. nov.) from Chengjiang, Yunnan. Acta Paleontol. Sin. 28:32–41. In Chinese; English summary.

Hou, X., J. Chen, and H. Lu. 1989. Early Cambrian new arthropods from Chengjiang, Yunnan. Acta Paleontol. Sin. 28:42–57. In Chinese; English summary.

Hou, X., L. Ramsköld, and J. Bergström. 1991. Composition and preservation of the Chengjiang fauna: A Lower Cambrian soft-bodied biota. Zool. Scr. 20:395–411.

Houbrick, R. S., W. Stürmer, and E. L. Yochelson. 1988. Rare Mollusca from the Lower Devonian Hunsruck Slate of southern Germany. Lethaia 21:395–402.

Hsu, K. J., H. Oberhänsli, J. Y. Gao, S. Shu, C. Haihong, and U. Krähenbühl. 1985. "Strangelove Ocean" before the Cambrian explosion. Nature 316:809–811.

Huelsenbeck, J. P. 1995. The performance of phylogenetic methods in simulation. Syst. Biol. 44:17–48.

Huelsenbeck, J. P., and D. M. Hillis. 1993. Success of phylogenetic methods in the four-taxon case. Syst. Biol. 42:247–264.

Huelsenbeck, J. P., and B. Rannala. 1997. Maximum likelihood estimation of topology and node times using stratigraphic data. Paleobiology 23:174–180.

Hughes, C. P. 1975. Redescription of *Burgessia bella* from the Middle Cambrian Burgess Shale, British Columbia. Fossils and Strata 4:415–435.

Hummon, W. D. 1982. Gastrotricha. Pp. 857–863 in S. P. Parker, ed., Synopsis and Classification of Living Organisms, vol. 1. McGraw-Hill, New York.

Humphries, S., T. Humphries, and J. Sano. 1977. Organization and polysaccharides of sponge aggregation factor. J. Supramol. Struct. 7:339–351.

Hutchinson, G. E. 1961. The biologist poses some problems. Pp. 85–94 in M. Sears, ed., Oceanography. American Association for the Advancement of Science Pub. 67, Washington, DC.

Hutter, H., and R. Schnabel. 1995. Specification of anterior-posterior differences within the AB lineage in the *C. elegans* embryo: A polarising induction. Development 121:1559–1568.

Hyde, W. T., T. J. Crowley, S. K. Baum, and W. R. Peltier. 2000. Neoproterozoic "snowball Earth" simulations with a coupled climate/ice-sheet model. Nature 405:425–429.

Hyman, L. H. 1940. The Invertebrates. Vol. 1, Protozoa through Ctenophora. McGraw-Hill, New York.

———. 1951a. The Invertebrates. Vol. 2, Platyhelminthes and Rhynchocoela. McGraw-Hill, New York.

———. 1951b. The Invertebrates. Vol. 3, Acanthocephala, Aschelminthes, and Entoprocta: The Pseudocoelomate Bilateria. McGraw-Hill, New York.

———. 1955. The Invertebrates. Vol. 4, Echinodermata. McGraw-Hill, New York.

———. 1959. The Invertebrates. Vol. 5, Smaller Coelomate Groups. McGraw-Hill, New York.

———. 1967. The Invertebrates. Vol. 6, Mollusca 1. McGraw-Hill, New York.

Ijima, I. 1901. Studies on the Hexactinellida: Contribution 1. J. Coll. Sci. Imp. Univ. Tokyo 15:1–299.

Inglis, H. G. 1985. Evolutionary waves: Patterns in the origins of animal phyla. Aust. J. Zool. 33:153–178.

Inoue, I. 1958. Studies on the life history of *Chordodes japonensis*, a species of Gordiacea: 1, The development and structure of the larva. Jpn. J. Zool. 12:203–218.

Israelsson, O. 1997. *Xenoturbella*'s molluscan relatives and molluscan embryogenesis. Nature 390:32.

Ivanov, A. V. 1960. Pogonophores. Fauna SSSR, n. s., 75. Akad. Nauk SSSR, Moscow. In Russian.

———. 1963. Pogonophora. Academic Press, London.

———. 1973. *Trichoplax adherens:* A phygocytella-like animal. Zool. Zh. 52:1117–1131. In Russian.

Ivantsov, A. Yu. 1990. New data on the ultrastructure of sabellitids (Pogonophora?). Paleontol. Zh. 1990:125–128. In Russian.

Jablonski, D. 1986a. Background and mass extinctions: The alternation of macroevolutionary regimes. Science 231:129–133.

———. 1986b. Causes and consequences of mass extinctions: A comparative approach. Pp. 183–229 in D. K. Elliott, ed., Dynamics of Extinctions. Wiley, New York.

———. 1986c. Larval ecology and macroevolution in marine invertebrates. Bull. Mar. Sci. 39:565–587.

———. 1987. Heritability at the species level: Analysis of geographic ranges of Cretaceous mollusks. Science 238:360–363.

———. 1993. The tropics as a source of evolutionary novelty through geological time. Nature 364:142–144.

———. 1994. Extinctions in the fossil record. Philos. Trans. R. Soc. London B 344:11–17.

———. 1995. Extinctions in the fossil record. Pp. 25–44 in J. H. Lawton and R. M. May, eds., Extinction Rates. Oxford Univ. Press, Oxford.

Jablonski, D., and D. J. Bottjer. 1988. Onshore-offshore evolutionary patterns in post-Paleozoic echinoderms: A preliminary analysis. Pp. 81–90 in R. D. Burke, P. V. Mladenov, P. Lambert, and R. L. Parsley, eds., Echinoderm Biology: Proceedings of the 6th International Echinoderm Conference. Balkema, Rotterdam.

———. 1990. Onshore-offshore trends in marine invertebrate evolution. Pp. 21–75 in R. M. Ross and W. D. Allmon, eds., Causes of Evolution: A Paleontological Perspective. Univ. Chicago Press, Chicago.

Jablonski, D., S. Lidgard, and P. D. Taylor. 1997. Comparative ecology of bryozoan radiations: Origin of novelties in cyclostomes and cheilostomes. Palaios 12:505–523.

Jablonski, D., and R. A. Lutz. 1983. Larval ecology of marine benthic invertebrates: Paleobiological implications. Biol. Rev. 58:21–89.

Jablonski, D., J. J. Sepkoski, Jr., D. J. Bottjer, and P. M. Sheehan. 1983. Onshore-offshore patterns in the evolution of Phanerozoic shelf communities. Science 222:1123–1125.

Jackson, J. B. C. 1974. Biogeographic consequences of eurytopy and stenotopy among marine bivalves and their evolutionary significance. Am. Nat. 108:541–560.

Jackson, J. B. C., et al. 2001. Historical overfishing and the recent collapse of coastal ecosystems. Science 293:629–638.

Jacobs, D. K. 1990. Selector genes and the Cambrian radiation of the Bilateria. Proc. Nat. Acad. Sci. USA 87:4406–4410.

Jacobs, D. K., S. E. Lee, M. N. Dawson, J. L. Staton, and K. A. Raskoff. 1998. The history of development through the evolution of molecules: Gene trees, hearts, eyes, and dorsoventral inversion. Pp. 323–355 in R. DeSalle and B. Schierwater, eds., Molecular Approaches to Ecology and Evolution. Birkhäuser Verlag, Basel.

Jacobs, D. K., C. G. Wray, C. J. Wedeen, R. Kostriken, R. DeSalle, J. L. Staton, R. D. Gates, and D. R. Lindberg. 2000. Molluscan engrailed expression, serial organization, and shell evolution. Evol. Dev. 2:340–347.

Jaekel, O. 1902. Über verschiedene Wege phylogenetischer Entwicklung. Fischer, Jena.

———. 1918. Über fragliche Tunikaten aus dem Perm Siciliens. Palaeontol. Z. 2:66–74.

Jägersten, G. 1955. On the early phylogeny of the Metazoa: The bilaterogastraea theory. Zool. Bidr. Uppsala 30:321–354.

———. 1972. Evolution of the Metazoan Life Cycle. Academic Press, London.

Jahn, B., and G. Miklos. 1988. The Eukaryotic Genome in Development and Evolution. Allen and Unwin, London.

Janussen, D., M. Steiner, and A. Maoyan. 2002. New well-preserved scleritomes of Chancelloridae from the Early Cambrian Yuanshan Formation (Chengjiang, China) and the Middle Cambrian Wheeler Shale (Utah, USA) and paleobiological implications. J. Paleontol. 76:596–606.

Janvier, P. 1999. Catching the first fish. Nature 402:21–22.

Jefferies, R. P. S. 1967. Some fossil chordates with echinoderm affinities. Symp. Zool. Soc. London 20:163–208.

———. 1968. The subphylum Calcichordata (Jefferies 1967): Primitive fossil chordates with echinoderm affinities. Bull. Br. Mus. (Nat. Hist.) Geol. 16:243–339.

———. 1979. The origin of the chordates: A methodological essay. Pp. 443–477 in M. R. House, ed., The Origin of Major Invertebrate Groups. Academic Press, London.

———. 1981. In defence of the calcichordates. Zool. J. Linn. Soc. 73:351–396.

———. 1986. The Ancestry of the Vertebrates. Cambridge Univ. Press, Cambridge.

———. 1990. The solute *Dendrocystoides scoticus* from the Upper Ordovician of Scotland and the ancestry of chordates and echinoderms. Palaeontology 33:631–679.

Jeffery, W. R., and B. J. Swalla. 1997. Tunicates. Pp. 331–364 in S. F. Gilbert and A. M. Raunio, eds., Embryology. Sinauer, Sunderland, MA.

Jeffreys, A. J. 1982. Evolution of globin genes. Pp. 157–176 in G. A. Dover and R. B. Flavell, eds., Genome Evolution. Academic Press, New York.

Jell, J. S. 1984. Cambrian cnidarians with mineralized skeletons. Palaeontogr. Am. 54:105–109.

Jenkins, R. J. F. 1984. Interpreting the oldest fossil cnidarians. Palaeontogr. Am. 54:95–104.

———. 1985. The enigmatic Ediacaran (late Precambrian) genus *Rangea* and related forms. Paleobiology 11:336–355.

———. 1989. The "supposed terminal Precambrian extinction event" in relation to the Cnidaria. Mem. S. Aust. Mus. 17:347–359.

———. 1992. Functional and ecological aspects of Ediacaran assemblages. Pp. 131–176 in J. H. Lipps and P. W. Signor, eds., Origin and Early Evolution of the Metazoa. Plenum, New York.

Jenkins, R. J. F., P. S. Plummer, and K. C. Moriarty. 1981. Late Precambrian pseudofossils from the Flinders Ranges, South Australia. Trans. R. Soc. S. Aust. 105:67–83.

Jenner, R. A. 2001. Bilaterian phylogeny and uncritical recycling of morphological data sets. Syst. Biol. 50:730–742.

Jenner, R. A. 2002. Boolean logic and character state identity: Pitfalls of character coding in metazoan characters. Contrib. Zool. 71:67–91.

Jenner, R. A., and F. R. Schram. 1999. The grand game of metazoan phylogeny: Rules and strategies. Biol. Rev. 74:121–142.

Jennings, H. S. 1896. The early development of *Asplanchna herrickii* de Guerne. Bull. Mus. Comp. Zool. 30:1–117.

Jensen, S. 1997. Trace fossils from the Lower Cambrian Mickwitzia sandstone, south-central Sweden. Fossils and Strata 42:1–111.

Jensen, S., J. G. Gehling, and M. L. Droser. 1998. Ediacara-type fossils in Cambrian sediments. Nature 393:567–569.

Jeong, H., B. Tombor, Z. N. Oltvai, and A.-L. Barabási. 2000. The large-scale organization of metabolic networks. Nature 407:651–654.

Jeram, A. J., P. A. Selden, and D. Edwards. 1990. Land animals in the Silurian: Arachnids and myriapods from Shropshire, England. Science 250:658–661.

Jespersen, A., and J. Lützen. 1988. Ultrastructure and morphological interpretation of the circulatory system of nemertines (phylum Rhynchocoela). Vidensk. Medd. Dan. Naturhist. Foren. 147:47–66.

Jin, Y., and H. Wang. 1992. Revision of the Lower Cambrian brachiopod *Heliomedusa* Sun and Hou, 1987. Lethaia 25:35–49.

Johansson, K. E. 1937. Über *Lamellisabella zachsi* und ihre systematische Stellung. Zool. Anz. 117:23–26.

Jondelius, U., I. Ruiz-Trillo, J. Baguñà, and M. Riutort. 2002. The Nemertodermatida are basal bilaterians and not members of the Platyhelminthes. Zool. Scr. 31:201–215.

Jones, B. 1990. Tunicate spicules and their syntaxial overgrowths: Examples from the Pleistocene Ironshore Formation, Great Cayman, British West Indies. Can. J. Earth Sci. 27:525–532.

Jones, D., and I. Thompson. 1977. Echiura from the Pennsylvanian Essex fauna of northern Illinois. Lethaia 10:317–326.

Jones, H. D., and E. R. Trueman. 1970. Locomotion of the limpet, *Patella vulgata* L. J. Exp. Biol. 52:201–216.

Jones, M. L. 1981. *Riftia pachyptila,* new genus, new species: The vestimentiferan worm from the Galápagos Rift geothermal vents (Pogonophora). Proc. Biol. Soc. Wash. 93:1295–1313.

———. 1985. Vestimentiferan pogonophores: Their biology and affinities. Pp. 327–342 in S. Conway Morris, J. D. George, R. Gibson, and H. M. Platt, eds., The Origins and Relationships of Lower Invertebrates. Oxford Univ. Press, Oxford.

Jones, P. J., and K. G. McKenzie. 1980. Queensland Middle Cambrian Bradoriida (Crustacea): New taxa, paleobiogeography, and biological affinities. Alcheringa 4:203–225.

Joysey, K. A. 1959. Probable cirripede, phoronoid, and echiuroid burrows within a Cretaceous echinoid test. Palaeontology 1:397–400.

Kaletta, T., H. Schnabel, and R. Schnabel. 1997. Binary specification of the embryonic lineage in *Caenorhabditis elegans*. Nature 390:294–298.

Kanegae, Y., A. T. Tavares, J. C. I. Belmonte, and I. M. Verma. 1998. Role of Rel/NF$_K$B transcription factors during the outgrowth of the vertebrate limb. Nature 392:611–614.

Kapp, H. 1991. Morphology and anatomy. Pp. 5–17 in Q. Bone, H. Kapp, and A. C. Pierrot-Buls, eds., The Biology of Chaetognaths. Oxford Univ. Press, Oxford.

———. 2000. The unique embryology of Chaetognatha. Zool. Anz. 239:263–266.

Karling, T. G. 1974. On the anatomy and affinities of the turbellarian orders. Pp. 1–16 in N. W. Riser and M. P. Morse, eds., Biology of the Turbellaria. McGraw-Hill, New York.

Katayama, T., H. Wada, H. Furuya, N. Satoh, and M. Yamamoto. 1995. Phylogenetic position of the dicyemid Mesozoa inferred from 18S rDNA sequences. Biol. Bull. 189:81–90.

Katz, M. J. 1983. Comparative anatomy of the tunicate tadpole, *Ciona intestinalis*. Biol. Bull. 164:1–27.

Kaufman, A. J., A. H. Knoll, M. A. Semikhatov, J. P. Grotzinger, S. B. Jacobsen, and W. Adams. 1997. Isotopes, ice ages, and terminal Proterozoic earth history. Proc. Nat. Acad. Sci. USA 94:6600–6605.

Kent, M. L., K. B. Andree, J. L. Bartholomew, M. El-Matboull, S. S. Desser, R. H. Devlin, S. W. Feist, R. P. Hedrick, R. W. Hoffmann, J. Khattra, S. L. Hallett, R. J. G. Lester, M. Longshaw, O. Palenzuela, M. E. Siddall, and C. Xiao. 2001. Recent advances in our knowledge of the Myxozoa. J. Eukaryot. Microbiol. 48:395–413.

Kent, M. L., L. Margolis, and J. O. Corliss. 1994. The demise of a class of protists: Taxonomic and nomenclatural revisions proposed for the protist phylum Myxozoa Grassé, 1970. Can. J. Zool. 72:932–937.

Kenyon, C., and B. Wang. 1991. A cluster of *Antennapedia*-class homeobox genes in a nonsegmented animal. Science 253:516–517.

Kerk, D., A. Gee, M. Standish, P. O. Wainright, A. S. Drum, R. A. Elston, and M. L. Sogin. 1995. The rosette agent of chinook salmon (*Oncorhynchus tshawytscha*) is closely related to chanoflagellates, as determined by the phylogenetic analyses of its small ribosomal subunit RNA. Mar. Biol. 122:187–192.

Kerszberg, M., and L. Wolpert. 1998. The origin of the Metazoa and the egg: A role for cell death. J. Theor. Biol. 193:535–537.

Kidwell, S. M. 1993. Patterns of time-averaging in the shallow marine fossil record. Pp. 275–300 in S. M. Kidwell and A. K. Behrensmeyer, Taphonomic Approaches to Time Resolution in Fossil Assemblages. Short Course in Paleontology 6. Paleontological Society.

———. 2001. Preservation of species abundance in marine death assemblages. Science 294:1091–1094.

Kidwell, S. M., and A. K. Behrensmeyer, eds. 1993. Taphonomic Approaches to Time Resolution in Fossil Assemblages. Short Course in Paleontology 6. Paleontological Society.

Kidwell, S. M., and D. W. J. Bosence. 1991. Taphonomy and time-averaging of marine shelly faunas. Pp. 115–209 in P. A. Allison and D. E. G. Briggs, eds., Taphonomy: Releasing the Data Locked in the Fossil Record. Plenum, New York.

Kidwell, S. M., and K. W. Flessa. 1995. The quality of the fossil record: Populations, species, and communities. Ann. Rev. Ecol. Syst. 26:269–299.

Kim, J. 1993. Improving the accuracy of the phylogenetic estimation by combining different methods. Syst. Biol. 42:331–340.

Kim, J., W. Kim, and C. W. Cunningham. 1999. A new perspective on lower metazoan relationships from 18S rDNA sequences. Mol. Biol. Evol. 16:423–427.

Kimura, M. 1968. Evolutionary rate at the molecular level. Nature 217:624–626.

———. 1983. The Neutral Theory of Molecular Evolution. Cambridge Univ. Press, Cambridge.

Kinchin, I. M. 1994. The Biology of Tardigrades. Portland Press, London.

King, J. L., and T. H. Jukes. 1969. Non-Darwinian evolution: Random fixation for selectively neutral alleles. Science 164:788–798.

King, N., and S. B. Carroll. 2001. A receptor tyrosine kinase from choanoflagellates: Molecular insights into early animal evolution. Proc. Nat. Acad. Sci. USA 98:15032–15037.

King, N., G. T. Hittinger, and S. B. Carroll. 2003. Evolution of key cell signaling and adhesion protein families predates animal origins. Science 301:361–363.

Kirschvink, J. L. 1992. Late Proterozoic low-latitude global glaciation: The snowball Earth. Pp. 51–52 in J. W. Schopf and C. Klein, eds., The Proterozoic Biosphere: A Multidisciplinary Approach. Cambridge Univ. Press, Cambridge.

Kitching, I. J., P. L. Forey, C. J. Humphries, and D. M. Williams. 1998. Cladistics. 2nd ed. Oxford Univ. Press, Oxford.

Knauss, E. B. 1979. Indication of an anal pore in Gnathostomulida. Zool. Scr. 8:181–186.

Knight, J. B. 1952. Primitive fossil gastropods and their bearing on gastropod classification. Smithson. Misc. Coll. 114:1–55.

Knoll, A. H. 1992a. Biological and biogeochemical preludes to the Ediacaran radiation. Pp. 53–84 in J. H. Lipps and P. W. Signor, eds., Origin and Early Evolution of the Metazoa. Plenum, New York.

———. 1992b. The early evolution of eukaryotes: A geological perspective. Science 256:622–628.

———. 1996. Breathing room for early animals. Nature 382:111–112.

Knoll, A. H., and S. B. Carroll. 1999. Early animal evolution: Emerging views from comparative biology and geology. Science 284:2129–2137.

Knoll, A. H., and S.-H. Xiao. 1999. On the age of the Doushantuo Formation. Acta Micropaleont. Sin. 16:225–236.

Kobayashi, M., H. Furuya, and P. W. H. Holland. 2000. Dicyemids are higher animals. Nature 401:762.

Kobluk, D. R. 1981. Lower Cambrian cavity-dwelling endolithic boring sponges. Can. J. Earth Sci. 18:972–980.

Koehl, M. A. R. 1976. Mechanical design in sea anemones. Pp. 23–31 in G. O. Mackie, ed., Coelenterate Ecology and Behavior. Plenum, New York.

———. 1977. Mechanical diversity of connective tissue of the body wall of sea anemones. J. Exp. Biol. 69:107–125.

———. 1982. Mechanical design of spicule-reinforced connective tissue: Stiffness. J. Exp. Biol. 98:239–267.

———. 1995. Fluid flow through hair-bearing appendages: Feeding, smelling, and swimming at low and intermediate Reynolds numbers. Pp. 157–182 in C. P. Ellington and T. J. Pedley, eds., Biological Fluid Dynamics. Society for Experimental Biology, Symposium 49. Company of Biologists, Cambridge.

Kojima, S. 1998. Paraphyletic status of Polychaeta suggested by phylogenetic analysis based on the amino acid sequences of elongation factor 1-alpha. Mol. Phylogenet. Evol. 9:255–261.

Kourakis, M. J., and M. Q. Martindale. 2000. Combined-method phylogenetic analysis of Hox and ParaHox genes of the Metazoa. J. Exp. Zool. 288:175–191.

Kozloff, E. N. 1972. Some aspects of development in *Echinoderes* (Kinorhyncha). Trans. Am Microsc. Soc. 91:119–130.

Kozlowski, R. 1948. Les graptolithes et quelques nouveaux groupes d'animaux du Tremadoc de la Pologne. Palaeontol. Pol. 3:1–235.

Kozur, H. 1970. Fossile Hirudinea aus dem Oberjura von Bayern. Lethaia 3:225–232.

Kristensen, R. M. 1983. Loricifera: A new phylum with Aschelminthes characters from the meiobenthos. Z. Zool. Syst. Evolutionforsch. 21:163–180.

———. 1991. Loricifera. Pp. 351–375 in F. W. Harrison and E. E. Ruppert, eds., Microscopic Anatomy of Invertebrates, vol. 4, Aschelminthes. Wiley-Liss, New York.

Kristensen, R. M., and R. P. Higgins. 1991. Kinorhyncha. Pp. 377–404 in F. W. Harrison and E. E. Ruppert, eds., Microscopic Anatomy of Invertebrates, vol. 4, Aschelminthes. Wiley-Liss, New York.

Kristensen, R. M., and A. Nørrevang. 1977. On the fine structure of *Rastrognathia macrostoma* gen. et sp. n. placed in Rastrognathidae fam. n. (Gnathostomulida). Zool. Scr. 6:27–41.

Krumlauf, R. 1994. *Hox* genes in vertebrate development. Cell 78:191–201.

Kruse, P. D. 1996. Hyolith guts in the Cambrian of northern Australia: Turning hyolithomorphs upside down. Lethaia 29:213–217.

Kuhn, K., B. Streit, and B. Schierwater. 1996. Homeobox genes in the cnidarian *Eleutheria dichotoma:* Evolutionary implications for the origin of *Antennapedia*-class (HOM/Hox) genes. Mol. Phylogenet. Evol. 6:30–38.

Kumar, S., and S. B. Hedges. 1998. A molecular timescale for vertebrate evolution. Nature 392:917–920.

Kume, M., and K. Dan, eds. 1968. Invertebrate Embryology. NOLIT Publishing House, Belgrade.

Kuznetsov, A. P., V. V. Maslennikov, V. V. Zaidov, and L. P. Zonenshain. 1990. Fossil hydrothermal vent fauna in Devonian sulfide deposits of the Uralian ophiolites. Deep-Sea Newsletter 17:9–10.

Lafay, B., N. Boury-Esnault, J. Vacelet, and R. Christen. 1992. An analysis of partial 28S ribosomal RNA sequences suggests early radiations of sponges. Biosystems 28:139–151.

Lafay, B., A. B. Smith, and R. Christen. 1995. A combined morphological and molecular approach to the phylogeny of asteroids (Asteroidea: Echinodermata). Syst. Biol. 4:190–208.

Lake, J. A. 1987. Rate-independent technique for analysis of nucleic acid sequences: Evolutionary parsimony. Mol. Biol. Evol. 4:167–191.

———. 1989. Origin of the eukaryotic nucleus determined by rate-invariant analyses of ribosomal RNA genes. Pp. 87–101 in B. Fernholm, K. Bremer, and H. Jörnvall, eds., The Hierarchy of Life. Elsevier Scientific, Amsterdam.

———. 1990. Origin of the Metazoa. Proc. Nat. Acad. Sci. USA 87:763–766.

———. 1991. The order of sequence alignment can bias the selection of tree topology. Mol. Biol. Evol. 8:378–385.

Lammert, V. 1991. Gnathostomulida. Pp. 19–39 in F. W. Harrison and E. E. Rupert, eds., Anatomy of Invertebrates, vol. 4, Aschelminthes. Wiley-Liss, New York.

Land, J. van der, and A. Nørrevang. 1977. Structure and relationships of the *Lamellibrachia* (Annelida, Vestimentifera). K. Dan. Vidensk. Selsk. Biol. Skr. 21:1–102.

Lande, R. 1986. The dynamics of peak shifts and the pattern of morphologic evolution. Paleobiology 12:343–354.

Landing, E., S. A. Bowring, K. L. Davidek, S. R. Westrop, G. Geyer, and W. Heldmaier. 1998. Duration of the Early Cambrian: U-Pb ages of volcanic ashes from Avalon and Gondwana. Can. J. Earth Sci. 35:329–338.

Landing, E., and S. R. Westrop, eds. 1998. Avalon 1997: The Cambrian standard. N.Y. State Mus. Bull. 492.

Lang, A. 1882. Der Bau von Gunda segmentata und die Verwandtschaft der Platyhelminthen mit Coelenteraten und Hirudineen. Mitt. Zool. Stat. Neapel. 3:187–251.

Lankester, E. R. 1870. On the use of the term homology in modern zoology, and the distinction between homogenetic and homoplastic agreements. Ann. Mag. Nat. Hist. 6:34–43.

Lanzavecchia, G., M. de Eguileor, and R. Valvassori. 1988. Muscles. Pp. 71–88 in W. Westheide and C. O. Hermans, eds., The Ultrastructure of Polychaeta. Fischer, New York.

Laval, M. 1971. Ultrastructure et mode de nutrition du choanoflagellé *Salpingoeca pelagica* sp. nov.: Comparaison avec les choanocytes des spongiaires. Protohistologica 8:325–336.

Leadbeater, B. S. C. 1983. Life-history and ultrastructure of a new marine species of *Proterospongia* (Choanoflagellida). J. Mar. Biol. Assoc. U.K. 63:135–160.

———. 1985. Class 2, Zoomastigophorea Calkins; Order 1, Choanoflagellida Kent, 1880. Pp. 106–116 in J. J. Lee, S. H. Hutner, and E. C. Bovee, eds., An Illustrated Guide to the Protozoa. Allen Press, Lawrence, KS.

Lechner, M. 1966. Untersuchungen zur Embryonalenwicklung des Rädertieres *Asplancha girodi* de Guerne. Roux Arch. Entwicklungsmech. Org. 157:117– 173.

Ledger, P., and W. C. Jones. 1977. Spicule formation in the calcareous sponge *Sycon ciliatum*. Cell Tissue Res. 181:553–567.

Lee, D. L., and W. D. Biggs. 1990. Two- and three-dimensional locomotion of the nematode *Nippostrongylus brasiliensis*. Parasitology 101:301–308.

Lefebvre, B. 2003. Functional morphology of stylophoran echinoderms. Palaeontology 46:511–555.

Lemche, H. 1957. A new living deep-sea mollusc of the Cambro-Devonian class Monoplacophora. Nature 179:413–416.

Lemche, H., and K. Wingstrand. 1959. The anatomy of *Neopilina galatheae* Lemche, 1957 (Mollusca, Tryblidiacea). Galatea Rep. 3:1–63.

Lesh-Laurie, G. E., and P. E. Suchy. 1991. Cnidaria: Scyphozoa and Cubozoa. Pp. 185–266 in F. W. Harrison and J. A. Westfall, eds., Microscopic Anatomy of Invertebrates, vol. 2. Wiley-Liss, New York.

Lester, S. M. 1985. *Cephalodiscus* sp. (Hemichordata: Pterobranchia): Observations of functional morphology, behavior, and occurrence in shallow water around Bermuda. Mar. Biol. 85:263–268.

Levins, R. 1968. Evolution in Changing Environments. Princeton Univ. Press, Princeton, NJ.

Levinton, J. 1988. Genetics, Paleontology, and Macroevolution. Cambridge Univ. Press, Cambridge.

Levitan, D. 1998. Does Bateman's principle apply to broadcast-spawning organisms? Egg traits influence in situ fertilization rates among congeneric sea urchins. Evolution 52:1043–1056.

Lewis, E. B. 1978. A gene complex controlling segmentation in *Drosophila*. Nature 276:565–570.

———. 1985. Regulation of the genes of the bithorax complex in *Drosophila*. Cold Spring Harbor Symp. Quant. Biol. 50:155–164.

Leys, S. P., and B. N. Degnan. 2002. Embryogenesis and metamorphosis in a haplosclerid demosponge: Gastrulation and transdifferentiation of larval ciliated cells to choanocytes. Invert. Biol. 121:171–189.

Leys, S. P., and G. O. Mackie. 1994. Cytoplasmic streaming in the hexactinellid sponge *Rhabdocalyptus dawsoni* (Lambe, 1873). Pp. 417–423 in R. W. M. van Soest, T. M. G. van Kempen, and J. D. Braekman, eds., Sponges in Time and Space. Balkema, Rotterdam.

Li, C. W., J. Y. Chen, and T. E. Hua. 1998. Precambrian sponges with cellular structures. Science 279:879–882.

Light, S. F., R. I. Smith, F. A. Pitelka, D. P. Abbott, and M. Weesner. 1970. Intertidal Invertebrates of the Central California Coast. Univ. California Press, Berkeley and Los Angeles.

Lindberg, D. R., and W. F. Ponder. 1996. An evolutionary tree for the Mollusca: Branches or roots? Pp. 67–75 in J. Taylor, ed., Origin and Evolutionary Radiation of the Mollusca. Oxford Univ. Press, Oxford.

Lindstrom, M. 1974. The conodont apparatus as a food-gathering mechanism. Palaeontology 17:729–744.

Linnaeus, C. 1758. Systema Naturae. Laurentii Salvii, Holmiae.

Linsley, R. M., and W. M. Kier. 1984. The Paragastropoda: A proposal for a new class of Paleozoic Mollusca. Malacologia 25:241–254.

Lipps, J. H. 1992. Origin and early evolution of Foraminifera. Pp. 3–9 in Studies in Benthic Foraminifera. Tokai Univ. Press, Sendai.

Lissmann, H. W. 1945a. The mechanism of locomotion in gastropod molluscs: 1, Kinematics. J. Exp. Biol. 21:58–69.

———. 1945b. The mechanism of locomotion in gastropod molluscs: 2, Kinetics. J. Exp. Biol. 22:37–50.

Little, C. T. S., R. J. Herrington, V. V. Maslennikov, N. J. Morris, and V. V. Zaykov. 1997. Silurian hydrothermal-vent community from the southern Urals, Russia. Nature 385:146–148.

Littlewood, D. T. J., and R. A. Bray, eds. 2001. Interrelationships of the Platyhelminthes. Taylor and Francis, London.

Littlewood, D. T. J., P. D. Olson, M. J. Telford, E. A. Herniou, and M. Riutort. 2001. Elongation factor 1-alpha sequences alone do not assist in resolving the position of the Acoela within the Metazoa. Mol. Biol. Evol. 18:437–442.

Littlewood, D. T. J., K. Rohde, R. A. Bray, and E. A. Herniou. 1999. Phylogeny of the Platyhelminthes and the evolution of parasitism. Biol. J. Linn. Soc. 68:257–287.

Littlewood, D. T. J., K. Rohde, and K. A. Clough. 1999. The interrelationships of all major groups of Platyhelminthes: Phylogenetic evidence from morphology and molecules. Biol. J. Linn. Soc. 66:75–114.

Littlewood, D. T. J., A. B. Smith, K. A. Clough, and R. H. Emson. 1997. The interrelationships of the echinoderm classes: Morphological and molecular evidence. Biol. J. Linn. Soc. 61:409–438.

Littlewood, D. T. J., M. J. Telford, K. A. Clough, and K. Rohde. 1998. Gnathostomulida: An enigmatic metazoan phylum from both morphological and molecular perspectives. Mol. Phylogenet. Evol. 9:72–79.

Logan, G. A., J. M. Hayes, G. B. Hieshima, and R. E. Summons. 1995. Terminal Proterozoic reorganization of biogeochemical cycles. Nature 366:53–56.

Long, J. A., and S. A. Stricker. 1991. Brachiopoda. Pp. 1–35 in A. C. Giese, J. S. Pearse, and V. B. Pearse, eds., Reproduction of Invertebrates, vol. 6, Lophophorates and Echinoderms. Boxwood Press, Pacific Grove, CA.

Loomis, W. F., and M. E. Gilpin. 1986. Multigene families and vestigial sequences. Proc. Nat. Acad. Sci. USA 83:2143–2147.

Lorenzen, S. 1985. Phylogenetic aspects of pseudocoelomate evolution. Pp. 210–233 in S. Conway Morris, J. D. George, R. Gibson, and H. M. Platt, eds., The Origins and Relationships of Lower Invertebrates. Clarendon Press, Oxford.

Lowe, C. J., L. Issel-Tarver, and G. A. Wray. 2002. Gene expression and larval evolution: Changing roles of *distal-less* and *orthodenticle* in echinoderm larvae. Evol. Dev. 4:111–123.

Lowe, C. J., and G. A. Wray. 1997. Radical alterations in the roles of homeobox genes during echinoderm evolution. Nature 389:718–721.

Lowe, C. J., M. Wu, A. Salic, L. Evans, E. Lander, N. Stang-Thoman, C. E. Gruber, J. Gerhart, and M. Kirschner. 2003. Anteroposterior patterning in hemichordates and the origins of the chordate nervous system. Cell 113:853–865.

Lowenstam, H. A., and S. Weiner. 1989. On Biomineralization. Oxford Univ. Press, New York.

Ludwig, M. Z., C. Bergman, N. H. Patel, and M. Kreitman. 2000. Evidence for stabilizing selection in a eukaryotic enhancer element. Nature 403:564–567.

Lundin, K., and W. Sterrer. 2001. The Nemertodermatida. Pp. 24–27 in D. T. J. Littlewood and R. A. Bray, eds., Interrelationships of the Platyhelminthes. Taylor and Frances, London.

Lundin, L.-G. 1999. Gene duplications in early metazoan evolution. Cell Dev. Biol. 10:523–530.

Lüter, C. 2000. The origin of the coelom in Brachiopoda and its phylogenetic significance. Zoomorphology 120:15–28.

———. 2001. Brachiopod larval setae: A key to the phylum's ancestral life cycle? Pp. 46–55 in C. H. C. Brunton, L. R. M. Cocks, and S. L. Long, eds., Brachiopods Past and Present. Taylor and Francis, London.

Lynch, M. 1999. The age and relationships of the major animal phyla. Evolution 53:319–325.

Lynch, M., and J. C. Conery. 2000. The evolutionary fate and consequences of duplicate genes. Science 297:1151–1155.

MacArthur, R. H. 1969. Patterns of communities in the tropics. Biol. J. Linn. Soc. 1:19–30.

MacArthur, R. H., and E. O. Wilson. 1967. The Theory of Island Biogeography. Princeton Univ. Press, Princeton, NJ.

Mackey, L. Y., B. Winnepenninckx, R. DeWachter, T. Backeljau, P. Emschermann, and J. R. Garey. 1996. 18S rRNA suggests that Entoprocta are protostomes, unrelated to Ectoprocta. J. Mol. Evol. 42:552–559.

Mackie, G. O. 1963. Siphonophores, bud colonies, and superorganisms. Pp. 329–337 in E. C. Dougherty, ed., The Lower Metazoa: Comparative Biology and Phylogeny. Univ. California Press, Berkeley and Los Angeles.

Mackie, G. O., and C. L. Singla. 1983. Studies on hexactinellid sponges: 1, Histology of *Rhabdocalypterus dawsoni* (Lambe, 1873). Philos. Trans. R. Soc. London B 301: 365–400.

Maggenti, A. R. 1970. System analysis and nematode phylogeny. J. Nematol. 2:7–15.

Malakhov, V. V. 1994. Nematodes: Structure, Development, Classification, and Phylogeny. Smithsonian Institution Press, Washington, DC.

Mallatt, J., J. Chen, and N. D. Holland. 2003. Comment on "A new species of Yunnanozoan with implications for deuterostome evolution." Science 300:1372.

Mannervik, M., Y. Nibu, H. Zhang, and M. Levine. 1999. Transcriptional coregulators in development. Science 284:606–609.

Manton, S. M. 1950. The evolution of arthropodan locomotory mechanisms: Part 1, The locomotion of *Peripatus*. Zool. J. Linn. Soc. 41:529–570.

———. 1973. The evolution of arthropodan locomotory mechanisms: Part 2, Habits, morphology, and evolution of the Uniramia (Onychophora, Myriapoda, Hexapoda) and comparisons with Arachnida, together with a functional review of uniramian musculature. Zool. J. Linn. Soc. 53:257–375.

———. 1977. The Arthropoda. Clarendon Press, Oxford.

Manton, S. M., and D. T. Anderson. 1979. Polyphyly and the evolution of arthropods. Pp. 269–321 in M. R. House, ed., The Origin of Major Invertebrate Groups. Academic Press, London.

Manuel, M., and Y. Le Parco. 2000. Homeobox gene diversification in the calcareous sponge, *Sycon raphanus*. Mol. Phylogenet. Evol. 17:97–107.

Marcus, E. 1929. Zur Embryologie der Tardigraden. Zool. Jb. Abt. Anat. 50:333–384.

———. 1958. On the evolution of animal phyla. Q. Rev. Biol. 33:24–58.

Marek, L. 1967. The class Hyolitha in the Caradoc of Bohemia. Sb. Geol. Ved. Rada P Paleontol. 9:51–113.

Marek, L., and E. L. Yochelson. 1976. Aspects of the biology of Hyolitha (Mollusca). Lethaia 9:65–81.

Margulis, L. 1981. Symbiosis in Cell Evolution. Freeman, San Francisco.

———. 1984. Early Life. Jones and Bartlett, Boston.

———. 1993. Symbiosis in Cell Evolution: Microbial Communities in the Archean and Proterozoic Eons. Freeman, New York.

Margulis, L., and K. V. Schwartz. 1982. Five Kingdoms: An Illustrated Guide to the Phyla of Life on Earth. Freeman, San Francisco.

Mariscal, R. N. 1975. Entoprocta. Pp. 1–41 in A. C. Giese and S. Pearse, eds., Reproduction of Marine Invertebrates, vol. 2. Academic Press, New York.

Marshall, C. R. 1990. Confidence intervals on stratigraphic ranges. Paleobiology 16:1–10.

———. 1992a. Character analysis and the integration of molecular and morphological data in an understanding of sand dollar phylogeny. Mol. Biol. Evol. 9:309–322.

———. 1992b. Substitution bias, weighted parsimony, and amniote phylogeny as inferred from 18S rRNA sequences. Mol. Biol. Evol. 9:370–373.

Marshall, C. R., E. C. Raff, and R. A. Raff. 1994. Dollo's law and the death and resurrection of genes. Proc. Nat. Acad. Sci. USA 91:12283–12287.

Martin, A. W., F. M. Harrison, M. J. Huston, and D. M. Stewart. 1958. The blood volumes of some representative molluscs. J. Exp. Biol. 35:260–279.

Martin, J. W. 1992. Branchiopoda. Pp. 25–224 in F. W. Harrison and A. G. Humes, eds., Microscopic Anatomy of Invertebrates, vol. 9, Crustacea. Wiley-Liss, New York.

Martin, M. W., D. V. Grazhdankin, S. A. Bowring, D. A. D. Evans, M. A. Fedonkin, and J. L. Kirschvink. 2000. Age of Neoproterozoic bilatarian [*sic*] body and trace fossils, White Sea, Russia: Implications for metazoan evolution. Science 288: 841–845.

Martin, V. 1997. Cnidarians, the jellyfish, and hydras. Pp. 57–86 in S. F. Gilbert and A. M. Raunio, eds., Embryology. Sinauer, Sunderland, MA.

Martindale, M. Q., and J. Q. Henry. 1995. Diagonal development: Establishment of the anal axis in the ctenophore *Mnemiopsis leidyi*. Biol. Bull. 189:190–192.

———. 1997a. Ctenophorans, the comb jellies. Pp. 87–111 in S. F. Gilbert and A. M. Raunio, eds., Embryology. Sinauer, Sunderland, MA.

———. 1997b. Reassessing embryogenesis in the Ctenophora: The inductive role of e_1 micromeres in organizing ctene row formation in the "mosaic" embryo, *Mnemiopsis leidyi*. Development 124:1999–2006.

Martinez, P., J. P. Rast, C. Arenas-Mena, and E. H. Davidson. 1999. Organization of an echinoderm *Hox* gene cluster. Proc. Nat. Acad. Sci. USA 96:1469–1474.

Mayr, E. 1982. The Growth of Biological Thought: Diversity, Evolution, and Inheritance. Belknap Press, Cambridge, MA.

———. 1987. The ontological status of biological species. Biol. Philos. 2:145–166.

McConnaughey, B. H. 1963. The Mesozoa. Pp. 151–165 in E. C. Dougherty, ed., The Lower Metazoa: Comparative Biology and Phylogeny. Univ. California Press, Berkeley and Los Angeles.

McGinnis, W., M. Levine, E. Hafen, A. Kuroiwa, and W. J. Gehring. 1984. A conserved DNA sequence in homeotic genes of the *Drosophila Antennapedia* and *bithorax* complexes. Nature 308:428–433.

McHugh, D. 1997. Molecular evidence that echiurans and pogonophorans are derived annelids. Proc. Nat. Acad. Sci. USA 94:8006–8009.

McIlroy, D., and G. A. Logan. 1999. The impact of bioturbation on infaunal ecology and evolution during the Proterozoic-Cambrian transition. Palaios 14:58–72.

McKinney, F. K., and J. B. C. Jackson. 1989. Bryozoan Evolution. Univ. Chicago Press, Chicago.

McKinney, M. L. 1987. Taxonomic selectivity and continuous variation in mass and background extinctions of marine taxa. Nature 325:143–145.

McKinney, M. L., and K. J. McNamara. 1991. Heterochrony: The Evolution of Ontogeny. Plenum, New York.

McMenamin, M. A. S. 1986. The garden of Ediacara. Palaios 1:178–182.

McMenamin, M. A. S., and D. L. S. McMenamin. 1990. The Emergence of Animals. Columbia Univ. Press, New York.

McShea, D. W. 1991. Complexity and evolution: What everybody knows. Biol. Philos. 6:303–324.

———. 2001. The hierarchical structure of organisms: A scale and documentation of a trend in the maximum. Paleobiology 27:405–423.

Medawar, P. 1974. A geometric model of reduction and emergence. Pp. 57–63 in F. J. Ayala and T. Dobzhansky, eds., Studies in the Philosophy of Biology. Univ. California Press, Berkeley and Los Angeles.

Medina, M., A. G. Collins, J. D. Silberman, and M. L. Sogin. 2001. Evaluating hypotheses of basal animal phylogeny using complete sequences of large and small subunit rRNA. Proc. Nat. Acad. Sci. USA. 98:9707–9712.

Meier, S., and D. S. Packard. 1984. Morphogenesis of the cranial segments and distribution of neural crest in the embryos of the snapping turtle, *Chelydra serpentina*. Dev. Biol. 102:309–323.

Meixner, J. 1925. Beitrag zur Morphologie und zum System der Turbellaria-Rhabdocoela: 1, Der Kalyptorhynchia. Z. Morphol. Oekol. Tiere 3:255–343.

Melton, W. G., and H. W. Scott. 1973. Conodont-bearing animal from the Bear Gulch Limestone, Montana. Geol. Soc. Am. Spec. Pap. 141:31–65.

Mettam, C. 1971. Functional design and evolution of the polychaete *Aphrodite aculeata*. J. Zool. 63:489–514.

———. 1985. Functional constraints in the evolution of the Annelida. Pp. 297–309 in S. Conway Morris, J. D. George, R. Gibson, and H. M. Platt, eds., The Origins and Relationships of Lower Invertebrates. Oxford Univ. Press, Oxford.

Metz, R. 1998. Nematode trails from the Late Triassic of Pennsylvania. Ichnos 5: 303–308.

Meyer, A. 1933. Acanthocephala. Pp. 333–582 in Dr H. G. Bronn's Klassen und Ordnungen des Tierreichs 4. Akademische Verlags., Leipzig.

———. 1938. Die plasmodiale Entwicklung und Formbildung des Riesenkratzers (*Macracanthorhynchus hirudinaceus*) (Pallas). Zool. Abt. Jb. Anat. 62:111–172.

Meyer, K., and T. Bartolomaeus. 1996. Ultrastructure and formation of the hooked setae in *Owenia fusiformis* delle Chiaje, 1842: Implications for annelid phylogeny. Can. J. Zool. 74:2143–2153.

Mikulik, D. G., D. E. G. Briggs, and J. Kluessendorf. 1985a. A new exceptionally preserved biota from the Lower Silurian of Wisconsin, USA. Philos. Trans. R. Soc. London B 311:75–85.

———. 1985b. A Silurian soft-bodied fauna. Science 228:715–717.

Millar, R. H. 1966. Evolution in ascidians. Pp. 519–534 in H. Barnes, ed., Some Contemporary Studies in Marine Science. Allen and Unwin, London.

Miller, S. L. 1974. Adaptive design of locomotion and foot form in prosobranch gastropods. J. Exp. Mar. Biol. Ecol. 14:99–156.

Mills, C. 1984. Density is altered in hydromedusae and ctenophores in response to changes in salinity. Biol. Bull. 166:206–215.

Mokady, O., M. H. Dick, D. Lackschewitz, B. Schierwater, and L. W. Buss. 1998. Over one-half billion years of head conservation? Expression of an *ems* class gene in *Hydractinia symbiolongicarpus* (Cnidaria: Hydrozoa). Proc. Nat. Acad. Sci. USA 95:3673–3678.

Monge-Najera, J. 1995. Phylogeny, biogeography, and reproductive trends in the Onychophora. Zool. J. Linn. Soc. 114:21–60.

Mooi, R., and B. David. 1998. Evolution within a bizarre phylum: Homologies of the first echinoderms. Am. Zool. 38:965–974.

Mooi, R., B. David, and D. Marchand. 1994. Echinoderm skeletal homologies: Classical morphology meets modern phylogenetics. Pp. 87–95 in B. David, A. Guille, J. P. Féral, and M. Roux, eds., Echinoderms through Time (Echinoderms Dijon). Balkema, Rotterdam.

Moon, S. Y., C. B. Kim, S. R. Gelder, and W. Kim. 1996. Phylogenetic position of the aberrant branchiobdellidans and aphanoneurans within the Annelida as derived from 18S ribosomal RNA gene sequences. Hydrobiologia 334:229–236.

Moore, R. C. 1966. Treatise on Invertebrate Paleontology, pt. U, Echinodermata 3, vol. 1. Geological Society of America, Boulder, CO; Univ. Kansas Press, Lawrence.

Moores, E. M. 1991. Southwest U.S.–East Antarctic (SWEAT) connections: A hypothesis. Geology 19:425–428.

Morgan, C. I. 1982. Tardigrada. Pp. 731–740 in S. P. Parker, ed., Synopsis and Classification of Living Organisms, vol. 2. McGraw-Hill, New York.

Morgan, C. I., and P. E. King. 1976. British Tardigrades. Academic Press, New York.

Morris, P. J. 1993. The developmental role of extracellular matrix suggests a monophyletic origin of the kingdom Animalia. Evolution 47:152–165.

Morse, M. P., and P. D. Reynolds. 1996. Ultrastructure of the heart-kidney complex in smaller classes supports symplesiomorphy of molluscan coelomic characters. Pp. 89–97 in J. Taylor, ed., Origin and Evolutionary Radiation of the Mollusca. Oxford Univ. Press, Oxford.

Mourant, A. E. 1971. Transduction and skeletal evolution. Nature 231:486–487.

Moussa, M. T. 1970. Nematode fossil trails from the Green River Formation (Eocene) in the Uinta Basin, Utah. J. Paleontol. 44:304–307.

Mukouyama, Y.-S., D. Shin, S. Britsch, M. Taniguchi, and D. J. Anderson. 2002. Sensory nerves determine the pattern of arterial differentiation and blood vessel branching in the skin. Cell 109:693–705.

Müller, K. J. 1977. *Palaeobotryllus* from the Upper Cambrian of Nevada: A probable ascidian. Lethaia 10:107–118.

———. 1981. Zoological affinities of conodonts. Pp. 78–82 in R. A. Robison, ed., Treatise on Invertebrate Paleontology, pt. W, suppl 2. Geological Society of America, Boulder, CO: Univ. Kansas Press, Lawrence.

Müller, K. J., and I. Hinz-Schallreuter. 1993. Paleoscolecid worms from the Middle Cambrian of Australia. Palaeontology 36:549–592.

Müller, K. J., and Y. Nogami. 1971. Über den Feinbau der Conodonten. Mem. Fac. Sci. Kyoto Univ. Ser. Geol. Mineral. 38:1–88.

Müller, W. E. G. 1995. Molecular phylogeny of Metazoa (animals): Monophyletic origin. Naturwissenschaften 82:321–329.

———. 1998. Origin of Metazoa: Sponges as living fossils. Naturwissenschaften 85:11–25.

Mundy, W. P., P. D. Taylor, and J. P. Thorpe. 1981. A reinterpretation of phylactolaemate phylogeny. Pp. 185–190 in G. P. Larwood and C. Nielsen, eds., Recent and Fossil Bryozoa. Olsen and Olsen, Fredensborg.

Narbonne, G. M., and J. D. Aitken. 1990. Ediacaran fossils from the Sewki Brook area, Mackenzie Mountains, northwestern Canada. Palaeontology 33:945–980.

Narbonne, G. M., P. M. Myrow, and E. Landing. 1987. A candidate stratotype for the Precambrian-Cambrian boundary, Fortune Head, Burin Peninsula, southeastern Newfoundland. Can. J. Earth Sci. 24:1277–1293.

Near, T. J., J. R. Garey, and S. A. Nadler. 1998. Phylogenetic relationships of the Acanthocephala inferred from 18S ribosomal DNA sequences. Mol. Phylogenet. Evol. 10:287–298.

Nebelsick, M. 1993. Introvert, mouth cone, and nervous system of *Echinoderes capitatus* (Kinorhyncha, Cyclorhagida) and implications for the phylogenetic relationships of Kinorhyncha. Zoomorphology 113:211–232.

Neuhaus, B. 1994. Ultrastructure of alimentary canal and body cavity, ground pattern, and phylogenetic relationships of the Kinorhyncha. Microfauna Marina 9:61–156.

———. 1995. Postembryonic development of *Paracentrophyes praedictus* (Homalorhagida): Neoteny questionable among the Kinorhyncha. Zool. Scr. 24:179–192.

Newby, W. W. 1940. The embryology of the echiuroid worm *Urechis caupo*. Mem. Am. Philos. Soc. 16:1–219.

Nichols, D. 1969. Echinoderms. 4th ed. Hutchinson Univ. Library, London.

———. 1971. The phylogeny of the invertebrates. Pp. 362–381 in J. E. Smith, J. D. Carthy, G. Chapman, R. B. Clark, and D. Nichols, The Invertebrate Panorama. Weidenfeld and Nicolson, London.

Nielsen, C. 1971. Entoproct life-cycles and the entoproct-ectoproct relationship. Ophelia 9:209–341.

———. 1985. Animal phylogeny in the light of the Trochaea theory. Biol. J. Linn. Soc. 23:243–299.

———. 1987. Structure and function of metazoan ciliary bands and their phylogenetic significance. Acta Zool. 68:205–262.

———. 1991. The development of the brachiopod *Crania (Neocrania) anomala* (O. F. Muller) and its phylogenetic significance. Acta Zool. 72:7–28.

———. 1995. Animal Evolution: Interrelationships of the Living Phyla. Oxford Univ. Press, Oxford.

———. 1998. Origin and evolution of animal life cycles. Biol. Rev. 73:125–155.

———. 2001. Animal Evolution: Interrelationships of the Living Phyla. 2nd ed. Oxford Univ. Press, Oxford.

Nielsen, C., and A. Jespersen. 1997. Entoprocta. Pp. 13–43 in F. W. Harrison and R. M. Woollacott, eds., Microscopic Anatomy of Invertebrates, vol. 13, Lophophorates, Entoprocta, and Cycliophora. Wiley-Liss, New York.

Nielsen, C., and A. Nørrevang. 1985. The Trochaea theory: An example of life cycle phylogeny. Pp. 28–41 in S. Conway Morris, J. D. George, R. Gibson, and H. M. Platt, eds., The Origin and Relationships of Lower Invertebrate Groups. Oxford Univ. Press, Oxford.

Nielsen, C., and H. U. Riisgård. 1998. Tentacle structure and filter-feeding in *Crisia eburnea* and other cyclostomatous bryozoans, with a review of upstream-collecting mechanisms. Mar. Ecol. Prog. Ser. 168:163–186.

Nielsen, C., N. Scharff, and D. Eibye-Jacobsen. 1996. Cladistic analyses of the animal kingdom. Biol. J. Linn. Soc. 57:385–410.

Nigg, E. A. 1995. Cyclin-dependent protein kinases: Key regulators of the eukaryote cell cycle. BioEssays 17:471–480.

Nislow, C. 1994. Cellular dynamics during the early development of an articulate brachiopod, *Terebratalia transversa*. Pp. 118–128 in W. H. Wilson, Jr., S. A. Stricker, and G. L. Shinn, eds., Reproduction and Development of Marine Invertebrates. Johns Hopkins Univ. Press, Baltimore.

Nogrady, T. 1982. Rotifera. Pp. 865–872 in S. P. Parker, ed., Synopsis and Classification of Living Organisms, vol. 1. McGraw-Hill, New York.

Norén, M., and U. Jondelius. 1997. *Xenoturbella*'s molluscan relatives. Nature 390:31–32.

Norris, R. D. 1989. Cnidarian taphonomy and affinities of the Ediacaran biota. Lethaia 22:381–393.

Norris, R. E. 1982. Choanoflagellida. P. 497 in S. P. Parker, ed., Synopsis and Classification of Living Organisms, vol. 1. McGraw-Hill, New York.

Nursall, J. R. 1962. On the origins of the major groups of animals. Evolution 16:118–123.

Nyholm, K.-G. 1943. Zur Entwicklung und Entwicklungsbiologie der Ceriantharien und Aktinaria. Zool. Bidr. Uppsala 22:87–248.

Odorico, D. M., and D. J. Miller. 1997. Internal and external relationships of the Cnidaria: Implications of primary and predicted secondary structure of the 5'-end of the 23S-like rDNA. Proc. R. Soc. London B 264:77–82.

Oelofsen, B. W., and J. G. Loock. 1981. A fossil cephalochordate from the Early Permian of South Africa. So. Afr. J. Sci. 77:178–180.

Oliver, W. A., and A. G. Coates. 1987. Phylum Cnidaria. Pp. 140–193 in R. Boardman, A. H. Cheetham, and A. J. Rowell, eds., Fossil Invertebrates. Blackwell Scientific, Palo Alto, CA.

Ono, K., H. Suga, N. Iwabe, K. Kuma, and T. Miyata. 1999. Multiple protein tyrosine phosphatases in sponges and explosive gene duplication in the early evolution of animals before the parazoan-eumetazoan split. J. Mol. Evol. 48:654–662.

Öpik, A. A. 1968. Ordian (Cambrian) Crustacea Bradoriida of Australia. Aust. Bur. Miner. Resour. Geol. Geophys. Bull. 103:1–44.

Owen, R. 1843. Lectures on the Comparative Anatomy and Physiology of the Invertebrate Animals. Longman, Brown, Green and Longmans, London.

———. 1860. Palaeontology, or a Systematic Summary of Extinct Animals and Their Geological Relations. Adam and Charles Black, Edinburgh.

Palmer, D., and R. B. Rickards. 1991. Graptolites: Writing in the Rocks. Boydell Press, Woodbridge, U.K.

Panchen, A. L. 1992. Classification, Evolution, and the Nature of Biology. Cambridge Univ. Press, Cambridge.

Panganiban, G., S. M. Irvine, C. Lowe, H. Roehl, L. S. Corley, B. Sherbon, J. K. Grenier, J. F. Fallon, J. Kimble, M. Walker, G. A. Wray, B. J. Swalla, M. Q. Martindale, and S. B. Carroll. 1997. The origin and evolution of animal appendages. Proc. Nat. Acad. Sci. USA 94:5162–5166.

Pantin, C. F. A. 1950. Locomotion in British terrestrial nemertines and planarians: With a discussion on the identity of *Rhynchodemus bilineatus* (Mecznikow) in Britain, and on the name *Fasciola terrestris* O. F. Muller. Proc. Linn. Soc. London 162:23–37.

———. 1966. Homology, analogy, and chemical identity in the Cnidaria. Pp. 1–16 in W. J. Rees, ed., The Cnidaria and Their Evolution. Academic Press, New York.

Parker, A. R. 1998. Colour in Burgess Shale animals and the effect of light on evolution in the Cambrian. Proc. R. Soc. London B 265:967–972.

Parker, S. P., ed. 1982. Synopsis and Classification of Living Organisms. 2 vols. McGraw-Hill, New York.

Parsley, R. L. 1988. Feeding and respiratory strategies in Stylophora. Pp. 347–361 in

C. R. C. Paul and A. B. Smith, eds., Echinoderm Phylogeny and Evolutionary Biology. Clarendon Press, Oxford.

―――. 1999. The Cincta (Homostelea) as blastozoans. Pp. 369–375 in M. D. Candia Carnivali and F. Bonasoro, eds., Echinoderm Research 1998. Balkema, Rotterdam.

Parsley, R. L., and L. W. Mintz. 1975. North American Paracrinoidea (Ordovician, Paracrinozoa, new, Echinodermata). Bull. Am. Paleontol. 68:6–112.

Patterson, C. 1987. Introduction. Pp. 1–22 in C. Patterson, ed., Molecules and Morphology in Evolution: Conflict or Compromise? Cambridge Univ. Press, Cambridge.

―――. 1988. Homology in classical and molecular biology. Mol. Biol. Evol. 5:603–625.

―――. 1989. Phylogenetic relations of major groups: Conclusions and prospects. Pp. 471–488 in B. Fernholm, K. Bremer, and J. Jörnvall, eds., The Hierarchy of Life. Elsevier Scientific, Amsterdam.

Patton, S. J., G. N. Luke, and P. W. H. Holland. 1998. Complex history of a chromosomal paralogy region: Insights from amphioxus aromatic amino acid hydroxylase genes and insulin-related genes. Mol. Biol. Evol. 15:1373–1380.

Paul, C. R. C. 1977. Evolution of primitive echinoderms. Pp. 123–157 in A. Hallam, ed., Patterns of Evolution. Elsevier, Amsterdam.

―――. 1982. The adequacy of the fossil record. Pp. 75–117 in K. A. Joysey and A. E. Friday, eds., Problems of Phylogenetic Reconstruction. Academic Press, London.

―――. 1990. Thereby hangs a tail. Nature 348:680–681.

―――. 1992. The recognition of ancestors. Hist. Biol. 6:239–250.

Paul, C. R. C., and A. B. Smith. 1984. The early radiation and phylogeny of echinoderms. Biol. Rev. 59:443–481.

Pawlowski, J., J.-I. Montoya-Burgos, J. F. Fahrni, J. Wuest, and L. Zaninetti. 1996. Origin of the Mesozoa inferred from 18S rRNA gene sequences. Mol. Biol. Evol. 13:1128–1132.

Pearse, V., J. Pearse, M. Buchsbaum, and R. Buchsbaum. 1987. Living Invertebrates. Blackwell Scientific, Palo Alto, CA; Boxwood Press, Pacific Grove, CA.

Pébusque, M.-J., F. Coulier, D. Birnbaum, and P. Pontarotti. 1998. Ancient large-scale genome duplications: Phylogenetic and linkage analyses shed light on chordate genome evolution. Mol. Biol. Evol. 15:1145–1159.

Peel, J. S. 1991a. The classes Tergomya and Helcionelloida and early molluscan evolution. Bull. Groenl. Geol. Unders. 161:11–65.

―――. 1991b. Functional morphology of the class Helcionelloida nov., and the early evolution of the Mollusca. Pp. 157–177 in A. M. Simonetta and S. Conway Morris, eds., The Early Evolution of Metazoa and the Significance of Problematic Taxa. Cambridge Univ. Press, Cambridge.

Percival, E. 1944. A contribution to the life history of the brachiopd Terebratella inconspicua Sowerby. Trans. R. Soc. N.Z. 74:1–23.

Perkins, F. O. 1991. "Sporozoa": Apicomplexa, Microsporidia, Haplosporidia, Paramyxea, Myxosporidia, and Actinosporidia. Pp. 261–331 in F. W. Harrison and J. O. Corliss, eds., Microscopic Anatomy of Invertebrates, vol. 1, Protozoa. Wiley-Liss, New York.

Person, P. 1983. Invertebrate cartilages. Cartilage 1:31–57.

Peterson, K. J., R. A. Cameron, and E. H. Davidson. 1997. Set-aside cells in maximal indirect development: Evolutionary and developmental significance. BioEssays 19:623–631.

———. 2000. Bilaterian origins: Significance of new experimental observations. Dev. Biol. 219:1–17.

Peterson, K. J., R. A. Cameron, K. Tagawa, N. Satoh, and E. H. Davidson. 1999. A comparative molecular approach to mesodermal patterning in basal deuterostomes: The expression pattern of *Brachyury* in the enteropneust hemichordate *Ptychodera flava*. Development 126:85–95.

Peterson, K. J., and D. J. Eernisse. 2001. Animal phylogeny and the ancestry of bilaterians: Inferences from morphology and 18S rDNA gene sequences. Evol. Dev. 3:170–197.

Pflug, H. D. 1970. Zur Fauna der Nama-Schichten in sudwest-Afrika: 1, Pteridinia: Bau und systematische Zugerhorigkeit. Palaeontographica Abt. A 134:226–262.

Philip, G. M. 1979. Carpoids: Echinoderms or chordates? Biol. Rev. 54:439–471.

Philippe, H., A. Chenuil, and A. Adoutte. 1994. Can the Cambrian explosion be inferred through molecular phylogeny? Development 120, suppl.: 15–25.

Pianka, E. R. 1976. Competition and niche theory. Pp. 114–141 in R. M. May, ed., Theoretical Ecology: Principles and Applications. Saunders, Philadelphia.

Pinker, S. 1995. The Language Instinct. Harper Perennial, New York.

Piper, D. J. W. 1972. Sediments of the Middle Cambrian Burgess Shale, Canada. Lethaia 5:169–175.

Podar, M., S. H. D. Haddock, M. Sogin, and G. R. Harbison. 2001. A molecular phylogenetic framework for the phylum Ctenophora using 18S rRNA genes. Mol. Phylogenet. Evol. 21:218–230.

Poinar, G. O., Jr. 1977. Fossil nematodes from Mexican amber. Nematologica 25:232–238.

———. 1981. Fossil dauer rhabditoid nematodes. Nematologica 27:466–467.

———. 1996. Fossil velvet worms in Baltic and Dominican amber: Onychophoran evolution and biogeography. Science 273:1370–1371.

Poinar, G. O., Jr., and C. Ricci. 1992. Bdelloid rotifers in Dominican amber: Evidence for parthenogenetic continuity. Experientia 48:408–410.

Pojeta, J., Jr. 1980. Molluscan phylogeny. Tulane Stud. Geol. Paleontol. 16:55–80.

———. 1987. Phylum Hyolitha. Pp. 436–444 in R. S. Boardman, A. H. Cheetham, and A. J. Rowell, eds., Fossil Invertebrates. Blackwell Scientific, Palo Alto, CA.

Pojeta, J., Jr., B. Runnegar, N. J. Morris, and N. D. Newell. 1972. Rostroconchia: A new class of bivalved mollusks. Science 177:264–267.

Pollard, S. L., and P. W. H. Holland. 2000. Evidence for 14 homeobox gene clusters in human genome ancestry. Curr. Biol. 10:1059–1062.

Ponder, W. F., and D. R. Lindberg. 1997. Towards a phylogeny of gastropod molluscs: An analysis using morphological characters. Zool. J. Linn. Soc. 119:1–183.

Popov, L. Xe. 1992. The Cambrian radiation of brachiopods. Pp. 399–423 in J. H. Lipps and P. W. Signor, eds., Origin and Early Evolution of the Metazoa. Plenum, New York.

Potswald, H. E. 1981. Abdominal segment formation in *Spirorbis moerchi* (Polychaeta). Zoomorphology 97:225–245.

Powell, C. M. 1995. Are Neoproterozoic glacial deposits preserved on the margins of Laurentia related to the fragmentation of two supercontinents? Geology 23:1053–1054.

Powers, T. P., J. Hogan, Z. Ke, K. Dymbrowski, X. Wang, F. H. Collins, and T. C. Kaufman. 2000. Characterization of the Hox cluster from the mosquito *Anopheles gambiae* (Diptera: Culicidae). Evol. Dev. 2:311–325.

Pridmore, P. A., R. E. Barwick, and R. S. Nicoll. 1997. Soft anatomy and the affinities of conodonts. Lethaia 29:317–328.

Ptashne, M., and A. Gann. 1997. Transcriptional activation by recruitment. Nature 386:569–577.

Purnell, M. A. 1995. Microwear on conodont elements and macrophagy in the first vertebrates. Nature 374:798–800.

Quiring, R., U. Waldorf, U. Kloter, and W. J. Gehling. 1994. Homology of the *eyeless* gene of *Drosophila* to the *small eye* gene in mice and *Aniridia* in humans. Science 265:785–789.

Raff, E. C., E. M. Popodi, B. J. Sly, F. R. Turner, J. T. Villinski, and R. A. Raff. 1999. A novel ontogenetic pathway in hybrid embryos between species with different modes of development. Development 126:1937–1945.

Raff, R. A. 1996. The Shape of Life. Univ. Chicago Press, Chicago.

Raff, R. A., and T. C. Kaufman. 1983. Embryos, Genes, and Evolution. Macmillan, New York.

Raff, R. A., C. R. Marshall, and J. M. Turbeville. 1994. Using DNA sequences to unravel the Cambrian radiation of the animal phyla. Ann. Rev. Ecol. Syst. 25:351–375.

Ragan, M. A., C. L. Goggins, R. J. Cawthorn, L. Cerenius, A. V. C. Jamieson, S. M. Plourdes, T. G. Rand, K. Soderhall, and R. R. Gutell. 1996. A novel clade of protistan parasites near the animal-fungal divergence. Proc. Nat. Acad. Sci. USA 93:11907–11912.

Raikova, O. I., A. Falleni, and J. L. Justine. 1997. Spermiogenesis in *Paratomella rubra* (Platyhelminthes, Acoela): Ultrastructural, immunocytochemical, cytochemical studies, and phylogenetic implications. Acta Zool. 78:295–307.

Raikova, O. I., M. Reuter, U. Jondelius, and M. K. S. Gustafsson. 2000. The brain of the Nermertodermatida (Platyhelminthes) as revealed by anti-5HT and anti-FMRFamide immunostainings. Tissue and Cell 32:358–365.

Raikova, O. I., M. Reuter, and J.-L. Justine. 2001. Contributions to the phylogeny and systematics of the Acoelomorpha. Pp. 13–23 in D. T. J. Littlewood and R. A. Bray, eds., Interrelationships of the Platyhelminthes. Taylor and Francis, London.

Raikova, O. I., M. Reuter, E. A. Kotikova, and M. K. S. Gustafsson. 1998. A commisssural brain! The pattern of 5-HT immunoreactivity in Acoela (Platyhelminthes). Zoomorphology 118:69–77.

Ramazzotti, G. 1972. Il phylum Tardigrada. Mem. Ist. Ital. Idrobiol. 28:1–732.

Ramsköld, L. 1992. Homologies in Cambrian Onychophora. Lethaia 25:443–460.

Ramsköld, L., and J. Chen. 1998. Cambrian lobopodians: Morphology and phylogeny. Pp. 107–150 in G. D. Edgecombe, ed., Arthropod Fossils and Phylogeny. Columbia Univ. Press, New York.

Ramsköld, L., J. Chen, G. D. Edgecombe, and G. Zhou. 1996. Preservational folds simulating tergite junctions in tegopeltid and naraoiid arthropods. Lethaia 29:15–20.

Ramsköld, L., and X. Hou. 1991. New early Cambrian animal and onychophoran affinities of enigmatic metazoans. Nature 351:225–228.

Rattenbury, J. C. 1954. The embryology of *Phoronopsis viridus*. J. Morphol. 95: 289–349.

Raup, D. M. 1975. Taxonomic diversity estimation using rarefaction. Paleobiology 1:333–342.

———. 1979. Biases in the fossil record of species and genera. Carnegie Museum Nat. Hist. Bull. 13:85–91.

———. 1985. Mathematical models of cladogenesis. Paleobiology 11:42–52.

Raup, D. M., and G. E. Boyajian. 1988. Patterns of generic extinction in the fossil record. Paleobiology 14:109–125.

Raup, D. M., S. J. Gould, T. J. M. Schopf, and D. S. Simberloff. 1973. Stochastic models of phylogeny and the evolution of diversity. J. Geol. 81:525–542.

Raup, D. M., and L. G. Marshall. 1980. Variation between groups in evolutionary rates: A statistical test of significance. Paleobiology 6:9–23.

Raup, D. M., and J. J. Sepkoski, Jr. 1982. Mass extinctions in the marine fossil record. Science 215:1501–1503.

Ravasz, E., A. L. Somera, D. A. Mongru, Z. N. Oltavai, and A.-L. Barabási. 2002. Hierarchical organization of modularity in metabolic networks. Science 297:1551–1555.

Redecker, D., R. Kodner, and L. E. Graham. 2000. Glomalean fungi from the Ordovician. Science 289:1920–1921.

Reed, C. G., and R. A. Cloney. 1977. Brachiopod tentacles: Ultrastructure and functional significance of the connective tissue and myoepithelial cells in Terebratalia. Cell Tissue Res. 185:17–42.

Rees, W. J. 1966. The evolution of the Hydrozoa. Pp. 199–221 in W. J. Rees, ed., The Cnidaria and Their Evolution. Academic Press, London.

Regnell, G. 1966. Edrioasteroids. Pp. 136–173 in R. C. Moore, ed., Treatise on Invertebrate Paleontology, pt. U, vol. 1. Geological Society of America, Boulder, CO; Univ. Kansas Press, Lawrence.

Reineck, H.-E., W. F. Gutmann, and G. Hertweck. 1967. Das Schlickgebiet südlich Helgoland als Beispiel rezenter Schelfablagerungen. Senckenb. Lethaea 48:219–275.

Reiswig, H. M. 1979. Histology of Hexactinellida (Porifera). Pp. 173–180 in C. Levi and N. Boury-Esnault, eds., Biologie des Spongiaires. Colloques Internationaux du Centre National de la Recherche Scientifique no. 291. Editions du Centre National de la Recherche Scientifique, Paris.

Reiswig, H. M., and G. O. Mackie. 1983. Studies on hexactinellid sponges; 3, The taxonomic status of Hexactinellida within the Porifera. Philos. Trans. R. Soc. London B 301:419–428.

Reitner, J., and D. Mehl. 1995. Early Paleozoic diversification of sponges: New data and evidences. Geol. Palaontol. Mitt. Innsbruck 20:335–347.

Remane, A. 1952. Die Grundlagen des natürlich Systems der vergleichenden Anatomie und der Phylogenetik. Geest und Portig, Leipzig.

———. 1954. Die Geschichte der Tiere. Pp. 340–422 in G. Herberer, ed., Die Evolution der Organismen, vol. 2. 2nd ed. Fischer, Stuttgart.

———. 1963. The enterocelic origin of the celom. Pp. 78–90 in E. C. Dougherty, ed., The Lower Metazoa. Univ. California Press, Berkeley and Los Angeles.

Rensch, B. 1947. Neuere Probleme der Abstammungslehre. Ferdinand Enke, Stuttgart.

Reverberi, G. 1971. Ctenophores. Pp. 85–103 in G. Reverberi, ed., Experimental Embryology of Marine and Fresh-Water Invertebrates. North-Holland, Amsterdam.

Rhoads, D. C., and J. W. Morse. 1971. Evolutionary and ecological significance of oxygen-deficient marine basins. Lethaia 4:413–428.

Rhodes, F. H. T., and R. L. Austin. 1981. Natural assemblages of elements: Interpretation and taxonomy. Pp. 68–78 in R. A. Robison, ed., Treatise on Invertebrate Paleontology, pt. W, suppl. 2, Conodonta. Geological Society of America, Boulder, CO; Univ. Kansas Press, Lawrence.

Rhyu, M. S., and J. A. Knoblich. 1995. Spindle orientation and asymmetric cell fate. Cell 82:523–526.

Rice, M. E. 1969. Possible boring structures of sipunculids. Am. Zool. 9:803–812.

———. 1982. Sipuncula. Pp. 67–69 in S. P. Parker, ed., Synopsis and Classification of Living Organisms. McGraw-Hill, New York.

———. 1985. Sipuncula: Developmental evidence for phylogenetic inference. Pp. 274–296 in S. Conway Morris, J. D. George, R. Gibson, and H. M. Platt, eds., The Origins and Relationships of Lower Invertebrates. Oxford Univ. Press, Oxford.

Richards, R. J. 1992. The structure of narrative explanation in history and biology. Pp. 19–53 in N. H. Nitecki and D. V. Nitecki, eds., History and Evolution. State Univ. New York Press, Albany.

Rickards, R. B., A. J. Chapman, and J. T. Temple. 1984. *Rhabdopleura hollandi:* A new pterobranch hemichordate from the Silurian of the Llandovery district, Powys, Wales. Proc. Geol. Assoc. 95:23–28.

Rickards, R. B., and B. A. Stait. 1984. *Psigraptus:* Its classification, evolution, and zooid. Alcheringa 8:101–111.

Ricklefs, R. E. 1990. Ecology. 3rd ed. Freeman, San Francisco.

Ridley, M. 1986. Evolution and Classification: The Reformation of Cladism. Longman Science and Technology, Harlow, U.K.

Riedl, R. J. 1969. Gnathostomulida from America. Science 163:445–452.

Rieger, R. M. 1974. A new group of Turbellaria-Typhloplanoida with a proboscis and its relationship to Kalyptorhynchia. Pp. 23–62 in N. W. Riser and M. P. Morse, eds., Biology of the Turbellaria. McGraw-Hill, New York.

———. 1981. Morphology of the Turbellaria at the ultrastructural level. Hydrobiologia 84:213–229.

———. 1985. The phylogenetic status of the acoelomate organisation within the Bilateria: A histological perspective. Pp. 101–122 in S. Conway Morris, J. D. George, R. Gibson, and H. M. Platt, eds., The Origins and Relationships of Lower Invertebrates. Clarendon Press, Oxford.

———. 1986. Uber den Ursprung der Bilateria: Die Bedeutung der Ultrastrukturforschung für ein neues Verstehen der Metazoenevolution. Verh. Dtsch. Zool. Ges. 79:31–50.

Rieger, R. M., and M. Mainitz. 1977. Comparative fine structure and study of the body wall in Gnathostomulida and their phylogenetic position between Platyhelminthes and Aschelminthes. Z. Zool. Syst. Evolutionsforsch. 15:9–35.

Rieger, R. M., and S. Tyler. 1995. Sister-group relationship of Gnathostomulida and Rotifera-Acanthocephala. Invertebr. Biol. 114:186–188.

Rieger, R. M., S. Tyler, J. P. S. Smith, III, and G. E. Rieger. 1991. Platyhelminthes: Turbellaria. Pp. 7–140 in F. W. Harrison and B. J. Bogitsh, eds., Microscopic Anatomy of Invertebrates, vol. 3, Platyhelminthes and Nemertina. Wiley-Liss, New York.

Riemann, F. 1977. Causal aspects of nematode evolution: Relations between structure, function, habitat, and evolution. Mikrofauna Meeresboden 61:217–230.

Rigby, J. K. 1978. Porifera of the Middle Cambrian Wheeler Shale from the Wheeler Amphitheater, House Range in western Utah. J. Paleontol. 52:1325–1345.

Riutort, M., K. G. Field, R. A. Raff, and J. Baguñà. 1993. 18S rRNA sequences and phylogeny of Platyhelminthes. Biochem. Syst. Ecol. 21:71–77.

Robison, R. A. 1969. Annelids from the Middle Cambrian of Utah. J. Paleontol. 43:1169–1173.

———. 1990. Earliest-known uniramous arthropod. Nature 343:163–164.

Robison, R. A., and R. L. Kaesler. 1987. Phylum Arthropoda. Pp. 205–269 in R. S. Boardman, A. H. Cheetham, and A. J. Rowell, eds., Fossil Invertebrates. Blackwell Scientific, Palo Alto, CA.

Robison, R. A., and J. Sprinkle. 1969. Ctenocystoidea: New class of primitive echinoderms. Science 166:1512–1514.

Rodgers, J. 1969. Gnathostomulida: Is there a fossil record? Science 164:855–856.

Rodrigues-Trelles, F., R. Tarrio, and F. J. Ayala. 2001. Erratic overdispersion of three molecular clocks: GPDH, SOD, and XDH. Proc. Nat. Acad. Sci. USA 98:11404–11410.

Rolfe, W. D. I., F. R. Schram, G. Pacaud, D. Sotty, and S. Secretan. 1982. A remarkable Stephanian biota from Montceau-les-Mines, France. J. Paleontol. 56:426–428.

Ronshaugen, M., N. McGinnis, and W. McGinnis. 2002. Hox protein mutation and macroevolution of the insect body plan. Nature 415:914–917.

Rosenzweig, M. L. 1975. On continental steady states of species diversity. Pp. 121–140 in M. L. Cody and J. M. Diamond, eds., Ecology and Evolution of Communities. Belknap Press, Cambridge, MA.

Roth, V. L. 1991. Homology and hierarchies: Problems solved and unresolved. J. Evol. Biol. 4:167–194.

Rouse, G. W. 2000a. The epitome of hand waving? Larval feeding and hypotheses of metazoan phylogeny. Evol. Dev. 2:222–233.

———. 2000b. Polychaetes have evolved feeding larvae numerous times. Bull. Mar. Sci. 67:393–409.

Rouse, G. W., and K. Fauchald. 1995. The articulation of annelids. Zool. Scr. 24:269–301.

Rowe, F. W. E., A. N. Baker, and H. E. S. Clark. 1988. The morphology, development, and taxonomic status of Xyloplax Baker, Rowe, and Clark, 1986 (Echinodermata: Concentricycloidea), with the description of a new species. Philos. Trans. R. Soc. London B 233:431–459.

Rowell, A. J. 1977. Early Cambrian brachiopods from the southwestern Great Basin of California and Nevada. J. Paleontol. 51:68–85.

———. 1982. The monophyletic origin of the Brachiopoda. Lethaia 15:299–307.

Rowell, A. J., and N. E. Caruso. 1985. The evolutionary significance of Nisusia sulcata, an early articulate brachiopod. J. Paleontol. 59:1227–1242.

Rozanov, A. Yu. 1984. The Precambrian-Cambrian boundary in Siberia. Episodes 7:20–24.

Rozanov, A. Yu., V. V. Missarzhevsky, N. A. Volkova, L. G. Voronova, I. N. Krylov, B. M. Keller, I. K. Korolyut, K. Lendzion, P. Miknia, N. G. Pykhova, and A. D. Sidorov. 1969. The Tommotian stage and the Cambrian lower boundary problem. Tr. Geol. Inst. Akad. Nauk SSSR 206:5–380. In Russian.

Rozanov, A. Yu., and A. Yu. Zhuravlev. 1992. The Lower Cambrian fossil record of the Soviet Union. Pp. 205–282 in J. H. Lipps and P. W. Signor, eds., Origin and Early Evolution of the Metazoa. Plenum, New York.

Rozov, S. N. 1984. Morphology, terminology, and systematic affinity of stenothecoids. Tr. Inst. Geol. Geofiz. Sibirsk. Ofd. Akad. Nauk SSSR 597:117–133. In Russian.

Rubin, G. M., et al. 2000. Comparative genomics of the eukaryotes. Science 287:2204–2215.

Rudwick, M. J. S. 1970. Living and Fossil Brachiopods. Hutchinson, London.

———. 1976. The Meaning of Fossils. 2nd ed. Science History Publishers, New York.

Ruiz, G., and D. R. Lindberg. 1989. A fossil record for trematodes: Extent and potential uses. Lethaia 22:431–438.

Ruiz-Trillo, I., J. Paps, M. Loukota, C. Ribera, U. Jondelius, J. Baguñà, and M. Riutort. 2002. A phylogenetic analysis of myosin heavy chain type II sequences corroborates that Acoela and Nemertodermatida are basal bilaterians. Proc. Nat. Acad. Sci. USA 99:11246–11251.

Ruiz-Trillo, I., M. Riutort, D. T. J. Littlewood, E. A. Herniou, and J. Baguñà. 1999. Acoel flatworms: Earliest extant bilaterian metazoans, not members of Platyhelminthes. Science 283:1919–1923.

Runnegar, B. 1982a. The Cambrian explosion: Animals or fossils? J. Geol. Soc. Aust. 29:395–411.

———. 1982b. Oxygen requirements, biology, and phylogenetic significance of the late Precambrian worm *Dickinsonia,* and the evolution of the burrowing habit. Alcheringa 6:223–239.

———. 1983. Molluscan phylogeny revisited. Mem. Assoc. Australas. Palaeontol. 1:121–144.

———. 1985. Collagen gene construction and evolution. J. Mol. Evol. 5:141–149.

———. 1992a. Proterozoic fossils of soft-bodied metazoans (Ediacara faunas). Pp. 999–1007 in J. W. Schopf, and C. Klein, eds., The Proterozoic Biosphere: A Multidisciplinary Study. Cambridge Univ. Press, Cambridge.

———. 1992b. Proterozoic metazoan trace fossils. Pp. 1009–1025 in J. W. Schopf and C. Klein, eds., The Proterozoic Biosphere: A Multidisciplinary Study. Cambridge Univ. Press, Cambridge.

———. 1996. Early evolution of the Mollusca: The fossil record. Pp. 77–87 in J. Taylor, ed., Origin and Evolutionary Radiation of the Mollusca. Oxford Univ. Press, Oxford.

Runnegar, B., and J. Pojeta, Jr. 1974. Molluscan phylogeny: The paleontological viewpoint. Science 186:311–317.

———. 1985. Origin and diversification of the Mollusca. Pp. 1–57 in E. R. Trueman and M. R. Clarke, eds., The Mollusca, vol. 10, Evolution. Academic Press, Orlando, FL.

Runnegar, B., J. Pojeta, Jr., N. J. Morris, J. D. Taylor, M. E. Taylor, and G. McClung. 1975. Biology of the Hyolitha. Lethaia 8:181–191.

Ruppert, E. E. 1991a. Gastrotricha. Pp. 41–109 in F. W. Harrison and E. E. Ruppert, eds., Microscopic Anatomy of Invertebrates, vol. 4, Aschelminthes. Wiley-Liss, New York.

———. 1991b. Introduction to the aschelminth phyla: A consideration of mesoderm, body cavities, and cuticle. Pp. 1–17 in F. W. Harrison and E. E. Ruppert, eds., Microscopic Anatomy of Invertebrates, vol. 4, Aschelminthes. Wiley-Liss, New York.

————. 1994. Evolutionary origin of the vertebrate nephron. Am. Zool. 34:542–553.

Ruppert, E. E., and R. D. Barnes. 1994. Invertebrate Zoology. 6th ed. Saunders, Fort Worth.

Ruppert, E. E., and K. J. Carle. 1983. Morphology of metazoan circulatory systems. Zoomorphology 103:193–208.

Ruppert, E. E., and P. R. Smith. 1988. The functional organization of filtration nephridia. Biol. Rev. 63:231–258.

Ruse, M. 1973. The Philosophy of Biology. Hutchinson, London.

Russell-Hunter, W. D. 1979. A Life of Invertebrates. Macmillan, New York.

Rutherford, S. L., and S. Lindquist. 1998. Hsp90 as a capacitor for morphological evolution. Nature 396:336–342.

Ruthmann, A. 1977. Cell differentiation, DNA content, and chromosomes of *Trichoplax adherens* F. E. Schulze. Cytobiologie 15:58–64.

Ryland, J. S. 1970. Bryozoans. Hutchinson, London.

Sachs, M. 1955. Observations on the embryology of an aquatic gastrotrich *Lepidodermella squammata* (Dujardin, 1841). J. Morphol. 96:473–495.

Sadler, L. A., and C. F. Brunk. 1992. Phylogenetic relationships and unusual diversity in histone H4 proteins within the *Tetrahymina pyriformis* complex. Mol. Biol. Evol. 9:70–85.

Sadler, P. M. 1981. Sediment accumulation rates and the completeness of stratigraphic sections. J. Geol. 89:569–584.

Sadler, P. M., and D. J. Strauss. 1990. Estimation of completeness of stratigraphic sections from empirical data and theoretical models. J. Geol. Soc. London 147:471–485.

Saitou, N., and M. Nei. 1987. The neighbor-joining method: A new method for reconstructing phylogenetic trees. Mol. Biol. Evol. 4:406–425.

Saló, E., J. Tauler, E. Jimenez, J. R. Bayascas, J. Gonzalez-Linares, J. Garcia-Fernàndez, and J. Baguñà. 2001. Hox and ParaHox genes in flatworms: Characterization and expression. Am. Zool. 41:652–663.

Salthe, S. N. 1985. Evolving Hierarchical Systems: Their Structure and Representation. Columbia Univ. Press, New York.

Salvini-Plawen, L. v. 1972. Zur Morphologie und Phylogenie der Mollusken. Z. Wiss. Zool. 184:205–394.

————. 1978. On the origin and evolution of the lower Metazoa. Z. Zool. Syst. Evolutionsforsch. 16:40–88.

————. 1980. A reconsideration of systematics in the Mollusca (phylogeny and higher classification). Malacologia 19:249–278.

————. 1982. A paedomorphic origin of the oligomerous animals? Zool. Scr. 11:77–81.

————. 1985. Early evolution and the primitive groups. Pp. 59–150 in E. R. Trueman and M. R. Clarke, eds., The Mollusca, vol. 10, Evolution. Academic Press, Orlando, FL.

————. 1991. Origin, phylogeny, and classification of the phylum Mollusca. Iberus 9:1–33.

Salvini-Plawen, L. v., and G. Steiner. 1996. Synapomorphies and plesiomorphies in higher classification of Mollusca. Pp. 29–51 in J. Taylor, ed., Origin and Evolutionary Radiation of the Mollusca. Oxford Univ. Press, Oxford.

Sanderson, M. J. 1997. A nonparametric approach to estimating divergence times in the absence of rate constancy. Mol. Biol. Evol. 14:1218–1231.

Sanderson, M. J., and L. Hufford. 1996. Homoplasy: The Recurrence of Similarity in Evolution. Academic Press, San Diego.

Sansom, I. J., M. P. Smith, H. A. Armstrong, and M. M. Smith. 1992. Presence of the earliest vertebrate hard tissues in conodonts. Science 256:1308–1311.

Satoh, N. 1994. Developmental Biology of Ascidians. Cambridge Univ. Press, Cambridge.

Savarese, M. 1992. Functional analysis of archaeocyathan skeletal morphology and its paleobiological implications. Paleobiology 18:464–480.

Schaeffer, B., M. K. Hecht, and N. Eldredge. 1972. Phylogeny and paleontology. Evol. Biol. 6:31–46.

Schäfer, W. 1972. Ecology and Palaeoecology of Marine Environments. Univ. Chicago Press, Chicago.

Schaffer, J. 1930. Die Stützgewebe. Pp. 1–390 in W. von Mollendorf, ed., Handbuch der microskopischen Anatomie des Menschen, vol. 2. Springer-Verlag, Berlin.

Scheltema, A. H. 1993. Aplacophora as progenetic aculiferans and the coelomate origin of mollusks as the sister taxon of Sipuncula. Biol. Bull. 184:57–78.

———. 1996. Phylogenetic position of Sipuncula, Mollusca, and the progenetic Aplacophora. Pp. 53–58 in J. Taylor, ed., Origin and Evolutionary Radiation of the Mollusca. Oxford Univ. Press, Oxford.

Schierenberg, E. 1997. Nematodes: The roundworms. Pp. 131–148 in S. F. Gilbert and A. M. Raunio, eds., Embryology. Sinauer, Sunderland, MA.

Schierwater, B., and K. Kuhn. 1998. Homology of Hox genes and the zootype concept in early metazoan evolution. Mol. Phylogenet. Evol. 9:375–381.

Schierwater, B., M. Murtha, M. Dick, F. H. Ruddle, and L. W. Buss. 1991. Homeoboxes in cnidarians. J. Exp. Zool. 260:413–416.

Schindel, D. E. 1980. Microstratigraphic sampling and the limits of stratigraphic resolution. Paleobiology 6:408–426.

———. 1982. Resolution analysis: A new approach to the gaps in the fossil record. Paleobiology 8:340–353.

Schlegel, M., J. Lom, A. Stechmann, D. Bernhard, D. Leipe, I. Dykova, and M. L. Sogin. 1996. Phylogenetic analysis of complete small subunit ribosomal RNA coding region of *Myxidium lieberkuehni:* Evidence that Myxozoa are Metazoa and related to the Bilateria. Arch. Protistenk. 147:1–9.

Schmidt, G. D. 1985. Development and life cycles. Pp. 273–305 in D. W. T. Crompton and B. B. Nickol, eds., Biology of the Acanthocephala. Cambridge Univ. Press, Cambridge.

Schmidt-Nielsen, K. 1984. Scaling: Why Is Animal Size So Important? Cambridge Univ. Press, Cambridge.

Schnabel, R. 1996. Pattern formation: Regional specification in the early *C. elegans* embryo. BioEssays 18:591–594.

Scholz, G. 1998. Cleavage, germ band formation, and head segmentation: The ground pattern of the Euarthropoda. Pp. 317–332 in R. A. Fortey and R. H. Thomas, eds., Arthropod Relationships. Chapman and Hall, London.

Schopf, J. W., and C. Klein, eds. 1992. The Proterozoic Biosphere: A Multidisciplinary Study. Cambridge Univ. Press, Cambridge.

Schopf, K. M., and T. K. Baumiller. 1998. A biomechanical approach to Ediacaran hypotheses: How to weed the Garden of Ediacara. Lethaia 31:89–97.

Schram, F. R. 1973. Pseudocoelomates and a nemertine from the Illinois Pennsylvanian. J. Paleontol. 47:985–989.

———. 1979. Worms of the Mississippian Bear Gulch Limestone of central Montana, USA. Trans. San Diego Soc. Nat. Hist. 19:107–120.

———. 1991. Cladistic analysis of metazoan phyla and the placement of fossil problematica. Pp. 35–46 in A. M. Simonetta and S. Conway Morris, eds., The Early Evolution of Metazoa and the Significance of Problematic Taxa. Cambridge Univ. Press, Cambridge.

Schubiger, G. 1968. Anlageplan, Determinationszustand, und Transdeterminations-leistungen der männlichen Vorderbeinscheibe von *Drosophila melanogaster*. Roux Arch. Entwicklungsmech. Org. 160:9–40.

Schumann, D. 1973. Mesodermale Endoskelette terebratulider Brachiopoden: 1. Palaeontol. Z. 47:77–103.

Schwann, T. 1839. Mikroscopische Untersuchungen über die Uebereinstimmung in der Structur und dem Wachsthum der Thiere und Pflanzen. Berlin. Translated by H. Smith, Microscopical Researches into the Accordance in the Structure of Growth of Animals and Plants, Kraus Reprint, New York, 1969.

Scott, H. W. 1973. New Conodontochordata from the Bear Gulch Limestone (Namurian, Montana). Mich. State Univ. Paleontol. Ser. 1:81–100.

Scott, M. P., and A. J. Weiner. 1984. Structural relationships among genes that control development: Sequence homology between the *Antennapedia, Ultrabithorax,* and *fushi tarazu* loci of *Drosophila*. Proc. Nat. Acad. Sci. USA 81:4115–4119.

Scrutton, C. T., and E. N. K. Clarkson. 1991. A new scleractinian-like coral from the Ordovician of the Southern Uplands, Scotland. Palaeontology 34:179–194.

Sedgwick, A. 1884. On the origin of metameric segmentation and some other morphological questions. Q. J. Microsc. Sci. 24:43–82.

Seilacher, A. 1955. Spuren und Fazies im Unterkambrium. Akad. Wiss. Lit. Mainz Abh. Math. Naturwiss. Kl. 10:11–143.

———. 1970. *Cruziana* stratigraphy of "non-fossiliferous" Paleozoic sandstones. Pp. 447–476 in T. P. Crimes and J. C. Harper, eds., Geol. J. Special Issue 3.

———. 1984. Late Precambrian and Early Cambrian Metazoa: Preservational or real extinctions? Pp. 159–168 in H. D. Holland and A. F. Trendall, eds., Patterns of Change in Earth Evolution. Springer-Verlag, Berlin.

———. 1989. Vendozoa: Organismic construction in the Proterozoic biosphere. Lethaia 22:229–239.

———. 1992. Vendobionta and Psammocorallia: Lost constructions of Precambrian evolution. J. Geol. Soc. London 149:607–613.

———. 1999. Precambrian trace fossils. Geol. Soc. Am. Abstr. Programs 330:A–147.

Seilacher, A., P. K. Bose, and F. Pfluger. 1998. Triploblastic animals more than 1 billion years ago: Trace fossil evidence from India. Science 282:80–83.

Seilacher, A., M. Meschede, E. W. Bolton, and H. Luginsland. 2000. Precambrian "fossil" *Vermiforma* is a tectograph. Geology 28:235–238.

Seimiya, M., H. Ishiguro, K. Miura, Y. Watanabe, and Y. Kurosawa. 1994. Homeobox-containing genes in the most primitive Metazoa, the sponges. Eur. J. Biochem. 221:219–225.

Seimiya, M., Y. Watanabe, and Y. Kurosawa. 1996. Analysis of POU-type homeobox genes in sponge. International Conference on Sponge Science, Otsu, Japan, abstract vol., p. 80.

Sepkoski, J. J., Jr. 1979. A kinetic model of Phanerozoic taxonomic diversity: 2, Early Phanerozoic families and multiple equilibria. Paleobiology 5:222–251.

———. 1981. A factor analytic description of the Phanerozoic marine fossil record. Paleobiology 7:36–53.

———. 1982. A compendium of fossil marine families. Milw. Public Mus. Contrib. Biol. Geol. 51:1–125.

———. 1988. Alpha, beta, or gamma: Where does all the diversity go? Paleobiology 14:221–234.

———. 2002. A compendium of fossil marine animal genera. Ed. D. Jablonski and M. Foote. Bull. Amer. Paleontol. 363:1–560.

Sepkoski, J. J., Jr., and A. Miller. 1985. Evolutionary faunas and the distribution of Paleozoic marine communities in space and time. Pp. 153–190 in J. W. Valentine, ed., Phanerozoic Diversity Patterns: Profiles in Macroevolution. Princeton Univ. Press, Princeton, NJ.

Sepkoski, J. J., Jr., and P. M. Sheehan. 1983. Diversification, faunal change, and community replacement during the Ordovician radiations. Pp. 673–717 in M. J. S. Tevesz and P. L. McCall, eds., Biotic Interactions in Recent and Fossil Benthic Communities. Plenum, New York.

Serafinski, W., and M. Strzelec. 1988. Molluscan phylogeny from the neontological viewpoint. Przeglad Zoologiczny 32:163–176. In Polish; English summary.

Sereno, P. C., and F. E. Novas. 1992. The complete skull and skeleton of an early dinosaur. Science 258:1137–1140.

Seslavinsky, I. B., and I. D. Maidanskaya. 2001. Global facies distributions from Late Vendian to Mid-Ordovician. Pp. 47–68 in A. Yu. Zhuravlev and R. Riding, eds., The Ecology of the Cambrian Radiation. Columbia Univ. Press, New York.

Sharman, A. C., and P. W. H. Holland. 1996. Conservation, duplication, and divergence of developmental genes during chordate evolution. Neth. J. Zool. 46:47–67.

Shaw, A. B. 1964. Time in Stratigraphy. McGraw-Hill, New York.

Shear, W. A. 1997. The fossil record and evolution of the Myriapoda. Pp. 211–219 in R. A. Fortey and R. H. Thomas, eds., Arthropod Relationships. Chapman and Hall, London.

Shelton, C. A., and S. A. Wasserman. 1993. Pelle encodes a protein kinase required to establish dorsoventral polarity in the Drosophila embryo. Cell 72:515–525.

Shinn, G. L. 1997. Chaetognatha. Pp. 103–220 in F. W. Harrison and E. E. Ruppert, eds., Microscopic Anatomy of Invertebrates, vol. 15, Hemichordata, Chaetognatha, and the Invertebrate Chordates. Wiley-Liss, New York.

Shu, D. 2003. A paleontological perspective of vertebrate origin. Chinese Sci. Bull. 48:725–735.

Shu, D., S. Conway Morris, X.-L. Zhang, L. Chen, Y. Li, and J. Han. 1999. A pipiscid-like fossil from the Lower Cambrian of south China. Nature 400:746–749.

Shu, D., S. Conway Morris, Z. F. Zhang, J. N. Liu, J. Han, L. Chen, S. L. Zhang, K. Yasui, and Y. Li. 2003. A new species of yunnanozoan with implications for deuterostome evolution. Science 299:1380–1384.

Shu, D., H.-L. Luo, S. Conway Morris, X.-L. Zhang, S.-X. Hu, L. Chen, J. Han, M. Zhu, and L.-Z. Chen. 1999. Lower Cambrian vertebrates from south China. Nature 402:42–46.

Shu, D., J. Vannier, H.-L. Luo, L. Chen, X.-L. Zhang, and S.-X. Hu. 1999. Anatomy and lifestyle of *Kunmingella* (Arthropoda, Bradoriida) from the Chengjiang fossil lagerstätte (Lower Cambrian; southwest China). Lethaia 32:279–298.

Shu, D.-G., S. Conway Morris, J. Han, L. Chen, K.-L. Zhang, H.-Q. Liu, Y. Li, and J.-N. Liu. 2001. Primitive deuterostomes from the Chengjiang lagerstätte (Lower Cambrian, China). Nature 414:419–424.

Shu, D.-G., S. Conway Morris, and X.-L. Zhang. 1996. A *Pikaia*-like chordate from the Lower Cambrian of China. Nature 384:157–158.

Shu, D.-G., X. Zhang, and L. Chen. 1996. Reinterpretation of *Yunnanozoon* as the earliest known hemichordate. Nature 380:428–430.

Shubin, N., C. Tabin, and S. Carroll. 1997. Fossils, genes, and the evolution of animal limbs. Nature 388:639–648.

Siddall, M. E., D. S. Martin, D. Bridge, S. S. Desser, and D. K. Cone. 1995. The demise of a phylum of protists: Phylogeny of Myxozoa and other parasitic Cnidaria. J. Parasitol. 81:961–967.

Siewing, R. 1980. Das Archicoelomatenkonzept. Zool. Jb. Abt. Syst. 103:439–482.

Signor, P. W. 1985. Real and apparent trends in species richness through time. Pp. 129–150 in J. W. Valentine, ed., Phanerozoic Diversity Patterns: Profiles in Macroevolution. Princeton Univ. Press, Princeton, NJ.

Signor, P. W., and G. J. Vermeij. 1994. The plankton and the benthos: Origins and early history of an evolving relationship. Paleobiology 20:297–319.

Simmen, M. W., S. Leitgeb, V. H. Clark, S. J. M. Jones, and A. Bird. 1998. Gene number in an invertebrate chordate, *Ciona intestinalis*. Proc. Nat. Acad. Sci. USA 95:4437–4440.

Simon, H. A. 1962. The architecture of complexity. Proc. Am. Philos. Soc. 106:462–482.

———. 1973. The organization of complex systems. Pp. 1–27 in H. H. Pattee, ed., Hierarchy Theory. Brazilier, New York.

Simpson, G. G. 1944. Tempo and Mode in Evolution. Columbia Univ. Press, New York.

———. 1953. The major features of evolution. Columbia Univ. Press, New York.

Simpson, T. L. 1984. The Cell Biology of Sponges. Springer-Verlag, New York.

Siveter, D. J., M. Williams, and D. Walossek. 2001. A phosphatocopid crustacean with appendages from the Lower Cambrian. Science 293:479–481.

Slack, J. M. W., D. W. H. Holland, and C. F. Graham. 1993. The zootype and the phylotypic stage. Nature 361:490–492.

Sleigh, M. A., J. R. Blake, and N. Liron. 1988. The propulsion of mucus by cilia. Am. Rev. Respir. Dis. 137:726–741.

Sloss, L. L. 1963. Sequences in the cratonic interior of North America. Geol. Soc. Am. Bull. 74:93–114.

Smith, A. B. 1984. Classification of the Echinodermata. Palaeontology 27:431–459.

———. 1988a. Fossil evidence for the relationships of extant echinoderm classes and their times of divergence. Pp. 85–97 in C. R. C. Paul and A. B. Smith, eds., Echinoderm Phylogeny and Evolutionary Biology. Clarendon Press, Oxford.

———. 1988b. To group or not to group: The taxonomic position of *Xyloplax*. Pp. 17–23 in R. D. Burke, P. V. Mladenov, P. Lambert, and R. L. Parsley, eds., Echinoderm Biology. Balkema, Rotterdam.

———. 1994. Systematics and the Fossil Record. Blackwell, Oxford.

Smith, A. B., R. Lafay, and R. Christen. 1992. Comparative variation of morphological and molecular evolution through geologic time: 28S ribosomal RNA versus morphology in echinoids. Philos. Trans. R. Soc. London B 338:365–382.

Smith, A. G. 2001. Paleomagnetically and tectonically based global maps for Vendian to Mid-Ordovician time. Pp. 11–46 in A. Yu. Zhuravlev and R. Riding, eds., The Ecology of the Cambrian Radiation. Columbia Univ. Press, New York.

Smith, J. P. S., III, and S. Tyler. 1985. The acoel turbellarians: Kingpins of metazoan evolution or a specialized offshoot? Pp. 123–142 in S. Conway Morris, J. D. George, R. Gibson, and H. M. Platt, eds., The Origins and Relationships of Lower Invertebrates. Clarendon Press, Oxford.

Smith, P. R. 1986. Development of the blood vascular system in *Sabellaria cementarium* (Annelida, Polychaeta): An ultrastructural investigation. Zoomorphology 106:67–74.

Smothers, J. F., C. D. von Dohlen, L. H. Smith, Jr., and R. D. Spall. 1994. Molecular evidence that the myxozoan protists are metazoans. Science 265:1719–1721.

Sneath, P. H. A. 1964. Comparative biochemical genetics in bacterial taxonomy. Pp. 565–583 in C. A. Leone, ed., Taxonomic Biochemistry and Serology. Ronald Press, New York.

Snodgrass, R. E. 1938. Evolution of the Annelida, Onychophora, and Arthropoda. Smithson. Misc. Coll. 97 (6):1–159.

Sogin, M. L. 1989. Evolution of eukaryotic microorganisms and their small subunit ribosomal RNAs. Am. Zool. 29:487–499.

———. 1991. Early evolution and the origin of eukaryotes. Curr. Opin. Genet. Dev. 1:457–463.

———. 1994. The origin of eukaryotes and evolution into major kingdoms. Pp. 181–192 in S. Bengtson, ed., Early Life on Earth. Columbia Univ. Press, New York.

Sokal, R. R., and P. H. A. Sneath. 1963. Principles of Numerical Taxonomy. Freeman and Co., San Francisco.

Sokolov, B. S. 1952. On the age of the ancient sedimentary cover of the Russian Platform. Izv. Akad. Nauk SSSR Ser. Geol. 5:121–131. In Russian.

———. 1972. Vendian and Early Cambrian Sabelliditida (Pogonophora) of the USSR. Proceedings of the 23rd International Geological Congress, 79–86.

Sokolov, B. S., and M. A. Fedonkin. 1984. The Vendian as the terminal system of the Precambrian. Episodes 7:12–19.

Solow, A. R., and W. Smith. 1997. On fossil preservation and the stratigraphic ranges of taxa. Paleobiology 23:271–277.

Southcott, R. V., and R. T. Lange. 1971. Acarine and other microfossils from the Maslin Eocene, South Australia. Rec. S. Aust. Mus. 16:1–21.

Southward, E. C. 1980. Regionation and metamerisation in Pogonophora. Zool. Jb. Abt. Anat. 103:264–275.

———. 1982. Bacterial symbionts in Pogonophora. J. Mar. Biol. Assoc. U.K. 62:889–906.

———. 1984. Pogonophora. Pp. 376–388 in J. Bereiter-Hahn, A. G. Matoltsy, and K. S. Richards, eds., Biology of the Integument, vol. 1, Invertebrates. Springer-Verlag, Berlin.

Spring, J., N. Yanze, C. Jo"sch, A. M. Middell, B. Winnenger, and V. Schmidt. 2002. Conservation of Brachyury, Met2, and Snail in the myogenic lineage of a jellyfish: A connection to the mesoderm of Bilateria. Dev. Biol. 244:372–384.

Sprinkle, J. 1973. Morphology and Evolution of Blastozoan Echinoderms. Museum of Comparative Zoology, Harvard Univ. Spec. Pub., Cambridge, MA.

———. 1980. An overview of the fossil record. Pp. 15–26 in T. W. Broadhead and J. A. Waters, Echinoderms. Studies in Geology 3. Univ. Tennessee, Department of Geological Science, Knoxville.

Sprinkle, J., and D. Collins. 1998. Revision of Echmatocrinus from the Middle Cambrian Burgess Shale of British Columbia. Lethaia 31:269–358.

Sprinkle, J., and P. M. Kier. 1987. Phylum Echinodermata. Pp. 550–611 in R. S. Boardman, A. H. Cheetham, and A. J. Rowell, eds., Fossil Invertebrates. Blackwell Scientific, Palo Alto, CA.

Stach, T., and J. M. Turbeville. 2002. Phylogeny of Tunicata inferred from molecular and morphological characters. Mol. Phylogenet. Evol. 25:408–428.

Stachowitsch, M. 1992. The Invertebrates: An Illustrated Glossary. Wiley-Liss, New York.

Stanley, G. D. 1986. Chondrophorine hydrozoans as problematic fossils. Pp. 68–86 in A. Hoffman and M. H. Nitecki, eds., Problematic Fossil Taxa. Oxford Univ. Press, New York.

Stanley, G. D., and W. Sturmer. 1983. The first fossil ctenophore from the Lower Devonian of West Germany. Nature 303:518–520.

———. 1987. A new fossil ctenophore discovered by X-rays. Nature 327:61–63.

Stanley, S. M. 1968. Post-Paleozoic adaptive radiation of infaunal bivalve mollusks: A consequence of mantle fusion and siphon formation. J. Paleontol. 42:214–229.

———. 1973. An ecological theory for the sudden origin of multicellular life in the late Precambrian. Proc. Nat. Acad. Sci. USA 70:1486–1489.

———. 1976. Ideas on the timing of metazoan diversification. Paleobiology 2:209–219.

———. 1979. Macroevolution: Principles and Processes. Freeman, San Francisco.

———. 1990. The general correlation between rate of speciation and rate of extinction: Fortuitous causal links. Pp. 103–127 in R. M. Ross and W. D. Allmon, eds., Causes of Evolution: A Paleontological Perspective. Univ. Chicago Press, Chicago.

Stanley, S. M., P. W. Signor, III, S. Lidgard, and A. F. Karr. 1981. Natural clades differ from "random" clades: Simulations and analyses. Paleobiology 7:115–127.

Stanojevic, D., S. Small, and M. Levine. 1991. Regulation of a segmentation stripe by overlapping activators and repressors in the Drosophila embryo. Science 254:1385–1387.

Stasek, C. R. 1972. The molluscan framework. Pp. 1–44 in M. Florkin and B. J. Sheer, eds., Chemical Zoology, vol. 7. Academic Press, New York.

Stauffer, H. 1924. Die Lokomotion der Nematoden. Zool. Jb. 49:119–130.

Stebbing, A. R. D., and P. N. Dilly. 1972. Some observations on living Rhabdopleura compacta (Hemichordata) from Plymouth. J. Mar. Biol. Assoc. U.K. 50:209–221.

Steiner, G., and H. Dreyer. 2002. Cephalopoda and Scaphopoda are sister taxa: an evolutionary scenario. Zoology 105 (suppl. 5): 95.

Steiner, G., and L. v. Salvini-Plawen. 2001. Acaenoplax: Polychaete or mollusc? Nature 414:601–602.

Steiner, M., D. Mehl, J. Reitner, and B.-D. Erdtmann. 1993. Oldest entirely preserved sponges and other fossils from the lowermost Cambrian and a new facies reconstruction of the Yangtze Platform (China). Berliner Geowiss. Anh., ser. E, 9:293–329.

Stent, G. S. 1985. The role of cell lineage in development. Philos. Trans. R. Soc. London B 312:3–19.

Stephen, A. C., and S. J. Edmonds. 1972. The Phyla Sipuncula and Echiura. British Museum (Natural History), London.

Stern, D. L. 2000. Evolutionary developmental biology and the problem of variation. Evolution 54:1079–1091.

Sterrer, W. 1972. Systematics and evolution within the Gnathostomulida. Syst. Zool. 21:151–173.

Sterrer, W., M. Mainitz, and R. M. Rieger. 1985. Gnathostomulida: Enigmatic as ever. Pp. 181–199 in S. Conway Morris, J. D. George, R. Gibson, and H. M. Platt, eds., The Origins and Relationships of Lower Invertebrates. Clarendon Press, Oxford.

Stolc, A. 1899. Actinomyxidies: Nouveau groupe de Mesozoaires parent des Myxosporidies. Bull. Int. Acad. Sci. Boheme 22:1–12.

Stone, J. R., and G. A. Wray. 2001. Rapid evolution of cis-regulatory sequences via local point mutations. Mol. Biol. Evol. 18:1764–1770.

Storch, V. 1979. Contributions of comparative ultrastructural research to problems of invertebrate evolution. Am. Zool. 19:637–645.

———. 1991. Priapulida. Pp. 333–350 in F. W. Harrison and E. E. Ruppert, eds., Microscopic Anatomy of Invertebrates, vol. 4, Aschelminthes. Wiley-Liss, New York.

Storch, V., R. P. Higgins, and M. P. Morse. 1989. Internal anatomy of Meiopriapulus fijiensis (Priapulida). Trans. Am. Microsc. Soc. 108:245–261.

Storch, V., and H. Ruhberg. 1993. Onychophora. Pp. 11–56 in F. W. Harrison and M. E. Rice, eds., Microscopic Anatomy of Invertebrates, vol. 12, Onychophora, Chilopoda, and Lesser Protostomata. Wiley-Liss, New York.

Størmer, L. 1959. Arthropoda: General features. Pp. 3–16 in R. C. Moore, ed., Treatise on Invertebrate Paleontology, pt. O, Arthropoda 1. Geological Society of America, Boulder, CO; Univ. Kansas Press, Lawrence.

———. 1963. Gigantoscorpio willsi, a new scorpion from the Lower Carboniferous of Scotland and its associated preying microorganisms. Skrifter utgitt av Det Norska Videnskaps-Akademi i Oslo I, Math-Nuturv, Klasse Ny Serie, no. 8:1–171.

Strathmann, R. R. 1978. Progressive vacating of adaptive types during the Phanerozoic. Evolution 32:894–906.

Strauss, D., and P. M. Sadler. 1989a. Classical confidence intervals and Bayesian probability estimates for ends of local taxon ranges. Math. Geol. 21:411–427.

———. 1989b. Stochastic models for the completeness of stratigraphic sections. Math. Geol. 21:37–59.

Streidter, G. F., and R. G. Northcutt. 1991. Biological hierarchies and the concept of homology. Brain Behav. Evol. 38:177–189.

Stunkard, H. W. 1954. The life-history and systematic relations of the Mesozoa. Q. Rev. Biol. 29:220–244.

———. 1982. Mesozoa. Pp. 853–855 in S. P. Parker, ed., Synopsis and Classification of Living Organisms, vol. 1. McGraw-Hill, New York.

Suga, H., M. Koyanagi, D. Hoshiyama, K. Ono, N. Iwabe, K. Kuma, and T. Miyata. 1999. Extensive gene duplication in the early evolution of animals before the parazoan-eumetazoan split demonstrated by G proteins and protein tyrosine kinases from sponge and *Hydra*. J. Mol. Evol. 48:646–653.

Sun, W., and X. Hou. 1987. Early Cambrian worms from Chengjiang, Yunnan, China: *Maotianshania* gen. nov. Acta Paleontol. Sin. 26:299–305. In Chinese; English summary.

Sundberg, P., J. M. Turbeville, and S. Lindh. 2001. Phylogenetic relationships among higher nemertean (Nemertea) taxa inferred from 18S rDNA sequences. Mol. Phylogenet. Evol. 20:327–334.

Sutcliffe, O. E., W. H. Südkamp, and R. P. S. Jefferies. 2000. Ichnological evidence on the behaviour of mitrates: Two trails associated with the Devonian mitrate *Rhenocystis*. Lethaia 33:1–12.

Sutton, M. D., D. E. G. Briggs, D. J. Siveter, and D. J. Siveter. 2001. An exceptionally preserved vermiform mollusc from the Silurian of England. Nature 410:461–463.

Swalla, B. J., C. B. Cameron, L. S. Corley, and J. R. Garey. 2000. Urochordates are monophyletic within the deuterostomes. Syst. Biol. 49:52–64.

Swalla, B. J., K. W. Makabe, N. Satoh, and W. R. Jeffery. 1993. Novel genes expressed differentially in ascidians with alternate modes of development. Development 119:307–318.

Swedmark, B. 1964. The interstitial fauna of marine sand. Biol. Rev. 39:1–42.

Sweet, W. C. 1988. The Conodonta: Morphology, Taxonomy, Paleoecology, and Evolutionary History of a Long-Extinct Animal Phylum. Clarendon Press, Oxford.

Swofford, D. L., and G. J. Olsen. 1995. Phylogeny reconstruction. Pp. 411–501 in D. M. Hillis and C. Moritz, eds., Molecular Systematics. Sinauer, Sunderland, MA.

Swofford, D. L., G. J. Olsen, P. J. Waddel, and D. M. Hillis. 1996. Phylogenetic inference. Pp. 407–514 in D. M. Hillis, C. Moritz, and B. K. Mable, eds., Molecular Systematics, 2nd ed. Sinauer, Sunderland, MA.

Szaniawski, H. 1982. Chaetognath grasping spines recognized among Cambrian protoconodonts. J. Paleontol. 56:806–810.

Szaniawski, H., and S. Bengtson. 1993. Origin of euconodont elements. J. Paleontol. 67:640–654.

Tautz, D., and K. J. Schmid. 1998. From genes to individuals: Developmental genes and the generation of the phenotype. Philos. Trans. R. Soc. London B 353:231–240.

Taylor, F. J. R. 1974. Implications and extensions of the serial endosymbiosis theory of the origin of eukaryotes. Taxon 23:229–258.

Taylor, P. D. 1981. Functional morphology and evolutionary significance of differing modes of tentacle eversion in marine bryozoans. Pp. 235–247 in G. P. Larwood and C. Nielsen, eds., Recent and Fossil Bryozoa. Olsen and Olsen, Fredensborg.

Taylor, P. D., and G. B. Curry. 1985. The earliest known fenestrate bryozoan, with a short review of Lower Ordovician Bryozoa. Palaeontology 28:147–158.

Technau, U. 2001. *Brachyury,* the blastopore, and the evolution of the mesoderm. BioEssays 23:788–794.

Telford, M. J. 2000. Turning Hox "signatures" into synapomorphies. Evol. Dev. 2:360–364.

Telford, M. J., E. A. Herniou, R. B. Rusell, and D. T. J. Littlewood. 2000. Changes in mitochondrial genetic codes as phylogenetic characters: Two examples from the flatworms. Proc. Nat. Acad. Sci. USA 97:11359–11364.

Telford, M. J., and P. W. H. Holland. 1993. The phylogenetic affinities of the chaetognaths: A molecular analysis. Mol. Biol. Evol. 10:660–676.

Termier, H., and G. Termier. 1968. Évolution et Biocinese: Les Invertébrés dans l'Histoire du Monde Vivant. Masson et Cie, Paris.

Tessier, G. 1931. Étude expérimentale du développement de quelques hydraires. Ann. Sci. Nat. Ser. Bot. Zool. 14:5–60.

Teuchert, G. 1968. Zur Fortpflanzung und Entwicklung der Macrodasyoidea (Gastrotricha). Z. Morph. Tiere 63:343–418.

Theusen, E. V. 1991. The tetrodotoxin venom of chaetognaths. Pp. 55–60 in Q. Bone, H. Kapp, and A. C. Pierrot-Buls, eds., The Biology of Chaetognaths. Oxford Univ. Press, Oxford.

Thomas, M. B., and N. C. Edwards. 1991. Cnidaria: Hydrozoa. Pp. 91–183 in F. W. Harrison and J. A. Westfall, eds., Microscopic Anatomy of Invertebrates, vol. 2. Wiley-Liss, New York.

Thompsen, H. A. 1976. Studies on marine choanoflagellates: 2, Fine-structural observations on some silicified choanoflagellates from the Isefjord (Denmark), including the description of two new species. Norw. J. Bot. 23:33–51.

Thompson, D. W. 1942. On Growth and Form. 2nd ed. Cambridge Univ. Press, Cambridge.

Thompson, E., and D. S. Jones. 1980. A possible onychophoran from the Middle Pennsylvanian Mazon Creek beds of northern Illinois. J. Paleontol. 54:588–596.

Thorson, G. 1936. The larval development, growth, and metabolism of arctic bottom invertebrates compared with those of other seas. Medd. Groenl. 100 (b):1–155.

Todd, J. A., and P. D. Taylor. 1992. The first fossil entoproct. Naturwissenschaften 79:311–314.

Tomarev, S. I., P. Callaerts, L. Kos, R. Zinovieva, G. Halder, W. Gehring, and J. Piatigorsky. 1997. Squid Pax-6 and eye development. Proc. Nat. Acad. Sci. USA 94:2421–2426.

Towe, K. M. 1970. Oxygen-collagen priority and the metzoan fossil record. Proc. Nat. Acad. Sci. USA 65:781–788.

———. 1981. Biochemical keys to the emergence of complex life. Pp. 297–306 in J. Billingham, ed., Life in the Universe. MIT Press, Cambridge, MA.

Tranter, P. R. G., D. N. Nicholson, and D. Kinchington. 1982. A description of spawning and post-gastrula development of the cool temperate coral, *Caryophyllia smithi*. J. Mar. Biol. Assoc. U.K. 62:845–854.

Trueman, E. R. 1975. The Locomotion of Soft-Bodied Animals. Edward Arnold, London.

Truex, R., and M. B. Carpenter. 1969. Human Neuroanatomy. 6th ed. Williams and Wilkins, Baltimore.

Turbeville, J. M. 1986. An ultrastructural analysis of coelomogenesis in the hoplonemertine *Prosorhochmus americanus* and the polychaete *Magelona* sp. J. Morphol. 187:51–60.

———. 1991. Nemertinea. Pp. 285–328 in F. W. Harrison and B. J. Bogitsh, eds., Microscopic Anatomy of Invertebrates, vol. 3, Platyhelminthes and Nemertinea. Wiley-Liss, New York.

Turbeville, J. M., K. G. Field, and R. A. Raff. 1992. Phylogenetic position of phylum Nemertini, inferred from 18S rRNA sequences: Molecular data as a test of morphological character homology. Mol. Biol. Evol. 9:235–249.

Turbeville, J. M., and E. E. Ruppert. 1983. Epidermal muscles and peristaltic burrowing in *Carinoma tremaphoros* (Nemertini): Correlates of effective burrowing without segmentation. Zoomorphology 103:103–120.

———. 1985. Comparative ultrastructure and the evolution of nemertines. Am. Zool. 25:53–71.

Turbeville, J. M., J. R. Schulz, and R. A. Raff. 1994. Deuterostome phylogeny and the sister group of the chordates: Evidence from molecules and morphology. Mol. Biol. Evol. 11:648–655.

Twitchett, R. J. 1996. The resting trace of an acorn-worm (class Enteropneusta) from the Lower Triassic. J. Paleontol. 70:128–131.

Tyler, S. 2001. The early worm: Origins and relationships of the lower flatworms. Pp. 3–12 in D. T. J. Littlewood and R. A. Bray, eds., Interrelationships of the Platyhelminthes. Taylor and Francis, London.

Tyler, S., and R. Rieger. 1975. Uniflagellate spermatozoa in Nemertoderma (Turbellaria), and their phylogenetic significance. Science 188:730–732.

Ubaghs, G. 1961. Sur la nature de l'organe appele tige ou pedoncule chez les carpoides Cornuta et Mitrata. Acad. Sci. Paris C. R. 253:2738–2740.

———. 1967a. Eocrinoidea. Pp. 455–495 in R. C. Moore, ed., Treatise on Invertebrate Paleontology, pt. S, Echinodermata 1. Geological Society of America, Boulder, CO; Univ. Kansas Press, Lawrence.

———. 1967b. General characters of Echinodermata. Pp. 3–60 in R. C. Moore, ed., Treatise on Invertebrate Paleontology, pt. S, Echinodermata 1. Geological Society of America, Boulder, CO; Univ. Kansas Press, Lawrence.

———. 1967c. Stylophora. Pp. 495–565 in R. C. Moore, ed., Treatise on Invertebrate Paleontology, pt. S, Echinodermata 1. Geological Society of America, Boulder, CO; Univ. Kansas Press, Lawrence.

———. 1978. Skeletal morphology of fossil crinoids. Pp. 58–216 in R. C. Moore, ed., Treatise on Invertebrate Paleontology, pt. T, Echinodermata 2, vol. 1. Geological Society of America, Boulder, CO; Univ. Kansas Press, Lawrence.

Ubaghs, G., and K. E. Caster. 1967. Homalozoans. Pp. 495–627 in R. C. Moore, ed., Treatise on Invertebrate Paleontology, pt. S, Echinodermata 1, vol. 2. Geological Society of America, Boulder, CO; Univ. Kansas Press, Lawrence.

Urbanek, A., and G. Mierzejewska. 1977. The fine structure of zooidal tubes in Sabelliditida and Pogonophora with reference to their affinity. Acta Palaeontol. Pol. 19:223–240.

Ushatinskaya, G. T. 1987. Unusual inarticulate brachiopods from the Lower Cambrian sequence of Mongolia. Paleontol. J. 2:59–66.

Vacelet, J., and N. Boury-Esnault. 1995. Carnivorous sponges. Nature 373:333–335.

Vagvolgyi, J. 1967. On the origin of the molluscs, the coelom, and coelomic segmentation. Syst. Zool. 16:153–168.

Vail, P. P. R., R. M. Mitchum, R. G. Todd, J. M. Widmier, S. Thompson, J. B. Songree, J. N. Bubb, and W. G. Hatlelid. 1977. Seismic stratigraphy and global changes of sea level. Am. Assoc. Pet. Geol. Mem. 26:49–212.

Valentine, J. W. 1968. Climatic regulation of species diversification and extinction. Geol. Soc. Am. Bull. 79:273–276.

———. 1969. Patterns of taxonomic and ecological structure of the shelf benthos during Phanerozoic time. Palaeontology 12:684–709.

———. 1973a. Coelomate superphyla. Syst. Zool. 22:97–102.

———. 1973b. Evolutionary Paleoecology of the Marine Biosphere. Prentice Hall, Englewood Cliffs, NJ.

———. 1975. Adaptive strategies and the origin of grades and ground-plans. Am. Zool. 15:391–404.

———. 1980. Determinants of diversity in higher taxonomic categories. Paleobiology 6:444–450.

———. 1986. Fossil record of the origin of Baupläne and its implications. Pp. 209–231 in D. M. Raup and D. Jablonski, eds., Patterns and Processes in the History of Life. Springer-Verlag, Berlin.

———. 1989a. Bilaterians of the Precambrian-Cambrian transition and the annelid-arthropod relationship. Proc. Nat. Acad. Sci. USA 86:2272–2275.

———. 1989b. How good was the fossil record? Clues from the Californian Pleistocene. Paleobiology 15:83–94.

———. 1990. The macroevolution of clade shape. Pp. 128–150 in R. M. Ross and W. D. Allmon, eds., Causes of Evolution: A Paleontological Perspective. Univ. Chicago Press, Chicago.

———. 1991. The sequence of bodyplans and locomotory systems during the Precambian-Cambrian transition. Pp. 389–397 in N. Schmidt-Kittler and K. Vogel, eds., Constructional Morphology and Evolution. Springer-Verlag, Berlin.

———. 1992a. Dickinsonia as a polypoid organism. Paleobiology 18:378–382.

———. 1992b. The macroevolution of phyla. Pp. 525–553 in J. H. Lipps and P. W. Signor, eds., Origin and Early Evolution of the Metazoa. Plenum, New York.

———. 1994. Late Precambrian bilaterians: Grades and clades. Proc. Nat. Acad. Sci. USA 91:6751–6757.

———. 1995. Why no new phyla after the Cambrian? Genome and ecospace hypotheses revisited. Palaios 10:190–194.

———. 1996. The evolution of complexity in metazoans. Pp. 327–362 in T. Riste and D. Sherrington, eds., Physics of Biomaterials: Fluctuation, Selfassembly, and Evolution. Kluwer Academic, Dordrecht.

———. 1997. Cleavage patterns and the topology of the metazoan tree of life. Proc. Nat. Acad. Sci. USA 94:8001–8005.

———. 1999. The evolution of diversity: Richness and disparity. Pp. 329–339 in J. Scotchmoor and D. A. Springer, eds., Evolution: Investigating the Evidence. Paleontology Society Special Publication 9.

———. 2000. Two genomic paths to the evolution of complexity in bodyplans. Paleobiology 26:522–528.

———. 2001. How were vendobiont bodies patterned? Paleobiology 27:425–428.

Valentine, J. W., S. M. Awramik, P. W. Signor, and P. M. Sadler. 1991. The biological explosion at the Precambrian-Cambrian boundary. Evol. Biol. 25:279–356.

Valentine, J. W., and C. A. Campbell. 1976. Genetic regulation and the fossil record. Am. Sci. 63:673–680.

Valentine, J. W., and A. G. Collins. 2000. The significance of moulting in ecdysozoan evolution. Evol. Dev. 2:152–156.

Valentine, J. W., A. G. Collins, and C. P. Meyer. 1994. Morphological complexity increase in metazoans. Paleobiology 20:131–142.

Valentine, J. W., and D. H. Erwin. 1987. Interpreting great developmental experiments: The fossil record. Pp. 71–107 in R. A. Raff and E. C. Raff, eds., Development as an Evolutionary Process. Liss, New York.

Valentine, J. W., T. C. Foin, and D. Peart. 1978. A provincial model of Phanerozoic marine diversity. Paleobiology 4:55–66.

Valentine, J. W., and H. Hamilton. 1997. Body plans, phyla, and arthropods. Pp. 1–9 in R. A. Fortey and R. H. Thomas, eds., Arthropod Relationships. Chapman and Hall, London.

Valentine, J. W., D. Jablonski, and D. H. Erwin. 1999. Fossils, molecules, and embryos: New perspectives on the Cambrian explosion. Development 126:851–859.

Valentine, J. W., and C. L. May. 1996. Hierarchies in biology and paleontology. Paleobiology 22:23–33.

Valentine, J. W., and E. M. Moores. 1972. Global tectonics and the fossil record. J. Geol. 80:167–184.

Valentine, J. W., B. H. Tiffney, and J. J. Sepkoski, Jr. 1991. Evolutionary dynamics of plants and animals: A comparative approach. Palaios 6:81–88.

Valentine, J. W., and T. D. Walker. 1986. Diversity trends within a model taxonomic hierarchy. Physica D 22:31–42.

Vance, R. R. 1973. On reproductive strategies in marine benthic invertebrates. Am. Nat. 107:339–352.

Van Cleve, H. J. 1948. Expanding horizons in the recognition of a phylum. J. Parasitol. 234:1–20.

van den Biggelaar, J. A. M. 1977. Development of dorsoventral polarity and mesentoblast determination in Patella vulgata. J. Morphol. 154:157–186.

van den Biggelaar, J. A. M., and P. Guerrier. 1979. Dorsoventral polarity and mesentoblast determination as concomitant results of cellular interactions in the mollusk Patella vulgata. Dev. Biol. 68:462–471.

van der Land, J. 1970. Systematics, zoogeography, and ecology of the Priapulida. Zool. Verh. Leiden 112:1–118.

Van de Vyver, G. 1980. A comparative study of the embryonic development of Hydrozoa athecata. Pp. 109–120 in P. Tardent and R. Tardent, eds., Developmental and Cellular Biology of Coelenterates. Elsevier/North-Holland, Amsterdam.

Vannier, J., and J.-Y. Chen. 2000. The Early Cambrian colonization of pelagic niches exemplified by Isoxys (Arthropoda). Lethaia 33:295–311.

Van Valen, L. M. 1985. A theory of origination and extinction. Evol. Theory 7:133–142.

Van Valen, L. M., and V. C. Maiorana. 1985. Patterns of origination. Evol. Theory 7:107–125.

van Wijhe, J. W. 1914. Studien uber Amphioxus: 1, Mund und Darmkanal wahrend der Metamorphose. Verh. K. Akad. Wet. Amst., sect. 2, 18:1–84.

Vawter, L., and W. M. Brown. 1993. Rates and patterns of base changes in the small subunit ribosomal RNA gene. Genetics 134:597–608.

Ventner, J. C., et al. 2001. The sequence of the human genome. Science 291:1304–1351.

Vermeij, G. J. 1978. Biogeography and Adaptation. Harvard Univ. Press, Cambridge, MA.

Vidal, G. 1997. Biodiversity, speciation, and extinction trends of Proterozoic and Cambrian phytoplankton. Paleobiology 23:230–246.

Viidik, A. 1972. Functional properties of collagenous tissues. Int. Rev. Connect. Tissue Res. 6:127–215.

Vogel, S. 1977. Current-induced flow through sponges *in situ*. Proc. Nat. Acad. Sci. USA 74:2069–2071.

———. 1981. Life in Moving Fluids: The Physical Biology of Flow. Willard Grant Press, Boston.

Voigt, E. 1938. Ein fossiler Saltenwurm (*Gordia tenuifibrosis*) aus der Eozanen Braunkohle des Geiseltales. Nova Acta Leopold., N. F., 5:53–56.

———. 1988. Preservation of soft tissues in the Eocene lignite of the Geiseltal near Halle/S. Cour. Forsch.-Inst. Senckenb. 107:325–343.

Von Baer, K. E. 1828. Entwicklungsgeschichte der Thiere: Beobachtung und Reflexion. Bornträger, Königsberg.

Voronov, D. A., and Yu. V. Panchin. 1998. Cell lineage in marine nematode *Enoplus brevis*. Development 125:143–150.

Voronov, D. A., Yu. V. Panchin, and S. E. Spiridonov. 1998. Nematode phylogeny and embryology. Nature 395:28.

Vrba, E. S., and N. Eldredge. 1984. Individuals, hierarchies, and processes: Towards a more complete evolutionary theory. Paleobiology 10:146–171.

Wada, H., and N. Satoh. 1994. Details of the evolutionary history from invertebrates to vertebrates, as deduced from the sequences of 18S rDNA. Proc. Nat. Acad. Sci. USA 91:1801–1804.

Wade, M. 1969. Medusae from uppermost Precambrian or Cambrian sandstones, central Australia. Palaeontology 12:351–365.

———. 1972a. *Dickinsonia*: Polychaete worms from the late Precambrian Ediacara fauna, South Australia. Mem. Queensland Mus. 16:171–190.

———. 1972b. Hydrozoa and Scyphozoa and other medusoids from the Precambrian Ediacara fauna, South Australia. Palaeontology 15:197–225.

Waggoner, B. M. 1996. Phylogenetic hypotheses of the relationships of arthropods to Precambrian and Cambrian problematic fossil taxa. Syst. Biol. 45:190–222.

———. 1999. Biogeographic analyses of the Ediacara biota: A conflict with paleotectonic reconstructions. Paleobiology 25:440–458.

Waggoner, B. M., and G. O. Poinar, Jr. 1993. Habrotrochid rotifers from Dominican amber. Experientia 49:354–357.

Wagner, A. 1994. Evolution of gene networks by gene duplications: A mathematical model and its implications on genome organization. Proc. Nat. Acad. Sci. USA 91:4387–4391.

Wagner, G. 1989. The biological homology concept. Ann. Rev. Ecol. Syst. 20:51–69.

Wagner, P. J. 1995a. Stratigraphic tests of cladistic hypotheses. Paleobiology 21:153–178.

———. 1995b. Testing evolutionary constraint hypotheses with early Paleozoic gastropods. Paleobiology 21:248–272.

———. 1997. Patterns of morphologic diversification among the Rostrochonchia. Paleobiology 23:115–150.

———. 1998. A likelihood approach for estimating phylogenetic relationships among fossil taxa. Paleobiology 24:430–449.

———. 1999. Phylogenetics of the earliest anisostrophically coiled gastropods. Smithson. Contrib. Paleobiol. 88:1–154.

———. 2000. Exhaustion of morphologic character states among fossil taxa. Evolution 54:365–386.

Wainright, P. O., G. Hinkle, M. L. Sogin, and S. K. Stickel. 1993. Monophyletic origins of the Metazoa: An evolutionary link with fungi. Science 260:340–342.

Walker, K. R., and R. K. Bambach. 1971. The significance of fossil assemblages from fine-grained sediments: Time-averaged communities. Geol. Soc. Am. Abstr. Programs 24:7.

Walker, T. D. 1984. The evolution of diversity in an adaptive mosaic. Ph.D. thesis, Univ. California, Santa Barbara.

Walker, T. D., and J. W. Valentine. 1984. Equilibrium models of evolutionary species diversity and the number of empty niches. Am. Nat. 124:887–899.

Wallace, B. 1963. Genetic diversity, genetic uniformity, and heterosis. Can. J. Genet. Cytol. 5:239–253.

Wallace, H. R. 1968. The dynamics of nematode movement. Ann. Rev. Phytopath. 6:91–114.

Wallace, R. L., C. Ricci, and G. Melone. 1996. A cladistic analysis of pseudocoelomate (aschelminth) morphology. Invertebr. Biol. 115:104–112.

Walossek, D., and K. J. Müller. 1998. Cambrian "Orsten"-type arthropods and the phylogeny of Crustacea. Pp. 139–153 in R. A. Fortey and R. H. Thomas, eds., Arthropod Relationships. Chapman and Hall, London.

Walossek, D., K. J. Müller, and R. M. Kristensen. 1994. A more than half a billion years old stem-group tardigrade from Siberia. Abstracts of the Sixth International Symposium on Tardigrada, Cambridge.

Walossek, D., J. E. Repetski, and K. J. Müller. 1994. An exceptionally preserved arthropod, *Heymonsicambria taylori* n. sp. (Arthropoda incertae sedis: Pentastomida), from Cambrian-Ordovician boundary beds of Newfoundland, Canada. Can. J. Earth Sci. 31:1664–1671.

Wang, B. B., M. M. Müller-Immergluck, J. Austin, N. Tamar Robinson, A. Chisholm, and C. Kenyon. 1993. A homeotic gene cluster patterns the anteroposterior body axis of *C. elegans*. Cell 74:29–42.

Ward, P. D., and P. W. Signor, III. 1983. Evolutionary tempo in Jurassic and Cretaceous ammonites. Paleobiology 9:183–198.

Warén, A., and S. Gofas. 1996. A new species of Monoplacophora, redescription of the genera *Veleropilina* and *Rokopella,* and new information on three species of the class. Zool. Scr. 25:215–232.

Warner, B. G., and R. Chengalath. 1988. Holocene fossil *Habrotrocha angusticollis* (Bdelloidea: Rotifera) in North America. J. Paleolimnol. 1:141–147.

Watrous, L. E., and Q. D. Wheeler. 1981. The outgroup comparison method of character analysis. Syst. Zool. 30:1–11.

Webb, J. E., and M. B. Hill. 1958. The ecology of Lagos Lagoon: 4, On the reactions of *Branchiostoma nigeriense* to its environment. Philos. Trans. R. Soc. London B 241:355–391.

Webb, M. 1964. The posterior extremity of *Siboglinum fiordicum* (Pogonophora). Sarsia 15:33–36.

Weill, R. 1938. L'interprétation des Cnidosporidies et la valeur taxonomique de leur cnidome: Leur cycle comparé à la phase larvaire des Narcomeduses Cuninides. Trav. Sta. Zool. Wimereaux 13:727–744.

Welch, D. B. M. 2000. Evidence from a protein-coding gene that acanthocephalans are rotifers. Invertebr. Biol. 119:17–26.

Welch, D. B. M., and M. Meselson. 2000. Evidence for the evolution of bdelloid rotifers without sexual reproduction or genetic exchange. Science 288:1211–1215.

Welsch, U., and V. Storch. 1976. Comparative Animal Cytology and Histology. Univ. Washington Press, Seattle.

Werner, B. 1955. Anatomie, Entwicklung, und Biologie des Veligers. Helgolander Wissenschaft. Meeresuntersuchungen 5:169–217.

West, L., and D. Powers. 1993. Molecular phylogenetic position of hexactinellid sponges in relation to the Protista and Demospongiae. Mol. Mar. Biol. Biotechnol. 2:71–75.

Weygoldt, P. 1979. Significance of later embryonic stages and head development in arthropod phylogeny. Pp. 107–135 in A. P. Gupta, ed., Arthropod Phylogeny. Van Nostrand Reinhold, New York.

Wheeler, W. C. 1998. Molecular systematics and arthropods. Pp. 9–32 in G. D. Edgecombe, ed., Arthropod Fossils and Phylogeny. Columbia Univ. Press, New York.

Wheeler, W. C., P. Cartwright, and C. Y. Hayashi. 1993. Arthropod phylogeny: A combined approach. Cladistics 9:1–39.

White, J., and S. Strome. 1996. Cleavage plane specification in *C. elegans:* How to divide the spoils. Cell 84:195–198.

Whittaker, J. R. 1997. Cephalochordates: The lancelets. Pp. 365–381 in S. F. Gilbert and A. M. Raunio, eds., Embryology. Sinauer, Sunderland, MA.

Whittington, H. B. 1971. Redescription of *Marella splendens* (Trilobitoidea) from the Burgess Shale, Middle Cambrian, British Columbia. Geol. Surv. Can. Bull. 209:1–24.

———. 1974. *Yohoia* Walcott and *Plenocaris* n. gen.: Arthropods from the Burgess Shale, Middle Cambrian, British Columbia. Geol. Surv. Can. Bull. 231:1–21.

———. 1975. The enigmatic animal *Opabinia regalis,* Middle Cambrian, Burgess Shale, British Columbia. Philos. Trans. R. Soc. London B 271:1–43.

———. 1977. The Middle Cambrian trilobite *Naraoia,* Burgess Shale, British Columbia. Philos. Trans. R. Soc. London B 280:409–443.

———. 1978. The lobopod animal *Aysheaia pedunculata* Walcott, Middle Cambrian, Burgess Shale, British Columbia. Philos. Trans. R. Soc. London B 284:165–197.

———. 1980. Exoskeleton, moult stage, appendage morphology, and habits of the Middle Cambrian trilobite *Olenoides serratus.* Palaeontology 23:171–204.

———. 1985a. The Burgess Shale. Yale Univ. Press, New Haven.

————. 1985b. *Tegopelte gigas,* a second soft-bodied trilobite from the Burgess Shale, British Columbia. J. Paleontol. 59:1251–1274.

Wicken, J. S. 1979. The generation of complexity in evolution: A thermodynamic and information-theoretic discussion. J. Theor. Biol. 77:349–365.

Wilbur, K. M. 1964. Shell formation and regeneration. Pp. 243–282 in K. M. Wilbur and C. M. Yonge, eds., Physiology of Mollusca. Academic Press, New York.

Wiley, E. O. 1980. Is the evolutionary species fiction? A consideration of classes, individuals, and historical entities. Syst. Zool. 29:76–80.

————. 1981. Phylogenetics. Wiley, New York.

Wilkins, A. S. 2002. The Evolution of Developmental Pathways. Sinauer, Sunderland, MA.

Williams, A. 1965. Stratigraphic distribution. Pp. 237–250 in R. C. Moore, ed., Treatise on Invertebrate Paleontology, pt. H, Brachiopoda, vol. 1. Geological Society of America, Boulder, CO; Univ. Kansas Press, Lawrence.

Williams, A., S. J. Carlson, C. H. C. Brunton, L. E. Holmer, and L. Popov. 1996. A supra-ordinal classification of the Brachiopoda. Philos. Trans. R. Soc. London B 351:1171–1193.

Williams, A., and A. J. Rowell. 1965a. Brachiopod anatomy. Pp. 6–57 in R. C. Moore, ed., Treatise on Invertebrate Paleontology, pt. H, Brachiopoda, vol. 1. Geological Society of America, Boulder, CO; Univ. Kansas Press, Lawrence.

————. 1965b. Classification. Pp. 214–237 in R. C. Moore, ed., Treatise on Invertebrate Paleontology, pt. H, Brachiopoda, vol. 1. Geological Society of America, Boulder, CO; Univ. Kansas Press, Lawrence.

Williams, G. C. 1966. Adaptation and Natural Selection. Princeton Univ. Press, Princeton, NJ.

————. 1992. Natural Selection: Domains, Levels, and Challenges. Oxford Univ. Press, New York.

————. 1997. Preliminary assessment of the phylogeny of Pennatulacea (Anthozoa: Octocoralia), with a reevaluation of Ediacaran frond-like fossils, and a synopsis of the history of evolutionary thought regarding the sea pens. Pp. 497–509 in J. C. den Hartog, ed., Proceedings of the Sixth International Conference on Coelenterate Biology, 1995. Nationaal Naturhistorisch Museum, Leiden.

Williams, M., D. J. Siveter, and J. S. Peel. 1996. *Isoxys* (Arthropoda) from the Early Cambrian Sirius Passet lagerstätte, north Greenland. J. Paleontol. 70:947–954.

Williams, P. L., and W. M. Fitch. 1990. Phylogeny determination using dynamically weighted parsimony method. Methods Enzymol. 183:615–626.

Williamson, D. I. 1982. Larval morphology and diversity. Pp. 43–110 in L. G. Abele, ed., The Biology of Crustacea, vol. 2, Embryology, Morphology, and Genetics. Academic Press, New York.

Willmer, P. 1990. Invertebrate Relationships: Patterns in Animal Evolution. Cambridge Univ. Press, Cambridge.

Wills, M. A. 1998. Cambrian and Recent disparity: The picture from priapulids. Paleobiology 24:177–199.

Wills, M. A., D. E. G. Briggs, and R. A. Fortey. 1994. Disparity as an evolutionary index: A comparison of Cambrian and Recent arthropods. Paleobiology 20:93–130.

————. 1997. Evolutionary correlates of arthropod tagmosis: Scrambled legs. Pp. 57–65 in R. A. Fortey and R. H. Thomas, eds., Arthropod Relationships. Chapman and Hall, London.

Wills, M. A., D. E. G. Briggs, R. A. Fortey, and M. Wilkinson. 1995. The significance of fossils in understanding arthropod evolution. Verh. Dtsch. Zool. Ges. 88:203–215.

Wills, M. A., D. E. G. Briggs, R. A. Fortey, M. Wilkinson, and P. H. A. Sneath. 1998. An arthropod phylogeny based on fossil and recent taxa. Pp. 33–105 in G. D. Edgecombe, ed., Arthropod Fossils and Phylogeny. Columbia Univ. Press, New York.

Wilson, A. C., H. Ochman, and E. M. Prager. 1987. Molecular time scale for evolution. Trends Genet. 3:241–247.

Wilson, E. B. 1892. The cell lineage of *Nereis*. J. Morphol. 6:361–480.

Winchell, C. J., J. Sullivan, C. B. Cameron, B. J. Swalla, and J. Mallatt. 2002. Evaluating hypotheses of deuterostome phylogeny and chordate evolution with new LSU and SSU ribosomal DNA data. Mol. Biol. Evol. 19:762–766.

Wingstrand, K. G. 1972. Comparative spermatology of a pentastomid, *Raillietiella hemidactyli*, and a brachiuran crustacean, *Argulus foliaceous,* with a discussion of pentastomid relationships. K. Dan. Vidensk. Selsk. Biol. Skr. 19:1–72.

————. 1985. On the anatomy and relationships of recent Monoplacophora. Galathea Report 16:1–94.

Winnepenninckx, B., T. Backeljau, and R. DeWachter. 1995. Phylogeny of protostome worms derived from 18S rRNA sequences. Mol. Biol. Evol. 12:641–649.

————. 1996. Investigation of molluscan phylogeny on the basis of 18S rRNA sequences. Mol. Biol. Evol. 13:1306–1317.

Winnepenninckx, B., T. Backeljau, and R. M. Kristensen. 1998. Relations of the new phylum Cycliophora. Nature 393:636–638.

Winnepenninckx, B., T. Backeljau, L. Y. Mackey, J. M. Brooks, R. DeWachter, S. Kumar, and J. R. Garey. 1995. 18S rRNA data indicate that Aschelminthes are polyphyletic in origin and consist of at least three distinct clades. Mol. Biol. Evol. 12:1132–1137.

Winnepenninckx, B., Y. Van de Peer, and T. Backeljau. 1998. Metazoan relationships on the basis of 18S rRNA sequences: A few years later. Am. Zool. 38:888–906.

Withers, T. H. 1926. Catalogue of the Machaeridia. British Museum (Natural History), London.

Woese, C. R. 1987. Bacterial evolution. Microbiol. Rev. 51:221–271.

Wolf, K., and M. E. Markiw. 1984. Biology contravenes taxonomy in the Myxozoa: New discoveries show alternation of invertebrate and vertebrate hosts. Science 225:1449–1452.

Wolpert, L., R. Beddington, J. Brockes, T. Jessell, P. Lawrence, and E. Meyerowitz. 1998. Principles of Development. Current Biology, London; Oxford Univ. Press, Oxford.

Wood, R. 1999. Reef Evolution. Oxford Univ. Press, Oxford.

Wood, R. A., J. P. Grotzinger, and J. A. D. Dickson. 2002. Proterozoic modular biomineralized metazoan from the Nama Group, Namibia. Science 296:2383–2386.

Wood, V., et al. 2002. The genome sequence of *Schizosaccharomyces pombe*. Nature 415:871–880.

Wood, W. B., ed. 1988. The nematode *Caenorhabditis elegans*. Cold Spring Harbor Monograph 17. Cold Spring Harbor Laboratories, Cold Spring, NY.

Woodger, J. H. 1952. From biology to mathematics. Br. J. Philos. Sci. 3:1–21.

Wray, G. A. 1997. Echinoderms. Pp. 309–329 in S. F. Gilbert and A. M. Raunio, eds., Embryology. Sinauer, Sunderland, MA.

Wray, G. A., and A. E. Bely. 1994. The evolution of echinoderm development is driven by several distinct factors. Development 1994 suppl.: 97–106.

Wray, G. A., J. S. Levinton, and L. H. Shapiro. 1996. Molecular evidence for deep Precambrian divergences among metazoan phyla. Science 274:568–573.

Wray, G. A., and R. A. Raff. 1989. Evolutionary modification of cell lineage in the direct-developing sea urchin *Heliocidaris erythrogramma*. Dev. Biol. 132:458–470.

———. 1990. Novel origins of lineage founder cells in the direct-developing sea urchin *Heliocidaris erythrogramma*. Dev. Biol. 141:41–54.

Wright, A. D. 1979. Brachiopod radiation. Pp. 235–252 in M. R. House, ed., The Origin of Major Invertebrate Groups. Academic Press, London.

Wright, K. W. 1991. Nematoda. Pp. 111–195 in F. W. Harrison and E. E. Ruppert, eds., Microscopic Anatomy of Invertebrates, vol. 4, Aschelminthes. Wiley-Liss, New York.

Wulfert, J. 1902. Die embryonalentwicklung von *Gonothyraea loveni* Allm. Z. Wiss. Zool. 71:296–327.

Xiao, S., and A. H. Knoll. 2000. Phosphatized animal embryos from the Neoproterozoic Doushantuo Formation at Weng'an, Guizhou, south China. J. Paleontol. 74:767–788.

Xiao, S., X. Yuan, and A. H. Knoll. 2000. Eumetazoan fossils in terminal Proterozoic phosphorites? Proc. Nat. Acad. Sci. USA 97:13684–13689.

Xiao, S., X. Yuan, M. Steiner, and A. H. Knoll. 2002. Microscopic carbonaceous compressions in a terminal Proterozoic shale: A systematic reassessment of the Miaohe biota, south China. J. Paleontol. 76:347–376.

Xiao, S., Y. Zhang, and A. H. Knoll. 1998. Three-dimensional preservation of algae and animal embryos in a Neoproterozoic phosphorite. Nature 391:553–558.

Yochelson, E. L. 1969. Stenothecoida: A proposed new class of Cambrian Mollusca. Lethaia 2:49–62.

Young, C. M., E. Vazquez, A. Metaxas, and P. A. Tyler. 1996. Embryology of vestimentiferan tube worms from deep-sea methane/sulphide seeps. Nature 381:514–516.

Yuh, C.-H., H. Bolouri, and E. H. Davidson. 1998. Genomic *cis*-regulatory logic: Experimental and computational analysis of a sea urchin gene. Science 279:1896–1902.

Yuh, C.-H., and E. H. Davidson. 1996. Modular *cis*-regulatory organization of *Endo16*, a gut-specific gene of the sea urchin embryo. Development 122:1069–1082.

Yuh, C.-H., A. Ransick, P. Martinez, R. J. Britten, and E. H. Davidson. 1994. Complexity and organization of DNA-protein interactions in the 5′ regulatory region of an endoderm-specific marker gene in the sea urchin embryo. Mech. Dev. 47:165–186.

Yurchenko, P. D., and G. C. Ruben. 1987. Basement membrane structure *in situ*: Evidence for lateral associations in the type IV collagen network. J. Cell Biol. 105:2259–2268.

Zaykov, V. V., and V. V. Maslennikov. 1987. Sea-bottom sulfide structure in massive sulfide deposits of the Urals. Dokl. Akad. Nauk SSSR 293:60–62. In Russian.

Zhang, A. 1987. Fossil appendicularians in the Early Cambrian. Sci. Sin. B 30:888–896.

Zhang. X.-L., J. Han, Z.-F. Zhang, H.-Q. Liu, and D.-G. Shu. 2003. Reconsideration of the supposed Naraoiid larvae from the Early Cambrian Chengjiang lagerstätte, south China. Palaeontology 46:447–465.

Zhao, Y.-L., J.-L. Yuan, Y.-Z. Huang, J.-R. Mao, Y. Qian, Z.-H. Zhang, and X.-Y. Gong. 1994. Middle Cambrian Kaili fauna in Taijiang, Guizhou. Acta Paleontol. Sin. 33:263–271. In Chinese; English summary.

Zhao, Y.-L., J.-L. Yuan, M.-Y. Zhu, R.-D. Yang, Q.-J. Guo, Q. Yi, Y.-Z. Huang, and Y. Pan. 1999. A progress report on research on the early Middle Cambrian Kaili biota, Guizhou, PRC. Acta Paleontol. Sin. 38, suppl.: 10–14.

Zhuravlev, A. Yu. 1986. Radiocyathids. Pp. 35–44 in A. Hoffman and M. H. Nitecki, eds., Problematic Fossil Taxa. Oxford Univ. Press, New York.

Zhuravlev, A. Yu., and R. Riding, eds. 2001. The Ecology of the Cambrian Radiation. Columbia Univ. Press, New York.

Zhuravleva, I. T. 1960. Archaeocyaths of the Siberian Platform. Akademiya Nauk SSSR, Moscow. In Russian.

———. 1970a. Marine faunas and Lower Cambrian stratigraphy. Am. J. Sci. 269:417–445.

———. 1970b. Porifera, Sphinctozoa, Archaeocyathi: Their connections. Pp. 41–59 in W. G. Fry, ed., The Biology of the Porifera. Symposia of the Zoological Society of London 25. Academic Press, London.

Zimmer, R. L. 1980. Mesoderm proliferation and formation of the protocoel and metacoel in early embryos of *Phoronis vancouverensis* (Phoronida). Zool. Jb. Abt. Anat. 103:219–233.

———. 1997. Phoronids, brachiopods, and bryozoans: The lophophorates. Pp. 279–305 in S. F. Gilbert and A. M. Raunio, eds., Embryology. Sinauer, Sunderland, MA.

Zrzavy, J., S. Mihulka, P. Kepka, A. Bezdek, and D. Tietz. 1998. Phylogeny of the Metazoa based on morphological and 18S ribosomal DNA evidence. Cladistics 14:249–285.

Zuckerkandl, E., and L. Pauling. 1965. Molecules as documents of evolutionary history. J. Theor. Biol. 8:357–366.

Index